Luminescent Metal Oxides

The focus of the book is to explore metal oxides exhibiting a high optical transmittance as applicable in the field of light-emitting diodes (LEDs), photo-catalysts, and so forth. It provides exposure to structural and chemical parameters of optically active metal oxides as a phosphor, innovative and currently demanded synthesis methods, and their proper characterization. It further covers applications such as optical thermometry, scintillation, anti-counterfeit, solid-state lighting and spectral modifier for solar cells, VUV application, and long persistent light emission phenomenon.

Features:

- Reviews selection of structurally and functionally active materials for effective synthesis of metal oxides
- Exclusively covers large number of areas of applications of the luminescent metal oxides
- Cover various aspects of metal oxide research including synthesis and applications
- Includes chapters on synthesis-related predictions using machine learning
- Discusses radiation dosimetry and bio-imaging aspects

This book is aimed at researchers and graduate students in materials science and phosphor technology.

Luminescent Metal Oxides
Materials to Technologies

Edited by
S.V. Moharil, N.S. Bajaj and P.K. Tawalare

CRC Press
Taylor & Francis Group
Boca Raton London New York

CRC Press is an imprint of the
Taylor & Francis Group, an **informa** business

Designed cover image: © Shutterstock

First edition published 2024
by CRC Press
6000 Broken Sound Parkway NW, Suite 300, Boca Raton, FL 33487-2742

and by CRC Press
4 Park Square, Milton Park, Abingdon, Oxon, OX14 4RN

CRC Press is an imprint of Taylor & Francis Group, LLC

ISBN: 9781032415611 (hbk)
ISBN: 9781032432205 (pbk)
ISBN: 9781003366232 (ebk)

DOI: 10.1201/9781003366232

Typeset in Times
by codeMantra

Contents

Preface

Materials are a crucial part of human advancement since the inception of life on earth. With the passage of time, innumerable new materials have been explored as well as developed and the search for new innovative materials continues smartly. Keeping in mind the enormous perspectives of various classes of oxide materials, this book aims at providing an inclusive collection of works across the breadth of oxide materials research at cutting-edge interface of materials science with physics, chemistry, biology, and engineering. The main focus to explore metal oxides is because of its fascinating properties like electrical properties of metal oxides as insulators and semiconductors to metals and even superconductors. Moreover, metal oxides that exhibit a high optical transmittance along with high electrical conductivity, referred to as transparent conductive oxides, have also gained immense attraction in the field of solar cells, gas sensors, field emitters, LEDs, photo-catalysts, piezoelectric nano-generators and nano-optoelectronic devices. This book compiles the reach of oxide materials for various device applications with brief reviews. The main highlight of the book is to provide exposure to metal oxides over the other ceramics as they have several advantages such as thermal, chemical, and physical properties when compared with other ceramic materials.

In this book, we have discussed the following factors:

- The structural and chemical parameters of optically active metal oxides as phosphors.
- The suitable innovative and currently demanded synthesis methods for production of metal oxides on a large scale in crystal as well as thin film form
- Brief about modern ways of proper characterization of materials based on their applications.
- A tremendous literature survey on optically active metal oxides based on their cutting-edge applications in the research and innovations.
- The book inclusively covers applications of luminescent metal oxide such as optical thermometry, scintillation, anti-counterfeit, solid-state lighting and spectral modifier for solar cells, and long persistent light emission phenomenon and many more.

About the Editors

S.V. Moharil obtained his master's degree in physics from the Department of Physics, RTM Nagpur University, Nagpur, in 1974. He was awarded a Ph.D. in Physics by the same University in 1977 and the highest degree of D.Sc. in 1984. He has secured several prestigious fellowships like Research Associates of UGC and CSIR during his research career. He has a long experience of more than 45 years in research and teaching. He worked as a teacher at the Department of Physics, RTM Nagpur University, from 1977 to 2014. His area of expertise includes material synthesis and characterization, luminescence (PL, TL and OSL), and radiation dosimeters. He has 320 publications in international journals to his credit, and 36 students have obtained a doctorate under his guidance. He is also an inventor with five patents. He is a reviewer for many reputed publishers.

N.S. Bajaj obtained his master's degree from the Vidhyabharti College, Amravati, affiliated to SGB Amravati University, Amravati, in 2008 and his Ph.D. (Physics) from the same university in 2014. He served from 2010 to 2015 as a UGC-Project Fellow at SGB Nagpur University, in projects sanctioned under the supervision of Dr. S. K. Omanwar. He also served his services to various colleges at the same university as a visiting lecturer. In 2015 he accepted to join as Assistant Professor at Toshniwal Arts, Commerce and Science College, Sengaon, affiliated to SRTM University, Nanded. During his short career, he has achieved many milestones in the field of research and received a couple of international awards and reorganization. Recently he was awarded the most precious Marathwada Bhushan award. Dr. Bajaj has expertise in the synthesis of phosphor and phosphor applications for radiation dosimeters. He has excellent publications in indexed journals. Recently in the past 2 years he has published many books, the special one being *Borate Phosphor Processing to Applications* published with CRC Press. Recently Government of Australia and Government of South Africa granted patents to his innovations.

P.K. Tawalare obtained her master's degree in physics from Department of Physics, SGB Amravati University, Amravati, in 2012. She was awarded a Ph.D. degree in Physics by the same University in 2019. Her area of expertise includes material synthesis and characterization, luminescence, and solar cells. She has 15 publications in international journals to her credit. She has presented papers at six national and international conferences. She has been a reviewer for many research journals. Presently she is working as an Assistant Professor in Jagadamba Mahavidyalaya, Achalpur City, District Amravati. In the past 2 years, she has published many book chapters, the special one being *Borate Phosphor Processing to Applications* published with CRC Press, as she has contributed two main chapters in a book. She has one patent to her credit granted by the Government of South Africa.

Contributors

D.P. Awade
Department of Physics
G. N. Khalsa College
Mumbai, India

S.R. Dhakate
Advanced Carbon Products and Metrology
CSIR – National Physical Laboratory
New Delhi, India

S.P. Dhale
Department of Physics
Anand Niketan College
Anandwan, India

S.R. Gosavi
Material Research Laboratory
C.H.C. Arts, S.G.P. Commerce, and B.B.J.P.
    Science College
Nandurbar, India

R.A. Joshi
Department of Physics
Toshniwal Arts, Commerce and Science College
Hingoli, India

R.K. Joshi
Department of First Year Engineering
Dr. D.Y.Patil Institute of Technology
Pune, India

K.A. Koparkar
Department of Physics
M. S. P. Arts, Science and K. P. T. Commerce
    College
Manora, India

R.G. Korpe
Department of Physics
Shri. Shivaji Science College
Amravati, India

P.P. Lohe
Department of Physics and Electronics
Jhulelal Institute of Technology
Nagpur, India

G.B. Nair
Department of Physics
University of the Free State
Bloemfontein, Republic of South Africa

V.R. Raikwar
Department of Physics
Ramniranjan Jhunjhunwala College
    of Arts, Science and Commerce
Mumbai, India

S.G. Revankar
Department of Physics
Priyadarshani Bhagwati College of
    Engineering
Nagpur, India

V.S. Singh
CSIR – National Physical Laboratory
New Delhi, India

A.D. Sontakke
Guest Researcher, CMI, Utrecht University
Netherland

S.J. Tamboli
Department of Physics
Faculty of Natural and Agricultural Sciences
Bloemfontein, Republic of South Africa

P.K. Tawalare
Department of Physics
Jagadamba Mahavidyalaya
Achalpur, India

V.G. Thakare
Department of Physics
Shri Shivaji Arts, Commerce and Science
    College
Akola, India

P.V. Tumram
Department of Physics
Amolakchand Mahavidyalaya
Yavatmal, India

N.S. Ugemuge
Department of Physics
Anand Niketan College
Anandwan, India

1 Metal Oxide-Based Luminescent Technologies
Past and Future

S.P. Dhale and N. S. Ugemuge
Anand Niketan College

CONTENTS

1.1 INTRODUCTION

Lanthanide-doped metal oxides are ideal materials for the development of multifunctional luminescent materials. This is due to the luminescence of lanthanide ions, which is used for a large number of diverse functions, and host materials may have their own functional abilities. It is well known that lanthanide luminescence can be used for scintillator [1,2], radiation dosimetry materials [3,4], phototherapy [5], vacuum ultraviolet (VUV) application [6], solid-state lighting (SSL) [7], spectral modifier for solar cell application [8], solid-state lasers [9], optical sensing [10,11], antimicrobial technology [12,13], bioimaging and biomedical imaging [14–16], optical thermometry [17], and anticounterfeit applications [18,19].

DOI: 10.1201/9781003366232-1

1

Several studies have concentrated on metal oxide nanoparticles (NPs) because of their distinguishing catalytic, electronic, and optical properties [20,21]. Luminescence alters its characteristics in the presence of a chemical category or as a response to physical stimuli acting as high-energy radiation, magnetic and electric fields, temperature, etc.

Metal oxides act as bulk grown-up luminescent materials, which can be categorized into two major categories: metal activator-doped metal oxide phosphors (e.g. Y_2O_3:Eu^{3+}) and defect-related materials [22]. Synthesis of these materials is possible with several routes namely solid-state reaction, combustion synthesis, sol–gel synthesis, and hydrothermal synthesis.

Metal oxide-based luminescent material includes phosphates, borates, silicates, and aluminate. The oxide-type materials have good chemical stability and mechanical strength, but the phonon energy is huge and the preparation is complex [23–25]. These materials possess several characteristics such as stability, high luminescent efficiency, non-linear properties, high UV transparency, less sintering temperature, high mechanical resistance, large quantum yields, and high chemical and thermal stability [26–29]. The detailed properties of these materials are discussed in the following sections.

1.1.1 BORATES

Borates as luminescence hosts were found to be good materials because of their easy fabrication, good thermal stability, high luminescent efficiency, and low synthetic temperatures [30–33]. Due to their remarkable properties of rare earth-doped alumina, borate attracted much interest such as non-linear properties, high UV transparency, exceptional optical damage threshold, and their ability to withstand the vulgar condition in vacuum discharge screens or lamps [34–36].

In the structure of borate, boron atoms have three oxygen atoms by trigonal sp^2 bonds accepted pyramidal BO_3 or planar structure or four oxygen atoms by tetragonal sp^3 bonds giving formation of tetrahedral BO_4 structure [37,38] and that the PO_4, BO_4, or BO_3 and tetrahedra divide corners and assemble infinite networks or chains, and these are the main structural characteristic of them [39].

1.1.1.1 Borophosphates

The high-temperature technique has constructed a handful of metal borophosphates, such as $XBPO_5$, where X=Ca, Sr, or Ba [40–43], $X_3BP_3O_{12}$, where X = Ba or Pb [40,44], $Na_5B_2P_3O_{13}$, $Co_5BP_3O_{14}$, and X_3BPO_7, where X=Mg or Zn [39,45].

$XBPO_5$ (X=Ca, Sr, Ba) being an adjunct of stillwellite type materials, which are crystallized in the trigonal system, having space group $P3_121$, with three formula units per cell, X^{2+} ions are tenfold coordinated using O^{2-} ions in the form of C_2 symmetry. The common corners are actually connected by anionic units of BO_4^{5-} and PO_4^{3-} tetrahedra. Predominantly, the positioning of BO_4^{5-} and PO_4^{3-} was link branched chains [46].

According to A.M. Srivastava, the structure has periodic positioning of tetrahedrally coordinated BO_4 and PO_4 groups and X^{2+} ions are ninefold coordinated [47]. The central three solitary chains of BO_4 tetrahedra rush surely parallel to [0 0 1], the BO_4 units were basically associated with the terminal PO_4 tetrahedra [41,46,48,49]. Furthermore, borates have more complex groups apart from the above simple groups such as B_3O_6, B_3O_7, and $(BO_3)_n$ [38,50] anhydrous structure of borophosphate compounds. $MBPO_5$ (M=Ca, Sr, Ba) was first obtained by Bauer with the thermal method [51,52]. Gozel reported the synthesis of $CaBPO_5$ by the conventional solid-state method [38].

1.1.1.2 Aluminoborates

Plasma display panels (PDPs) are the most approving large-shaped flat panel information display devices, which have pixels consisting of small gas-discharge cells [53–58]. Aluminoborate compounds seem to be good for thermal stability and high luminescent efficiency [59–62]. The rare earth ion-doped aluminoborates attracted more attention due to their non-linear properties and high UV transparency. Borates have an exceptional optical damage threshold and capacity to resist the harsh state in vacuum discharge screens or lamps [63].

1.1.1.3 Borosilicates

The LaBSiO$_5$ compound exhibits excellent hydrolytic and thermal stability and is considered to be an effective luminescent host [64–67]. It has a space group of P3$_1$ and shares a trigonal crystal structure. a=6.874 Å, b=6.874 Å, c=6.717 Å, Z=3, and V=274.87 Å^3 are the corresponding lattice parameters [68]. Xue et al. [69] reported the photo-luminescent properties of red-emitting LaBSiO$_5$:Eu^{3+} phosphors [67]. In this structure, six-coordination La^{3+} ion with O^{2-} ions and SiO$_4$ and BO$_4$ are interconnected by corner sharing to form a six-membered ring [66,68]. LaBSiO$_5$:Eu^{3+} co-doping with Bi^{3+} and Sm^{3+} ions much enhances red emission intensity [67], because low cost and water resistance, excellent thermal and chemical stability, high quantum efficiency, and silicate-based phosphors have been widely investigated [69–74].

For the single-crystal structure analysis, LaBSiO$_5$ crystal has approximately 0.15×0.075×0.075 mm^3 dimensions [66]. They exhibit 3D structures composed of BO$_4$ and RO$_4$ tetrahedra and LaO$_n$ (n = 10 for LaBSiO$_5$ and n=9 for LaBGeO$_5$) polyhedra. The BO$_4$ tetrahedron connects each other by corner sharing forming infinite helical chains along the c (and 31) axis, and each SiO$_4$ tetrahedron connects two neighbouring BO$_4$ tetrahedra by sharing O atoms. The helical chains are further linked together by La atoms forming the final 3D frameworks. The frequency-dependent optical properties of LaBRO$_5$ (R=Si and Ge) are calculated based on the crystal structures. It is clearly shown that the absorption area regions of LaBGeO$_5$ and LaBSiO$_5$ are from UV areas, at about 243 and 220 nm, respectively, which indicates the large band gaps [75]. Inspection has shown that during the last two decapods the area of the emission band changes from the UV range to red in various strontium borates (SrAl$_2$B$_2$O$_7$) of Eu^{2+} luminescence, depending on the host lattice [76–80].

In particular, as an alternative to conventional incandescent and fluorescent lamps, white light-emitting diodes (LEDs) have high potential utilities because of their superior advantages such as less energy consumption, high efficiency, easy maintenance, environmental benefit, and long life-time [69,81,82]. The electronic arrangement of metal oxides, specifically those depending upon metals and having d and f electrons, possesses a constant challenge to an ideologue, addressing the electronic structure property interconnection in these materials [83]. Firstly, Blasse reported the violet emission of MBPO$_5$:Eu^{2+} (M=Ca^{2+}, Sr^{2+}, Ba^{2+}) phosphor synthesized in H$_2$/N$_2$ reducing atmosphere [84]. Moreover, the phosphor SrBPO$_5$:Eu^{2+} showed efficient photo-stimulated luminescence at 390 nm by illumination with 635 nm laser light after neutron irradiation; thus, it is a possible neutron imaging material [46,85].

1.1.2 Silicates

Silicate matrix has been studied as a marvellous host because of its low cost, good chemical stability, high photoluminescence intensity, water resistance property, and predominantly very strong absorption in the near UV region [86–91]. Rare earth actuates light-emitting inorganic materials that have future applications in displays, lighting, and optical communication fields consisting of PDPs, cathode ray tubes, field emission displays, fibre amplifiers, lamps, etc. [86,92,93]. Silica NPs are multifaceted platforms with numerous intrinsic features such as low toxicity. Proper design and derivatization yield particularly stable colloids, even in physiological conditions, and provide them with multiple functions. Defect-related luminescent materials were mainly based on silica and silica-based materials before time, including SiO$_2$ gels [94,95], silicate carboxylate [96], silica nanotubes [97], SiO$_2$ spheres [98,99], molecular sieves [100,101], organic/inorganic hybrid silicone (Si) [102–104], and SiO$_2$ glass [105,106] [107–112]. The PL quantum product of these materials lies between 20% and 45% under UV (365 nm) excitation having a PL lifetime of less than 10 ns [22]. They have developed novel multicoloured phosphorescent CdSiO$_3$ doped and co-doped with transition metal ions [113].

Cuimiao Zhang et al. reported defect associated with luminescent materials and their applications, emission properties, and synthesis [22]. Sara Bonacchi et al. reported luminescence of silica NPs with extending frontier brightness [114]. Devender Singh et al. reported the optical characterization of Eu^{3+}-doped MLSiO$_4$ (M=Sr, Ba, Ca, and L=Mg) luminescent materials for display devices [86]. Carolina M. Abreu et al. reported the colour control of the persistent luminescence of cadmium silicate doped with transition metals by solid-state method [113].

1.1.3 ALUMINATE

For doping with transition and rare earth ions, alumina is examined as a rising host material to get an efficient luminescent approach [115]. It gives an effective atomic number nearer to that of the natural skeleton (Z_{eff} = 11.6–13.8) and becomes a good option for environmental and medical dosimetry studies [116–119]. To learn the luminescence characteristics of the synthesized SrAl$_2$O$_4$:Dy^{3+} phosphor, excitation peaks were noticed in the range of 200–400 nm and emission peaks were observed in between the range of 400 and 700 nm. This phosphor has two phases that are a raised temperature hexagonal phase (b-phase) and a lowering temperature monoclinic phase (a-phase). The five dominant peaks in the XRD patterns are located at 35.11, 29.98, 29.27, 28.46, and 19.96, and these essential peaks were recognized as the (0 1 1), (2 1 1), (2 2 0), (2 1 1), and (0 3 1) planes that indicate the crystalline SrAl$_2$O$_4$ shape [120]. From these materials, zinc aluminate (ZnAl$_2$O$_4$), with Fd3m space group having a spinel shape, also offers various advantages, such as high chemical and thermal stability, hydrophobic behaviour, low sintering temperature, high mechanical resistance, and high quantum yields [26–29].

Nano-ZnAl$_2$O$_4$ synthesized by a microwave method using metal nitrates and extract of aloe vera solution as precursors was first time reported by C. Ragupathi et al. [119]. Ishwar Prasad Sahu et al. reported a study of white light-emitting phosphor of dysprosium-doped strontium aluminate through the combustion method [120].

1.1.4 PHOSPHATE

Phosphates consisting of calcium and rare earth metals have captivated continuous attention because of their beneficial applications such as scintillation materials, fluorescence materials, and laser materials [121–123]. Calcium phosphate (Ca$_3$(PO$_4$)$_2$) is usually seen as a host compound for various derivatives consisting of alkaline and rare earth metals jointly. It has two stable forms, i.e. α- and β-modifications [123].

When a small quantity of alkali metal or alkaline earth metal ions (Ca^{2+}, Mg^{2+}, Sr^{2+}, Ba^{2+}, Na$^+$, K$^+$, or Li$^+$) were doped into the BPO$_4$ host lattice, then the luminescent intensities were enhanced [22].

Anees A. Ansari et al. reported that GdPO$_4$:Eu^{3+} (core) and GdPO$_4$:Eu, LaPO$_4$ (core/shell) nanorods (NRs) were successfully synthesized through urea-based co-precipitation method at ambient state followed by coating with amorphous silica shell by the sol–gel chemical synthesis [124]. Cuimiao Zhang et al. reported defect associated with luminescent materials and their applications, emission properties, and synthesis [22]. A. Terebilenko et al. discuss the synthesis of calcium phosphate by solid-state reaction and luminescence properties of it [125].

Several researchers have studied various metal oxide phosphors till date for different applications such as moderate synthesis temperature, high luminescence, low thermal degradation, and great X colour coordinate [126]. Metal oxides are widely used in many technological applications, such as coating, catalysis, electrochemistry, optical fibres, and sensors [127]. Metal oxide sensors have been utilized for several decades for their low cost and simplicity [128].

This kind of material and technology had a great impact on human life. At present, several studies are going on in these areas leading to optical applications such as human-centric lighting and plant growth. Metal oxide micro- and nano-crystallites that are enclosed by unusual high-energy facets exhibit enhanced performance in photodegradation, water splitting, gas sensing, catalysis, luminescence, and antibiosis [129].

Many types of NPs have been shown to be useful for such purposes. There are some metals and metal oxides, which are used as micronutrients for plant growth. When mixed with soil, they can be absorbed by plants. On a similar path, the respective NPs of metals and metal oxides can also be useful for plant growth. The advantage of using NPs of corresponding metals and metal oxide lies in their larger effect with a minimum amount. These NPs can promote plant growth. On the contrary, if they are being used beyond certain concentrations, they can affect the soil ecosystem and thus restrict the growth of soil microorganisms and the plants itself. Nano-SiO_2 improves seed germination and nutrient availability to maize plants [130]. It has been reported that the application of SiO_2 NPs to plants improves the photosynthetic rate by improving the activity of carbonic anhydrase enzyme (that supplies CO_2 to the Rubisco) and synthesis of photosynthetic pigments [131,132]. Similarly, the use of nano-anatase TiO_2 enhances photosynthesis by stimulating the Rubisco activity that could eventually increase the growth rate of plants [133,134]. Similarly, lettuce plants were challenged for nutrients on the application of CuO NPs [135–137].

Metal oxide NPs possess different physiochemical properties such as surface, optical, thermal, and electrical properties over their native bulk compounds [138]. The cellular stress, damage to cell membrane, and higher extent deactivation of the cellular defence systems were reported upon exposure to metal oxide NPs such as Al_2O_3 NPs, ZrO_2 NPs, ZnO NPs, and TiO_2 NPs [139]. A few other metal oxide NPs also shown a prominent impact on growth and physiology of plant [140].

1.2 APPLICATIONS OF METAL OXIDE LUMINESCENT MATERIALS

1.2.1 SCINTILLATOR

The characteristics of the Li_2O-B_2O_3 glass scintillator inspected in this study can be used for the detection of neutrons [141], energy transfer from Gd^{3+} to Ce^{3+} is efficient and plays an important role in the scintillation process. Also, in addition to increasing the glass density, Lu_2O_3 and La_2O_3 appear and can increase the scintillation light yields. The properties of these scintillating glasses make them potentially useful in high-energy physics experiments and gamma-ray imaging. The ratio of the scintillation and Cherenkov light can be easily tuned to a desired level by adjusting the Ce^{3+} doping level [142]. Vasilii M. Khanin et al. disclosed high-quality scintillators for medical imaging, mainly for computed tomography (CT) and positron-emission tomography (PET), where afterglow adversely affects image quality [14].

1.2.2 RADIATION DOSIMETRY MATERIALS

Occupational disclosure of ionizing radiation can occur in a range of medical institutions, educational and research establishments, nuclear fuel cycle facilities, and industries. Satisfactory diagnosis of radiation is requisites for the secure and acceptable utilization of radiation, nuclear energy, and radioactive materials. Various metal oxides are considered to be in use as sensitive elements for gas, temperature detectors, and humidity. For the extensive interpretation of the physical properties of radiation on metal oxides, materials under the influence of radiation exposure are essential for the worthwhile design of dosimeters [143]. The detection of radiation was carried out on the basis of the fact that structural, optical, and electrical properties of the materials switch upon the effect of gamma radiation [15,16].

MgO has also attracted the radiation dosimetry industry because of having a comparatively simple structure, a $Z_{eff} = 10.8$, low effective atomic number [144], and $E_g = 7.8\,eV$, wide band gap [1,2,145,146]. Although interest in this material was due to its potential use as a UV dosimeter and neutron dosimetry [3,147], there has been a tendency against its use as a commercial dosimetric material due to its radiation damage properties [148] regarded as a cost-effective alternative for room temperature real-time gamma-radiation dosimetry [149].

1.2.3 PHOTOTHERAPY

In 1903, Niels Finsen was honoured with the Nobel Prize for the implementation of red light and UV light to diagnose smallpox spots and cutaneous tuberculosis, which was known as the origination of "phototherapy" (Figure 1.1) [150,151].

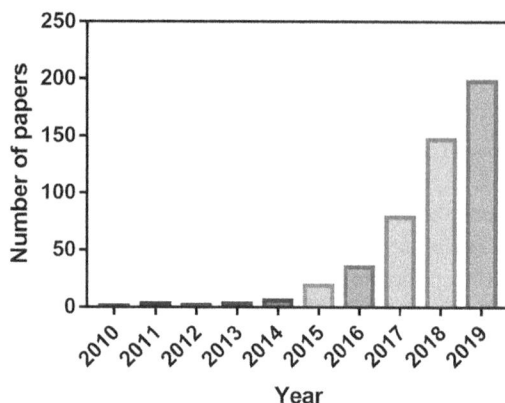

FIGURE 1.1 In the last decade, the number of publications that concentrated under the title of phototherapy using a metal–organic framework based on a Web of Science search conducted on March 31, 2020 [152].

Today, phototherapy is globally used to treat numerous diseases, such as cancer [153], vitiligo, psoriasis [154], atopic dermatitis, and acne vulgaris [155–157].

Phototherapy demonstrates scientific medicine in which basic research and latest technologies have grown up in parallel with clinical demand [4]. Intriguingly, WO x < 3 has attracted much more interest in cancer therapy, because of its absorption of near-infrared (NIR) light, which is essential for phototherapy. Photodynamic therapy (PDT) and photothermal therapy (PTT) have attracted immense interest in cancer therapy in view of their non-invasive nature [158–160].

1.2.4   VUV Application

The supported catalysts are more captivating among the transition metal oxides, due to their better catalytic stability and lower cost. The supports include molecular sieve [161], Al_2O_3 [162–164], TiO_2 [165], and SiO_2. This technology has a large development in air pollution control [166]. The application of monochromatic in several broadband crystals and VUV radiation is more useful for studying methods of defect formation in several broadband crystals since it also allows spectral dependence measurements of the production ability of the radiation to influence defects [167]. VUV photons are also used to a very large extent in PDPs and mercury-free fluorescent lamps. In these applications, phosphors are used to convert VUV excimer radiation to visible light [168].

1.2.5   Solid-State Lighting

SSL using LEDs is poised to reduce this value by at least 50%, so that lighting will then use less than one-tenth of all electricity generated. LED lighting will provide reductions of at least 10% in fuel consumption and carbon dioxide emissions from power stations within the next 5–10 years. Even greater reductions are likely on a 10- to 20-year timescale [169].

A set of Eu^{2+}-activated Li_2SrSiO_4 orange-yellow phosphor compositions reveal enormous emission under 400–470 nm excitation and are prepared by solid-state reaction, and their luminescence properties are investigated as a function of activator concentration (Eu^{2+}). This feature makes these phosphor compositions very attractive for application in LEDs. Li_2SrSiO_4:Eu^{2+} is a potential phosphor for SSL using InGaN (420 nm) to produce white light [170]. This new application of EPD of phosphor for white SSL is promising and offers easy tuning of emissive colour by varying phosphor blend compositions [171]. Colloidal quantum dots (QDs) emit bright, pure, and tuneable colours of light, making them excellent candidates for colour centres in next-generation display and SSL technologies [5,172]. The benefits of SSL in terms of energy savings, reduction in environmental pollution and new functionalities are reviewed. Solid-state light sources are in the process of profoundly changing the way we generate light for general lighting and other applications [173].

1.2.6 Spectral Modifier for Solar Cell Application

Considering the availability of a wide range of ternary metal oxides and their tuneable properties, the investigation of their applications in solar cells is interesting. $BaSnO_3$ (BSO) is also an interesting ternary metal oxide for photovoltaic applications [174]. Solar cells are devices containing semiconductors that exhibit the photovoltaic effect [175]. This technology is able to collect and convert solar radiation photons to electric energy, which then becomes readily available to supply a load or, when produced in surplus, to be stored for future uses. In theory, 1 hour of sunlight is more than enough for a whole year of global consumption.

Spectral modification is used to convert or shift the incident solar spectrum to better match the absorbed spectrum of the semiconductor to enhance the conversion efficiency of the solar cell. In 1995, Gibart et al. reported the first application of UC on solar cells. In 2002, Trupke et al. realized that DC materials can be used for increasing the efficiency of solar cells [176,177].

1.2.7 Solid-State Lasers

All this happened after the first invention of the lasing phenomenon by Snitzer in Nd^{3+}-doped glasses [178]. Nd^{3+} ion possesses a ladder-like energy-level structure (e.g. $^4I_{11/2}$ level lies about ~2,000 cm^{-1} above the $^4I_{9/2}$ ground state). It gives lasing emissions in the UV, visible (blue, green, orange, and red), and NIR regions following a four-level scheme when it is pumped at 808 nm ($^4I_{9/2}/^4F_{5/2}$, $^2H_{9/2}$) and 885 nm ($^4I_{9/2}/^4F_{3/2}$) laser diodes (diode-pumped solid-state laser (DPSSL) systems) [6,7,179]. Laser-based processing has been utilized in applications ranging from communications to material processing [180].

1.2.8 Optical Sensing

The optical properties of metal oxides are rich with opportunities as luminescence, dielectric function changes, and optically active dopants are all sensitive to environmental changes and can be utilized in a sensing device. The UV emission has been sought to be optimized for UV optical devices, and visible fluorescence has been primarily explored for sensing applications and visible light phosphors [181].

Metallic and other conducting NPs are used in optical sensor applications due to the phenomenon of localized surface plasmon resonance (LSPR) and the optical excitation of surface charge modes, which is strongly dependent upon the near-field environment of the NP. For optical sensing applications in high-temperature and chemically harsh environments, Au NPs provide several benefits: a strong LSPR response (typically in the visible region of the spectrum), resistance to chemical activity, and stability at high temperature [182].

The advantages of optical sensors include their sensitivity, selectivity, linearity, accuracy, limit of detection, dynamic range, reproducibility, and resolution [183]. The sensing potential of metal oxide (CuO and ZnO) NPs could be used for sensing metal ion (Li^+ and Ag^+) contamination from marine water samples through optical sensor devices [21]. A broadband increase in transmittance is also observed upon reducing gas exposure that competes with the near-IR absorption for intermediate wavelengths and may be potentially useful for optical sensing applications as well [184].

In optical sensors, changes/shifts in a wavelength of an optical output signal depend on an external stimulus, such as temperature, strain, stress, acceleration, bio-coating, or chemical environment. They are very sensitive, allow remote and distributed sensing, can be used in harsh environments, and are immune to electromagnetic interference [185].

1.2.9 Antimicrobial Technology

These features confer unique chemical and physical properties to MeO-NPs through which they interact with biological systems [186]. The different properties of MeONPs are the major contributors to antimicrobial activity. The alkalinity of the calcium oxide (CaO) and magnesium oxide (MgO) NP surface is a significant component in conferring antimicrobial activity [187]. MeO-NP antimicrobials may extend to wastewater treatment, food packaging, surface coatings on prosthetic and medical devices, dental implants, and textiles [188]. Cobalt oxide (Co_3O_4) NPs (Co_3O_4-I) showed antimicrobial activity on the two tested bacteria [189]. A broad spectrum of antimicrobial activity was revealed from their efficacy on both Gram-positive and Gram-negative bacteria by NiO NPs [190]. ZnO NPs showed antimicrobial effects against common pathogenic microorganisms [191]. The antimicrobial activity of metals such as silver (Ag), copper (Cu), gold (Au), titanium (Ti), and zinc (Zn), each having various properties, potencies, and spectra of activity, has been known and applied for centuries [192]. Metal NPs such as Ag, silver oxide (Ag_2O), titanium dioxide (TiO_2), Si, copper oxide (CuO), zinc oxide (ZnO), Au, CaO, and MgO were identified to exhibit antimicrobial activity. According to the literature, Ag NPs are the most popular inorganic NPs used as antimicrobial agents. The antimicrobial application of Ag additives is widely beneficial in the various injection-moulded plastic products, textiles, and coating-based usages [10]. Ag NPs also possess a range of biomedical applications [192]. It has been revealed that Ag NPs show a high antimicrobial activity comparable to that of its ionic form [193]. It has also been demonstrated that Ag NPs are potential antimicrobial agents against drug-resistant bacteria [194]. According to the literature, the antibacterial action of Ag NPs results from damage to the bacteria's outer membrane [195]. It has also been known that Ag ions interact with di-sulphide or sulfhydryl groups of enzymes that lead to disruption of metabolic processes, which in turn cause cell death [10]. In addition, the non-toxicity of Si NPs offers their use as antimicrobial agents in biomedical applications. We believe that the development of simple and low-cost inorganic antimicrobial agents such as metal and metal oxide NPs as an alternative to traditional antibiotics might be promising for the future of pharmaceutics and medicine [196].

New strategies to target metal toxicity can aid in overcoming the concerns associated with utilizing metal-based NPs as an antimicrobial agent in the clinical setting [197].

1.2.10 BIOIMAGING AND BIOMEDICAL IMAGING

More recently, metal oxide NPs have been used in a variety of biomedical applications due to their high surface area and enhanced magnetic and catalytic activity [11,198]. Many biomedical applications of metal oxides involve surface modification with biocompatible coatings, fluorescence tags, or biological molecules (e.g. antibodies and receptors) for targeted biological response and imaging purposes [199,200]. Natural or synthetic polymeric macromolecules have been used most frequently as NP coatings for biomedical applications [201].

ZnO NPs have exhibited promising biomedical applications based on their anticancer, antibacterial, antidiabetic, anti-inflammatory, drug delivery, and bioimaging activity [202]. These properties allow for its use in many biomedical applications such as gene delivery, bioimaging, drug delivery, cell labelling, and hyperthermia [203].

Iron oxide NPs (IONPs) are excellent candidates for biomedical imaging because of unique characteristics such as enhanced colloidal stability and excellent in vivo biocompatibility. Among NPs, iron oxide (Fe_3O_4) NPs (IONPs) have been widely utilized in biomedical applications due to instinctive characteristics, such as biocompatibility, good colloidal stability after appropriate surface coatings, high relaxation profiles, and tuneable surface modifications [204] IONPs have been widely explored for uses in biomedical techniques [205], especially as MRI contrast agents (Figure 1.2) [206].

Anticancer and antibacterial properties have been manipulated by the coupling of metal oxides such as MgO, NiO, and ZnO, which are the best candidates for biomedical applications. Furthermore, CdO is present in numerous cosmetics and anti-tan creams. Therefore, CdO can be regarded as chemically well matched with the body moiety. Furthermore, it is a suitable choice for biomedical applications because of its excellent optical features such as fluorescence and high resolution. Due to the rare optical, chemical, photoelectrochemical, and electrical properties, it has been demonstrated

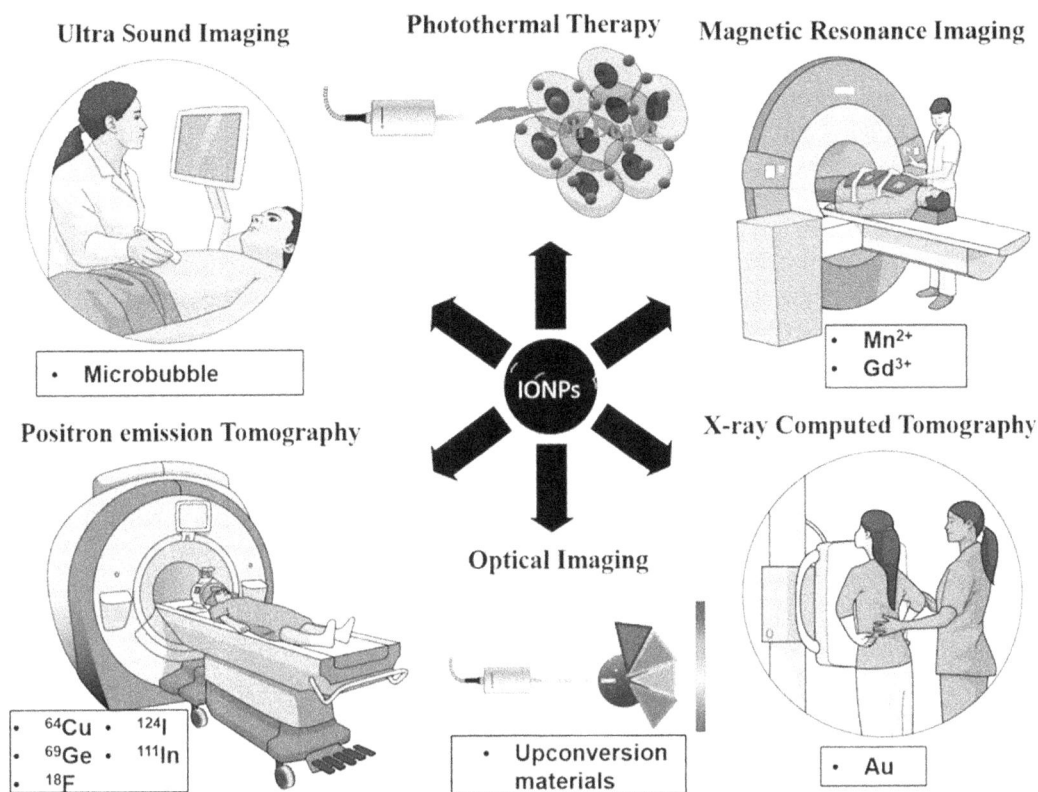

FIGURE 1.2 Schematic illustration of the biomedical application of IONPs [207].

that many heavy metals, including cadmium, can destroy cancer cells at minimum concentrations. The key mechanism of CdO NPs' influence on cancer cells is thought to be deoxyribonucleic acid (DNA) and protein damage and the destruction of the cell wall. Along with cadmium, some other heavy metals, such as Cu and Co, also have anticancer characteristics. The thought-provoking fact about the anticancer features of CdO nanostructures is that they are not harmful to human and mammalian cells.

Among several treatments for cancer therapy, very few nanomedicine-based formulations have been permitted for clinical usage and various other formulations are underneath medical experiments [208]. Anteneh Marelign Beyene et al. reported biomedical applications of curcumin nanoformulations with metal oxide nanomaterials [209].

1.2.11 OPTICAL THERMOMETRY

To measure body temperature, there is always required direct contact, whereas due to contactless functioning, immediate response, non-invasive operations, and inertness to the strong magnetic or electric field of optical thermometry, it is gaining popularity nowadays [210,211]. Moreover, applications in different areas consist of optical thermometer applications such as corrosive environments, electrically or magnetically hazardous conditions such as aerodynamics applications, electric power stations, cold storage facilities, and coal or other mines [12,13,212–214].

For example, the phonon-assisted energy transfer between Eu^{3+} and Tb^{3+} ions has been utilized in optical thermometry [215–219]. Inorganic optical thermometers show great potential in non-contact temperature sensing due to the excellent repeatability of inorganic materials [220].

1.2.12 Anticounterfeit

We have done a survey of the recent progress in anticounterfeiting applications using new collections of optical nanomaterials as security inks and point out that UCNPs are one of the most promising candidates for high-security-level anticounterfeiting applications [221,222]. Si NPs are advantageous in terms of their stability and biosafety, making them promising candidates in anticounterfeiting application. In recent years, there has been growing interest in developing dye-sensitized UCNPs for anticounterfeiting applications [223,224]. For example, Nd^{3+}- and Yb^{3+}-co-doped UCNPs with IR 808 dyes as sensitizers on the surface have been shown to be highly responsive to the excitation wavelength under scanning excitation in the range of 730–980 nm [225]. Anticounterfeiting ink is a direct and facile form of transforming a variety of NPs into real-world applications. Before they are widely accepted by different printing technologies [226], recent advances in the use of metal nanomaterials for optical anticounterfeit labels may offer a multiplexed approach to security tags that can be easily fabricated, offer large coding capacity, and be interrogated throughout the supply chain and by the end user [227].

Applications of luminescence attract more attention towards alkaline earth borophosphate lattices trigger with rare earth ions, because of its. Ce^{3+}-activated scintillation phosphors are extensively applicable for a range of purposes such as security inspection, nuclear medical imaging, and high-energy physics. High scintillation light yield, fast decay time, and high density used in these applications are the mass-desired properties for this type of phosphor [228]. The Ce^{3+} ion has PL excitation at around 230–350 nm and gives an emission at around 350–500 nm, and broadband emissions (375 nm) are noticed in all prepared borophosphate-based phosphors. From all these, we can say that the emission from Ce^{3+} ions has emission peak in the ultraviolet region. The PL emission spectra of Ce^{3+} ions in $SrBPO_5$ and $CaBPO_5$ phosphors may be useful for scintillation applications after a detailed study [204].

A large extent of experimental work has been carried out to probe the electronic shape of the host materials [229] and numerous dopants [230]. Strontium borophosphates are light sensitive, and they can be used in solar energy research [231]. However, in the solid-state rechargeable lithium-ion battery, lithium borophosphate (Li_2OBPO_4) is used as an electrolyte alkaline earth metal borophosphate powder <10 μm diameter (especially $CaBPO_4$) and is used for the corrosion protection of metal surface by evenly spreading in a polymer binder [42].

To enhance the red emission, the Eu^{3+} intensity of rare earth or transition metal ions is generally co-doped as sensitizers [232,233]. Luminescence thermometry is the most developed approach among optical thermometry, leaving apart general pyrometry. Lanthanide-doped salts and metal oxides were among the first investigated probes and are commonly the most choice materials [234]. The Mn(II), Co(II), Ni(II), and Li(I) metals were selected due to their significant achievement in the future of green energy and their estimated order in various associated applications such as energy-storing lithium-ion batteries [235]. Deep eutectic solvent phase has a potential interest in luminescence properties for specific applications consisting of organic LEDs [236] or metal sensing [237,238], and nanophosphors doped with rare earth ion have drawn special interest because of their applications in display technologies, biological fluorescence imaging and detection [239,240], lasers [2,241], IR quantum counters [242], and solar cells [18,19,243,244]. Moreover, further investigation into the alkaline earth borosilicate system had been a field of interest for its potential applications such as low-cost optical connectors, optical amplifiers, and phosphor materials [245]. Also, the applications of low-temperature co-fired ceramics (LTCC) of alkaline earth borosilicate in wireless communication and microwave products have recently immerged [19,246].

1.3 CONCLUSION

Metal oxide-based luminescent technologies play a vital role in various optical applications discussed in this chapter. Also, in the future there must be a game changer in the field of optical materials such as in applications including human-centric lighting and anticounterfeiting. Synthesis and characterizations of oxide materials play a vital role in technological developments. In the future, there is huge scope to synthesize existing and novel oxide-based materials.

REFERENCES

1. Oliveira, L. C., Doull, B. A and Yukihara, E. G. (2013). Investigations of MgO: Li, Gd thermally and optically stimulated luminescence. *Journal of Luminescence*, 137, 282–289, https://doi.org/10.1016/j.jlumin.2013.01.018.

2. Oliveira, L. C. D., Yukihara, E. G and Baffa, O. (2019). Lanthanide-doped MgO: A case study on how to design new phosphors for dosimetry with tailored luminescent properties. *Journal of Luminescence*, 209, 21–30, https://doi.org/10.1016/j.jlumin.2019.01.015.

3. Ritz, V. H. and Attix, F. H. (1973). Fast-neutron dosimetry using F centers in MgO. *Applied Physics Letters*, 23(3), 166–168, https://doi.org/10.1063/1.1654846.

4. Grossweiner, L. I., Grossweiner, J. B. and Rogers, B. G. (2005). *The Science of Phototherapy: An Introduction* (pp. 1–328). New York: Springer, https://doi.org/10.1007/1-4020-2885-7.

5. Shirasaki, Y., Supran, G. J., Bawendi, M. G and Bulović, V. (2013). Emergence of colloidal quantum-dot light-emitting technologies. *Nature Photonics*, 7(1), 13–23, https://doi.org/10.1038/nphoton.2012.328.

6. Skrzypczak, U., Seifert, G. and Schweizer, S. (2015). Highly efficient and broadband upconversion of NIR sunlight with neodymium-doped glass ceramics. *Advanced Optical Materials*, 3(4), 541–545, https://doi.org/10.1002/adom.201400512.

7. Lakshminarayana, G., Kaky, K. M., Baki, S. O., Lira, A., Meza-Rocha, A. N., Falcony, C. and Mahdi, M. A. (2018). Nd3+-doped heavy metal oxide based multicomponent borate glasses for 1.06 µm solid-state NIR laser and O-band optical amplification applications. *Optical Materials*, 78, 142–159, https://doi.org/10.1016/j.optmat.2018.02.011.

8. Shin, S. S., Yang, W. S., Noh, J. H., Suk, J. H., Jeon, N. J., Park, J. H. and Seok, S. I. (2015). High-performance flexible perovskite solar cells exploiting Zn2SnO4 prepared in solution below 100° C. *Nature Communications*, 6(1), 7410, https://doi.org/10.1038/ncomms8410.

9. Jha, A., Richards, B., Jose, G., Teddy-Fernandez, T., Joshi, P., Jiang, X. and Lousteau, J. (2012). Rare-earth ion doped TeO2 and GeO2 glasses as laser materials. *Progress in Materials Science*, 57(8), 1426–1491, https://doi.org/10.1016/j.pmatsci.2012.04.003.

10. Egger, S., Lehmann, R. P., Height, M. J., Loessner, M. J. and Schuppler, M. (2009). Antimicrobial properties of a novel silver-silica nanocomposite material. *Applied and Environmental Microbiology*, 75(9), 2973–2976, https://doi.org/10.1128/AEM.01658-08.

11. Donaldson, K. and Seaton, A. (2007). The Janus faces of nanoparticles. *Journal of Nanoscience and Nanotechnology*, 7(12), 4607–4611, https://doi.org/10.1166/jnn.2007.18113.

12. Shinde, S. L. and Nanda, K. K. (2013). Wide-range temperature sensing using highly sensitive green-luminescent ZnO and PMMA-ZnO film as a non-contact optical probe. *Angewandte Chemie International Edition*, 52(43), 11325–11328, https://doi.org/10.1002/anie.201302449.

13. Senapati, S. and Nanda, K. K. (2017). Red emitting Eu: ZnO nanorods for highly sensitive fluorescence intensity ratio based optical thermometry. *Journal of Materials Chemistry C*, 5(5), 1074–1082, https://doi.org/10.1039/C6TC04296A.

14. Khanin, V. M., Venevtsev, I., Chernenko, K., Tukhvatulina, T., Rodnyi, P. A., Spoor, S. and Meijerink, A. (2020). Influence of 3d transition metal impurities on garnet scintillator afterglow. *Crystal Growth and Design*, 20(5), 3007–3017, https://doi.org/10.1021/acs.cgd.9b01660.

15. Arshak, A., Arshak, K., Korostynska, O and Zleetni, S. (2003, October). Review of various gamma radiation dosimeters based on thin and thick films of metal oxides and polymer materials. In *2003 IEEE Nuclear Science Symposium. Conference Record (IEEE Cat. No. 03CH37515)* (Vol. 1, pp. 78–82). IEEE, doi: 10.1109/NSSMIC.2003.1352002.

16. Arshak, K., Arshak, A., Zleetni, S. and Korostynska, O. (2004). Thin and thick films of metal oxides and metal phthalocyanines as gamma radiation dosimeters. *IEEE Transactions on Nuclear Science*, 51(5), 2250–2255, doi: 10.1109/TNS.2004.834718.

17. Auzel, F. (2004). Upconversion and anti-stokes processes with f and d ions in solids. *Chemical Reviews*, 104(1), 139–174, https://doi.org/10.1021/cr020357g.

18. Su, L. T., Karuturi, S. K., Luo, J., Liu, L., Liu, X., Guo, J. and Tok, A. I. Y. (2013). Photon upconversion in hetero-nanostructured photoanodes for enhanced near-infrared light harvesting. *Advanced Materials*, 25(11), 1603–1607, https://doi.org/10.1002/adma.201204353.

19. Aarts, L., Van der Ende, B. M. and Meijerink, A. (2009). Downconversion for solar cells in NaYF 4: Er, Yb. *Journal of Applied Physics*, 106(2), 023522, https://doi.org/10.1063/1.3177257.

20. Falcaro, P., Ricco, R., Yazdi, A., Imaz, I., Furukawa, S., Maspoch, D. and Doonan, C. J. (2016). Application of metal and metal oxide nanoparticles@ MOFs. *Coordination Chemistry Reviews*, 307, 237–254, https://doi.org/10.1016/j.ccr.2015.08.002.

21. Maruthupandy, M., Zuo, Y., Chen, J. S., Song, J. M., Niu, H. L., Mao, C. J. and Shen, Y. H. (2017). Synthesis of metal oxide nanoparticles (CuO and ZnO NPs) via biological template and their optical sensor applications. *Applied Surface Science*, 397, 167–174, https://doi.org/10.1016/j.apsusc.2016.11.118

22. Zhang, C. and Lin, J. (2012). Defect-related luminescent materials: Synthesis, emission properties and applications. *Chemical Society Reviews*, 41(23), 7938–7961, https://doi.org/10.1039/C2CS35215J.

23. Elrafei, S., Kandas, I., Shehata, N. and Samir, E. (2018). Efficiency improvement of up-conversion process of plasmonic-enhanced Er-doped-NaYF 4 nanoparticles under IR excitation. *Optics Express*, 26(19), 25492–25506, https://doi.org/10.1364/OE.26.025492.

24. Hassairi, M. A., Dammak, M., Zambon, D., Chadeyron, G. and Mahiou, R. (2017). Red-green-blue upconversion luminescence and energy transfer in Yb3+/Er3+/Tm3+ doped YP5O14 ultraphosphates. *Journal of Luminescence*, 181, 393–399, https://doi.org/10.1016/j.jlumin.2016.09.054.

25. Ritter, B., Haida, P., Fink, F., Krahl, T., Gawlitza, K., Rurack, K. and Kemnitz, E. (2017). Novel and easy access to highly luminescent Eu and Tb doped ultra-small CaF 2, SrF 2 and BaF 2 nanoparticles-structure and luminescence. *Dalton Transactions*, 46(9), 2925–2936, https://doi.org/10.1039/C6DT04711D.

26. Pandey, R., Gale, J. D., Sampath, S. K. and Recio, J. M. (1999). Atomistic simulation study of spinel oxides: Zinc aluminate and zinc gallate. *Journal of the American Ceramic Society*, 82(12), 3337–334, https://doi.org/10.1111/j.1151-2916.1999.tb02248.x.

27. Wrzyszcz, J., Zawadzki, M., Trawczyński, J., Grabowska, H. and Miśta, W. (2001). Some catalytic properties of hydrothermally synthesised zinc aluminate spinel. *Applied Catalysis A: General*, 210(1–2), 263–269, https://doi.org/10.1016/S0926-860X(00)00821-8.

28. Mathur, S., Veith, M., Haas, M., Shen, H., Lecerf, N., Huch, V. and Jilavi, M. (2001). Single-source sol-gel synthesis of nanocrystalline ZnAl2O4: Structural and optical properties. *Journal of the American Ceramic Society*, 84(9), 1921–1928, https://doi.org/10.1111/j.1151-2916.2001.tb00938.x.

29. Yang, Y., Sun, X. W., Tay, B. K., Wang, J. X., Dong, Z. L. and Fan, H. M. (2007). Twinned Zn2TiO4 spinel nanowires using ZnO nanowires as a template. *Advanced Materials*, 19(14), 1839–1844, https://doi.org/10.1002/adma.200700299.

30. Gawande, A. B., Sonekar, R. P. and Omanwar, S. K. (2014). Synthesis and enhancement of luminescence intensity by co-doping of M+ (M= Li, Na, K) in Ce³⁺ doped strontium haloborate. *Optical Materials*, 36(7), 1143–1145, https://doi.org/10.1016/j.optmat.2014.02.017.

31. Gawande, A. B., Sonekar, R. P. and Omanwar, S. K. (2014). Combustion synthesis and energy transfer mechanism of Bi3+→ Gd3+ and Pr3+→ Gd3+ in YBO3. *Combustion Science and Technology*, 186(6), 785–791, https://doi.org/10.1080/00102202.2013.878708.

32. He, H., Fu, R., Song, X., Wang, D. and Chen, J. (2008). White light-emitting Mg0. 1Sr1. 9SiO4: Eu2+ phosphors. *Journal of Luminescence*, 128(3), 489–493, https://doi.org/10.1016/j.jlumin.2007.09.023.

33. Sonekar, R. P., Omanwar, S. K., Moharil, S. V., Dhopte, S. M., Muthal, P. L. and Kondawar, V. K. (2007). Combustion synthesis of narrow UVB emitting rare earth borate phosphors. *Optical Materials*, 30(4), 622–625, https://doi.org/10.1016/j.optmat.2007.02.016.

34. Marzouk, M. A. and Abdel-Hameed, S. A. M. (2019). Crystallization and photoluminescent properties of Eu, Gd, Sm, Nd co-doped SrAl2B2O7 nanocrystals phosphors prepared by glass-ceramic technique. *Journal of Luminescence*, 205, 248–257, https://doi.org/10.1016/j.jlumin.2018.09.019.

35. Yang, Y., Bao, A., Lai, H., Tao, Y. and Yang, H. (2009). Luminescent properties of SrAl2B2O7: Ce3+, Tb3+. *Journal of Physics and Chemistry of Solids*, 70(10), 1317–1321, https://doi.org/10.1016/j.jpcs.2009.06.012.

36. Chen, X., Jiang, L., Zhou, L., He, M., Li, J., Wu, A. and Chen, H. (2020). Up-conversion luminescence properties and energy transfer mechanism of SrAl2B2O7: Er3+@ Yb3+ phosphors. *Journal of Alloys and Compounds*, 840, 155489, https://doi.org/10.1016/j.jpcs.2009.06.012.

37. Sologubenko, A. V., Jun, J., Kazakov, S. M., Karpinski, J. and Ott, H. R. (2002). Thermal conductivity of single-crystalline MgB 2. *Physical Review B*, 66(1), 014504, https://doi.org/10.1103/PhysRevB.66.014504.

38. Gul, G. Ç. and Kurtulus, F. (2016). Rietveld refinement, optical and photoluminescence properties of blue emitting phosphors RE (Y, Er, Gd, La, Nd)-doped CaBPO5. *Optik*, 127(24), 11674–11680, https://doi.org/10.1016/j.ijleo.2016.09.097.

39. Pan, S., Wu, Y., Fu, P., Wang, G., Zhang, G., Li, Z. and Chen, C. (2001). Flux growth of single crystals of BaBPO5. *Chemistry Letters*, 30(7), 628–629, https://doi.org/10.1246/cl.2001.628.

40. Bauer, H. (1965). Über eine Reihe isotyper Erdalkaliboratphosphate und-arsenate vom Typus 2 MeO· X2O5· B2O3. *Zeitschrift für anorganische und allgemeine Chemie*, 337(3–4), 183–190, https://doi.org/10.1002/zaac.19653370311.

41. Kniep, R., Gözel, G., Eisenmann, B., Röhr, C., Asbrand, M. and Kizilyalli, M. (1994). Borophosphates-A neglected class of compounds: Crystal structures of Mii [BPO5](Mii= Ca, Sr) and Ba3 [BP$_3$O$_{12}$]. *Angewandte Chemie International Edition in English*, 33(7), 749–751, https://doi.org/10.1002/anie.199407491.

42. Baykal, A., Kizilyalli, M., Gözel, G. and Kniep, R. (2000). Synthesis of strontium borophosphate, SrBPO5 by solid state and hydrothermal methods and characterisation. *Crystal Research and Technology: Journal of Experimental and Industrial Crystallography*, 35(3), 247–254, https://doi.ohttps://doi.org/10.1002/1521-4079(200003)35:rg/10.1002/1521-4079(200003)35:3<247::AID-CRAT247>3.0.CO;2–9.

43. Bauer, H. (1966). Über eine Reihe isotyper Erdalkaliboratphosphate und-arsenate vom Typus 2MeO· X2O5· B2O3. II. Die Verbindungen 2BaO· P2O5· B2O3 und 2BaO· As2 O5· B2O3. *Zeitschrift für anorganische und allgemeine Chemie*, 345(5–6), 225–229, https://doi.org/10.1002/zaac.19663450502.

44. Park, C. H. and Bluhm, K. (1995). Synthese und Kristallstruklur von Triblei-Diphosphato-Borat-Phosphat, eine Verbindung mit einem 1∞[(PO4) 2BPO4] 6– Anion/synthesis and crystal structure of trilead-diphosphato-borate-phosphate, a compound with a 1∞[(PO4) 2BPO4] 6–anion. *Zeitschrift für Naturforschung B*, 50(11), 1617–1622, https://doi.org/10.1515/znb-1995-1107.

45. Liebertz, J. and Stähr, S. (1982). Zur Existenz und Einkristallzüchtung von Zn3BPO7 und Mg3BPO7? *Zeitschrift für Kristallographie-Crystalline Materials*, 160(1–4), 135–138, https://doi.org/10.1524/zkri.1982.160.14.135.

46. Liang, H., Shi, J., Su, Q., Zhang, S. and Tao, Y. (2005). Spectroscopic properties of Ce3+ doped MBPO5 (M= Ca, Sr, Ba) under VUV excitation. *Materials Chemistry and Physics*, 92(1), 180–184, https://doi.org/10.1016/j.matchemphys.2005.01.012.

47. Srivastava, A. M. (1998). Luminescence of divalent bismuth in M2+ BPO5 (M2+= Ba2+, Sr2+ and Ca2+). *Journal of Luminescence*, 78(4), 239–243, https://doi.org/10.1016/S0022-2313(98)00010-6.

48. Shi, Y., Liang, J., Zhang, H., Liu, Q., Chen, X., Yang, J. and Rao, G. (1998). Crystal structure and thermal decomposition studies of barium borophosphate, BaBPO5. *Journal of Solid State Chemistry*, 135(1), 43–51, https://doi.org/10.1006/jssc.1997.7588.

49. Pan, S., Wu, Y., Fu, P., Zhang, G., Li, Z., Du, C. and Chen, C. (2003). Growth, structure and properties of single crystals of SrBPO5. *Chemistry of Materials*, 15(11), 2218–2221, https://doi.org/10.1021/cm020878k.

50. Toraman, Ö. Y. and Depçi, T. (2007). Applications of microwave pretreatment in coal. *Scientific Mining Journal*, 46(3), 43–53, https://www.mining.org.tr/en/pub/issue/32494/361193.

51. Schaber, P. M., Colson, J., Higgins, S., Thielen, D., Anspach, B. and Brauer, J. (2004). Thermal decomposition (pyrolysis) of urea in an open reaction vessel. *Thermochimica Acta*, 424(1–2), 131–142, https://doi.org/10.1016/j.tca.2004.05.018.

52. Martínez-García, A., Ortiz, M., Martínez, R., Ortiz, P. and Reguera, E. (2004). The condensation of furfural with urea. *Industrial Crops and Products*, 19(2), 99–106, https://doi.org/10.1016/j.indcrop.2003.07.004.

53. Wang, Y., Endo, T., Xie, E., He, D. and Liu, B. (2004). Luminescence properties of Ca4GdO (BO3) 3: Eu in ultraviolet and vacuum ultraviolet regions. *Microelectronics Journal*, 35(4), 357–361, https://doi.org/10.1016/S0026-2692(03)00245-3.

54. Hayakawa, T., Selvan, S. T. and Nogami, M. (2000). Influence of adsorbed CdS nanoparticles on 5D0→7FJ emissions in Eu3+-doped silica gel. *Journal of Luminescence*, 87, 532–534, https://doi.org/10.1016/S0022-2313(99)00280-X.

55. Kwon, I. E., Yu, B. Y., Bae, H., Hwang, Y. J., Kwon, T. W., Kim, C. H. and Kim, S. J. (2000). Luminescence properties of borate phosphors in the UV/VUV region. *Journal of Luminescence*, 87, 1039–1041, https://doi.org/10.1016/S0022-2313(99)00532-3.

56. Okazaki, C., Shiiki, M., Suzuki, T. and Suzuki, K. (2000). Luminance saturation properties of PDP phosphors. *Journal of Luminescence*, 87, 1280–1282, https://doi.org/10.1016/S0022-2313(99)00575-X.

57. Jüstel, T., Krupa, J. C. and Wiechert, D. U. (2001). VUV spectroscopy of luminescent materials for plasma display panels and Xe discharge lamps. *Journal of Luminescence*, 93(3), 179–189, https://doi.org/10.1016/S0022-2313(01)00199-5.

58. Zhang, X., Zhang, J., Liang, L. and Su, Q. (2005). Luminescence of SrGdGa3O7: RE3+ (RE= Eu, Tb) phosphors and energy transfer from Gd3+ to RE3+. *Materials Research Bulletin*, 40(2), 281–288, https://doi.org/10.1016/j.materresbull.2004.10.011.

59. Ruan, S. K., Zhou, J. G., Zhong, A. M., Duan, J. F., Yang, X. B. and Su, M. Z. (1998). Synthesis of Y3Al5O12: Eu3+ phosphor by sol-gel method and its luminescence behavior. *Journal of Alloys and Compounds*, 275, 72–75, https://doi.org/10.1016/S0925-8388(98)00276-X.

60. Kim, C. H., Kwon, I. E., Park, C. H., Hwang, Y. J., Bae, H. S., Yu, B. Y. and Hong, G. Y. (2000). Phosphors for plasma display panels. *Journal of Alloys and Compounds*, 311(1), 33–39, https://doi. org/10.1016/S0925-8388(00)00856-2.

61. Lo, C. L., Duh, J. G., Chiou, B. S., Peng, C. C. and Ozawa, L. (2001). Synthesis of Eu3+-activated yttrium oxysulfide red phosphor by flux fusion method. *Materials Chemistry and Physics*, 71(2), 179–189, https://doi.org/10.1016/S0254-0584(01)00279-6.

62. Ronda, C. R. (1997). Recent achievements in research on phosphors for lamps and displays. *Journal of Luminescence*, 72, 49–54, https://doi.org/10.1016/S0022-2313(96)00374-2.

63. Yang, Y., Ren, Z., Tao, Y., Cui, Y. and Yang, H. (2009). Eu3+ emission in SrAl2B2O7 based phosphors. *Current Applied Physics*, 9(3), 618–621, https://doi.org/10.1016/j.cap.2008.05.015.

64. Leonyuk, N. I., Belokoneva, E. L., Bocelli, G., Righi, L., Shvanskii, E. V., Henrykhson, R. V. and Kozhbakhteeva, D. E. (1999). High-temperature crystallization and X-ray characterization of Y2SiO5, Y2Si2O7 and LaBSiO5. *Journal of Crystal Growth*, 205(3), 361–367, https://doi.org/10.1016/ S0022-0248(99)00233-X.

65. Leonyuk, N. I., Belokoneva, E. L., Bocelli, G., Righi, L., Shvanskii, E. V., Henrykhson, R. V. and Kozhbakhteeva, D. E. (1999). Crystal growth and structural refinements of the Y2SiO5, Y2Si2O7 and LaBSiO5 single crystals. *Crystal Research and Technology: Journal of Experimental and Industrial Crystallography*, 34(9), 1175–1182, https://doi.org/10.1002/ (SICI)1521-4079(199911)34:9<1175::AID-CRAT1175>3.0.CO;2–2.

66. Chi, L., Chen, H., Zhuang, H. and Huang, J. (1997). Crystal structure of LaBSiO5. *Journal of Alloys and Compounds*, 252(1–2), L12–L15, https://doi.org/10.1002/bio.4301.

67. Wang, Z., Cheng, P., Liu, Y., Zhou, Y., Zhou, Q. and Guo, J. (2014). Novel red phosphors LaBSiO5 co-doped with Eu3+, Al3+ for near-UV light-emitting diodes. *Optical Materials*, 37, 277–280, https://doi. org/10.1016/j.optmat.2014.06.006.

68. Wang, Z., Cheng, P., He, P., Liu, Y., Zhou, Y. and Zhou, Q. (2015). Luminescent properties and energy transfer in the green phosphors LaBSiO5: Tb3+, Ce3+. *Luminescence*, 30(6), 719–722, https://doi. org/10.1002/bio.2810.

69. Xue, Y. N., Xiao, F. and Zhang, Q. Y. (2011). Enhanced red light emission from LaBSiO5: Eu3+, R3+ (R= Bi or Sm) phosphors. *Spectrochimica Acta Part A: Molecular and Biomolecular Spectroscopy*, 78(2), 607–611, https://doi.org/10.1016/j.saa.2010.11.030.

70. Wang, J., Zhang, M., Zhang, Q., Ding, W. and Su, Q. (2007). The photoluminescence and thermoluminescence properties of novel green long-lasting phosphorescence materials Ca 8 Mg (SiO 4) 4 Cl 2: Eu 2+, Nd 3+. *Applied Physics B*, 87, 249–254, https://doi.org/10.1007/s00340-007-2590-1.

71. Hirai, T. and Kondo, Y. (2007). Preparation of Y2SiO5: Ln3+ (Ln= Eu, Tb, Sm) and Gd9. 33 (SiO4) 6O2: Ln3+ (Ln= Eu, Tb) phosphor fine particles using an emulsion liquid membrane system. *The Journal of Physical Chemistry C*, 111(1), 168–174, https://doi.org/10.1021/jp064655j.

72. Ye, S., Wang, C. H. and Jing, X. P. (2008). Photoluminescence and Raman spectra of double-perovskite Sr2Ca (Mo/W) O6 with A-and B-site substitutions of Eu3+. *Journal of the Electrochemical Society*, 155(6), J148, doi: 10.1149/1.2898897.

73. Cooke, D. W., Lee, J. K., Bennett, B. L., Groves, J. R., Jacobsohn, L. G., McKigney, E. A. and Hong, K. S. (2006). Luminescent properties and reduced dimensional behavior of hydrothermally prepared Y 2 Si O 5: Ce nanophosphors. *Applied Physics Letters*, 88(10), 103108, https://doi.org/10.1063/1.2183737.

74. Zhang, Q. Y., Pita, K., Ye, W. and Que, W. X. (2002). Influence of annealing atmosphere and temperature on photoluminescence of Tb3+ or Eu3+-activated zinc silicate thin film phosphors via sol-gel method. *Chemical Physics Letters*, 351(3–4), 163–170, https://doi.org/10.1016/S0009-2614(01)01370-7.

75. Li, L., Jing, Q., Yang, Z., Su, X., Lei, B. H., Pan, S. and Zhang, J. (2015). Effect of the tetrahedral groups on the optical properties of LaBRO5 (R= Si and Ge): A first-principles study. *Journal of Applied Physics*, 118(11), 113104, https://doi.org/10.1063/1.4930224.

76. Machida, K., Adachi, G and Shiokawa, J. (1979). Luminescence properties of Eu (II)-borates and Eu2+-activated Sr-borates. *Journal of Luminescence*, 21(1), 101–110, https://doi. org/10.1016/0022-2313(79)90038-3.

77. Schipper, W. J., Van der Voort, D., Van den Berg, P., Vroon, Z. A. E. P and Blasse, G. (1993). The luminescence of europium in strontium borates. *Materials Chemistry and Physics*, 33(3–4), 311–317, https:// doi.org/10.1016/0254-0584(93)90080-6.

78. Diaz, A and Keszler, D. A. (1996). Red, green and blue Eu2+ luminescence in solid-state borates: A structure-property relationship. *Materials Research Bulletin*, 31(2), 147–151, https://doi. org/10.1016/0025-5408(95)00182-4.

79. Diaz, A and Keszler, D. A. (1997). Eu2+ luminescence in the borates X2Z (BO3) 2 (X= Ba, Sr; Z= Mg, Ca). *Chemistry of Materials*, 9(10), 2071–2077, https://doi.org/10.1021/cm9700817.

80. Lucas, F., Jaulmes, S., Quarton, M., Le Mercier, T., Guillen, F and Fouassier, C. (2000). Crystal structure of SrAl2B2O7 and Eu2+ luminescence. *Journal of Solid State Chemistry*, 150(2), 404–409, https://doi.org/10.1006/jssc.1999.8616.

81. Nishida, T., Ban, T. and Kobayashic, N. (2003). High 郢 color 郢 rendering light sources consisting of a 350 郢 nm ultraviolet light 郢 emitting diode and three 郢 basal 郢 color phosphors. *Applied Physics Letters*, 82(22), 3817, https://doi.org/10.1063/1.1580649.

82. Zhang, Q. Y., Yang, C. H. and Pan, Y. X. (2007). Enhanced white light emission from GdAl3 (BO3) 4: Dy3+, Ce3+ nanorods. *Nanotechnology*, 18(14), 145602, doi: 10.1088/0957-4484/18/14/145602.

83. Slater, J. C. (1974). *The Self-consistent Field for Molecules and Solids: Quantum Theory of Molecules and Solids*. McGraw-Hill Book Company, https://doi.org/10.5796/electrochemistry.72.54.

84. Blasse, G., Bril, A. and De Vries, J. (1969). Luminescence of alkaline-earth borate-phosphates activated with divalent europium. *Journal of Inorganic and Nuclear Chemistry*, 31(2), 568–570, https://doi.org/10.1016/0022-1902(69)80502-6.

85. Sakasai, K., Katagiri, M., Toh, K., Takahashi, H., Nakazawa, M. and Kondo, Y. (2002). A SrBPO 5: Eu 2+ storage phosphor for neutron imaging. *Applied Physics A*, 74, s1589–s1591, https://doi.org/10.1007/s003390101255.

86. Singh, D., Sheoran, S. and Singh, J. (2018). Optical characterization of Eu 3+ doped MLSiO 4 (M= Ca, Sr, Ba and L= Mg) phosphor materials for display devices. *Journal of Materials Science: Materials in Electronics*, 29, 294–302, https://doi.org/10.1007/s10854-017-7916-0.

87. Sahu, I. P., Chandrakar, P., Baghel, R. N., Bisen, D. P., Brahme, N. and Tamrakar, R. K. (2015). Luminescence properties of dysprosium doped calcium magnesium silicate phosphor by solid state reaction method. *Journal of Alloys and Compounds*, 649, 1329–1338, https://doi.org/10.1016/j.jallcom.2015.06.011.

88. Singh, D., Sheoran, S., Tanwar, V. and Bhagwan, S. (2017). Optical characteristics of Eu (III) doped MSiO 3 (M= Mg, Ca, Sr and Ba) nanomaterials for white light emitting applications. *Journal of Materials Science: Materials in Electronics*, 28, 3243–3253, https://doi.org/10.1007/s10854-016-5914-2.

89. Singh, D., Sheoran, S., Bhagwan, S. and Kadyan, S. (2016). Optical characteristics of sol-gel derived M3SiO5: Eu3+ (M= Sr, Ca and Mg) nanophosphors for display device technology. *Cogent Physics*, 3(1), 1262573, https://doi.org/10.1016/j.optmat.2022.112945.

90. Singh, D. and Sheoran, S. (2016). Synthesis and luminescent characteristics of M 3 Y 2 Si 3 O 12: Eu 3+(M= Ca, Mg, Sr and Ba) nanomaterials for display applications. *Journal of Materials Science: Materials in Electronics*, 27, 12707–12718, https://doi.org/10.1007/s10854-016-5405-5.

91. Singh, D., Sheoran, S. and Tanwar, V. (2017). Europium doped silicate phosphors: Synthetic and characterization techniques. *Advanced Materials Letters*, 8(5), 656–672, doi: 10.5185/amlett.2017.7011.

92. Xiaojun, Z. H. O. U., Xiong, Z., Hao, X. U. E., Yisen, L. I. N. and Chunxiao, S. O. N. G. (2015). Hydrothermal synthesis and photoluminescent properties of Li2Sr0. 996SiO4: Pr3+ 0.004 phosphors for white-LED lightings. *Journal of Rare Earths*, 33(3), 244–248, https://doi.org/10.1016/S1002-0721(14)60410-5.

93. Singh, J., Baitha, P. K. and Manam, J. (2015). Influence of heat treatment on the structural and optical properties of SrGd2O4: Eu3+ phosphor. *Journal of Rare Earths*, 33(10), 1040–1050, https://doi.org/10.1016/S1002-0721(14)60524-X.

94. Cordoncillo, E., Guaita, F. J., Escribano, P., Philippe, C., Viana, B. and Sanchez, C. (2001). Blue emitting hybrid organic-inorganic materials. *Optical Materials*, 18(3), 309–320, https://doi.org/10.1016/S0925-3467(01)00170-7.

95. Lin, J. and Baerner, K. (2000). Tunable photoluminescence in sol-gel derived silica xerogels. *Materials Letters*, 46(2–3), 86–92, https://doi.org/10.1016/S0167-577X(00)00147-6.

96. Green, W. H., Le, K. P., Grey, J., Au, T. T. and Sailor, M. J. (1997). White phosphors from a silicate-carboxylate sol-gel precursor that lack metal activator ions. *Science*, 276(5320), 1826–1828, doi: 10.1126/science.276.5320.1826.

97. Jang, J. Y. O. N. G. S. I. K. and Yoon, H. Y. E. O. N. S. E. O. K. (2004). Novel fabrication of size-tunable silica nanotubes using a reverse-microemulsion-mediated sol-gel method. *Advanced Materials*, 16(9–10), 799–802, https://doi.org/10.1002/adma.200306567.

98. Jakob, A. M. and Schmedake, T. A. (2006). A novel approach to monodisperse, luminescent silica spheres. *Chemistry of Materials*, 18(14), 3173–3175, https://doi.org/10.1021/cm060664t.

99. Kong, D., Zhang, C., Xu, Z., Li, G., Hou, Z. and Lin, J. (2010). Tunable photoluminescence in monodisperse silica spheres. *Journal of Colloid and Interface Science*, 352(2), 278–284, https://doi.org/10.1016/j.jcis.2010.08.054.

100. Gimon-Kinsel, M. E., Groothuis, K. and Balkus Jr, K. J. (1998). Photoluminescent properties of MCM-41 molecular sieves. *Microporous and Mesoporous Materials*, 20(1–3), 67–76, https://doi.org/10.1016/S1387-1811(97)00004-8.

101. Zhang, J. and Lin, J. (2004). Comparative study on the photoluminescent properties of siliceous MCM-41 with silica particles and xerogels. *Microporous and Mesoporous Materials*, 75(1–2), 115–120, https://doi.org/10.1016/j.micromeso.2004.07.015.

102. Carlos, L. D., Sá Ferreira, R. A., Pereira, R. N., Assunção, M. and de Zea Bermudez, V. (2004). White-light emission of amine-functionalized organic/inorganic hybrids: Emitting centers and recombination mechanisms. *The Journal of Physical Chemistry B*, 108(39), 14924–14932, https://doi.org/10.1021/jp049052r.

103. Fu, L., Sá Ferreira, R. A., Silva, N. J. O., Carlos, L. D., de Zea Bermudez, V. and Rocha, J. (2004). Photoluminescence and quantum yields of urea and urethane cross-linked nanohybrids derived from carboxylic acid solvolysis. *Chemistry of Materials*, 16(8), 1507–1516, https://doi.org/10.1021/cm035028z.

104. Hayakawa, T., Hiramitsu, A. and Nogami, M. (2003). White light emission from radical carbonyl-terminations in Al 2 O 3-SiO 2 porous glasses with high luminescence quantum efficiencies. *Applied Physics Letters*, 82(18), 2975–2977, https://doi.org/10.1063/1.1569038.

105. Yoldas, B. E. (1992). Thermochemically induced photoluminescence in sol-gel-derived oxide networks. *Journal of Non-crystalline Solids*, 147, 614–620, https://doi.org/10.1016/S0022-3093(05)80686-6.

106. Chen, Z., Wang, Y., He, H., Zou, Y., Wang, J. and Li, Y. (2005). Mechanism of intense blue photoluminescence in silica wires. *Solid State Communications*, 135(4), 247–250, https://doi.org/10.1016/j.ssc.2005.04.027.

107. Balagurov, L. A., Leiferov, B. M., Petrova, E. A., Orlov, A. F. and Panasenko, E. M. (1996). Influence of water and alcohols on photoluminescence of porous silicon. *Journal of Applied Physics*, 79(9), 7143–7147, https://doi.org/10.1063/1.361484.

108. Nishikawa, H., Watanabe, E., Ito, D., Sakurai, Y., Nagasawa, K. and Ohki, Y. (1996). Visible photoluminescence from Si clusters in γ-irradiated amorphous SiO2. *Journal of Applied Physics*, 80(6), 3513–3517, https://doi.org/10.1063/1.363223.

109. Nishikawa, H., Shiroyama, T., Nakamura, R., Ohki, Y., Nagasawa, K. and Hama, Y. (1992). Photoluminescence from defect centers in high-purity silica glasses observed under 7.9-eV excitation. *Physical Review B*, 45(2), 586, https://doi.org/10.1103/PhysRevB.45.586.

110. Awazu, K. and Kawazoe, H. (1990). O2 molecules dissolved in synthetic silica glasses and their photochemical reactions induced by ArF excimer laser radiation. *Journal of Applied Physics*, 68(7), 3584–3591, https://doi.org/10.1063/1.346318.

111. Ayers, M. R. and Hunt, A. J. (1997). Visibly photoluminescent silica aerogels. *Journal of Non-crystalline Solids*, 217(2–3), 229–235, https://doi.org/10.1016/S0022-3093(97)00126-9.

112. Augustine, B. H., Irene, E. A., He, Y. J., Price, K. J., McNeil, L. E., Christensen, K. N. and Maher, D. M. (1995). Visible light emission from thin films containing Si, *O, N and H*. *Journal of Applied Physics*, 78(6), 4020–4030, https://doi.org/10.1063/1.359925.

113. Abreu, C. M., Silva, R. S., Valerio, M. E. and Macedo, Z. S. (2013). Color-control of the persistent luminescence of cadmium silicate doped with transition metals. *Journal of Solid State Chemistry*, 200, 54–59, https://doi.org/10.1016/j.jssc.2012.11.031.

114. Bonacchi, S., Genovese, D., Juris, R., Montalti, M., Prodi, L., Rampazzo, E. and Zaccheroni, N. (2011). Luminescent silica nanoparticles: Extending the frontiers of brightness. *Angewandte Chemie International Edition*, 50(18), 4056–4066, https://doi.org/10.1002/anie.201004996.

115. Rai, R. K., Upadhyay, A. K., Kher, R. S. and Dhoble, S. J. (2012). Mechanoluminescence, thermoluminescence and photoluminescence studies on Al2O3: Tb phosphors. *Journal of Luminescence*, 132(1), 210–214, https://doi.org/10.1016/j.jlumin.2011.08.003.

116. Noh, A. M., Amin, Y. M., Mahat, R. H. and Bradley, D. A. (2001). Investigation of some commercial TLD chips/discs as UV dosimeters. *Radiation Physics and Chemistry*, 61(3–6), 497–499, https://doi.org/10.1016/S0969-806X(01)00313-9.

117. Makhov, V. N., Lushchik, A., Lushchik, C. B., Kirm, M., Vasil'Chenko, E., Vielhauer, S. and Aleksanyan, E. (2008). Luminescence and radiation defects in electron-irradiated Al2O3 and Al2O3: Cr. *Nuclear Instruments and Methods in Physics Research Section B: Beam Interactions with Materials and Atoms*, 266(12–13), 2949–2952, https://doi.org/10.1016/j.nimb.2008.03.145.

118. Bos, A. J. J. (2001). High sensitivity thermoluminescence dosimetry. *Nuclear Instruments and Methods in Physics Research Section B: Beam Interactions with Materials and Atoms*, 184(1–2), 3–28, https://doi.org/10.1016/S0168-583X(01)00717-0.

119. Ragupathi, C., Kennedy, L. J. and Vijaya, J. J. (2014). A new approach: Synthesis, characterization and optical studies of nano-zinc aluminate. *Advanced Powder Technology*, 25(1), 267–273, https://doi.org/10.1016/j.apt.2013.04.013.

120. Sahu, I. P., Bisen, D. P., Brahme, N., Tamrakar, R. K. and Shrivastava, R. (2015). Luminescence studies of dysprosium doped strontium aluminate white light emitting phosphor by combustion route. *Journal of Materials Science: Materials in Electronics*, 26(11), 8824–8839, https://doi.org/10.1007/s10854-015-3563-5.

121. Lecointre, A., Bessière, A., Viana, B., Benhamou, R. A. and Gourier, D. (2010). Thermally stimulated luminescence of Ca3 (PO4) 2 and Ca9Ln (PO4) 7 (Ln= Pr, Eu, Tb, Dy, Ho, Er, Lu). *Radiation Measurements*, 45(3–6), 273–276, https://doi.org/10.1016/j.radmeas.2010.02.008.

122. Wang, J., Zhang, Z., Zhang, M., Zhang, Q., Su, Q. and Tang, J. (2009). The energy transfer from Eu2+ to Tb3+ in Ca10K (PO4) 7 and its application in green light emitting diode. *Journal of Alloys and Compounds*, 488(2), 582–585, https://doi.org/10.1016/j.jallcom.2008.09.088.

123. Huang, Y., Zhao, W., Cao, Y., Jang, K., Lee, H. S., Cho, E. and Yi, S. S. (2008). Photoluminescence of Eu3+-doped triple phosphate Ca8MgR (PO4) 7 (R= La, Gd, Y). *Journal of Solid State Chemistry*, 181(9), 2161–2164, https://doi.org/10.1016/j.jssc.2008.04.044.

124. Ansari, A. A., Labis, J. P. and Manthrammel, M. A. (2017). Designing of luminescent GdPO4: Eu@ LaPO4@ SiO2 core/shell nanorods: Synthesis, structural and luminescence properties. *Solid State Sciences*, 71, 117–122, https://doi.org/10.1016/j.solidstatesciences.2017.07.012.

125. Terebilenko, A., Terebilenko, K., Slobodyanik, N. and Zubar, E. (2014, May). Synthesis and luminescence properties of calcium phosphates. In *International Conference on Oxide Materials for Electronic Engineering-fabrication, properties and applications (OMEE-2014)* (pp. 168–169). IEEE, doi: 10.1109/OMEE.2014.6912394.

126. Kharabe, V. R., Oza, A. H. and Dhoble, S. J. (2015). Photoluminescence characteristics of Ce and Eu activated MBPO5 (M= Sr, Ca) phosphors. *Optik*, 126(23), 4544–4547, https://doi.org/10.1016/j.ijleo.2015.08.090.

127. Pacchioni, G. (2000). Ab initio theory of point defects in SiO 2. Defects in SiO 2 and related dielectrics. *Science and Technology*, 161–195, https://doi.org/10.1007/978-94-010-0944-7_5.

128. Rao, V. H., Prasad, P. S., Babu, M. M., Rao, P. V., Santos, L. F., Raju, G. N. and Veeraiah, N. (2017). Luminescence properties of Sm3+ ions doped heavy metal oxide tellurite-tungstate-antimonate glasses. *Ceramics International*, 43(18), 16467–16473, https://doi.org/10.1016/j.ceramint.2017.09.028.

129. Kuang, Q., Wang, X., Jiang, Z., Xie, Z. and Zheng, L. (2014). High-energy-surface engineered metal oxide micro-and nanocrystallites and their applications. *Accounts of Chemical Research*, 47(2), 308–318, https://doi.org/10.1021/ar400092x.

130. Suriyaprabha, R., Karunakaran, G., Yuvakkumar, R., Rajendran, V. and Kannan, N. (2012). Silica nanoparticles for increased silica availability in maize (Zea mays. L) seeds under hydroponic conditions. *Current Nanoscience*, 8(6), 902–908, https://doi.org/10.2174/157341312803989033.

131. Siddiqui, M. H. and Al-Whaibi, M. H. (2014) Role of nano-SiO2 in germination of tomato (Lycopersicum esculentum Mill.) seeds. *Saudi Journal of Biological Sciences*, 21, 13–17, https://doi.org/10.1016/j.sjbs.2013.04.005.

132. Siddiqui, M. H., Al-Whaibi, M. H., Faisal, M. and Al Sahli, A. A. (2014) Nano-silicon dioxide mitigates the adverse effects of salt stress on Cucurbita pepo L. *Environmental Toxicology and Pharmacology*, 33, 2429–2437, https://doi.org/10.1002/etc.2697.

133. Gao, F., Hong, F., Liu, C., Zheng, L., Su, M., Wu, X. and Yang, P. (2006). Mechanism of nano-anatase TiO 2 on promoting photosynthetic carbon reaction of spinach: Inducing complex of rubisco-rubisco activase. *Biological Trace Element Research*, 111, 239–253, https://doi.org/10.1385/BTER:111:1:239.

134. Chand, N. and Siddiqui, N. (2012). Improvement in thermo mechanical and optical properties of in situ synthesized PMMA/TiO2 nanocomposite. *Composite Interfaces*, 19(1), 51–58, https://doi.org/10.1080/09276440.2012.688423.

135. Trujillo-Reyes, J., Majumdar, S., Botez, C. E., Peralta-Videa, J. R. and Gardea-Torresdey, J. L. (2014). Exposure studies of core-shell Fe/Fe3O4 and Cu/CuO NPs to lettuce (Lactuca sativa) plants: Are they a potential physiological and nutritional hazard? *Journal of Hazardous Materials*, 267, 255–263, https://doi.org/10.1016/j.jhazmat.2013.11.067.

136. Zhao, L., Peralta-Videa, J. R., Rico, C. M., Hernandez-Viezcas, J. A., Sun, Y., Niu, G. and Gardea-Torresdey, J. L. (2014). CeO2 and ZnO nanoparticles change the nutritional qualities of cucumber (Cucumis sativus). *Journal of Agricultural and Food Chemistry*, 62(13), 2752–2759, https://doi.org/10.1021/jf405476u.

137. Kanwar, M. K., Sun, S., Chu, X. and Zhou, J. (2019). Impacts of metal and metal oxide nanoparticles on plant growth and productivity. In *Nanomaterials and Plant Potential* (pp. 379–392). Cham: Springer, https://doi.org/10.1007/978-3-030-05569-1_15.

138. Santos, C.A.D., Seckler, M.M., Ingle, A.P. et al. (2014). Silver nanoparticles: Therapeutical uses, toxicity and safety issues. *Journal of Pharmaceutical Sciences*, 103(7), 1931–1944, https://doi.org/10.1002/jps.24001.

139. Czyżowska, A. and Barbasz, A. (2019). Effect of ZnO, TiO2, Al2O3 and ZrO2 nanoparticles on wheat callus cells. *Acta Biochimica Polonica*, 66(3), 365–370, https://doi.org/10.18388/abp.2019_2836.

140. Panakkal, H., Gupta, I., Bhagat, R. and Ingle, A. P. (2021). Effects of different metal oxide nanoparticles on plant growth. *Nanotechnology in Plant Growth Promotion and Protection: Recent Advances and Impacts*, 259–282, https://doi.org/10.1002/9781119745884.ch13.

141. Arshak, K. and Korostynska, O. (2004). Thick film oxide diode structures for personal dosimetry application. *Sensors and Actuators A: Physical*, 113(3), 319–323, https://doi.org/10.1016/j.sna.2004.01.050.

142. Wang, Q., Yang, B., Zhang, Y., Xia, H., Zhao, T. and Jiang, H. (2013). High light yield Ce3+-doped dense scintillating glasses. *Journal of Alloys and Compounds*, 581, 801–804, https://doi.org/10.1016/j.jallcom.2013.07.181.

143. Arshak, K., Korostynska, O. and Fahim, F. (2003). Various structures based on nickel oxide thick films as gamma radiation sensors. *Sensors*, 3(6), 176–186, https://doi.org/10.3390/s30600176.

144. Oliveira, L. C., Doull, B. A. and Yukihara, E. G. (2013). Investigations of MgO: Li, Gd thermally and optically stimulated luminescence. *Journal of Luminescence*, 137, 282–289, https://doi.org/10.1016/j.jlumin.2013.01.018.

145. Kawano, N., Kato, T., Okada, G., Kawaguchi, N. and Yanagida, T. (2017). Optical, scintillation and dosimeter properties of MgO: Tb translucent ceramics synthesized by the SPS method. *Optical Materials*, 73, 364–370, https://doi.org/10.1016/j.optmat.2017.08.025.

146. Guckan, V., Altunal, V., Ozdemir, A. and Yegingil, Z. (2020). Optically stimulated luminescence of MgO: Na, Li phosphor prepared using solution combustion method. *Journal of Alloys and Compounds*, 835, 155253, https://doi.org/10.1016/j.jallcom.2020.155253.

147. Dolgov, S., Kärner, T., Lushchik, A., Maaroos, A., Mironova-Ulmane, N. and Nakonechnyi, S. (2002). Thermoluminescence centres created selectively in MgO crystals by fast neutrons. *Radiation Protection Dosimetry*, 100(1–4), 127–130, https://doi.org/10.1093/oxfordjournals.rpd.a005828.

148. Guckan, V., Altunal, V. O. L. K. A. N., Ozdemir, A., Kurt, K. and Yegingil, Z. E. H. R. A. (2021). Cu, Li and K activated MgO: A metal oxide thermoluminescent synthesized using solution combustion technique for dosimetry. *Journal of Luminescence*, 230, 117751, https://doi.org/10.1016/j.jlumin.2020.117751.

149. Arshak, K. and Korostynska, O. (2004). Preliminary studies of properties of oxide thin/thick films for gamma radiation dosimetry. *Materials Science and Engineering: B*, 107(2), 224–232, https://doi.org/10.1016/j.mseb.2003.11.014.

150. Dolmans, D. E., Fukumura, D. and Jain, R. K. (2003). Photodynamic therapy for cancer. *Nature Reviews Cancer*, 3(5), 380–387, https://doi.org/10.1038/nrc1071.

151. Lan, G., Ni, K., Veroneau, S. S., Luo, T., You, E. and Lin, W. (2019). Nanoscale metal-organic framework hierarchically combines high-Z components for multifarious radio-enhancement. *Journal of the American Chemical Society*, 141(17), 6859–6863, https://doi.org/10.1021/jacs.9b03029.

152. Zheng, Q., Liu, X., Zheng, Y., Yeung, K. W., Cui, Z., Liang, Y. and Wu, S. (2021). The recent progress on metal-organic frameworks for phototherapy. *Chemical Society Reviews*, 50(8), 5086–5125, https://doi.org/10.1039/D1CS00056J.

153. Morton, C. A., Brown, S. B., Collins, S., Ibbotson, S., Jenkinson, H., Kurwa, H. and Rhodes, L. E. (2002). Guidelines for topical photodynamic therapy: Report of a workshop of the British Photodermatology Group. British Journal of Dermatology, 146(4), 552–567, https://doi.org/10.1046/j.1365-2133.2002.04719.x.

154. Diffey, B. L. (1980). Ultraviolet radiation physics and the skin. *Physics in Medicine and Biology*, 25(3), 405, doi: 10.1088/0031-9155/25/3/001.

155. Pei, S., Inamadar, A. C., Adya, K. A. and Tsoukas, M. M. (2015). Light-based therapies in acne treatment. *Indian Dermatology Online Journal*, 6(3), 145, doi: 10.4103/2229-5178.156379.

156. Hession, M. T., Markova, A. and Graber, E. M. (2015). A review of hand-held, home-use cosmetic laser and light devices. *Dermatologic Surgery*, 41(3), 307–320, doi: 10.1097/DSS.0000000000000283.

157. Lan, G., Ni, K. and Lin, W. (2019). Nanoscale metal-organic frameworks for phototherapy of cancer. *Coordination Chemistry Reviews*, 379, 65–81, https://doi.org/10.1016/j.ccr.2017.09.007.

158. (a) Guo, C., Yin, S., Yu, H., Liu, S., Dong, Q., Goto, T. and Sato, T. (2013). Photothermal ablation cancer therapy using homogeneous Cs x WO 3 nanorods with broad near-infra-red absorption. *Nanoscale*, 5(14), 6469–6478, https://doi.org/10.1039/C3NR01025B; (b) Kalluru, P., Vankayala, R., Chiang, C. S. and Hwang, K. C. (2013). Photosensitization of singlet oxygen and in vivo photodynamic therapeutic effects mediated by PEGylated W18O49 nanowires. *Angewandte Chemie International Edition*, 52(47), 12332–12336.

159. Chen, Z., Wang, Q., Wang, H., Zhang, L., Song, G., Song, L. and Zhao, D. (2013). Ultrathin PEGylated W18O49 nanowires as a new 980 nm-laser-driven photothermal agent for efficient ablation of cancer cells in vivo. *Advanced Materials*, 25(14), 2095–2100, https://doi.org/10.1002/anie.201307358.

160. Huang, Z. F., Song, J., Pan, L., Zhang, X., Wang, L. and Zou, J. J. (2015). Tungsten oxides for photocatalysis, electrochemistry and phototherapy. *Advanced Materials*, 27(36), 5309–5327, https://doi.org/10.1002/adma.201501217.

161. Einaga, H., Teraoka, Y. and Ogata, A. (2013). Catalyticoxidationofbenzeneby ozone over manganese oxides supported on US Y zeolite. *Journal of Catalysis*, 305, 227–237. doi: 10.1016/j.jcat.2013.05.016

162. Einaga, H. and Futamura, S. (2004b). Catalytic oxidation of benzene with ozone over alumina-supported manganese oxides. *Journal of Catalysis*, 227, 304–312. doi: 10.1016/j.jcat.2004.07.029

163. Einaga, H. and Futamura, S. (2005). Oxidation behavior of cyclohexane on alumina-supported manganese oxides with ozone. *Applied Catalysis B: Environmental*, 60, 49–55. doi: 10.1016/j.apcatb.2005.02.017.

164. Konova, P., Stoyanova, M., Naydenov, A., Christoskova, S. and Mehandjiev, D. (2006). Catalytic oxidation of VOCs and CO by ozone over alumina supported cobalt oxide. *Applied Catalysis A: General*, 298, 109–114, doi: 10.1016/j.apcata.2005.09.027.

165. Radhakrishnan, R. and Oyama, S. T. (2001). Ozone decomposition over manganese oxides upportedon ZrO2 and TiO2: A kinetic study using insitu laser Raman spectroscopy. *Journal of Catalysis*, 199, 282–290, doi: 10.1006/jcat.2001.3167

166. Huang, H., Lu, H., Huang, H., Wang, L., Zhang, J. and Leung, D. Y. (2016). Recent development of VUV-based processes for air pollutant degradation. *Frontiers in Environmental Science*, 4, 17, https://doi.org/10.3389/fenvs.2016.00017.

167. Kristianpoller, N., Weiss, D. and Chen, R. (2005). Defects induced in fluorides and oxides by VUV radiation. *Physica Status Solidi (c)*, 2(1), 409–412, https://doi.org/10.1002/pssc.200460195.

168. Kogelschatz, U., Esrom, H., Zhang, J. Y. and Boyd, I. W. (2000). High-intensity sources of incoherent UV and VUV excimer radiation for low-temperature materials processing. *Applied Surface Science*, 168(1–4), 29–36, https://doi.org/10.1016/S0169-4332(00)00571-7.

169. Humphreys, C. J. (2008). Solid-state lighting. *MRS Bulletin*, 33(4), 459–470, https://doi.org/10.1557/mrs2008.91.

170. Saradhi, M. P. and Varadaraju, U. V. (2006). Photoluminescence studies on Eu2+-activated Li2SrSiO4 a potential orange-yellow phosphor for solid-state lighting. *Chemistry of Materials*, 18(22), 5267–5272, https://doi.org/10.1021/cm061362u.

171. Han, J. K., Choi, J. I., Piquette, A., Hannah, M., Anc, M., Galvez, M. and McKittrick, J. (2012). Phosphor development and integration for near-UV LED solid state lighting. *ECS Journal of Solid State Science and Technology*, 2(2), R3138, doi: 10.1149/2.014302jss.

172. Supran, G. J., Shirasaki, Y., Song, K. W., Caruge, J. M., Kazlas, P. T., Coe-Sullivan, S. and Bulović, V. (2013). QLEDs for displays and solid-state lighting. *MRS Bulletin*, 38(9), 703–711, https://doi.org/10.1557/mrs.2013.181.

173. Schubert, E. F., Kim, J. K., Luo, H. and Xi, J. Q. (2006). Solid-state lighting-a benevolent technology. *Reports on Progress in Physics*, 69(12), 3069, doi: 10.1088/0034-4885/69/12/R01.

174. Shin, S. S., Lee, S. J. and Seok, S. I. (2019). Metal oxide charge transport layers for efficient and stable perovskite solar cells. *Advanced Functional Materials*, 29(47), 1900455, https://doi.org/10.1002/adfm.201900455.

175. Pearce, J. M. (2002). Photovoltaics-a path to sustainable futures. *Futures*, 34(7), 663–674, https://doi.org/10.1016/S0016-3287(02)00008-3.

176. Trupke, T., Green, M. A. and Würfel, P. (2002). Improving solar cell efficiencies by down-conversion of high-energy photons. *Journal of Applied Physics*, 92(3), 1668–1674, https://doi.org/10.1063/1.1492021.

177. Lian, H., Hou, Z., Shang, M., Geng, D., Zhang, Y. and Lin, J. (2013). Rare earth ions doped phosphors for improving efficiencies of solar cells. *Energy*, 57, 270–283, https://doi.org/10.1016/j.energy.2013.05.019.

178. Snitzer, E. J. P. R. L. (1961). Optical maser action of Nd+ 3 in a barium crown glass. *Physical Review Letters*, 7(12), 444, https://doi.org/10.1103/PhysRevLett.7.444.

179. Kesavulu, C. R., Kim, H. J., Lee, S. W., Kaewkhao, J., Wantana, N., Kaewnuam, E. and Kaewjaeng, S. (2017). Spectroscopic investigations of Nd3+ doped gadolinium calcium silica borate glasses for the NIR emission at 1059 nm. *Journal of Alloys and Compounds*, 695, 590–598, https://doi.org/10.1016/j.jallcom.2016.11.002

180. Zhang, Y. L., Chen, Q. D., Xia, H. and Sun, H. B. (2010). Designable 3D nanofabrication by femtosecond laser direct writing. *Nano Today*, 5(5), 435–448, https://doi.org/10.1016/j.nantod.2010.08.007.

181. Joy, N. A. and Carpenter, M. A. (2013). Optical sensing methods for metal oxide nanomaterials. *Metal Oxide Nanomaterials for Chemical Sensors*, 365–394, https://doi.org/10.1007/978-1-4614-5395-6_12.

182. Wuenschell, J. K., Jee, Y., Lau, D. K., Yu, Y. and Ohodnicki Jr, P. R. (2020). Combined plasmonic Au-nanoparticle and conducting metal oxide high-temperature optical sensing with LSTO. *Nanoscale*, 12(27), 14524–14537, https://doi.org/10.1039/D0NR03306E.

183. Ozbay, E. (2006). Plasmonics: Merging photonics and electronics at nanoscale dimensions. *Science*, 311(5758), 189–193, doi: 10.1126/science.1114849.

184. Ohodnicki Jr, P. R., Wang, C. and Andio, M. (2013). Plasmonic transparent conducting metal oxide nanoparticles and nanoparticle films for optical sensing applications. *Thin Solid Films*, 539, 327–336, https://doi.org/10.1016/j.tsf.2013.04.145.

185. Haick, H. (2007). Chemical sensors based on molecularly modified metallic nanoparticles. *Journal of Physics D: Applied Physics*, 40(23), 7173, doi: 10.1088/0022-3727/40/23/S01.

186. Singh, R. and Nalwa, H. S. (2011). Medical applications of nanoparticles in biological imaging, cell labeling, antimicrobial agents and anticancer nanodrugs. *Journal of Biomedical Nanotechnology*, 7(4), 489–503, https://doi.org/10.1166/jbn.2011.1324.

187. Sawai, J., Himizu, K. and Yamamoto, O. (2005). Kinetics of bacterial death by heated dolomite powder slurry. *Soil Biology and Biochemistry*, 37(8), 1484–1489, https://doi.org/10.1016/j.soilbio.2005.01.011.

188. Raghunath, A. and Perumal, E. (2017). Metal oxide nanoparticles as antimicrobial agents: Promise for the future. *International Journal of Antimicrobial Agents*, 49(2), 137–152, https://doi.org/10.1016/j.ijantimicag.2016.11.011.

189. Ghosh, T., Dash, S. K., Chakraborty, P., Guha, A., Kawaguchi, K., Roy, S. and Das, D. (2014). Preparation of antiferromagnetic Co_3O_4 nanoparticles from two different precursors by pyrolytic method: In vitro antimicrobial activity, https://doi.org/10.1039/C3RA47769J.

190. Rakshit, S., Ghosh, S., Chall, S., Mati, S. S., Moulik, S. P. and Bhattacharya, S. C. (2013). Controlled synthesis of spin glass nickel oxide nanoparticles and evaluation of their potential antimicrobial activity: A cost effective and eco friendly approach. *RSC Advances*, 3(42), 19348–19356, https://doi.org/10.1039/C3RA42628A.

191. Palanikumar, L., Ramasamy, S. N. and Balachandran, C. (2014). Size-dependent antimicrobial response of zinc oxide nanoparticles. *IET Nanobiotechnology*, 8(2), 111–117, https://doi.org/10.1049/iet-nbt.2012.0008.

192. Malarkodi, C., Rajeshkumar, S., Paulkumar, K., Vanaja, M., Gnanajobitha, G. and Annadurai, G. (2014). Biosynthesis and antimicrobial activity of semiconductor nanoparticles against oral pathogens. *Bioinorganic Chemistry and Applications*, https://doi.org/10.1155/2014/347167.

193. Jo, Y. K., Kim, B. H. and Jung, G. (2009). Antifungal activity of silver ions and nanoparticles on phyto-pathogenic fungi. *Plant Disease*, 93(10), 1037–1043, https://doi.org/10.1094/PDIS-93-10-1037.

194. Allahverdiyev, A. M., Abamor, E. S., Bagirova, M. and Rafailovich, M. (2011). Antimicrobial effects of TiO2 and Ag2O nanoparticles against drug-resistant bacteria and leishmania parasites. *Future Microbiology*, 6(8), 933–940, https://doi.org/10.2217/fmb.11.78.

195. Lok, C. N., Ho, C. M., Chen, R., He, Q. Y., Yu, W. Y., Sun, H. and Che, C. M. (2006). Proteomic analysis of the mode of antibacterial action of silver nanoparticles. *Journal of Proteome Research*, 5(4), 916–924, https://doi.org/10.1021/pr0504079.

196. Dizaj, S. M., Lotfipour, F., Barzegar-Jalali, M., Zarrintan, M. H. and Adibkia, K. (2014). Antimicrobial activity of the metals and metal oxide nanoparticles. *Materials Science and Engineering*: C, 44, 278–284, https://doi.org/10.1016/j.msec.2014.08.031.

197. Gold, K., Slay, B., Knackstedt, M and Gaharwar, A. K. (2018). Antimicrobial activity of metal and metal-oxide based nanoparticles. *Advanced Therapeutics*, 1(3), 1700033, https://doi.org/10.1002/adtp.201700033.

198. Meng, H., Chen, Z., Xing, G., Yuan, H., Chen, C., Zhao, F. and Zhao, Y. (2007). Ultrahigh reactivity provokes nanotoxicity: Explanation of oral toxicity of nano-copper particles. *Toxicology Letters*, 175(1–3), 102–110, https://doi.org/10.1016/j.toxlet.2007.09.015.

199. Gao, J., Gu, H. and Xu, B. (2009). Multifunctional magnetic nanoparticles: Design, synthesis and biomedical applications. *Accounts of Chemical Research*, 42(8), 1097–1107, https://doi.org/10.1021/ar9000026.

200. Fang, C. and Zhang, M. (2009). Multifunctional magnetic nanoparticles for medical imaging applications. *Journal of Materials Chemistry*, 19(35), 6258–6266, https://doi.org/10.1039/B902182E.

201. Andreescu, S., Ornatska, M., Erlichman, J. S., Estevez, A. and Leiter, J. C. (2012). Biomedical applications of metal oxide nanoparticles. *Fine Particles in Medicine and Pharmacy*, 57–100, https://doi.org/10.1007/978-1-4614-0379-1_3.

202. Jiang, J., Pi, J. and Cai, J. (2018). The advancing of zinc oxide nanoparticles for biomedical applications. *Bioinorganic Chemistry and Applications*, https://doi.org/10.1155/2018/1062562.

203. McNamara, K. and Tofail, S. A. (2017). Nanoparticles in biomedical applications. *Advances in Physics: X*, 2(1), 54–88, https://doi.org/10.1080/23746149.2016.1254570

204. Barrow, M., Taylor, A., Murray, P., Rosseinsky, M. J. and Adams, D. J. (2015). Design considerations for the synthesis of polymer coated iron oxide nanoparticles for stem cell labelling and tracking using MRI. *Chemical Society Reviews*, 44(19), 6733–6748, https://doi.org/10.1080/23746149.2016.1254570.

205. Lee, S. J., Muthiah, M., Lee, H. J., Lee, H. J., Moon, M. J., Che, H. L. and Park, I. K. (2012). Synthesis and characterization of magnetic nanoparticle-embedded multi-functional polymeric micelles for MRI-guided gene delivery. *Macromolecular Research*, 20, 188–196, https://doi.org/10.1007/s13233-012-0023-4.

206. Lee, N. and Hyeon, T. (2012). Designed synthesis of uniformly sized iron oxide nanoparticles for efficient magnetic resonance imaging contrast agents. *Chemical Society Reviews*, 41, 2575–2589, https://doi.org/10.1039/C1CS15248C.

207. Pillarisetti, S., Uthaman, S., Huh, K. M., Koh, Y. S., Lee, S. and Park, I. K. (2019). Multimodal composite iron oxide nanoparticles for biomedical applications. *Tissue Engineering and Regenerative Medicine*, 16, 451–465, https://doi.org/10.1007/s13770-019-00218-7.

208. Kannan, K., Radhika, D., Sadasivuni, K. K., Reddy, K. R. and Raghu, A. V. (2020). Nanostructured metal oxides and its hybrids for photocatalytic and biomedical applications. *Advances in Colloid and Interface Science*, 281, 102178, https://doi.org/10.1016/j.cis.2020.102178.

209. Beyene, A. M., Moniruzzaman, M., Karthikeyan, A. and Min, T. (2021). Curcumin nanoformulations with metal oxide nanomaterials for biomedical applications. *Nanomaterials*, 11(2), 460, https://doi.org/10.3390/nano11020460.

210. Park, Y., Koo, C., Chen, H. Y., Han, A. and Son, D. H. (2013). Ratiometric temperature imaging using environment-insensitive luminescence of Mn-doped core-shell nanocrystals. *Nanoscale*, 5(11), 4944–4950, https://doi.org/10.1039/C3NR00290J.

211. Cui, Y., Xu, H., Yue, Y., Guo, Z., Yu, J., Chen, Z. and Chen, B. (2012). A luminescent mixed-lanthanide metal-organic framework thermometer. *Journal of the American Chemical Society*, 134(9), 3979–3982, https://doi.org/10.1021/ja2108036.

212. Brites, C. D., Lima, P. P., Silva, N. J., Millán, A., Amaral, V. S., Palacio, F. and Carlos, L. D. (2012). Thermometry at the nanoscale. *Nanoscale*, 4(16), 4799–4829, https://doi.org/10.1039/C2NR30663H.

213. Jaque, D. and Vetrone, F. (2012). Luminescence nanothermometry. *Nanoscale*, 4(15), 4301–4326, https://doi.org/10.1039/C2NR30764B.

214. Balabhadra, S., Debasu, M. L., Brites, C. D., Nunes, L. A., Malta, O. L., Rocha, J. and Carlos, L. D. (2015). Boosting the sensitivity of Nd 3+-based luminescent nanothermometers. *Nanoscale*, 7(41), 17261–17267, https://doi.org/10.1039/C5NR05631D.

215. Brites, C. D., Lima, P. P., Silva, N. J., Millán, A., Amaral, V. S., Palacio, F. and Carlos, L. D. (2010). A luminescent molecular thermometer for long-term absolute temperature measurements at the nanoscale. *Advanced Materials*, 22(40), 4499–4504, https://doi.org/10.1002/adma.201001780.

216. Zheng, S., Chen, W., Tan, D., Zhou, J., Guo, Q., Jiang, W. and Qiu, J. (2014). Lanthanide-doped NaGdF 4 core-shell nanoparticles for non-contact self-referencing temperature sensors. *Nanoscale*, 6(11), 5675–5679, https://doi.org/10.1039/C4NR00432A.

217. Han, Y. H., Tian, C. B., Li, Q. H. and Du, S. W. (2014). Highly chemical and thermally stable luminescent Eu x Tb 1− x MOF materials for broad-range pH and temperature sensors. *Journal of Materials Chemistry C*, 2(38), 8065–8070, https://doi.org/10.1039/C4TC01336K.

218. Wang, Z., Ananias, D., Carné-Sánchez, A., Brites, C. D., Imaz, I., Maspoch, D., … Carlos, L. D. (2015). Lanthanide-organic framework nanothermometers prepared by spray-drying. *Advanced Functional Materials*, 25(19), 2824–2830, https://doi.org/10.1002/adfm.201500518.

219. Gao, Y., Huang, F., Lin, H., Zhou, J., Xu, J. and Wang, Y. (2016). A novel optical thermometry strategy based on diverse thermal response from two intervalence charge transfer states. *Advanced Functional Materials*, 26(18), 3139–3145, https://doi.org/10.1002/adfm.201505332.

220. Wang, Q., Liao, M., Lin, Q., Xiong, M., Mu, Z. and Wu, F. (2021). A review on fluorescence intensity ratio thermometer based on rare-earth and transition metal ions doped inorganic luminescent materials. *Journal of Alloys and Compounds*, 850, 156744, https://doi.org/10.1016/j.jallcom.2020.156744.

221. Huang, K., Idris, N. M. and Zhang, Y. (2016). Engineering of lanthanide-doped upconversion nanoparticles for optical encoding. *Small*, 12(7), 836–852, https://doi.org/10.1002/smll.201502722.

222. Shikha, S., Salafi, T., Cheng, J. and Zhang, Y. (2017). Versatile design and synthesis of nano-barcodes. *Chemical Society Reviews*, 46(22), 7054–7093, doi: 10.1039/C7CS00271H.

223. Wu, X., Lee, H., Bilsel, O., Zhang, Y., Li, Z., Chen, T. and Han, G. (2015). Tailoring dye-sensitized upconversion nanoparticle excitation bands towards excitation wavelength selective imaging. *Nanoscale*, 7(44), 18424–18428, https://doi.org/10.1039/C5NR05437K.

224. Lee, J., Yoo, B., Lee, H., Cha, G. D., Lee, H. S., Cho, Y. and Kim, D. H. (2017). Ultra-wideband multi-dye-sensitized upconverting nanoparticles for information security application. *Advanced Materials*, 29(1), 1603169, https://doi.org/10.1002/adma.201603169.

225. G. Chen, J. Damasco, H. Qiu, W. Shao, T. Y. Ohulchanskyy, R. R. Valiev, X. Wu, G. Han, Y. Wang, C. Yang, H. Ågren and P. N. Prasad (2015) *Nano Letters*, 15, 7400, https://doi.org/10.1021/acs.nanolett.5b02830.

226. Ren, W., Lin, G., Clarke, C., Zhou, J. and Jin, D. (2020). Optical nanomaterials and enabling technologies for high-security-level anticounterfeiting. *Advanced Materials*, 32(18), 1901430, https://doi.org/10.1002/adma.201901430.

227. Smith, A. F. and Skrabalak, S. E. (2017). Metal nanomaterials for optical anti-counterfeit labels. *Journal of Materials Chemistry C*, 5(13), 3207–3215, https://doi.org/10.1039/C7TC00080D.

228. Van Eijk, C. W. (2001). Inorganic-scintillator development. Nuclear Instruments and Methods in Physics Research Section A: Accelerators, Spectrometers, *Detectors and Associated Equipment*, 460(1), 1–14, https://doi.org/10.1016/S0168-9002(00)01088-3.

229. Kröger, F. A. and Nachtrieb, N. H. (1964). The chemistry of imperfect crystals. *Physics Today*, 17(10), 66, https://doi.org/10.1063/1.3051186.

230. Schön, M. (1948). Zur Kinetik des Leuchtens von Sulfidphosphoren mit mehreren Aktivatoren. *Annalen der Physik*, 438(1), 333–342, https://doi.org/10.1002/andp.19484380142.

231. Bulur, E. N. V. E. R., Goeksu, H. Y., Wieser, A., Figel, M. and Oezer, A. M. (1996). Thermoluminescence properties of fluorescent materials used in commercial lamps. *Radiation Protection Dosimetry*, 65(1–4), 373–379, https://doi.org/10.1093/oxfordjournals.rpd.a031665.

232. Mahalley, B. N., Dhoble, S. J., Pode, R. B. and Alexander, G. (2000). Photoluminescence in GdVO4: Bi3+, Eu3+ red phosphor. *Applied Physics A*, 70, 39–45, https://doi.org/10.1007/s003390050008.

233. Lin, H., Yang, D., Liu, G., Ma, T., Zhai, B., An, Q., … Pun, E. Y. B. (2005). Optical absorption and photoluminescence in Sm3+-and Eu3+-doped rare-earth borate glasses. *Journal of Luminescence*, 113(1–2), 121–128, https://doi.org/10.1016/j.jlumin.2004.09.115.

234. Dramićanin, M. D. (2016). Sensing temperature via downshifting emissions of lanthanide-doped metal oxides and salts. A review. *Methods and applications in fluorescence*, 4(4), 042001, doi: 10.1088/2050-6120/4/4/042001.

235. Herrington, R. (2021). Mining our green future. *Nature Reviews Materials*, 6(6), 456–458, https://doi.org/10.1038/s41578-021-00325-9.

236. Accorsi, G., Listorti, A., Yoosaf, K. and Armaroli, N. (2009). 1, 10-Phenanthrolines: Versatile building blocks for luminescent molecules, materials and metal complexes. *Chemical Society Reviews*, 38(6), 1690–1700, https://doi.org/10.1039/B806408N.

237. Carter, K. P., Young, A. M. and Palmer, A. E. (2014). Fluorescent sensors for measuring metal ions in living systems. *Chemical Reviews*, 114(8), 4564–4601, https://doi.org/10.1021/cr400546e.

238. Crema, A. P., Schaeffer, N., Bastos, H., Silva, L. P., Abranches, D. O., Passos, H. and Coutinho, J. A. (2023). New family of Type V eutectic solvents based on 1, 10-phenanthroline and their application in metal extraction. *Hydrometallurgy*, 215, 105971, https://doi.org/10.1016/j.hydromet.2022.105971.

239. Yu, X., Li, M., Xie, M., Chen, L., Li, Y. and Wang, Q. (2010). Dopant-controlled synthesis of water-soluble hexagonal NaYF 4 nanorods with efficient upconversion fluorescence for multicolor bioimaging. *Nano Research*, 3, 51–60, https://doi.org/10.1007/s12274-010-1008-2.

240. Tu, D., Liu, L., Ju, Q., Liu, Y., Zhu, H., Li, R. and Chen, X. (2011). Time-resolved FRET biosensor based on amine-functionalized lanthanide-doped NaYF4 nanocrystals. *Angewandte Chemie International Edition*, 50(28), 6306–6310, https://doi.org/10.1002/anie.201100303.

241. Danger, T., Koetke, J., Brede, R., Heumann, E., Huber, G. and Chai, B. H. T. (1994). Spectroscopy and green upconversion laser emission of Er3+-doped crystals at room temperature. *Journal of Applied Physics*, 76(3), 1413–1422, https://doi.org/10.1063/1.357745.

242. Esterowitz, L., Noonan, J. and Bahler, J. (1967). Enhancement in a Ho3+-Yb3+ quantum counter by energy transfer. *Applied Physics Letters*, 10(4), 126–127, https://doi.org/10.1063/1.1754876.

243. Huang, X., Han, S., Huang, W. and Liu, X. (2013). Enhancing solar cell efficiency: The search for luminescent materials as spectral converters. *Chemical Society Reviews*, 42(1), 173–201, https://doi.org/10.1039/C2CS35288E.

244. Wang, Y., Wang, X., Mao, Y. and Dorman, J. A. (2022). Impact of Sc3+-modified Local Site Symmetries on Er3+ Ion upconversion luminescence in Y2O3 nanoparticles. *The Journal of Physical Chemistry C*, 126(28), 11715–11722, https://doi.org/10.1021/acs.jpcc.2c00835.

245. Chang, C. R. and Jean, J. H. (1999). Crystallization kinetics and mechanism of low-dielectric, low-temperature, cofirable CaO-B2O3-SiO2 glass-ceramics. *Journal of the American Ceramic Society*, 82(7), 1725–1732, https://doi.org/10.1111/j.1151-2916.1999.tb01992.x.

246. Leow, T. Q., Hussin, R., Ibrahim, Z., Deraman, K., Shamsuri, W. N. W. and Lintang, H. O. (2015). Eu and Dy co-activated SrB2Si2O8 blue emitting phosphor: Synthesis and luminescence characteristics. *Sains Malaysiana*, 44(5), 753–760, https://doi.org/10.1002/bio.4301.

2 Modern Synthesis Techniques for Metal Oxides I

K. A. Koparkar
M. S. P. Arts, Science and K. P. T. Commerce College

R. G. Korpe
Shri. Shivaji Science College Amravati

CONTENTS

2.1 INTRODUCTION

Worldwide studies on metals, polymers, and ceramics during this century have resulted in the establishment of material science as a scientific discipline. A feature of those studies, particularly for metal oxide, is their interdisciplinary nature, and at present, chemistry is making an increasingly important contribution [1] to the research, development, and manufacture of metal oxide materials.

Chemistry has two major roles when applied to metal oxide and phosphors. The first is the synthesis of novel materials, usually in the form of powders. For example, the discovery in 1986 of $YBa_2Cu_3O_7$-x highlighted the role of chemical synthesis in the preparation of novel material. However, successful exploitation of metal oxide requires not only methods for their synthesis but also techniques for their fabrication into useful shapes, e.g., coatings, fibers, monolith ceramics, and powders with controlled particle size for a wide variety of applications. These applications include controlled porosity coatings for ceramic membrane, coating on window glass for selective transmission and reflection of solar radiations, optical fibers and fibers for lightweight thermal insulation, ceramic honeycombs for use in automotive catalytic converters, and un-aggregated powders for high-strength structural ceramic components.

DOI: 10.1201/9781003366232-2

The second chemical role is in the development of fabrication techniques for metal oxide shapes, for example, coatings, monoliths, and fibers. Synthesis and fabrication fall within the subject area of metal oxide processing that occupies an interface between conventional studies in chemistry and material science.

Processing is presently an important feature in universities and industrial environments, while chemical methods for metal oxide synthesis are under intense and increasing investigation. This is because, compared to conventional ceramic powder processing [2] using solid-state reaction (SSD) between powder reactants, these methods have the potential to yield metal oxides with tailored properties and with performance advantages over conventional materials. The conventional method for the preparation of metal oxide is first described together with the advantages and disadvantages of the technique.

In view of various phases of metal oxide applications, the preparation of some metal oxide by various synthesis methods differs in performance. As a result, the synthesis of a metal oxide is a very important starting step in material science. Spectroscopic characteristics of dopant ions, especially rare earths (REs), differ widely depending upon the host lattice used. This is because of local site symmetry offered to the RE ions, and correspondingly, the crystal field experienced by the ion is very much dependent on the environment surrounding the ion. In fact, 4f-5d transition and charge transfer state interactions can lead to emission ranging from Ultra violet (UV) to Infrared (IR) just by changing the host lattice, and in some cases, the emission is also dependent on the method used for its preparation.

There is increasing demand for modern techniques of metal oxide synthesis that impart superior properties compared with those attainable from conventional syntheses. A brief account of some of the methods used for the synthesis of phosphor materials is given as follows.

2.1.1 Conventional Method

2.1.1.1 Solid-State Reaction

The solid-state diffusion (SSD) reaction is the conventional method simply because it was widely used during the first half of the last century, for the synthesis [3,4] of multielement metal oxide powders. In this method, fine powders of the individual component oxides/carbonates are first mixed, milled, and calcined at elevated temperatures. Sometimes, the production of such oxides may require repeated grinding and calcining steps to achieve the desired phase. Relatively high temperatures are required for SSD, typically around 1,000°C–1,500°C because of limited diffusion during calcination, and this can result in the decomposition of the metal oxide product [5,6]. Other disadvantages of this method are the formation of undesirable phases, large particle sizes due to firing at high temperature, and poor chemical homogeneity, particularly when dopant oxides are introduced in small amounts during the synthesis of phosphors like $SrTiO_3$: Pr^{3+}. The as-synthesized particle size by this method is large for display applications; therefore, the powders must be ground into a finer powder. Particle size reduction by milling can introduce chemical impurities into the product. It is well known that mechanical crushing creates lattice defects, which, in turn, reduce the radiant efficiency of the phosphor. The preparation of single-phase materials is difficult with the SSD method. Hence, homogeneous doping at low activator concentration (2 mol%) has always been delicate; i.e., the product is inhomogeneous. A carefully controlled addition of dopant is important in this synthesis. The only advantage of this method is that the precursors are readily available. However, the method suffers from many disadvantages. Therefore, it becomes necessary to investigate simple, easy, and low-cost methods, which may overcome the following disadvantages, witnessed in the case of SSD method.

The disadvantages of SSD methods are as follows:

- Coarse grain size, powder agglomerates having nonhomogeneous properties due to limited diffusion lengths.
- Repeated cycles of heating and cooling followed after crushing the material in between these cycles.

- This added processing time and chemical impurity during crushing.
- Powders produced by this method are not suited for the fabrication of coatings and fibers because of nonuniform-sized particles.

Therefore it becomes necessary to explore simple, easy and low cost methods to overcome the disadvantages associated with the solid state diffusion process. The various novel and modern methods for the synthesis of metal oxides are described in this chapter.

2.1.2   MODERN METHODS OF SYNTHESIS

Chemical routes are attracting attention for metal oxide synthesis because some of them allow direct fabrication of coatings, fibers, and monoliths without powder intermediates. These routes have the potential to achieve the following:

1. Improved chemical homogeneity on the molecular scale, which is very important for electro-ceramics whose electrical functions are determined by the addition of a small amount of dopant oxides (e.g., semiconductors).
2. For the structural and single phase of metal oxides, improved mechanical properties such as strength can be obtained by the removal of powder aggregates since novel synthesis allows the preparation of non-aggregated powders.
3. Diffusion distances are reduced on calcination compared with conventional preparations owing to the mixing of components on the colloidal or molecular level that favors lower crystallization temperatures for multicomponent ceramics.

These potential advantages for improving the performance of materials have given rise to the increased application of chemistry, through material processing, for the development of metal oxides. A brief account of some of the novel and modern methods used for the synthesis of materials is as follows.

2.1.2.1   Co-Precipitation Method

In the precipitation process, the constituents are dissolved in a suitable solvent under pressure. This solution either is seeded or undergoes auto-precipitation. The problem can arise when two or more components are co-precipitated, and different species do not always deposit from the solution at the reaction pH. The difficulty in maintaining chemical homogeneity is serious as inhomogeneities have a deleterious effect on the properties of the material. The product is generally agglomerated. Therefore, grinding, dry milling, or wet milling with water or nonaqueous liquid is used for particle size reduction. These processes can introduce impurities from grinding media. Also, high temperatures are required for densification.

The co-precipitation method is used to prepare metal oxides through the formation of intermediate precipitates, usually hydrous oxides or oxalates, so that an intimate mixture of components is formed during precipitation, and chemical homogeneity is maintained on calcination. This method has been applied to synthesize a variety of oxides, e.g., Y_2O_3:Eu^{3+} [7–9]. Thus, a europium-doped yttrium oxalate precipitate was produced on the addition of oxalic acid to a mixed yttrium nitrate and europium nitrate solution under controlled conditions of pH, temperature, and reactant concentration. The precipitate was washed, dried, calcined, and milled to get the required metal oxide product (Figure 2.1).

In co-precipitation, careful control of solution conditions is required to precipitate all cations and thus maintain chemical homogeneity on the molecular scale. Water washing, solvent washing, and azeotropic distillation that are used for co-precipitated hydroxides can have a drastic effect on the mechanical properties of a sintered powder as they affect the degree of powder aggregation and need to be considered when developing a co-precipitation route to a ceramic powder. Another precipitation technique, not as widely reported as co-precipitation, involves the use of molten salts (Figure 2.2).

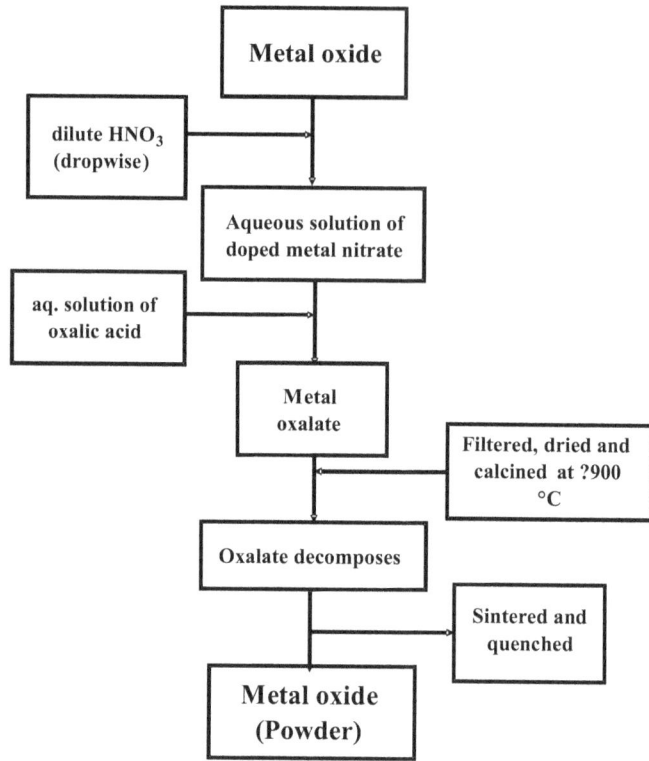

FIGURE 2.1   Flow chart of synthesis of phosphor by co-precipitation method-I.

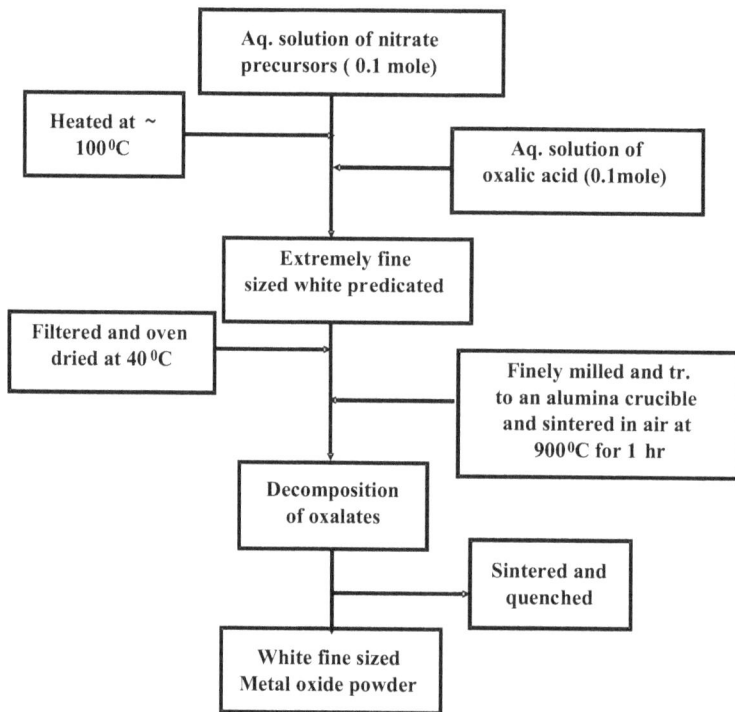

FIGURE 2.2   Flow chart of synthesis of phosphor by co-precipitation method-II.

2.1.2.2   Reaction in Molten Salts

The term molten salt refers [10] to the liquid state of compounds, which melt to give liquids displaying a degree of ionic properties. Alkali metal nitrates have relatively low melting points (Table 2.1), whereas even lower melting points are obtained in their eutectic mixtures [11,12]. Molten salt can behave as a solvent or as a reactant. Thus, in a nitrate melt acid–base reactions can occur according to the Lux–Flood formalism [13,14], whereby an acid is an oxide ion acceptor and a base is an oxide ion donor; nitrate ions are bases in this formalism. Nitrite melts are more basic than nitrate melts, whereas the addition of bases such as Na_2O_2, Na_2O, and NaOH to a nitrate melt increases its basicity. The starting materials for reactions in molten salts are inorganic compounds, in particular sulfates and chlorides that are blended with the alkali metal nitrates or nitrites as a powder mixture before heating to the reaction temperature. Thus, for the synthesis of Y_2O_3:Eu^{3+} phosphor, YCl_3 and $EuCl_3$ were mixed with an eutectic mixture of 43 mol% $LiNO_3$–57 mol% KNO_3 maintained at 400°C for 12 hours. The Y^{3+} and Eu^{3+} ions formed on the dissolution of these salts in the melt behave as acids. For this reaction, the eutectic was a solvent in which the reaction kinetics were considerably faster than in the conventional solid-state reaction.

Reactions in molten salts can be considered to fall within the general class of reactions in non-aqueous liquids and have not been widely applied for ceramic powder preparation even though the reactants are readily available. The nucleation, growth, and aggregation of ceramic powders during the preparation are of fundamental importance to the usefulness of the powders and are an area of continuing debate within activities in metal oxide processing. If non-aggregated powders can be obtained in melts for a wide range of compositions, then molten salt chemistry could, in the future, offer an attractive route for the synthesis of ceramic powders (Figure 2.3).

2.1.2.3   Combustion Synthesis

The synthesis of solids by the ceramic method (SSD method) is controlled by the diffusion of atoms and ionic species through reactants and products and thus requires repeated grinding, palletizing, and calcination of reactants (oxides or carbonates) for longer durations (than soft chemical routes) at high temperatures. The as-synthesized particle size by the ceramic method is too large to use in display applications; therefore, the powders must be ground into a finer powder. Attempts have recently been made to eliminate the diffusion control problems of solid synthesis [15]. Strongly exothermic reactions have been observed to sustain themselves and propagate in the form of a combustion wave until the reactants have been completely consumed. Booth [16] was the first to report such observations and to lay out the mathematical foundation for what has since been referred to as the self-propagating high-temperature synthesis (SHS) [17] or combustion synthesis and fire or furnace less synthesis [18,19]. The process makes use of highly exothermic redox chemical reactions between metals and nonmetals, metathetical (exchange) reactions between reactive compounds, and reactions involving redox compounds/mixtures. The term 'combustion' covers flaming (gas phase), smoldering (heterogeneous), and explosive reactions.

The combustion synthesis (CS) method has been successfully used in the preparation of a large number of technologically useful oxide (refractory oxides, magnetic, dielectric, semiconducting, insulators,

TABLE 2.1
Melting Points for Alkali Metal Nitrates and Eutectic Mixtures

Metal Nitrate	Melting Point (°C)
$LiNO_3$	225
$NaNO_3$	307
KNO_3	334
50 mol% $NaNO_3$–50 mol% KNO_3	220
43 mol% LiO_3–57 mol% KNO_3	132

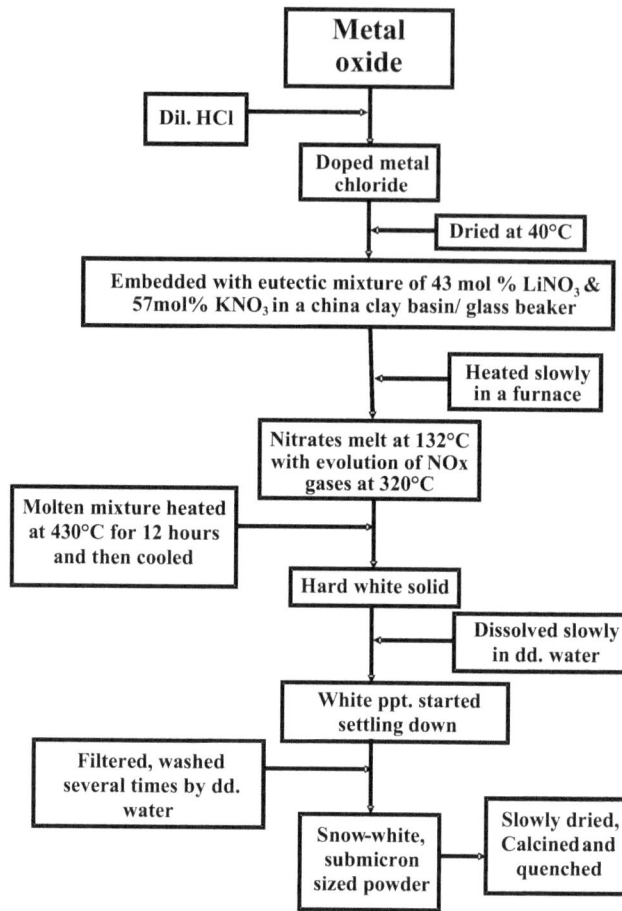

FIGURE 2.3 Flow chart of synthesis of phosphor by reaction in molten salts.

catalysts, sensors, phosphors, etc.) and non-oxide (carbides, borides, silicides, nitrides, etc.) materials. Important parameters that control the combustion synthesis are the particle size and shape of the reactants, ignition techniques and stoichiometric ratio, processing of reactant particles (green density, i.e., the density of the pellet before sintering), and adiabatic temperature (T_{ad}), which is a measure of the exothermicity of the reaction [20,21]. In this article, the recent trends in the SHS, solid-state metathesis (SSM), and combustion synthesis of oxide materials using redox compounds and mixtures are discussed.

2.1.2.4 Self-Propagating High-Temperature Synthesis (SHS)

Refractory materials such as ceramics, composites, and oxides can be synthesized by SHS methods. The process involves the ignition of the pellets prepared from fine powders of respective metals (fuels) and nonmetals (oxidizers) with a suitable heat source. A high ignition temperature (1,000°C–1,500°C) is required. After the ignition, the combustion reaction is self-propagating with an adiabatic temperature in the range of 1,500–3,000 K. The general chemical equation for the elemental combustion reaction can be represented as follows:

$$m\,X + n\,Y \rightarrow X_m\,Y_n$$

where X= metals (act as fuels) and Y= nonmetals (act as oxidizers).

The required ignition temperature can be achieved using laser radiations, a resistance heating coil, an electric arc, a chemical oven, and so on. Researchers are now trying to lower the ignition temperature and the use of metal oxides/halides instead of finely divided metal powders. To lower the ignition temperature, mechanical activation (ball milling) and field activation processes are used. Mechanical alloying has been successfully used to synthesize TiC, NbC, and their solid solutions using Ti, Nb, and graphite powders [22,23]. To activate low enthalpy of formation (ΔH_f) reactions, Munir and coworkers used the field activated (20 V) combustion synthesis. Using this method, ceramics and composites have been prepared [24]. A recent study [25] on the relationship between the field direction and wave propagation in activated combustion synthesis showed that the field applied in a direction perpendicular to wave propagation resulted in an enhancement of the wave velocity (which helps in the completion of the reaction and results in a decrease in the particle size of the product).

Thermite reactions that involve metallothermic reduction have been employed in the ceramic coating of pipes and preparation of composites. There are two types of thermite reactions. The first method involves the reduction of an oxide to the element, for example:

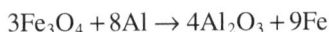

$$3Fe_3O_4 + 8Al \rightarrow 4Al_2O_3 + 9Fe$$

$$\Delta H^0{}_{298} = -3,400 \, KJ/mol$$

A modified thermite reaction called a 'centrifugal thermite reaction' has been used for coating the inner surface of pipes. By coupling SHS with a centrifugal process, a surface layer of alumina and an inner layer of Fe are formed in the pipe due to the difference in density between Fe and alumina. The composite pipe will have the strength and toughness of a metal and the corrosion and abrasion resistance of metal oxides. The second type of thermite process involves the reduction of an oxide to the element, which subsequently reacts with another element to form a refractory compound, for example

$$TiO_2 + B_2O_3 + 5Mg \rightarrow TiB_2 + 5MgO$$

These methods are of great importance in the synthesis of advanced metal oxides and composites [26] on account of the economic advantages of using cheaper oxide reactants. Merzhanov and his associates [27] have reported the synthesis of electronic engineering materials such as superconductors and ferroelectric and magnetic materials by SHS reaction using metal oxide and peroxide precursors. The properties of SHS-derived products are compared with those prepared by solid-state reaction. The advantages of SHS process are time- and energy-savings and an increase in the reactivity of products.

2.1.2.5 Combustion Synthesis of Oxide Materials Using Redox Compounds and Mixtures

A novel method of synthesis of simple and complex oxide materials is described. This method involves the use of combustible precursors (redox compounds) and redox mixtures. These are low-temperature (<500°C)-initiated gas-producing exothermic reactions and yield voluminous fine particle oxides in a few minutes. The combustion is smoldering type (flameless) and is accompanied by the evolution of gases resulting in fine, voluminous powder. However, the exothermicity of some precursors is not high enough to sustain combustion and the external heat source is required for the completion of the decomposition.

Precursors containing a carboxylate anion, hydrazide, hydrazine, or hydroxonium groups [28,29] are found to ignite at low temperatures (120°C–350°C) and decompose autocatalytically to yield fine particle and large surface area oxides. The high exothermicity ($T_{ad} = 10K$) of combustion is due to the oxidation of strong reducing halves such as COO^-, $N_2H_3{}^-$, N_2H_4, or N_2H_5 (present in the precursors) by atmospheric oxygen to CO_2, H_2O, and N_2. Using combustible precursors, nano-sized oxides such as Fe_2O_3 and ferrites, CeO_2, TiO_2, and Y_2O_3 can be synthesized. Besides magnetic oxides, this route has also reported a few ferroelectric titanates. Fine particles such as Fe_2O_3 and Fe_3O_4 and ferrites find use as recording materials and in the preparation of liquid magnets. Titania, ferrites, and cobaltites are all good catalysts.

Although the preparation of fine particle oxides by the above method of combustion of redox compounds is simple and attractive, it has certain limitations – (i) the preparation of precursors requires several days, (ii) the yield is only 20% of the precursors, and (iii) not all metals form complexes with the hydrazine carboxylate legends, and therefore, it is not possible to use this method to prepare high-temperature oxides such as chromites and alumina.

An efficient self-propagating combustion method for the synthesis of oxide materials is to use the combustible redox compounds and redox mixtures [30,31] (oxidizer–fuel). The stoichiometric compositions of redox mixtures used for the combustion synthesis are calculated based on the total oxidizing (O) and reducing (F) valencies of the components, which serves as the numerical coefficient for stoichiometric balance so that the equivalence ratio Φ_e, i.e., the ratio of oxidizing valency of metal nitrates to the reducing valency of fuel, is unity ($\phi_{e=}O/F=1$) and the energy released during combustion is maximum. According to the concepts used in propellant chemistry, the usual products of combustion are CO_2, H_2O, and N_2. The elements such as C, B, Zn, H, or any other metal are considered as reducing elements with valencies 4, 3, 2, and 1 (or valency of the metal ion in that compound), respectively, and oxygen as an oxidizing element having the valency of 2. The valency of nitrogen is taken as zero because of its conversion to molecular nitrogen.

The stoichiometric calculations for the proposed reaction are worked out first. The metal nitrates are weighed accordingly. The calculation of the amount of urea (fuel), required for each nitrate, was determined in the following way:

For a divalent oxidizer $M(NO_3)_2.xH_2O$ (where M is a metal) -
Oxidizing valency for M is   --------($+2 \times 1$) $=+2$
For O^{2-} is   --------($-2 \times 3) \times 2$ $=-12$
The net valency   -------- $=-10$
For fuel (urea), H_2N ---- C ----NH_2, the valency -------- $= +6.$
                                            $\parallel$
                                            O

Hence, the oxidizer-to-fuel valency ratio is 10:6. Therefore, one mole of oxidizer requires 10/6 moles of urea. Thus, depending upon the cation valency, +2 for $Ba(NO_3)_2$ and $Mg(NO_3)_2.6H_2O$, +3 for $Al(NO_3)_3.9H_2O$ and $Eu(NO_3)_3$, the multiplication factor is 10/6 and 15/6, respectively. Thus, the total weight of urea was calculated.

The host formation temperature for the thermal decomposition of precursors could be lowered by the addition of a calculated amount of either ammonium nitrate (NH_4NO_3) or ammonium perchlorate (NH_4ClO_4) or rapidly heating at 300°C. The addition of an oxidizer (NH_4NO_3 or NH_4ClO_4), while lowering the ignition temperature, increases the enthalpy of the reaction drastically, thus facilitating the crystallization of phosphor [32]. The stoichiometric amounts of ammonium nitrate or ammonium perchlorate and urea were calculated for the proposed reaction.

The Combustion Synthesis Involves Following Steps

a. Take stoichiometric compositions of oxidizer (e.g., metal nitrate and/or ammonium nitrate) and fuel (e.g., urea).
b. Crush and grind them for at least 30 minutes to form pasty material.
c. Transfer it to the borosilicate beaker and put it in the furnace maintained at 500°C.

The solution boils and undergoes dehydration followed by decomposition with the evolution of gases such as N_2, CO_2, H_2O, etc. The mixture then froths and swells, forming foam that ruptures with a flame on ignition of combustible gases, and glows to incandescence. During incandescence, the foam further swells to the capacity of the container. The flame temperature as measured by an optical pyrometer is around 1,600°C±20°C. The product of combustion is voluminous and foamy. The whole process completes in less than 5 minutes (Figure 2.4).

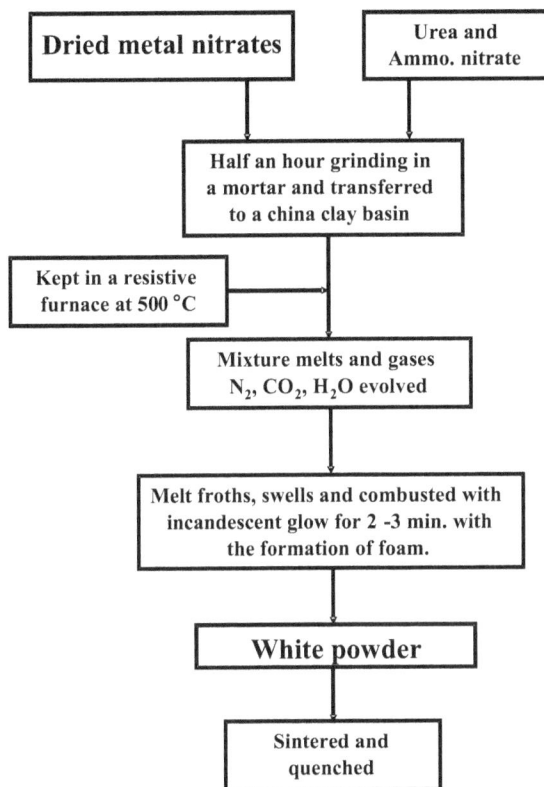

FIGURE 2.4   Flow chart of combustion synthesis using a redox mixture.

Patil K.C. and coworkers, and other research groups have carried out the synthesis of a variety of single and mixed oxides, aluminates, aluminosilicates, chromites, ferrites, ferroelectrics, zirconates, borates, vanadates [33,34], etc., using fuels such as urea (NH_2-CO-NH_2), carbohydrazide (CH_6N_4O), oxalyldihydrazide ($C_2H_6N_4O$), glycine ($C_2H_5NO_2$), 3Methylpyrazol-5 one ($C_4H_6N_2O$), and diformyl-hydrazine ($C_2H_4N_2O_2$). Glycine has been used to synthesize high Tc superconductors, magnates, and chromites. The solution combustion method has been used to yield nano-size TiO_2, ZrO_2, and pigments [35] (e.g., V^{4+} doped zircon blue pigment). The generation of gaseous products increases the surface area of the powders by creating micro- and nano-porous regions. Combustion-synthesized powders are generally more homogeneous, have fewer impurities, and have higher surface areas than powders prepared by conventional solid-state methods. The advantages of the solution combustion process over other combustion methods are as follows: (i) Being a solution process, it has control over the homogeneity and stoichiometry of the products; (ii) it is possible to incorporate desired impurity ions in the oxide hosts and prepare industrially useful materials such as pigments, phosphors [36,37], and high T_c cuprates and solid oxide fuel cell (SOFC); and (iii) process is simple, fast, and does not need any special equipment as in other SHS methods.

2.1.2.5.1 Preparation of Fuels

2.1.2.5.1.1 Preparation of Carbohydrazide (CH), CH_6N_4O   354 g of diethyl carbonate (3 moles) and 388 g of 99% hydrazine hydrate (6 moles) were placed in a 1-L round-bottom flask fitted with a thermometer. The reactants were partially miscible at first, the flask was then shaken well until a single phase was formed, and this was accompanied by the evolution of heat causing the temperature to rise to about 55°C. The flask was then connected through a standard taper joint, to a fractionating column filled with Raschig rings. A still head fitted with a thermometer and a water-cooled condenser was attached to the fractionating column. A heater regulated by a variable transformer was employed to heat the reaction mixture.

Distillation of the ethanol and water (byproducts) was quite rapid (5 mL/min) for the first 30 minutes but decreased as the reaction proceeded. Heating was continued for 4 hours during which around 350 mL of distillate was collected at a vapor temperature of 80°C–85°C. The liquor was cooled to 20°C and allowed to stand until the formation of carbohydrazide crystals [38]. The crystals were then separated by filtration and dried (yield = 165 g, 60%, m.p. = 153°C).

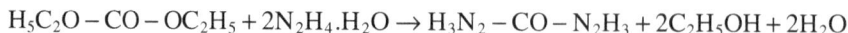

$$H_5C_2O - CO - OC_2H_5 + 2N_2H_4.H_2O \rightarrow H_3N_2 - CO - N_2H_3 + 2C_2H_5OH + 2H_2O$$

2.1.2.5.1.2 Preparation of N, N'-Diformylhydrazine (DFH) Hydrazine hydrate (50 mL, 1 mol) and formic acid (150 mL, 2 mol) were mixed and heated overnight in a steam bath, and the solvent was removed by heating under reduced pressure. Ethanol (100 mL) was added, and the solid separation was collected and air-dried (yield = 60%, m.p. = 160°C).

$$N_2H_4.H_2O + 2HCOOH \rightarrow OHC - HN - NH - CHO + 3H_2O$$

2.1.2.5.1.3 Preparation of Oxalyl Dihydrazide (ODH), $C_2H_6N_4O_2$ 146.14 g of diethyl oxalate (1 mol) was added dropwise to 100.122 g of hydrazine hydrate (2 mol) dissolved in 225 mL of double-distilled water in 1,000-mL beaker. The entire addition was carried out in an ice-cold bath with vigorous stirring. The white precipitate obtained was allowed to stand overnight, washed with ethanol, filtered, and dried (yield = 100.3 g, 85%).

$$H_5C_2O - CO - CO - OC_2H_5 + 2N_2H_4.H_2O \rightarrow H_3N_2 - CO - CO - N_2H_3 + 2C_2H_5OH + 2H_2O$$

2.1.2.5.1.4 Preparation of Tetraformal Trisazine (TFTA), $C_4H_{16}N_6O_2$ 100 mL of hydrazine hydrate (99%–100%) was cooled in an ice–salt mixture (temperature of the solvent being 0°C–5°C). To this ice-cold hydrazine hydrate, 175 mL of formaldehyde (37%–41%) was added dropwise from a burette. Throughout the addition, the temperature was maintained below 5°C. After the complete addition of HCHO, the mixture was concentrated in a water bath. When white precipitate began to appear, the solution was removed from the water bath and left for crystallization. A white lump of solid appeared, which was filtered, washed with ethanol, and stored in a desiccator. The mother liquor was again concentrated in a water bath (yield 70%, m.p. = 85°C).

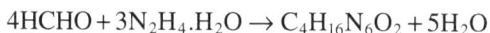

$$4HCHO + 3N_2H_4.H_2O \rightarrow C_4H_{16}N_6O_2 + 5H_2O$$

2.1.2.5.1.5 Preparation of 3Methylpyrazol-5 One (3MP$_5$O), $C_4H_6N_2O$ 3 Methylpyrazol-5 one (3MP5O, $C_4H_6N_2O$) was prepared by the dropwise addition of 1 mole of ethyl acetoacetate to 1 mole of hydrazine hydrate (99%–100%) cooled in an ice bath (0°C–5°C) according to the following equation:

$$CH_3COCH_2COOC_2H_5 + N_2H_4.H_2O \rightarrow C_4H_6N_2O + H_2O + C_2H_5OH$$

Combustion synthesis was successfully adopted for the synthesis of basic oxides; however, the most commonly available fuel is urea and oxidizer is ammonium nitrate. Both these compounds are not costly. In the research group, other coworkers have successfully established the combustion synthesis using various fuels but mostly urea for the synthesis of luminescent materials hosts such as aluminates, oxides, silicates, vanadates, and borates. The important well-known luminescent materials, which are commercially used and the same are synthesized in the laboratory using combustion, are listed in Table 2.2.

TABLE 2.2
Phosphors Synthesized by Combustion Method in Our Laboratory

Compound	Fuel	Emission Wavelength	Applications
Y_2O_3:Eu^{3+}	3MP5O, ODH, glycine	611 nm	Compact fluorescent lamps, color picture tubes
YVO_4:Eu^{3+}	3MP5O	619 nm	HPMV lamps
$(Ce, Tb)MgAl_{11}O_{19}$	Urea	541 nm	Compact fluorescent lamps
SrB_4O_7:Eu^{2+}	Urea and NH_4NO_3	370 nm	Sun tanning lamps
$BaMgAl_{10}O_{17}$:Eu^{2+}	Urea	450 nm	Compact fluorescent lamps
$BaMgAl_{10}O_{17}$:Eu, Mn	Urea	520 nm	Neon lamps
Zn_2SiO_4:Mn^{2+}	3MP5O	524 nm	Tricolor lamps, color picture tubes
$Sr_5(PO_4)_3Cl$:Eu^{2+}	Urea	445 nm	Compact fluorescent lamps
$(Gd, La)B_3O_6$:Bi	Urea and NH_4NO_3	311 nm	Psoriasis lamps
$Sr_2P_2O_7$:Eu	Urea and NH_4NO_3	420 nm	Photocopy lamps
$Y_3Al_5O_{12}$:Ce	Urea	525 nm	Deluxe lamps
CaO:Bi	Urea and NH_4NO_3	391 nm	Photoluminescent liquid crystal displays
$SrTiO_3$:Pr^{3+}	Urea and NH_4NO_3	617 nm	Vacuum fluorescent displays, low-voltage field-emission displays
MgB_4O_7 :Dy	Urea and NH_4NO_3	Main thermoluminescence (TL) peak at 163°C	TL dosimetry phosphor
MgB_4O_7 :Dy, Na	Urea and NH_4NO_3	Main TL peak at 190°C	Thermoluminescence dosimetry phosphor
$Li_2B_4O_7$:Cu	Urea and NH_4NO_3	Main TL peak at 210°C	Thermoluminescence dosimetry phosphor
$CeGdMgB_5O_{10}$	Urea and NH_4NO_3	311 nm	Phototherapy
$(CeGdTb)MgB_5O_{10}$:Mn	Urea and NH_4NO_3	616 nm	Tricolor lamps
$Ca_2B_5O_9Cl$:Eu^{2+}	Urea	455 nm	Photoluminescent liquid crystal displays

2.1.2.6   Sol–Gel Synthesis

The phrase sol–gel describes several types of processes in different areas of chemistry and material development, while the term 'gel' has been used to embrace a wide range of substances in systems as diverse as inorganic clays and oxides, phospholipids, disordered proteins, and three-dimensional or network polymers.

Colloid science is important for the successful application of chemistry to metal oxide synthesis. This is because powder preparation involves nucleation and growth of particles to a size often less than 1 μm, and thus, the powders are colloidal systems [39]. A 'colloid' is a suspension in which the dispersed phase is so small (~1–1,000 nm) that gravitational forces are negligible and interactions are dominated by short-range forces. In addition, powders are often handled in the form of colloidal dispersions, and this is illustrated by the synthesis method known as sol–gel processing of colloids [40]. A sol–gel preparation can be divided into five stages [41]. The starting material, for example, a metal salt, is converted in a chemical process to a dispersible oxide, which forms a colloidal dispersion (sol) in addition to dilute acid or H_2O. A 'Sol' is a colloidal suspension of solid particles in liquid. It may be produced from organic or inorganic precursors (e.g., nitrates or alkoxides).

The removal of H_2O and/or anions from the sol produces a stiff gel in the form of spheres, fibers, fragments, or coatings, and this transition is usually reversible. A 'gel' is a two-component system of a semisolid nature, which consists of continuous solid and fluid phases, of colloidal dimensions. Sol–gel processing is the preparation of phosphor materials by the preparation of the sol, gelation of the sol, and removal of the solvent. Calcination of the gel in air yields an oxide product after the decomposition of the salts. For the preparation of multicomponent oxides, sols are blended together before gelation and a component unavailable in sol form can be introduced as an electrolyte solution or oxide powder. Spherical sol–gel powders can be made by the following processes:

1. Dispersing a sol to an emulsion in an immiscible organic solvent capable of extracting H_2O from the sol, for example, 2-ethylhexanol. Gelation occurs during the dehydration process.
2. External gelation in which a sol is dispersed to an emulsion in a water-immiscible solvent and gelation is effected by the addition of a long-chain amine or NH_3 (g) to the solvent.
3. Internal gelation in which an ammonia donor such as hexamethylenetetramine or urea is added to the sol before emulsification and gelation occurs by the release of NH_3 (g) on warming solvent (Figures 2.5 and 2.6).

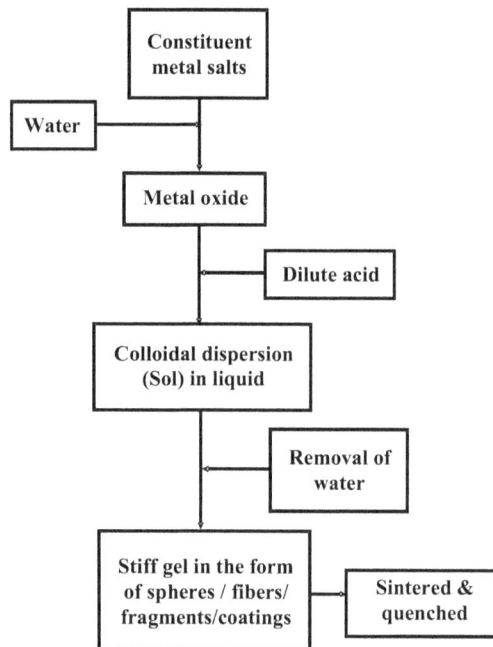

FIGURE 2.5 Flow chart of sol–gel processing of organometallic colloids.

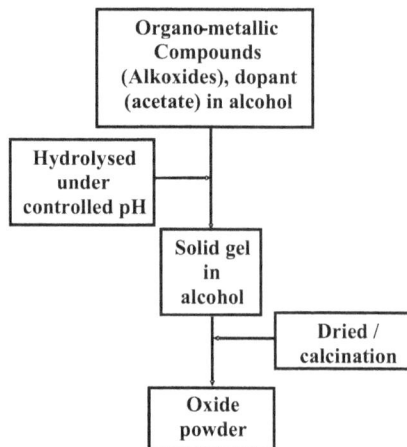

FIGURE 2.6 Flow chart of the sol–gel process for compounds.

The advantages of this sol–gel technique are good chemical homogeneity due to mixing components at the colloidal level and lower reaction temperatures, considerably lower than for the conventional powder mixing method. In addition, because it involves handling liquid feeds, small amounts of dopants can be readily introduced, while lower crystallization temperatures enable the preparation of phases that are unstable at high temperature. The hydrolysis of cations and inorganic polymerization in an aqueous solution is of fundamental importance to wet chemical methods of synthesis such as sol–gel in which sols are formed from cations that can undergo hydrolysis. The formation of monovalent hydrolysis products is represented by the following equation:

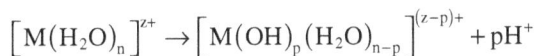

$$\left[M(H_2O)_n \right]^{z+} \rightarrow \left[M(OH)_p (H_2O)_{n-p} \right]^{(z-p)+} + pH^+$$

where n is the number of bound water molecules, p is the number of protons removed from the cation on hydrolysis, and z is the valency of the cation M, which can condense to polyvalent or polynuclear ions that can be colloidal.

However, sol–gel processing of alkoxides cannot compete economically with conventional routes for mass production due to the high cost of alkoxides.

Advantages
- Good chemical homogeneity due to mixing component at the colloidal level.
- Lower reaction temperatures (e.g., ThO_2-UO_2 spheres obtained at 1,150°C, whereas conventional powder mixing method produces at ≈1,700°C).
- As it involves handling liquid feeds, a small amount of dopants can be readily introduced, while lower crystallization temperatures enable the preparation of phase, which is unstable at high temperature.
- Because many alkoxides are liquids or volatile solids, they can be purified to form extremely pure oxide sources, which is important for electroceramic synthesis.
- The improved homogeneity is associated with lower crystallization temperatures than the use of colloids, often between 400°C and 800°C for gel-to-oxide conversion.

Disadvantages
- Alkoxides are relatively expensive compared with precursors for sol–gel processing of colloids.
- A limited range of them are commercially available.
- Their use involves a solvent-based process rather than a water-based process.

2.1.2.7 Stearic Acid Gel Method

Besides the advantages of sol–gel process of alkoxides, it is costly as compared to the conventional methods, and due to the high cost of alkoxides, the sol–gel processing by steric acid [42–46] provides an economical and effective route for the synthesis of luminescent materials.

In this sol–gel process, the precursors are added to the molten steric acid at 70°C for near about 30 minutes for the preparation of sol. After the complete dissolution of precursor, the solution is further heated near about 140°C and allowed to cool. The process of polymerization takes place due to the formation of metallic steroid and the gel is formed.

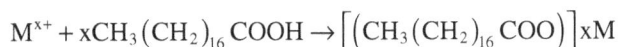

$$M^{x+} + xCH_3(CH_2)_{16} COOH \rightarrow \left[(CH_3(CH_2)_{16} COO) \right] xM$$

It is then dried further slowly to near about 230°C. Most gels are amorphous even after drying, but many crystallize when heated. The term aging is applied to the process of change in structure and properties after gelation. The dry gel is further heated till shining black resin or charcoal is formed. Shrinkage of gel, during either synthesis or drying, involves the deformation of the network and

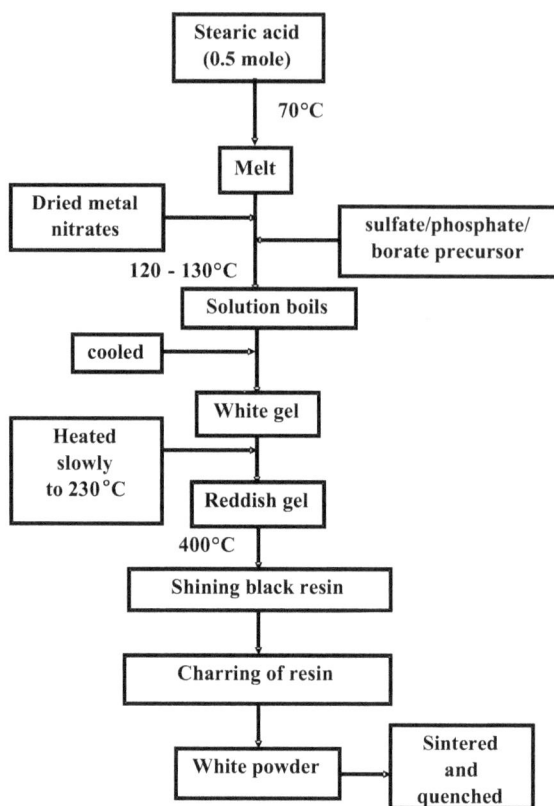

FIGURE 2.7 Flow chart of synthesis of phosphor by stearic acid sol–gel method.

transport of liquid through the pores. On further heating, the black resin or charcoal burns, yielding the fine-sized white crystalline powder (Figure 2.7).

This process offers a low-temperature route that gives high purity, homogeneous composition, and controlled particle size. It is energy-efficient and suited for compositions that are difficult to make by conventional methods.

2.1.2.8 Pechini and Citrate Gel Methods

In the Pechini method [47,48], polybasic chelates are formed between α-hydroxy carboxylic acids containing at least one hydroxy group, for example, citric acid and $HOC(CH_2CO_2H)_2.CO_2H$ with metallic ions.

$$M^{x+} + x\ HOC(CH_2CO_2H)_2.CO_2H \rightarrow \left[(CH_2COO)_2\ COH(COO)\right]xM$$

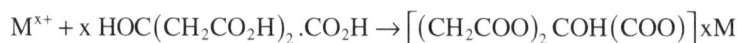

The chelate undergoes polyesterification on heating with a polyfunctional alcohol, for example, ethylene glycol, $HOCH_2CH_2OH$.

$$\left[(CH_2COO)_2\ COH(COO)\right]xM + 2xHOCH_2CH_2OH \rightarrow \left[(CH_2CH_2CH_2COO)_2\ COH(COO)\right]xM$$

Further heating produces a viscous resin, then a rigid transparent, glassy gel, and finally a fine oxide powder. This method is based on the esterification reaction between organic acid and alcohol (acid+alcohol → ester+water). Higher hydroxycarboxylic acid and polyfunctional alcohol yield the highly viscous polyester complex.

The advantages of the Pechini method are the ability to prepare complex compositions, good homogeneity through mixing at the molecular leveling solution, and control of stoichiometry. Low firing temperatures are required for the decomposition of the resin to the oxide, thus 650°C for $BaTiO_3$ compared with 1,000°C for the conventional solid-state reaction. This is a versatile method for the synthesis of multicomponent oxides, for example, $Pb_3MgNb_2O_9$ [49]. The citrate gel [50,51] method shares the advantages of the Pechini method with respect to chemical homogeneity and compositional control. In the citrate gel process, which was developed by Marcilly and coworkers (1970) [52], metal nitrate solutions are added to the citric acid solution, whose pH is increased to between 6 and 7.5 in order to dissolve insoluble citrates but not to precipitate metal hydroxides. Also for the synthesis [53] of $YBa_2Cu_3O_{7-x}$, metal nitrate solutions of Y, Ba, and Cu were added to the citric acid solution and the pH was raised to between 6.5 and 7.0 in order to dissolve insoluble barium citrate but not to precipitate metal hydroxides. The solutions, which contained polybasic chelates, were separately concentrated to a viscous resin and dried in transparent gels, which were pyrolyzed to fine powders. The citrate gel method is also a versatile method for the synthesis of multicomponent oxides, for example, $SrCo_{0.8}Fe_{0.2}O_3$-x [54]. Other organic acids containing at least one hydroxyl and one carboxylic group such as tartaric acid, lactic acid, and glycolic acid can be used.

A claimed advantage of many chemical syntheses is their improved homogeneity although evidence for this is not always available. However, a recent study using Raman and 13C NMR spectroscopy indicated that during the preparation [55] of $BaTiO_3$ by the Pechini method the coordination of Ba and Ti in the mixed metal complex remained almost unchanged on polymerization and that molecular-level mixing was retained at the resin stage and probably in the pyrolyzed resin. There is further scope for using characterization techniques for monitoring changes in homogeneity at different wide ranges of chemical syntheses.

A key advantage of polymer pyrolysis over the other methods is its ability to yield high-strength non-oxide fibers, and there is considerable scope for the chemical design of polymers suitable for conversion to fibers, coating, and monolithic ceramics.

The advantages of these methods are the ability to produce complex compositions, good homogeneity through mixing at the molecular level in solution, and control of stoichiometry. Low firing temperatures are required for decomposition of resin to the oxide. This method is primarily used to obtain oxide. However, it is extended to obtain the oxosalts (Figure 2.8).

2.1.2.9 Aldo-Keto Gel Method

The aldo-keto gel method is introduced and practiced for the first time for the synthesis of luminescent materials during this work. It is a novel and low-temperature synthesis method. The idea for this method originated from the glucose–fructose gel [56–58].

Glucose (dextrose) and fructose are the hexoses with aldehyde and ketonic functional groups, respectively (i.e., aldohexose and ketohexose, respectively). Also, glucose and fructose are optically active substances and are dextrorotatory and laevorotatory, respectively. Equal amounts (weights) of glucose and fructose were taken in a borosil beaker and dissolved in double-distilled water. This solution was slowly heated. It starts becoming thicker, and the gel is formed at near about 120°C. This led to the conclusion that the aldehyde and ketone are taken together in equal proportions and slowly heated to form the gel.

When the precursors are added to the liquid mixture of glucose and fructose and the solution is slowly heated, the intermolecular forces of attraction go on increasing due to the thermal energy being absorbed. Hence, the intermolecular distances go on decreasing and the viscosity of the mixture increases. Thus, polymerization starts. At a particular temperature, the viscosity increases sufficiently and the gelation starts with the evolution of gases. When the gelation and hence the polymerization are completed and the gel is allowed to cool, the viscosity is maximum. In this situation, the precursors achieve the molecular size.

When the gel is further heated slowly to a particular temperature, the formation of foam starts as the polymer burns into viscous resin followed by the evolution of gases. At this time, the viscosity of

```
                          ┌──────────────────┐
                          │   Citric acid    │
                          │   (0.5 mole)     │
                          └────────┬─────────┘
                                   │
                          ┌────────▼─────────┐
                          │ Nitrate precursor│
                          │  (0.1 mole each) │
        ┌──────────────┐  └────────┬─────────┘
        │ 1-2 drops of │───────────▶│
        │ acetic acid  │  ┌────────▼─────────┐
        └──────────────┘  │ Ethylene glycol  │
                          │    (1 mole)      │
     ┌────────────────┐   └────────┬─────────┘
     │ Stirring, slow │───────────▶│
     │ heating to 130°C│ ┌─────────▼──────────┐
     └────────────────┘  │ Highly viscous solution of │
                         │ metal citrate complexes    │
        ┌──────────────┐ └────────┬──────────┘
        │  Condensed   │──────────▶│
        │  at 130°C    │  ┌────────▼─────────┐
        └──────────────┘  │ Polyesterification│
                          │    /gelation     │
     ┌────────────────┐   └────────┬─────────┘
     │ Slow heating   │───────────▶│
     │ above 130°C    │  ┌────────▼─────────┐
     └────────────────┘  │Polymerized complex│
                         │ /resin formation  │
   ┌────────────────┐    └────────┬─────────┘
   │ Pyrolysis at 130°C│─────────▶│
   └────────────────┘   ┌────────▼─────────┐
                        │  Black mass with │
                        │ powder precursors│
  ┌────────────────┐    └────────┬─────────┘
  │ Pyrolysis at 600 to│────────▶│
  │    700°C       │    ┌────────▼─────────┐
  └────────────────┘    │  White powder    │
                        │   (phosphor)     │
                        └────────┬─────────┘
                        ┌────────▼─────────┐
                        │   Sintered and   │
                        │    quenched      │
                        └──────────────────┘
```

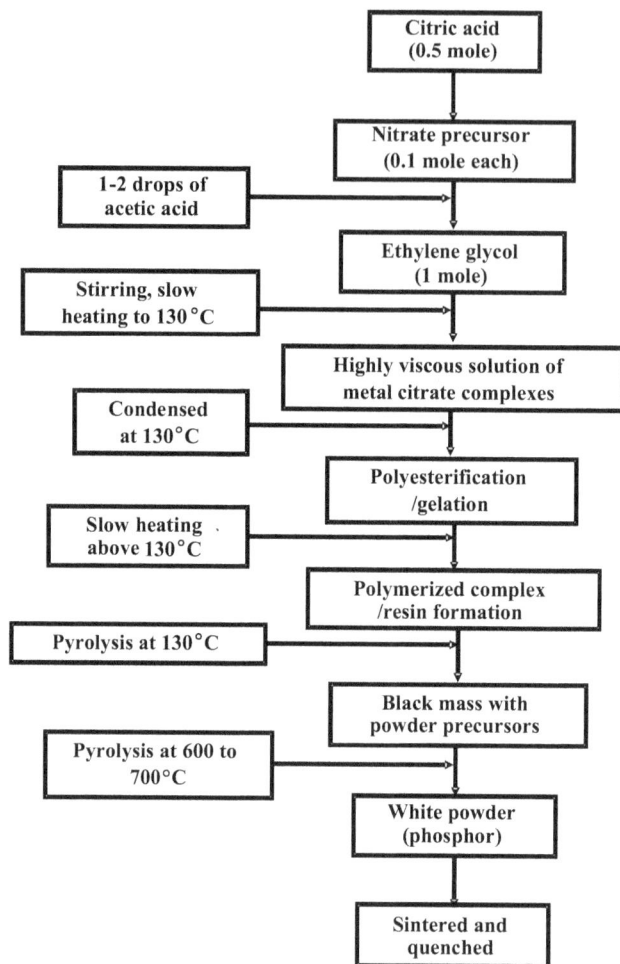

FIGURE 2.8 Flow chart of synthesis of phosphor by citrate gel/Pechini method.

gel increases enormously. When this resin is further heated, it starts burning and burns completely at a particular temperature leading to a fine crystalline phosphor.

Experimental procedure: Benzaldehyde (1 M) and acetone (1 M) (both of synthesis grade) were used together for gelation. The dried precursors (preferably nitrates) were weighed as per the stoichiometric proportions (0.1 mole of each) and finally powdered. Acetone (1 M) and benzaldehyde (1 M) (both of synthesis grade) were added to these nitrates. The pale yellow mixture obtained was stirred continuously and slowly heated to 120°C (near about) to form a reddish-brown gel.

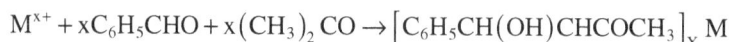

$$M^{x+} + xC_6H_5CHO + x(CH_3)_2 CO \rightarrow \left[C_6H_5CH(OH)CHCOCH_3 \right]_x M$$

The gel is further heated slowly to 300°C. Dark red foam (resin) was formed with the evolution of yellowish-brown fumes. A shining black resin was formed at 500°C. The resin was finally sintered at 800°C for 2 hours. The phosphor is obtained in powder form. This method was used for the synthesis of some of the phosphors (Figure 2.9).

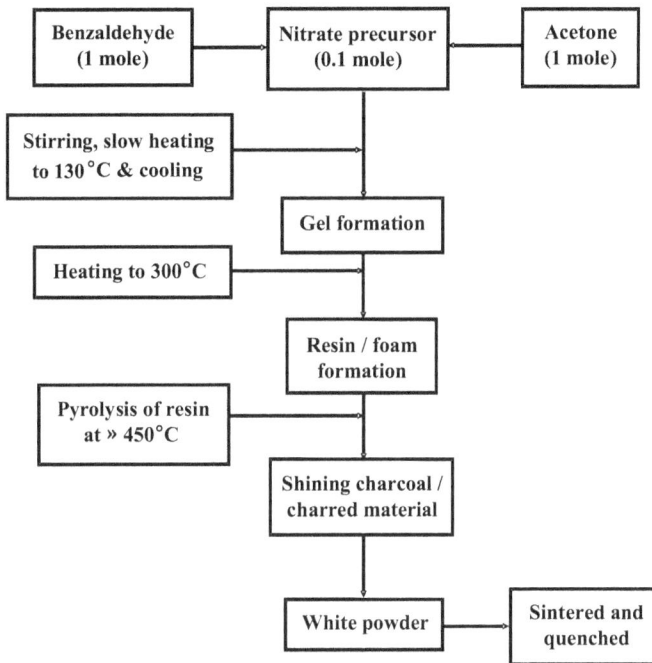

FIGURE 2.9   Flow chart of synthesis of phosphor by aldo-keto gel method.

2.2 CONCLUSIONS

Of all the methods reported, some are novel and modern synthesis methods that have their own advantages over each other in terms of above-discussed parameters such as particle sizes, morphology, required temperatures for synthesis, cost required for synthesis, and required time. It was found that the co-precipitation synthesis was found to be more prominent and effective due to a good morphological view with the required synthesis time is less.

Generally, SSD is used for the synthesis of commercial phosphor and most of the researchers usually prefer the CS method, which is used for the synthesis of phosphors, but both methods have some drawbacks. SSD requires stepwise repeated grinding and calcinating steps to get the pure phase of host. Repeated grinding may produce defects in the crystals of the prepared phosphor materials. Also, the crystalline size of the product is largely due to the sintering at high temperature for a long time. In CS method, fuel and oxidizer is required for synthesis and it is very difficult to maintain the fuel to oxidizer ratio. Stearic acid, Pechini, citrate gel methods, and aldo-keto gel method are modified version of the sol–gel method. In sol–gel method, costly organic chemical is used to form a gel, but in aldo-keto gel method, acetone and benzaldehyde are used for the gel formation. These organic chemicals are low cost and easily available in the laboratory.

REFERENCES

1. Segal, D. (1997). Chemical synthesis of ceramic materials, Paper 7/00881C; Received 6th February.
2. Moulson, A.J., and Herbert, J.M., *Electroceramics: Materials, Properties and Applications*, Chapman and Hall, London, 1990, p. 86
3. Segal, D.L., Materials science and technology: A comprehensive treatment. In *Processing of Ceramics*, vol. 17A, ed. R. J. Brook, VCH Publishers, Weinheim, 1996, p.69.
4. Moulson, J., and Herbert, J.M., *Electroceramics: Materials, Properties and Applications*, Chapman and Hall, London, 1990, p. 86.

5. Koparkar, K.A., Bajaj, N.S., and Omanwar, S.K. (2015). Exploring synthesis techniques for yttrium based phosphors. *Defect Diffus. Forum*, 361, 95–119. https://doi.org/10.4028/WWWSCIENTIFIC.NET/DDF.361.95.

6. Chu, M.S.H., and Rae, A.W.J.M., Ceramic transactions. In *Volume 49: Manufacture of Ceramic Components*, eds. B. Hiremath, A. Bruce and A. Ghosh, The American Ceramic Society, Westerville, OH, 1995, p. 65.

7. Ingale, N.B., Thakare, D.S., Omanwar, S.K., and Moharil S.V. (2002). Cost effective studies in Eu3+ activated yttrium host phosphor. *Proc. NSLA*, 9, 78–81.

8. Thakare, D.S., Ingale, N.B., and Omanwar, S.K. (2002). Novel synthesis of UV emitting phosphors. *Proc. NSNSM*, 1, 11–18.

9. Foo, Abdullah, A.Z., Amini Horri, B., and Salamatinia, B. (2018). Synthesis and characterisation of Y2O3 using ammonia oxalate as a precipitant in distillate pack co-precipitation process. *Ceram. Int.*, 44, 18693–18702. https://doi.org/10.1016/j.ceramint.2018.07.098.

10. Kerridge, D.H., *Inorg. Chemistry of Non-Aqueous Solvents*, ed. J.J. Lagowski, Academic Press, New York, 1978, voi. 5b, p. 269.

11. Segal, D. (1997). Chemical synthesis of ceramic materials. *J. Mater.*, 7, 1297–1305. doi: 10.1039/A700881C.

12. Pena-Pereira, F., and Calle, I., *Solvents and Eutectic Solvents, Encyclopedia of Analytical Science* (Third Edition), Academic Press, 2019, pp. 184–190, ISBN 9780081019849, doi: 10.1016/B978-0-12-409547-2.14020-X.

13. Lux, H. (1939). Zeitschrift fuer Elektrochemie und Angewandte Physikalische Chemie. *Elektrochem.*, 45, 303, ISSN: 0372–8323.

14. Mendoza-Mendoza, E., Montemayor, S.M., and Padmasree A.P. (2012). Molten salts synthesis and electrical properties of Sr and/or Mg doped Perovskite-type LaAlO$_3$ Powders. *J. Mat. Sci.*, 47, 6076. doi: 10.1007/s10853-012-6520-1.

15. Patil, K.C., Aruna S.T., and Ekambaram S., (1997). Combustion synthesis. *Curr. Opin. Solid State Mater. Sci.*, 2, 158–165. doi: 10.1016/S1359–0286(97)80060-5.

16. Booth, F. (1953). The theory of self-propagating exothermic reactions in solid systems. *Trans. Faraday Soc.*, 49, 272–281. doi: 10.1039/TF9534900272.

17. Philpot, K.A., Munir, Z.A., and Holt, J.B. (1987). An investigation of the synthesis of nickel aluminides through gasless combustion. *J. Mater. Sci.*, 22, 159–169. doi: 10.1007/BF01160566.

18. Moore, J.J., and Feng, H.J. (1995). Combustion synthesis of advanced materials-part I, reaction parameters. *Prog. Mater. Sci.*, 39, 243–273. doi: 10.1016/0079–6425(94)00011-5.

19. Moore, J.J., and Feng, H.J. (1995). Combustion synthesis of advanced materials- part II. Classification, applications and modeling. *Prog. Mater. Sci.*, 39, 275–316. doi: 10.1016/0079–6425(94)00012-3.

20. Merzhanov, A.G. (1993). Theory and practice of SHS, worldwide state of the art and the results. *Int. J. Self-Propag. High Temp Synth.*, , 19–39.

21. Subrahmanyam, J., and Vijaykumar, M. (1992). Self-propagating high temp synthesis. *J. Mater. Sci.*, 27, 6249–6273. doi: 10.1007/BF00576271.

22. Liu, Z.G., Ye, L.L., Guo, J.T., Li, G.S., and Hu, Z.Q. (1995). Self propogating high temperature synthesis of TiC and NbC by mechanical alloying. *J. Mater Res.*, 10, 3129–2135. doi: 10.1557/JMR.1995.3129.

23. Takacs, L. (1996). Ball milling-induced combustion in power mixtures containing titanium zirconium or hafnium. *J. Solid State Hem.*, 125, 75–84. doi: 10.1006/jssc.1996.0267.

24. Shon, I.J., and Munir, Z.A. (1995). Synthesis of MoSi$_2$-xNb and MoSi$_2$-yZrO$_2$ by the field activated combustion. *Mater. Sci. Eng.*, 202, 256–261. doi: 10.1016/0921–5093(95)09800-3.

25. Feng, A., and Munir, Z.A. (1996). Relationship between field direction and wave propogation in activated combustion synthesis. *J. Am. Ceram. Soc.*, 79, 2049–2058. doi: 10.1111/j.1151–2916.1996.tb08936.x.

26. Wang, L.L., Munir, Z.A., and Birch, J. (1995). Formation of MgO- B$_4$C composite via a thermite based combustion reaction. *J. Am. Ceram. Soc.*, 78, 756–764. doi: 10.1111/j.1151–2916.1995.tb08243.x.

27. Avakyan, P.B., Nersesyan, M.D., and Merzhanov, A.G. (1996). New materials for electronic engineering. *Am. Cer. Soc. Bull.*, 75, 50–55.

28. Gajapathy, D., and Patil, K.C. (1983). Mixed metal oxalate hydrazinites as compound precursors of spinnel ferrites. *Mat. Chem. Phys.*, 9, 423–438.

29. Patil, K.C., and Sekar, M.M.A. (1994). Synthesis, structure and reactivity of metal hydrazine carboxylates: Combustible precursors to the fine particle oxide materials. *Int. J. Self. Propag. High Temp. Synt.*, 3, 181–196.

30. Kingsley, J.J., and Patil, K.C. (1988). A novel combustion for the synthesis of fine particle α-alumina and related oxide materials. *Mat. Lett.*, 6, 427–432. doi: 10.1016/0167–577X(88)90045-6.

31. Patil, K.C. (1993). Advanced ceramics: Combustion, synthesis and properties. *Bul. Mat. Sci.*, 16, 533–541. doi: 10.1007/BF02757654.

32. Varma, A., Mukasyan, A.S., Rogachev, A.S., and Manukyan, K.V. (2016). Solution combustion synthesis of nanoscale materials. *Chem. Rev.*, 116, 14493–14586. doi: 10.1021/acs.chemrev.6b00279.

33. Suresh, K., and Patil, K.C., A recipe for an instant synthesis of fine particle oxide materials. In *Perspectives in Solid State Chemistry*, ed. K.J. Rao, Narosa Publishing House, New Delhi, 1995, pp. 376–388.

34. Aruna, S.T., and Patil, K.C. (1996). Synthesis and properties of nanosize titania. *J. Mater. Synth. Proc.*, 4, 175–179.

35. Patil, K.C., Ghosh, S., Aruna, S.T., and Ekambaram, S. (1996). Ceramic pigments: A solution combustion approach. *The Indian Potter*, 34, 1–9.

36. McKittrick, J., Shea, L.E., Bacalski, C.F., and Bosze, E.J. (1999). The influence of processing parameters on luminescent oxides produced by combustion synthesis. *Displays*, 19, 169–172. doi: 10.1016/S0141-9382(98)00046-8.

37. Kingsley, J.J., Suresh, K., and Patil, K.C. (1990). Combustion synthesis of fine- particle metal aluminates. *J. Mat. Sci.*, 25, 1305–1312. doi: 10.1007/BF00585441.

38. Patil, K.C., *Chemistry of Nanocrystalline Oxide Materials*, 2008. ISBN-13 978–981–279–314–0.

39. Everett, D.H., *Basic Principles of Colloid Science*, Royal Society of Chemistry, London, 1988. ISBN 0-85186-443-0.

40. Segal, D.L., *Chemical Synthesis of Advanced Ceramic Materials*, Cambridge University Press, Cambridge, 1989.

41. Segal, D.L. (1997). Chemical synthesis of ceramic materials. *J. Mater.*, 7, 1297–1305. doi: 10.1039/A700881C.

42. Zhang, W.F., Yin, Z., Zhang, M.S., Du, Z.L., and Chen, W.C. (1999). Roles of defects and grain sizes in photoluminescence of nanocrystalline SrTiO3. *J. Phys.: Cond. Mat.*, 11, 5655–5660. doi: 10.1088/0953-8984/11/29/312.

43. Meng, J.F., Zou, G.T., Cui, Q.L., Zhao, Y.N., and Zhu, Z.Q. (1994). Raman scattering from PbTiO$_3$ of various grain sizes at high hydrostatic pressures. *J. Phys.: Cond. Mat.*, 6, 6543. doi: 10.1088/0953-8984/6/32/015.

44. Ayyub, P., Palkar, V.R., Chattopadhyay, S., and Multani, M. (1995). Effect of crystal size reduction on lattice symmetry and cooperative properties. *Phys. Rev. B*, 51, 6135–6138. doi: 10.1103/PhysRevB.51.6135.

45. Qu, B.D., Evstigneev, M., Johnson, D.J., and Prince, R.H. (1998). Dielectric properties of BaTiO$_3$/SrTiO$_3$ multilayered thin films prepared by pulsed laser deposition. *Appl. Phys. Lett.*, 72, 1394. doi: 10.1063/1.121066.

46. Tsybeskov, L., Hirschman, K.D., Dasgupta, S.P., Zacharias, M., Fauchet, P.M., McCaffrey, J.P., and Lockwood, D.J. (1998). Nanocrystalline-silicon superlattice produced by controlled recrystallization. *Appl. Phys. Lett.*, 72, 43–45. doi: 10.1063/1.120640.

47. Pechini, M.P. (1967). Method of preparing lead and alkaline earth titanates and niobates and coating method using the same to form a capacitor. United States Patent Office, Patent 3,330,697.

48. Omanwar, S.K., Jaiswal, S.R., Bhatkar, V.B., and Koparkar, K.A. (2018). Comparative study of nanosized Al$_2$O$_3$ powder synthesized by sol-gel (citric and stearic acid) and aldo-keto gel method. *Optik*, 158, 1248–1254. doi: 10.1016/j.ijleo.2017.12.068.

49. Eror, N.G., and Anderson, H.U., Better ceramics through chemistry II. In *Mater. Res. Soc. Symp. Proc. 73*, eds. C.J. Brinker, D.E. Clark and D.R. Ulrich, Pittsburgh, 1986, p. 571.

50. Marcilly, C., Courty, P., and Delmon, B. (1970). Preparation of highly dispersed mixed oxides and oxide solid solutions by pyrolysis of amorphous organic precursors. *J. Am. Ceram. Soc.*, 53, 56–57. https://doi.org/10.1111/j.1151-2916.1970.tb12003.x.

51. Metlin, Y.G., and Tretyakov, Y.D. (1994). Chemical routes for preparation of oxide high-temperature superconducting powders and precursors for superconductive ceramics coatings and composites. *J. Mater. Chem.*, 4, 1659–1665. doi: 10.1039/JM9940401659.

52. Marcilly, C., Courty P., and Delmon, B. (1970). Preparation of highly dispersed mixed oxides and oxide solid solutions by pyrolysis of amorphous organic precursors. *J. Am. Ceram. Soc.*, 53, 56–57. https://doi.org/10.1111/j.1151-2916.1970.tb12003.x.

53. Blank, D.H.A., Kruidhof, H., and Flokstra, J. (1988). Preparation of YBa$_2$Cu$_3$O$_{7-\delta}$ by citrate synthesis and pyrolysis. *J. Phys. D: Appl. Phys.*, 21, 226. doi: 10.1088/0022-3727/21/1/036.

54. Lin, Y.S., and Zeng, Y. (1996). Catalytic properties of oxygen semipermeable perovskite-type ceramic membrane materials for oxidative coupling of methane. *J. Catal.*, 164, 220–231. doi: 10.1006/jcat.1996.0377.

55. Arima, M., Kakihana, M., Nakamura, Y., Yashima, M., and Yoshimura, M. (1996). Polymerized complex route to barium titanate powders using barium-titanium mixed-metal citric acid complex. *J. Am. Ceram. Soc.*, 79, 2847–2856. doi: 10.1111/j.1151–2916.1996.tb08718.x.

56. Omanwar, S.K., Jaiswal, S.R., Bhatkar, V.B., and Koparkar, K.A. (2018). Comparative study of nanosized Al_2O_3 powder synthesized by sol-gel (citric and stearic acid) and aldo-keto gel method. *Optik,* 158, 1248–1254. doi: 10.1016/j.ijleo.2017.12.068.

57. Koparkar, K.A., Korpe, R.G., Korpe, G.V., and Omanwar, S.K. (2021). Aldo-Keto gel synthesis and photoluminescence properties of YVO_4:Eu^{3+} microsphere. *Int. J. Scientif. Res. Sci. Technol.*, 8, 480–483, doi: 10.32628/IJSRST.RAMAN2181.

58. Koparkar, K.A., Bajaj, N., and Omanwar, S.K. (2015). Aldo-keto synthesis effect on Eu^{3+} fluorescence in YBO_3 compared with solid state diffusion. *J. Rare Earths*, 33, 486–490. doi: 10.1016/S1002–0721(14)60445–2.

3 Modern Synthesis Techniques for Metal Oxides II

S. R. Gosavi
C.H.C. Arts, S.G.P. Commerce, and B.B.J.P. Science College

R. A. Joshi
Toshniwal Arts, Commerce and Science College

CONTENTS

3.1 INTRODUCTION

From a scientific and technological point of view, metal oxides are considered the most important class of materials on account of their great variety of functional properties including chemical, semiconducting, piezoelectric, and optoelectronic. In the last few decades, metal oxides have been extensively studied for applications such as optoelectronics, piezoelectric applications, sensor devices, gas sensors for detection of a variety of combustible and toxic gases, photovoltaics, photocatalytic and photo-electrochromic devices, self-cleaning surfaces, thermochromic coatings, electromechanical devices, photocathode material in dye-sensitized solar cells, liquid crystal displays, transparent conducting electrodes, and memory devices [1–4].

DOI: 10.1201/9781003366232-3

Oxides in bulk form are usually stable and robust, with well-defined crystallographic structures [5]. However, generally fabricated in thin film form, these materials tend to display improvements in functional properties. Also, when metal oxides are utilized in applications such as windows for solar cell and memory devices, these materials must be grown in the form of smooth, compact thin films. Hence, recently, extensive research has been devoted to the growth of metal oxide thin films, which have been one of the most attractive themes in the field of research of chemistry, physics, and material science. Metal oxide thin films have attained great interest because of their excellent electrical conductivity, optical transmittance characteristics, nontoxic nature, large band gap, and low material cost, and they can be produced by a variety of methods as well [6,7]. Most metal oxide materials in thin film form have been identified as potential material for highly reproducible gas, humidity, and temperature sensor materials [8–14].

Generally, thin film synthesis methods can be classified into physical and chemical ones as shown in Figure 3.1. The process of deposition of thin film via physical methods is versatile technology, which can be used to deposit almost all kinds of materials. A wide variety of materials such as conductor materials are used in electronic circuits and devices, semiconducting materials are used for the application of dielectric and optical coatings, and superconductors can be deposited by physical methods. Though thin films with good quality and functionalized properties are prepared using physical methods, it is highly expensive and perhaps requires highly pure and large amount of material targets for thin film fabrication. These factors contribute to the increase in thin film fabrication cost. Also, in case of physical methods, doping is difficult during the growth process.

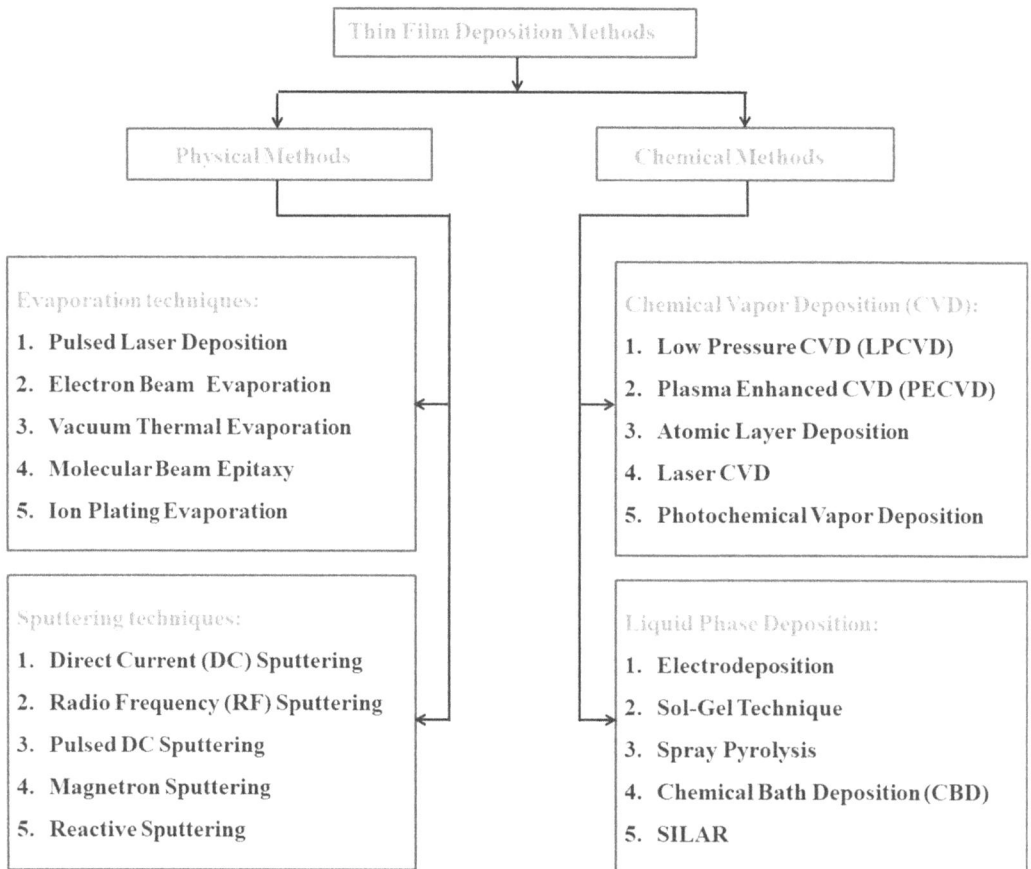

FIGURE 3.1   Classification of thin film deposition techniques.

Several metal oxides prepared by physical methods have been reported in recent years; however, the most investigated ones are centered on being low cost, nontoxic, and abundant. Zinc oxide (ZnO), titanium oxide (TiO_2), tungsten oxide (WO_3), copper oxide (CuO and Cu_2O), and tin oxide (SnO and SnO_2) have all these prerequisites, as well as some of them being environmentally friendly and easily synthesized by physical and chemical methods.

Since the need to produce device grade and good-quality thin films with low economical cost is necessary, chemical deposition techniques are widely used. Chemical methods are rather simple, cheap, and do not require vacuum or sophisticated instruments to accomplish the deposition of thin films. The chemical deposition is strongly dependent on the chemistry of solutions, pH value, temperature, viscosity, and so on. Hence, a comprehensive overview of the four different modern methods used for the synthesis of metal oxide thin films is discussed in the following sections of this chapter.

3.2 SYNTHESIS TECHNIQUES FOR METAL OXIDES

This chapter will report on the basic aspects of a few thin film deposition techniques, belonging to physical and chemical methods that are successfully utilized for the fabrication of thin films of metal oxides. In this chapter, we have discussed the experimental setup, preparative parameters, and advantages of particular deposition techniques. A review of the most popular methods conventionally applied for the fabrication of films of widely used metal oxides namely pulsed laser deposition (PLD), electron beam (e-beam) evaporation, successive ionic layer and adsorption (SILAR) method, and chemical bath deposition (CBD) method will be presented in the following sections.

3.3 PHYSICAL METHODS

3.3.1 Pulsed Laser Deposition

Since the invention of the pulsed ruby laser, PLD was used as an excellent method to grow high-quality thin films for the first time [15]. Since then, PLD is recognized as one of the most versatile techniques belonging to the physical vapor deposition (PVD) technique utilized for the growth of high-quality high-temperature superconducting thin films [16–18]. This PVD technique has been extensively used to fabricate thin films from a variety of functional materials such as metals, oxides, nitrides, carbides, and organic materials. PLD is simple in application and, therefore, is widely used in the fabrication of functional materials with different structural modifications such as amorphous, ultra-thin epitaxial single crystalline, polycrystalline, heterostructures, and nanocrystalline coatings [19,20]. Applications in the field of microelectronics, optoelectronics, optical industry, and other modern technologies require thin films of uniform thickness. Among the different deposition techniques, using the PLD technique not only the thickness but also phase purity and composition can be easily controlled by regulating the various growth parameters [21]. Hence, in the last few decades, PLD has emerged as a relatively simple and highly versatile technique for the deposition of epitaxial, stoichiometric thin films of simple materials, multielement complex compounds, multilayers, nanoparticles, and nanostructures.

3.3.1.1 Typical Experimental PLD Setup

In PLD, to fabricate uniform thin films on the substrate surface with good crystallinity, morphology, and stoichiometry, a high-power pulsed laser beam is focused inside a vacuum chamber to strike a target of the material that is to be deposited. During the deposition process, the following steps are very crucial [16,22]:

1. Laser absorption on the target of known composition and laser ablation of the target material and creation of plasma.
2. Dynamic of the plasma.

3. Deposition of the ablation material on the substrate.
4. Nucleation and growth of the film on the substrate surface.

Thus, it is an extremely simple technique, in which the high-power pulsed laser beam focuses on the surface of a target that is inside the vacuum chamber. The target material is vaporized by a laser beam. The vaporized species, containing atoms, ions, molecules, and electrons, is known as a laser-produced plasma plume and expands rapidly away from the target surface. Film growth occurs on a substrate upon which some of the plume material recondenses. This process can be performed in a high-vacuum environment or an environment with background gases such as oxygen. Oxygen is usually used for the oxide deposition to completely oxygenate the deposited thin film during the PLD process. If the deposition is made in a reactive gas and the obtained film has a composition different from that of the target, the name of the synthesis process is reactive PLD (RPLD). A schematic illustration of the PLD system is depicted in Figure 3.2. A PLD system primarily consists of the following three components [22]:

1. High-power laser source.
2. An ultra-high-vacuum chamber consisting of both the target and the substrate holder along with accessories to rotate the target and adjust the distance between the target and substrate.
3. Laser optics consists of a set of optical elements needed to guide the laser beam onto the target surface, including lenses, mirrors, and apertures.

Different kinds of laser sources are being used to ablate the target. The most common laser sources used in PLD ranges in output wavelength from the mid-infrared (CO_2 laser, i.e., 10.6 μm), through the near-infrared and visible (Nd-YAG laser, i.e., 1,064 and 532 nm), down into the ultraviolet. Currently, excimer lasers, which operate at several different UV wavelengths of 193 nm (ArF),

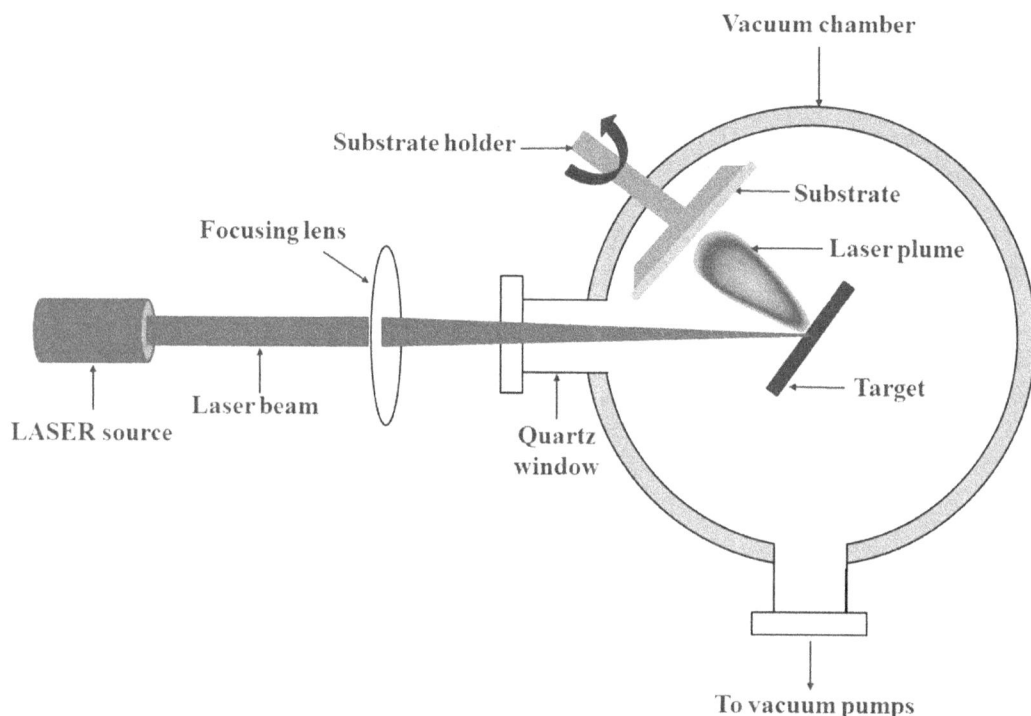

FIGURE 3.2 Schematic diagram of a typical pulsed laser deposition (PLD) setup.

248 nm (KrF), and 308 nm (XeCl), are used as laser sources in PLD. Excimer lasers receive considerable attention because most excimer lasers emit radiation in the range of interest of thin film fabrication [23].

In PLD, there are different processing parameters that play an important role during the deposition process, which controls the structural, morphological, and optoelectronic properties of the fabricated films. The following are the various parameters, which on adjustment can lead to desired properties of thin films and can affect the thin film quality [16,24,25]:

1. Wavelengths of laser
2. Pulsed duration
3. Repetition rate
4. Type of substrate
5. Substrate temperature: The morphology of the deposited thin films is also affected by the substrate temperature as surface temperature has a great influence on the nucleation density
6. The distance of the target to the substrate
7. Degree of ionization of the material separated from the target surface
8. Ambient gas pressure
9. Background pressure

3.3.1.2   Advantages of Thin Film Deposition Via Pulse Laser Deposition

PLD technique has significant advantages that distinguish it from other physical and chemical thin film deposition methods and provides special advantages for the growth of thin films of oxide materials. The following are the major advantages [16,26–29]:

1. In PLD, the deposition process is simple and flexible with great development potential with compatibility.
2. PLD is a versatile technique; i.e., a variety of materials, epitaxial films, multilayers, and heterostructures can be deposited in a wide variety of gases over a broad range of gas pressures.
3. PLD can be used to prepare a variety of thin film materials.
4. Any type of material can be ablated, and complex oxide compositions with high melting points can be easily deposited provided the target materials absorb the incident laser energy.
5. This method allows the growth of films under a highly reactive gas ambient over a wide range of pressure.
6. The precise chemical composition (stoichiometry) of a complex material can be reproduced with relative ease in the thin films fabricated by PLD.
7. PLD has a relatively high deposition rate. Also, the thickness of the deposited thin film can be controlled by the controlling repetition rate, and hence, the film prepared by PLD is uniform.
8. In PLD, the laser is used as an external energy source for plasma generation, so it is an energy-efficient and nonpolluting method of producing thin films.
9. The deposition process is very fast; i.e., high-quality thin films can be grown reliably in 10–15 minutes.
10. PLD does not require costly or corrosive precursors and large volume target.

3.3.1.3   Overview of Metal Oxide Thin Films Fabricated Using PLD

Herewith, the literature review is given in Table 3.1, which depicts the variation in the structure, morphology, and applications of the various metal oxide thin films prepared by PLD. Using the PLD technique, it is possible to deposit metal oxide films of desired film thickness, morphology, and stoichiometry by varying the parameters that are reviewed and presented in Table 3.1.

TABLE 3.1

Overview of metal oxide thin films fabricated using PLD

S. No	Oxide	Synthesis Parameters	Structure	Morphology	Application	Ref.
1.	Aluminum oxide (Al_2O_3)	Q-switched Nd:YAG laser Wavelength (λ)=1,064 nm Pulse duration=28 ns, 20 Hz Angle of incidence=45° Target used=Al_2O_3 crystal disk Chamber pressure=8×10^{-6} Torr Substrate: Si (1 0 0)	Amorphous	Splashed particles	Thermoluminescence	[2]
2.	Titanium oxide (TiO_x)	Pulsed Nd:YAG laser Wavelength (λ)=1,064 nm Pulse duration=10 ns, 50 Hz Pulse energy=0.6 J Target used=titanium Chamber pressure=2×10^{-3} Pa Substrate: molybdenum and steel	Mixture of monoclinic β-TiO_2 and amorphous TiO_x phase	Fine grains	Wear protection applications	[30]
3.	Titanium oxide (TiO_2)	Quadrupled Nd:YAG laser Wavelength (λ)=266 nm Pulse duration=7 ns, 10 Hz Target used=titanium Chamber pressure=10^{-6} Pa Gas=mixture of Ar and O_2 Substrate: silicon and aluminum	Amorphous	Porous nano- and mesostructure	Thermally sensitive substrates	[31]
4.	Vanadium oxide (VO_x)	Pulsed Nd:YAG laser Wavelength (λ)=355 nm Pulse width=19 ns, 10 Hz Target used=V_2O_5 Substrate=n-Si (1 0 0) Chamber pressure=0.1–0.5 mbar Gas=O_2	Tetragonal phase of VO_2 and V_3O_7	For 0.1 mbar, tetrahedral rods For 0.5 mbar, mesh-like rods	Electrochemistry and catalysis	[32]
5.	Manganese oxide (Mn_3O_4)	KrF excimer laser (Lambda Physik) Wavelength (λ)=248 nm Target used=metallic Mn Chamber pressure=200 mTorr Gas=oxygen Substrate: stainless steel	Tetragonal hausmannite phase of Mn_3O_4.	Octahedral-shaped grains	Microbatteries	[33]
6.	Iron oxide (Fe_2O_3)	Pulsed KrF excimer laser Wavelength (λ)=248 nm Repetition rate=2 Hz Energy density=2 J/cm² Substrate=sapphire Target=ablate magnetite Chamber pressure=more than 1×10^{-1} Pa Gas=oxygen	Hematite phase	Spherical grains	Magnetic material	[34]

(Continued)

TABLE 3.1 (*Continued*)

Overview of metal oxide thin films fabricated using PLD

S. No	Oxide	Synthesis Parameters	Structure	Morphology	Application	Ref.
7.	Cobalt oxide (Co_3O_4)	Nd:YAG laser Wavelength (λ) = 532 nm Repetition rate = 10 Hz Pulse energy = 0.6 J Substrate = silicon Target used = circular disk of cobalt boride Energy density = 20 and 30 J/cm^2 Chamber pressure = 10^{-5} mBar	For the as-deposited films, amorphous, and for air annealed films, cubic structure	Spherical particulates and urchin structures consisting of a core and nanowires	Micro- and nanoelectronic devices	[35]
8.	Nickel oxide (NiO)	KrF excimer laser Wavelength (λ) = 248 nm Repetition rate = 5 Hz Energy per pulse = 300 mJ Substrate = Cu foil and quartz Target = NiO ceramic target Chamber pressure = 5 Pa Gas = oxygen	Cubic	Irregularly rounded grains	Lithium-ion battery	[36]
9.	Copper oxide (CuO and Cu_2O)	Nd:YAG laser Wavelength (λ) = 532 nm Target used = copper oxide Energy per pulse = 100 mJ Chamber pressure = 1.33×10^{-5} mBar Gas = oxygen	For the as-deposited film, amorphous, and for annealed films, cubic (Cu_2O) and monoclinic (CuO) structure	Spherical grains	Next-generation semiconducting devices	[37]
10.	Zinc oxide (ZnO)	Krypton fluoride (KrF) laser Wavelength (λ) = 248 nm Pulse width = 15 ns, 20 Hz Target used = ZnO Energy per pulse = 100 mJ Substrate = Si wafer (1 0 0) Chamber pressure = 0.07 Torr Gas = oxygen	(0 0 2) c-axis orientation	Isolated "helices" and "posts"	UV and visible photoluminescence	[38]
11.	Zinc oxide (ZnO)	KrF excimer laser Wavelength (λ) = 248 nm Repetition rate = 10 Hz Target used = ZnO Energy density = 2 J/cm^2 Substrate = polyimide (Kapton) and polyethylene naphthalate (PEN) Chamber pressure = 1–2 mTorr Gas = oxygen	Hexagonal wurtzite structure	For ZnO film deposited on Kapton, parallel-chain-like texture, and for ZnO, film deposited on PEN, pit-like texture	Opto- and thermoelectrical applications	[39]

(Continued)

TABLE 3.1 (*Continued*)
Overview of metal oxide thin films fabricated using PLD

S. No	Oxide	Synthesis Parameters	Structure	Morphology	Application	Ref.
12.	Gallium oxide (Ga_2O_3)	KrF excimer laser Wavelength (λ)=248 nm Repetition rate=10 Hz Target used=Ga_2O_3 ceramic target Energy density=2.4 J/cm^2 Substrate=sapphire Chamber pressure=5×10^{-2} Torr Gas=oxygen	Amorphous and polycrystalline (depending on growth temperature)	Grains	Potential optoelectronic applications	[40]
13.	Molybdenum oxide (MoO_3)	ArF excimer laser (Lambda Physik, COMPex 102) Wavelength (λ)=193 nm Pulse width=12–16 ns Repetition rate=10 Hz Target used=MoO_3 Energy density=2.5 J/cm^2 Substrate=corning glass Medium=vacuum	Crystalline	Grains	Optical recording	[41]
14.	Silver oxide (Ag_4O_4)	Nd:YAG laser Wavelength (λ)=266 nm Pulse width=7 ns Energy density=1.6 J/cm^2 Target used=Ag Substrate=silicon Gas=pure argon and synthetic air Chamber pressure=4–150 Pa	Monoclinic	Columnar structure and nanoparticles	Catalysis and antibacterial coatings	[42]
15.	Cadmium oxide (CdO)	Nd:YAG laser Wavelength (λ)=1,064 nm Pulse width=5 ns Repetition rate=10 Hz Energy density=2 J/cm^2 Target used=CdO Medium=plasma Substrate: glass	Polycrystalline cubic	Quasi-spherical	Photoluminescence	[43]
16.	Tin oxide (SnO_2)	Pulsed KrF excimer laser Wavelength (λ)=248 nm Pulse width=12 ns Repetition rate=30 Hz Target used=pure tin oxide Energy density=6 J/cm^2 Substrate=alumina disk Gas=oxygen Chamber pressure=10^{-6} Torr	In vacuum, polycrystalline SnO_2 and amorphous SnO while at 150 mTorr or O_2, pure polycrystalline SnO_2 phase	Columnar structure with nodular feature	Gas sensing	[44]

(*Continued*)

TABLE 3.1 (*Continued*)
Overview of metal oxide thin films fabricated using PLD

S. No	Oxide	Synthesis Parameters	Structure	Morphology	Application	Ref.
17.	Tungsten oxide (WO_3)	KrF excimer laser (Lambda Physik LPX 220i) Wavelength (λ)=248 nm Pulse duration=25 ns Repetition rate=20 Hz Target used=disk made up of a mixture of metallic tungsten and boric acid powders Energy density=4.5 J/cm² Substrate=glass and Si Chamber pressure=1.5×10^{-2} mBar Gas=oxygen	Monoclinic	Flower shape	Photocatalysis	[45]
18.	Bismuth oxide (Bi_2O_3)	KrF excimer laser Wavelength (λ)=248 nm Pulse duration=25 ns Repetition rate=1 Hz Target used=Bi_2O_3 pellets Energy density=7 J/cm² Substrate=glass Chamber pressure=0.2 mBar Gas=oxygen	Mixed polycrystalline and amorphous	Nanocrystals	Photoluminescence	[46]

3.3.2 ELECTRON BEAM EVAPORATION

E-beam deposition is a PVD technique commonly used in the field of physics of thin film, because a large variety of materials can be rapidly and cleanly evaporated and deposited with minimum consumption of energy. E-beam deposition technique includes two modes [47] namely e-beam evaporation deposition method and e-beam-induced deposition. In this chapter, we mainly focus on the e-beam evaporation method, which is mainly utilized for growing thin films with high melting point materials or highly pure metals. The source material is evaporated with gas release and then transferred to the substrate for deposition to form a thin film or coating. This method is commonly used in the field of thin film processes, including semiconductors, optics, solar panels, glasses, and architectural glass, so that they have the required conductive, reflective, transmission, and electronic properties [48]. E-beam evaporation utilized a high-energy beam of electrons generated from a tungsten filament driven by electric and magnetic fields to heat the source material and transform it into the gaseous phase to be deposited on the substrate surface. This process occurs in a high-vacuum environment to form a thin film on the substrate surface by the free path. It is able to evaporate metals with high melting points, such as tungsten, tantalum, and graphite, and a very high rate of evaporation can be achieved with e-beam evaporation [49].

3.3.2.1 Typical Experimental Setup of Electron Beam Evaporation Method

E-beam evaporation deposition is a technique in which heat is produced by a high-energy e-beam bombarding the target materials placed in a water-cooling crucible and then deposited as a thin film on substrates. An intense e-beam is generated by an electron gun, which uses the thermionic emission of electrons produced by an incandescent filament. Emitted electrons are steered by a high-voltage potential and magnetic field to strike source material and vaporize it within a vacuum environment. The vapor will then travel and condense on the substrate forming a thin film. A schematic illustration of the e-beam evaporation system is depicted in Figure 3.3.

3.3.2.2 Advantages of Thin Film Deposition Via Electron Beam Evaporation Method

The e-beam evaporation technique has the following advantages over the other deposition techniques [47,50–52]:

1. E-beam evaporation is used to deposit a wide variety of materials with a high deposition rate.
2. This technique provides the required qualities in the deposited thin films, which are commonly used for optical thin film applications such as laser optics and solar panels to eyeglasses and architectural glass.
3. E-beam evaporation provides high material utilization efficiency compared with other PVD processes, thus reducing process costs.
4. E-beam evaporation is controllable, repeatable, and compatible with the use of an ion assist source to enhance the desired thin film performance characteristics.
5. During the deposition process, there is a direct transfer of energy to the target material using an e-beam, which makes suitability for this method to deposit metals with high melting points.
6. The source material is placed in the water-cooling crucible, which can avoid contamination in the deposits. Thus, the e-beam evaporation process may produce thin films with higher purity and better-controlled composition.
7. High density and homogeneity of the prepared films.

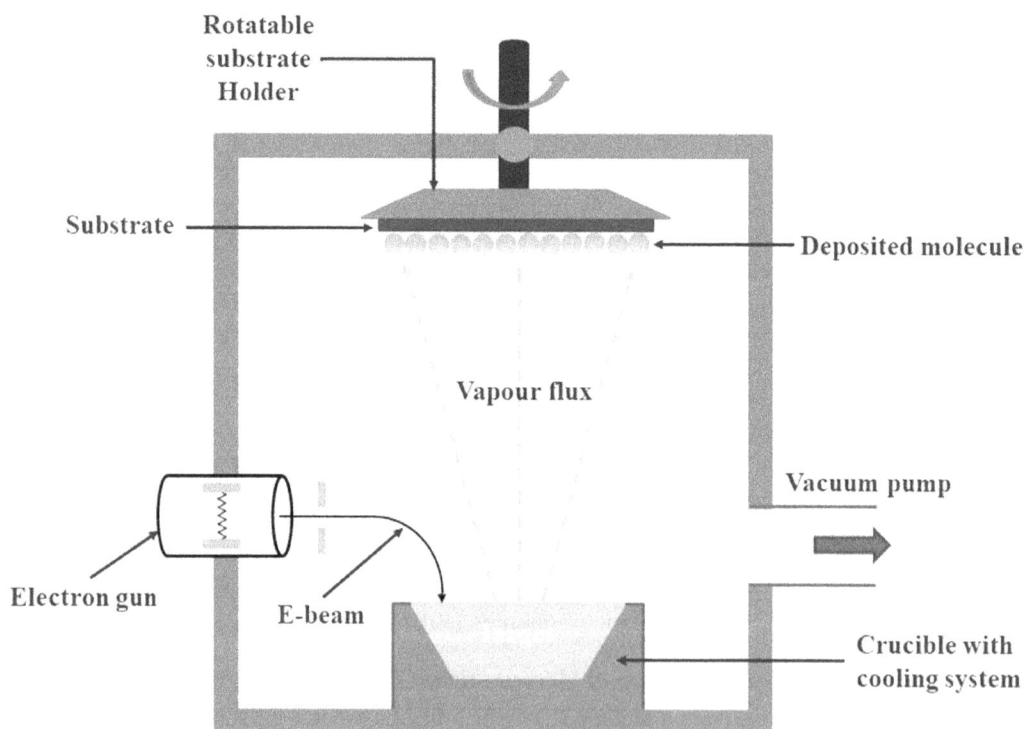

FIGURE 3.3   Illustration of the e-beam evaporation system.

3.3.2.3 Overview of Metal Oxide Thin Films Fabricated Using E-Beam Evaporation

Herewith, the literature review is given in Table 3.2, which depicts the variation in the structure, morphology, and applications of the various metal oxide thin films prepared by e-beam evaporation. E-beam evaporation is a relatively simple method utilized to prepare many different metal oxides in thin film forms using different deposition parameters, and some of them are summarized in Table 3.2.

TABLE 3.2

Overview of Metal Oxide Thin Films Fabricated Using E-Beam Evaporation

S. No.	Oxide	Synthesis Parameters	Structure	Morphology	Application	Ref
1.	Aluminum oxide (Al_2O_3)	Target: Al_2O_3 granules Substrate: n-Si (1 0 0) Base pressure: 2×10^{-6} mbar Deposition pressure: 2×10^{-4} mbar Substrate rotation speed: 20 rpm Deposition rate: 0.308 nm/s	–	Porous microstructure	Micro- and optoelectronic devices	[53]
2.	Titanium oxide (TiO_2)	Target: TiO_2 Substrate: n-Si (1 0 0) Chamber pressure: 1.1×10^{-6} mbar Deposition pressure: 4×10^{-5} mbar Deposition rate: 0.2 nm/s Substrate target distance: 14 cm	Anatase	Agglomerated nanoparticles	Magnetic material	[54]
3.	Vanadium oxide (VO_2)	Target: vanadium Substrate: c-type sapphire Chamber pressure: 10^{-4} Pa Chamber pressure: 8×10^{-2} Pa Accelerating voltage: 10 kV Deposition rate: 0.034–0.051 nm/s	Monoclinic	–	Optoelectronic switching devices	[55]
4.	Manganese oxide (MnO_2 and Mn_3O_4)	Target: MnO_2 pellet Substrate: Si (1 0 0) and stainless steel Base pressure: 2×10^{-6} Torr Chamber pressure: 4×10^{-4} Torr Substrate rotation speed: 25 rpm Deposition rate: 3.5 Å/s Substrate temperature: 300–343 K	For α-MnO_2, tetragonal and bixbyite Mn_2O_3 for sample heated at 973 K	Spherical particles	Electrochemical supercapacitor	[56]
5.	Nickel oxide (NiO)	Target: NiO commercial pellet Substrate: ITO Chamber pressure: 7×10^{-4} Pa Substrate rotation speed: 25 rpm Deposition rate: 6 nm/min	Cubic	Needle-shaped grains	Electrochromic properties	[57]
6.	Zinc oxide (ZnO)	Target: polycrystalline ZnO Substrate: glass Chamber pressure: Below 3×10^{-5} bar Deposition rate: 0.1–0.15 nm/s Substrate target distance: 10 cm Annealing temperature: 250°C–550°C	For the annealed film, hexagonal wurtzite structure	Spherical grains	Transparent conductive oxide	[58]

(Continued)

TABLE 3.2 (*Continued*)
Overview of Metal Oxide Thin Films Fabricated Using E-Beam Evaporation

S. No.	Oxide	Synthesis Parameters	Structure	Morphology	Application	Ref
7.	Molybdenum oxide (MoO_3)	Target: dry MoO_3 powder Substrate: glass and FTO-coated glass Deposition pressure: 1×10^{-5} bar Accelerating voltage: 5 kV Substrate temperature: RT, 100°C, 200°C	Orthorhombic phase	Needle-like crystallites	Electrochromic devices	[59]
8.	Cadmium oxide (CdO)	Target: CdO powder Substrate: glass and ITO-coated glass Chamber pressure: 2×10^{-6} bar Deposition rate: 8–10.6 Å/s	Cubic	Interconnected clusters	Optoelectronic applications	[51]
9.	Tin oxide (SnO_2)	Target: SnO_2 nano-powder Substrate: glass Chamber pressure: 5.5×10^{-6} mbar Deposition rate: 8–10.6 Å/s Annealing temperature: 400°C Annealing time: 3 hours	Tetragonal rutile	–	Thermistor applications	[60]
10.	Tungsten oxide (WO_3)	Target: tungsten oxide pellet Pellet size: 1 cm diameter and 0.14 cm thickness Substrate: glass and ITO Chamber pressure: 8.2×10^{-6} Torr Deposition rate: 2 Å/s	Monoclinic	–	Energy-saving system	[61]

3.4 CHEMICAL METHODS

Although physical methods are versatile technology and possess numerous benefits, they are highly expensive and perhaps require a large amount of material target. Also, these methods may require a larger investment than other thin film deposition processes. Since the need to produce device grade and good quality thin films with low economical cost is necessary, chemical deposition techniques are widely used. Chemical methods are rather simple, cheap, and do not require vacuum or sophisticated instruments to accomplish the deposition of thin films. The chemical deposition is strongly dependent on the chemistry of solutions, pH value, temperature, viscosity, and so on. The most common chemical deposition of metal oxide thin films is obtained via electrodeposition, spray pyrolysis, spin coating, sol–gel process, CBD, and SILAR. This section of the chapter is concerned only with SILAR and CBD techniques due to their ability to fabricate device quality thin films with low equipment requirement.

3.4.1 SUCCESSIVE IONIC LAYER ADSORPTION AND REACTION (SILAR) METHOD

The SILAR method, which is also known as the modified CBD (M-CBD) method, was first reported in 1985 by Ristov et al. [62] for the successful chemical deposition of Cu_2O thin film. In 1985, Nicolau has given the name SILAR to this method [63]. Since then, it has emerged as one of the chemical methods to deposit a variety of compound materials, i.e., metal chalcogenides and oxides in thin film form.

In the SILAR method, deposition proceeds via a layer-by-layer buildup of the film in the solution [64]. Essentially, the solutions can be in separate vessels, and then, the substrate is moved from one vessel to another. During the deposition process, the substrate is immersed in a solution containing metal cation, and ions of the precursor solution are adsorbed onto the surface of the substrate. The adsorption is due to attraction forces between ions in the solution and surface of the substrate [65]. In the following step, the substrate is rinsed in double-distilled water so that the layer, which is tightly stuck on the surface, remains on the substrate surface. In the next reaction step, the substrate is immersed in the anion precursor solution. The ions of the solution are diffused into the surface and react with the adsorbed cations. Ideally, this gives one monolayer of the deposits. In the final step, the substrate is rinsed to remove the ions from the diffusion layer. A second growth cycle can now commence. The rinsing steps are very important to avoid the formation of clusters of the semiconductor, rather than a film [64].

3.4.1.1 Typical Experimental Setup of SILAR Technique

The simplified schematic diagram of the experimental setup used for the deposition of thin films by the SILAR technique is shown in Figure 3.4.

The SILAR system consists of four beakers, two of which contain cationic and anionic precursor solutions and the other two beakers are filled with deionized water. The cationic precursor contains the solution of a salt of the cation, and the anionic precursor contains the solution of a salt of the anion.

A single SILAR immersion cycle involved the following four steps:

1. The cleaned substrate was immersed in the cationic precursor where cations of the compound to be grown had been adsorbed.
2. Loose cations from the substrate surface had been removed by rinsing it with distilled water.
3. The substrate was then immersed in the anionic precursor where cations react with anions to give a product.
4. Finally, again it is rinsed with water to remove loose material from the surface, which completes one SILAR deposition cycle. The process is then repeated as many times as needed to obtain the required thickness of the film, which can be utilized for desired applications.

During the deposition process via the SILAR method, the experimental conditions heavily affect the quality and the performance of the SILAR-coated thin films. The following

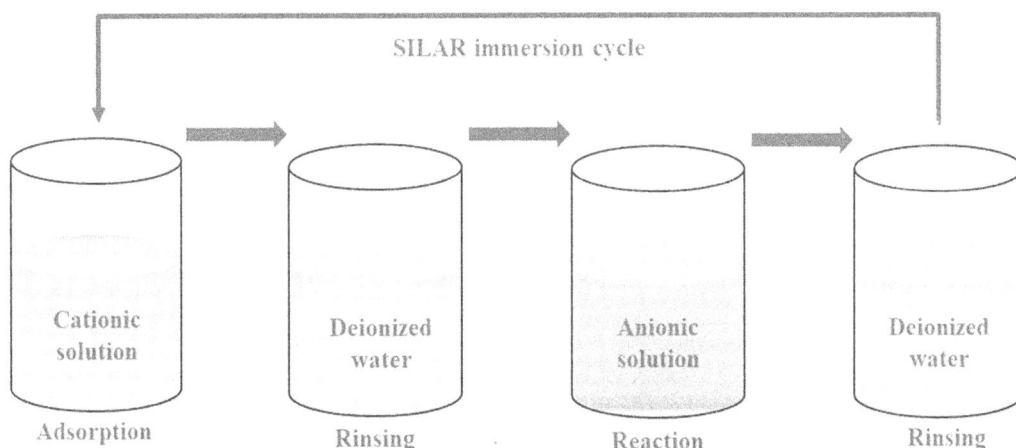

FIGURE 3.4   Schematic representation of the SILAR method [66].

parameters can affect the structural, morphological, and functional properties of the deposited metal oxide thin films:

1. Temperature
2. Type of precursor
3. Concentration and pH of solutions
4. Complexing agent
5. Adsorption time
6. Immersion time
7. Drying time
8. Number of immersion cycles

3.4.1.2 Advantages of Thin Film Deposition Via SILAR Method

Among the chemical route synthesis techniques, SILAR is one of the simplest and economically favorable chemical methods because of its following advantages [65,67–72]:

1. Since vacuum systems are not required, the SILAR deposition technique is simple, inexpensive, and convenient for large-area deposition.
2. Starting materials are commonly available and cheap.
3. The possibility to grow thin films has excellent physical properties at low temperature and normal pressure.
4. Film thickness and composition of a thin film can be controlled easily and accurately by varying the number of deposition cycles and immersion process.
5. Using SILAR, one can produce thin films of nanometer thickness with good stoichiometry.
6. It offers an extremely easy way to produce doped thin films.
7. Various kinds of substrates without special restrictions can be effectively coated using the SILAR technique.
8. Unlike physical methods, SILAR does not require high-quality target nor does it require a vacuum at any stage.
9. SILAR does not require any sophisticated instruments.
10. It can produce durable and adherent thin films of metal oxides.
11. In SILAR, the preparative parameters can be easily optimized and precipitate formation and wastage of material can be avoided.

3.4.1.3 Overview of Metal Oxide Thin Films Fabricated Using SILAR Method

Herewith, the literature review is given in Table 3.3, which depicts the properties and applications of different metal oxide thin films prepared by SILAR.

3.4.2 Chemical Bath Deposition (CBD) Method

During the last few decades, various materials such as metal chalcogenides and oxides in thin film form have been grown successfully using CBD. This method has been practiced since 1884. Nearly after 50 years, Bruckman used this process for the deposition of lead sulfide thin films using a chemical bath containing solutions of lead acetate, thiourea, and NaOH [84]. This method was limited to the deposition of PbS and PbSe for a long time. Later, this procedure was reviewed in 2000 [85]. Presently, it is a popular process to synthesize compound semiconductors and metal oxides such as ZnO, TiO_2, and CdO and ternary compounds in thin film form [62,85–87]. CBD has been specially characterized by its mass transport, which maintains the liquid phase as the mass transport media for the transportation of precursors from the source to the substrate. The representative schematic diagram showing the process of CBD of thin films is shown in Figure 3.5.

TABLE 3.3

Overview of Metal Oxide Thin Films Fabricated Using SILAR Method

S. No.	Oxide	Synthesis Parameters	Structure	Morphology	Application	Ref
1.	Titanium oxide (TiO_2)	Cationic precursor: 0.1 M Ti(III) Cl_3 and 30% HCl solution Anionic precursor: 0.01 M NaOH Reaction time: 10 seconds Rinsing time: 5 seconds Deposition temperature: RT Annealing temperature: 673 K Substrate: FTO-coated glass	Anatase phase	Spherical-sized grains	Photoelectrochemical cell	[73]
2.	Vanadium oxide (V_2O_5)	Cationic precursor: 0.2 M VCl_3 in HPLC H_2O+ HNO_3 maintained at 65°C Anionic precursor: 0.1% H_2O_2 pH: ~1 Reaction time: 20 seconds Rinsing time: 15 seconds Substrate: FTO-coated glass	Polycrystalline	Microporous	Electrochemical capacitors	[74]
3.	Manganese oxide (α-MnO_2 and γ-MnO_2)	Cationic precursor: 0.1 M $MnCl_2$ Anionic precursor: 1.0 M NaOH Reaction time: 60 seconds Drying time: 24 hours Drying temperature: 50°C Temperature: room temperature Substrate: stainless steel	Amorphous	Nanosheets	Pseudo-capacitor	[75]
4.	Iron oxide (α-Fe_2O_3)	Cationic precursor: 0.05 mol/L $FeCl_3 \cdot 6H_2O$ Anionic precursor: 0.001 mol/L NaOH Reaction time: 20 seconds Rinsing time: 20 seconds Annealing temperature: 773 K Substrate: glass	Rhombohedral hematite	Nano-grains	Antibacterial activity	[76]
5.	Cobalt oxide (Co_3O_4)	Cationic precursor: 0.4 M $CoCl_2$ complexed with liquor ammonia pH: ~12 Anionic precursor: 0.1% H_2O_2 Reaction time: 20 seconds Rinsing time: 15 seconds Substrate: copper substrate	Amorphous	Fine elongated particles	Supercapacitor	[67]
6.	Nickel oxide (NiO)	Cationic precursor: 0.05 M $NiSO_{4+}NH_4OH$ pH ~12±0.2 Anionic precursor: 2% H_2O_2 Reaction time: 30 seconds Reaction temperature: 333 K Annealing temperature: 573 K Substrate: stainless steel	Face-centered cubic	Nanoflakes sponge-like clusters	Supercapacitor	[77]

(Continued)

TABLE 3.3 (*Continued*)

Overview of Metal Oxide Thin Films Fabricated Using SILAR Method

S. No.	Oxide	Synthesis Parameters	Structure	Morphology	Application	Ref
7.	Copper oxide (CuO)	Cationic precursor: 0.1 M $CuSO_4.5H_2O+NH_4OH$ pH: ~11 Anionic precursor: hot deionized water maintained at 90°C Adsorption time: 30 seconds Reaction time: 30 seconds Substrate: glass	Monoclinic with tenorite phase	Grains	Photovoltaic applications	[78]
8.	Zinc oxide (ZnO)	Cationic precursor: 0.125 M ammonium zincate Anionic precursor: ethylene glycol solution maintained at 125°C Adsorption time: 5 seconds Reaction time: 5 seconds Rinsing time: 5 seconds Annealing temperature: 400°C Substrate: glass	Hexagonal wurtzite	Flower-like nanostructures with porous grains	Gas sensor	[79]
9.	Zinc oxide (ZnO)	Cationic precursor: ammonium zincate Anionic precursor: double-distilled water Annealing temperature: 300°C Annealing time: 1 hour Substrate: glass	Hexagonal wurtzite	Flower-like structures	Self-cleaning application	[80]
10.	Cadmium oxide (CdO)	Cationic precursor: 0.1 M $Cd(COOCH_3)_2.2H_2O$ and TEA pH: ~10 Solution temperature: 85°C Anionic precursor: distilled water Adsorption time: 20 seconds Rinsing time: 20 seconds Annealing temperature: 350°C Substrate: soda lime glass slides	Polycrystalline	Grains	Applications in different devices	[81]
11.	Tin oxide (SnO_2)	Cationic precursor: 0.05–0.15 M $SnCl4.5H_2O+TEA$ Anionic precursor: $H_2O_{2+}H_2O$ Immersion time: 20 seconds Rinsing time: 5 seconds Annealing temperature: 500°C Substrate: glass	Nanocrystalline	Spherical nano-size grains	Photosensor	[82]
12.	Bismuth oxide (Bi_2O_3)	Cationic precursor: 0.05M $Bi_2(NO_3)_3.5H_2O+TEA$ (30°C) Anionic precursor: double-distilled water (70°C) Immersion time: 10 seconds Rinsing time: 5 seconds Substrate: stainless steel Number of SILAR cycles: 80	Cubic δ-Bi_2O_3 phase	Interconnected nanoneedles	Supercapacitor	[83]

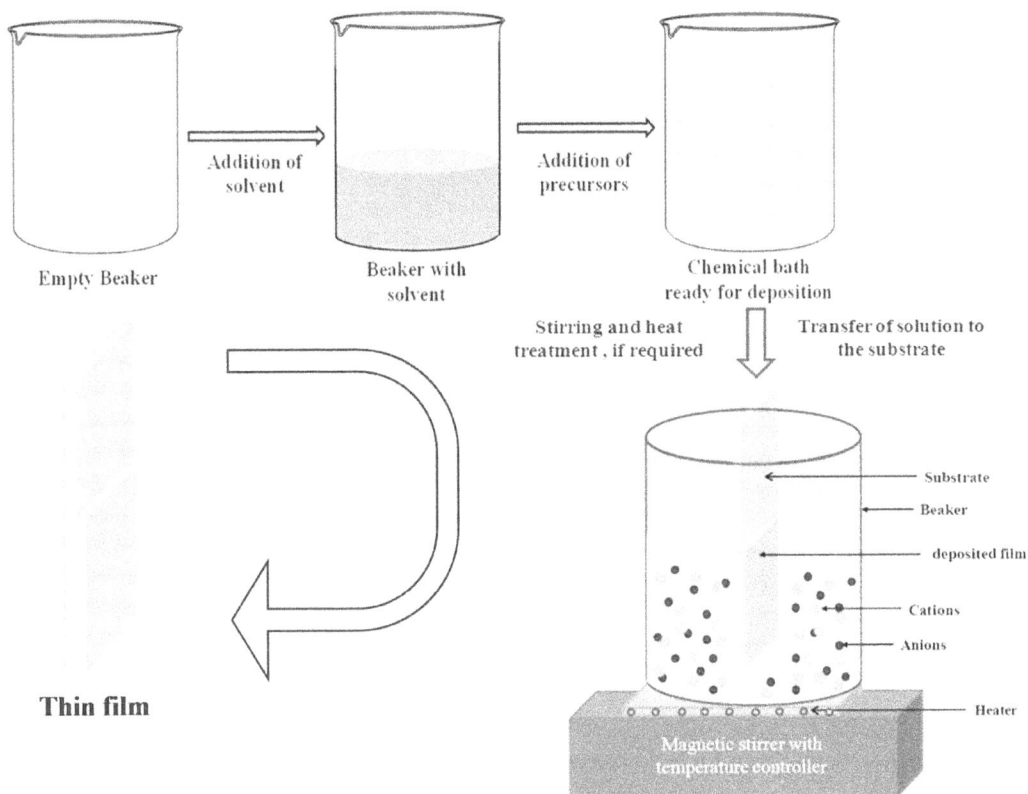

FIGURE 3.5   Basic aspects of chemical bath deposition.

In CBD, the chemical bath is prepared by mixing the proper proportion of cationic and anionic solutions in a beaker. Thereafter, the substrate is immersed in a solution containing the precursors. Substrates are then removed from the precursor solution and kept at a particular temperature after certain time intervals. Thin films are deposited on the substrate surface. Despite the growth mechanism, it is possible to control the properties of deposited thin films by varying the following deposition parameters:

1. Concentration of precursor solutions
2. Complexing agent
3. Deposition temperature and time
4. pH of precursor solutions
5. Nature of substrate

The most important requirement of the CBD technique is that the ionic product of the metal ion concentrations and the chalcogenide ion concentration must exceed the solubility product of the desired product [85].

3.4.2.1   Typical Experimental Setup of Chemical Bath Deposition

Thin films prepared using the CBD method are found to be of comparable quality to those prepared using sophisticated and expensive deposition techniques such as chemical spray pyrolysis, PLD, vacuum evaporation, and magnetron sputtering (Figure 3.6). Recently, CBD has emerged as one of the simple and cost-effective techniques for the deposition of ZnO thin films [88] as this method does not require sophisticated and expensive equipment.

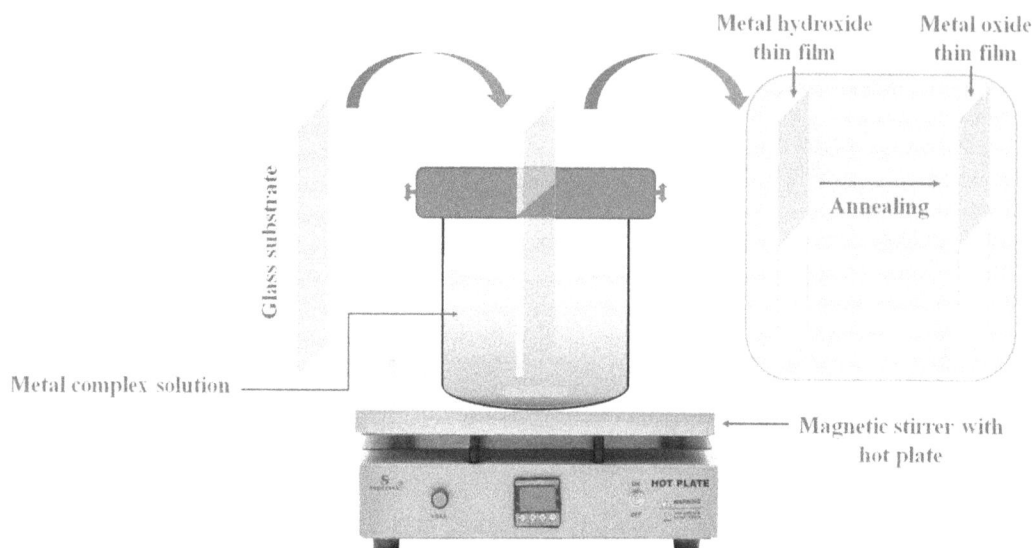

FIGURE 3.6 Illustration of the chemical bath deposition system.

3.4.2.2 Advantages of Thin Film Deposition Via Chemical Bath Deposition

This method offers many advantages over other well-known thin film deposition methods. The selection of this method for the fabrication of thin films arises from its low cost, ease of handling, possibility of large area deposition at room temperature, and inexpensive starting chemicals. Following are the major advantages [62,85,89]:

1. CBD is an attractive method for large-area deposition.
2. CBD method is simple and inexpensive and it does not require any sophisticated instruments.
3. Starting chemicals are easily available and cheap.
4. Film thickness can be controlled easily and accurately by changing the deposition time and temperature.
5. Various substrates without special restrictions can be effectively coated using the CBD technique.
6. Low operating temperature and it avoids oxidation and/or corrosion of metallic substrates.
7. The solution from which the deposition takes place always remains in touch with the substrates, and it results in pinhole-free and uniform deposits of complex shape and sizes.
8. In CBD, as the basic building blocks are ions instead of atoms, the stoichiometry of the deposits is easily achieved.
9. The preparative parameters are easily controllable, and slow film growth facilitates improved grain structure and better orientation.
10. By simply adding the dopant solution directly into the reaction bath, mixed/doped film structures could be obtained.
11. It uses very dilute solutions; hence, it offers minimum toxicity and occupational hazards.

3.4.2.3 Overview of Metal Oxide Thin Films Fabricated Using CBD Method

Table 3.4 shows representative examples of some of the most common precursors used today to deposit a few metal oxide thin films using CBD.

TABLE 3.4

Overview of Metal Oxide Thin Films Fabricated Using CBD Method

S. No.	Oxide	Synthesis Parameters	Structure	Morphology	Application	Ref
1.	Titanium oxide (TiO_2)	Precursors: $TiCl_3$ and 0.1 M NH_4Cl Deposition temperature: 55°C Deposition time: 3 hours Substrate: glass	Amorphous and anatase	Nanocrystalline grains	Photocatalytic activity	[90]
2.	Manganese oxide $(\beta\text{-}MnO_2)$	Precursors: 0.04 M $MnCl_2$, 4.0 M $NaBrO_3$, and 0.1 M HCl Deposition temperature: 60°C Deposition time: 12 hours Substrate: Ni foam substrate	Crystalline	Cross-linked round nanoplatelets	Supercapacitor	[91]
3.	Iron oxide $(\alpha\text{-}Fe_2O_3)$	Precursors: $FeCl_3.6H_2O$, distilled water Deposition temperature: 60°C–95°C Deposition time: 12 hours Substrate: FTO-coated glass	Rhombohedral	Nanorods	Photoelectrochemical cell	[92]
4.	Cobalt oxide (Co_3O_4)	Precursors: 0.1 M cobalt (II) Chloride and aqueous ammonia solution pH: ~12 Deposition temperature: 60°C–95°C Deposition time: 4–8 hours Substrate: glass	FCC cubic	Nanorods	Photovoltaic applications, electrochromic devices, supercapacitors, gas sensors, antireflection coatings, etc.	[93]
5.	Nickel oxide (NiO)	Precursors: 24 g Nickel sulfate hexahydrate, 4.0 g potassium persulfate, and aq. ammonia Substrate: ITO-coated glass pH: ~8.5–10 Deposition time: 30 minutes Annealing temperature: 400°C for 3 hours	Cubic	Nanoflakes	Smart windows	[94]
6.	Copper oxide (CuO)	Precursors: 0.1 M $CuCl_2 \cdot 2H_2O$ and ammonia solution Substrate: n-silicon glass Deposition temperature: 85°C	Monoclinic	Porous	Antireflection coating and solar energy harvesting	[95]
7.	Zinc oxide (ZnO)	Cationic precursor: 0.05 M zinc nitrate and 0.05 M hexamethylenetetramine Substrate: silicon and soda lime glass Deposition temperature: 90°C Deposition time: 3 hours	Hexagonal wurtzite	Nanowire	Electrical and optical applications	[96]
8.	Cadmium oxide (CdO)	Precursors: cadmium chloride and ethanolamine, and NaOH pH: ~13 Deposition temperature: 80°C Deposition time: 4 hours Substrate: glass	Hexagonal	Nanofiber	Gas sensors	[97]

(Continued)

TABLE 3.4 (*Continued*)
Overview of Metal Oxide Thin Films Fabricated Using CBD Method

S. No.	Oxide	Synthesis Parameters	Structure	Morphology	Application	Ref
9.	Tin oxide (SnO_2)	Precursors: 0.028 M tin chloride pentahydrate and deionized water Deposition temperature: 55°C Deposition time: 8–10 minutes (four cycles) Substrate: Schott Borofloat glass and quartz Annealing temperature: 400°C Annealing time: 1 hour	Orthorhombic	Overgrown crystallites	Transparent conducting oxide	
10.	Bismuth oxide (Bi_2O_3)	Precursors: 0.1 M $Bi(NO_3)_{3+}$ 3 mL TEA+ KOH pH: ~13.5 Deposition temperature: 60°C Deposition time: 2 hours Substrate: FTO glass Annealing temperature: 350°C Annealing time: 1 hour	Tetragonal	Nanoplates	Photoelectrocatalysis	

REFERENCES

1. Valerini, D., Caricato, A.P., Creti, A., Lomascolo M., Romano, F., Taurino, A., Tunno, T., & Martino, M. (2009). Morphology and photoluminescence properties of zinc oxide films grown by pulsed laser deposition. *Applied Surface Science*, 255, 9680–9683.
2. Fusi, M., Russo, V., Casari, C.S., Li Bassi, A., & Bottani, C.E. (2009). Titanium oxide nanostructured films by reactive pulsed laser deposition. *Applied Surface Science*, 255, 5334–5337.
3. Mazzi, A., Orlandi, M., Bazzanella, N., Popat, Y.J., Minati, L., Speranza, G., & Miotello, A. (2019). Pulsed laser deposition of nickel oxide films with improved optical properties to functionalize solar light absorbing photoanodes and very low overpotential for water oxidation catalysis. *Materials Science in Semiconductor Processing*, 97, 29–34.
4. Dolbec, R., El Khakani, M.A., Serventi, A.M., Trudeau, M., & Saint-Jacques, R.G. (2002). Microstructure and physical properties of nanostructured tin oxide thin films grown by means of pulsed laser deposition. *Thin Solid Films*, 419, 230–236.
5. Nunes, D., Pimentel, A., Santos, L., Barquinha, P., Pereira, L., Fortunato, E., & Martins, R. (2019). Structural, optical, and electronic properties of metal oxide nanostructures. In *Metal Oxide Nanostructures: Synthesis, Properties and Applications*, 59–102.
6. TulayHurma, T., & Caglar, M. (2020). Effect of anionic fluorine incorporation on structural, optical and electrical properties of ZnOnanocrystalline films. *Materials Science in Semiconductor Processing*, 110, 104949.
7. Aquí-Romero, F., Willars-Rodríguez, F.J., Chávez-Urbiola, I.R., & Ramírez-Bon, R. (2020). ZnO_2 films by successive ionic layer adsorption and reaction method and their conversion to ZnO ones for p-Si/ n-ZnO photodiode applications. *Semiconductor Science and Technology*, 35, 025012 (10 pp).
8. Nisha, R., Madhusoodanan, K.N., Vimalkumar, T.V., & Vijayakumar, K.P. (2015). Gas sensing application of nanocrystalline zinc oxide thin films prepared by spray pyrolysis. *Bulletin of Material Science*, 38(3), 583–591.
9. Bulakhe, R.N., & Lokhande, C.D. (2014). Chemically deposited cubic structured CdO thin films: Use in liquefied petroleum gas sensor. *Sensors and Actuators B: Chemical*, 200, 245–250.
10. Karunagaran, B., Uthirakumar, P., Chung, S.J., Velumani, S., & Suh, E.-K. (2007). TiO_2 thin film gas sensor for monitoring ammonia. *Materials Characterization*, 58, 680–684.

11. Hsu, N.-F., Chang, M., & Lin, C.-H. (2013). Synthesis of ZnO thin films and their application as humidity sensors. *Microsystem Technologies*, 19(11), 1737–1743.

12. Srivastava, R. (2012). Investigation on temperature sensing of nanostructured zinc oxide synthesized via oxalate route. *Journal of Sensor Technology*, 2, 8–12.

13. Boyadjiev, S., Georgieva, V., Vergov, L., Baji, Zs., Gáber, F., & Szilágyi, I.M. (2014). Gas sensing properties of very thin TiO_2 films prepared by atomic layer deposition (ALD). *Journal of Physics: Conference Series*, 559, 012013.

14. Mishra, S., Ghanshyam, C., Ram, N., Singh, S., Bajpai, R.P., & Bedi, R.K. (2002). Alcohol sensing of tin oxide thin film prepared by sol-gel process. *Bulletin of Material Science*, 25(3), 231–234.

15. Smith, H.M., & Turner, A.F. (1965). Vacuum deposited thin films using a ruby laser applied optics, 4(1), 147–148.

16. Ohnishi, T., Koinuma, H., & Lippmaa, M. (2006). Pulsed laser deposition of oxide thin films. *Applied Surface Science*, 252, 2466–2471.

17. Dijkkamp, D., Venkatesan, T., Wu, X.D., Shaheen, S.A., Jisrawi, N., Lee, Y.H., McLean, W.A., & Croft, M. (1987). Preparation of Y-Ba-Cu oxide superconductor thin films using pulsed laser evaporation from high T_c bulk material. *Applied Physics Letters*, 51, 619.

18. Venkatesan, T., Wu, X.D., Inam, A., & Wachtman, J.B. (1988). Observation of two distinct components during pulsed laser deposition of high T_c superconducting films. *Applied Physics Letters*, 52, 1193.

19. Chrisey, D., & Hubler, G. (1994). *Editors, Pulsed Laser Deposition of Thin Films*. New York: Wiley, 613 p.

20. Eason, R. Editor. (2006). *Pulsed Laser Deposition of Thin Films: Applications-Led Growth of Functional Materials*. New Jersey: Wiley, 682 p.

21. Yang, D. Editor. (2016). *Applications of Laser Ablation-Thin Film Deposition, Nanomaterial Synthesis and Surface Modification*. Intech, 150 p.

22. Masood, K.B., Kumar, P., Malik, M.A., & Singh, J. (2021). A comprehensive tutorial on the pulsed laser deposition technique and developments in the fabrication of low dimensional systems and nanostructures. *Emergent Materials*, 4, 737–754.

23. Delmdahl, R. (2020). Thin films for the future: Excimer lasers drive industrial-scale PLD manufacturing of innovative thin films. *PhotonicViews*, 17(6), 60–62.

24. Gupta, V., & Sreenivas, K. (2006). *Pulsed Laser Deposition of Zinc Oxide (ZnO) from Edited Book Entitled "Zinc Oxide Bulk. Thin Films and Nanostructures"*, C. Jagadish and S. Pearton (Editors). Elsevier Limited.

25. (a) Popescu, C., Dorcioman G., & Popescu, A.C. (2016). *Laser Ablation Applied for Synthesis of Thin Films: Insights into Laser Deposition Methods*. InTech. (b) Ogugua, S.N., Ntwaeaborwa O.M., & Swart, H.C. (2020). Latest development on pulsed laser deposited thin films for advanced luminescence applications. *Coatings*, 10, 1078.

26. Kumar, S., Saralch, S., Jabeen, U., & Pathak, D. (2020). *Metal Oxides for Energy Applications from Book Entitled "Colloidal Metal Oxide Nanoparticles"*. Elsevier Inc.

27. Lowndes, D.H., Geohegan, D.B., Puretzky, A.A., Norton, D.P., & Rouleau, C.M. (1996). Synthesis of novel thin-film materials by pulsed laser deposition. *Science*, 273, 898–903.

28. Villarreal-Barajas, J.E., Escobar-Alarc-on, L., Gonz-alez, P.R., Camps, E., & Barboza-Flores, M. (2002). Thermoluminescence properties of aluminum oxide thin films obtained by pulsed laser deposition. *Radiation Measurements*, 35, 355–359.

29. Lackner, J.M., Waldhauser, W., Ebner, R., Major, B., & Schöberl, T. (2004). Pulsed laser deposition of titaniumoxide coatings at room temperature-structural, mechanical and tribological properties. *Surface and Coatings Technology*, 180–181, 585–590.

30. Rama, N., & Ramachandra Rao, M.S. (2010). Synthesis and study of electrical and magnetic properties of vanadium oxide micro and nanosized rods grown using pulsed laser deposition technique. *Solid State Communications*, 150, 1041–1044.

31. Xia, H., Wan, Y., Yan, F., & Lu, L. (2014). Manganese oxide thin films prepared by pulsed laser deposition for thin film microbatteries. *Materials Chemistry and Physics*, 143, 720–727.

32. Guo O., Shi, W., Liu, F., Arita, M., Ikoma, Y., Saito, K., Tanaka, T., & Nishio, M. (2013). Effects of oxygen gas pressure on properties of iron oxide films grown by pulsed laser deposition. *Journal of Alloys and Compounds*, 552, 1–5.

33. Jadhav, H., Suryawanshi, S., More, M.A., & Sinha, S. (2018). Field emission study of urchin like nanostructured cobalt oxide films prepared by pulsed laser deposition. *Journal of Alloys and Compounds*, 744, 281–288.

34. Cao, L., Wang, D., & Wang, R. (2014). NiO thin films grown directly on Cu foils by pulsed laser deposition as anode materials for lithium ion batteries. *Materials Letters*, 132, 357–360.

35. Mistry, V.H., Mistry, B.V., Modi, B.P., & Joshi, U.S. (2017). Effect of annealing on pulse laser deposition grown copper oxide thin film. *AIP Conference Proceedings*, 1837, 040070.
36. Sun, Y.W., & Tsui, Y.Y. (2007). Production of porous nanostructured zinc oxide thin films by pulsed laser deposition. *Optical Materials*, 29, 1111–1114.
37. Tian, K., Tudu, B., & Tiwari, A. (2017). Growth and characterization of zinc oxide thin films on flexible substrates at low temperature using pulsed laser deposition. *Vacuum*, 146, 483–491.
38. Ou, S.-L., Wuu, D.-S. Fu, Y.-C., Liu, S.-P., Horng, R.-H., Liud, L., & Feng, Z.-C. (2012). Growth and etching characteristics of gallium oxide thin films by pulsed laser deposition. *Materials Chemistry and Physics*, 133, 700–705.
39. Aoki, T., Matsushita, T., Mishiro, K., Suzuki, A., & Okuda, M. (2008). Optical recording characteristics of molybdenum oxide films prepared by pulsed laser deposition method. *Thin Solid Films*, 517, 1482–1486.
40. Dellasega, D., Facibeni, A., Di Fonzo, F., Russo, V., Conti, C., Ducati, C., Casari, C.S., Li Bassi, A., Bottani, C.E. (2009). Nanostructured high valence silver oxide produced by pulsed laser deposition. *Applied Surface Science*, 255, 5248–5251.
41. Quiñones-Galván, J.G., Lozada-Morales, R., Jiménez-Sandoval S., Camps E., Castrejón-Sánchez V.H., Campos-González, E., Zapata-Torres M., Pérez-Centeno A., & Santana-Aranda, M.A. (2016). Physical properties of a non-transparent cadmium oxide thick film deposited at low fluence by pulsed laser deposition. *Materials Research Bulletin*, 76, 376–383.
42. El Khakani, M.A., Dolbec, R., Serventi, A.M., Horrillo, M.C., Trudeau, M., Saint-Jacques, R.G., Rickerby, D.G., & Sayago, I. (2001). Pulsed laser deposition of nanostructured tin oxide for gas sensing applications. *Sensors and Actuators B*, 77, 383–388.
43. Fendrich, M., Popat, Y., Orlandi, M., Quaranta, A., & Miotello, A. (2020). Pulsed laser deposition of nanostructured tungsten oxide films: A catalyst for water remediation with concentrated sunlight. *Materials Science in Semiconductor Processing*, 119, 105237.
44. Leontie, L., Caraman, M., Visinoiu, A., & Rusu, G.I. (2005). On the optical properties of bismuth oxide thin films prepared by pulsed laser deposition. *Thin Solid Films*, 473, 230–235.
45. Lin, Y., & Chen, X. Editors. (2016). *Advanced Nano Deposition Methods*, First Edition. Wiley-VCH Verlag GmbH & Co. KGaA.
46. https://www.syskey.com.tw/.
47. Yaroshevich, P.Y. (1976). Properties of tungsten-rhenium films produced by electron-beam evaporation and vacuum condensation. *Physics of Metals and Metallography*, 42(6), 68–71.
48. Moumen, A., Kumarage, G.C.W., & Comini, E. (2022). P-type metal oxide semiconductor thin films: Synthesis and chemical sensor applications. *Sensors*, 22(4), 1359.
49. Purohit, A., Chander, S., & Dhaka, M.S. (2017). Impact of annealing on physical properties of e-beam evaporated polycrystalline CdO thin films for optoelectronic applications. *Optical Materials*, 66, 512–518.
50. Keshavarzi, R., Mirkhani, V., Moghadam, M., Tangestaninejad, S., Baltork, I.M., Fallah, H.R., Dastjerdi, M.J.V., & Modayemzadeh, H.R. (2011). Preparation and characterization of indium zinc oxide thin films by electron beam evaporation technique. *Materials Research Bulletin*, 46, 615–620.
51. Kumar, P., V.SRS., Kumar, M., Kumari, N., & Sharma, A.L., (2020). Optical and morphological studies of aluminium oxide films fabricated at different leaning angles using ion assisted E-Beam deposition technique. *Optik – International Journal for Light and Electron Optics*, 222, 165376.
52. Mohanty, P., Kabiraj, D., Mandal, R.K., Kulriya, P.K., Sinha, A.S.K., & Rath, C. (2014). Evidence of room temperature ferromagnetism in argon/oxygen annealed TiO_2 thin films deposited by electron beam evaporation technique. *Journal of Magnetism and Magnetic Materials*, 355, 240–245.
53. Leroy, J., Bessaudou, A., Cosset, F., & Crunteanu, A. (2012). Structural, electrical and optical properties of thermochromic VO_2 thin films obtained by reactive electron beam evaporation. *Thin Solid Films*, 520, 4823–4825.
54. Sarkar, A., Satpati, A.K., Rao, P., & Kumar, S. (2015). Electron beam deposition of amorphous manganese oxide thin film electrodes and their predominant electrochemical properties. *Journal of Power Sources*, 284, 264–271.
55. Pereira, S., Gonçalves, A., Correia, N., Pinto, J., Pereira, L., Martins, R., & Fortunato, E. (2014). Electrochromic behavior of NiO thin films deposited by e-beam evaporation at room temperature. *Solar Energy Materials & Solar Cells*, 120, 109–115.
56. Varnamkhasti, M.G., Fallah, H.R., & Zadsar, M. (2012). Effect of heat treatment on characteristics of nanocrystallineZnO films by electron beam evaporation. *Vacuum*, 86, 871–875.

57. Sivakumar, R., Gopalakrishnan, R., Jayachandran, M., & Sanjeeviraja, C. (2007). Characterization on electron beam evaporated α-MoO$_3$ thin films by the influence of substrate temperature. *Current Applied Physics*, 7, 51–59.

58. Sohal, M.K., Singh, R.C., & Mahajan A. (2020). Electron beam evaporated tin oxide thin films for thermistor applications. *Materials Today: Proceedings*, 26(3), 4662–4665.

59. Evecan, D., & Zayim, E. (2019). Highly uniform electrochromic tungsten oxide thin films deposited by e-beam evaporation for energy saving systems. *Current Applied Physics*, 19, 198–203

60. Ristov, M., Sinadinovski, G.J., & Grozdanov, I. (1985). Chemical deposition of Cu$_2$O thin films. *Thin Solid Films*, 123, 63–67.

61. Nicolau, Y.F. (1985). Solution deposition of thin solid compound films by a successive ionic-layer adsorption and reaction process. *Applications of Surface Science*, 22–23(2), 1061–1074.

62. Hodes, G. (2003). *Chemical Solution Deposition of Semiconductor Films*. New York: Marcel Dekker, Inc.

63. Pathan, H.M., & Lokhande, C.D. (2004). Deposition of metal chalcogenide thin films by successive ionic layer adsorption and reaction (SILAR) method. *Bulletin of Material Science*, 27(2), 85–111.

64. Jellal, I., Nouneh, K., Jedryka, J., Chaumont, D., & Naja, J. (2020). Non-linear optical study of hierarchical 3D Al doped ZnOnanosheet arrays deposited by successive ionic adsorption and reaction method. *Optics and Laser Technology*, 130, 106348.

65. Ghos, B.C., Farhad, S.F.U., Patwary, Md A.M., Majumder, S., Hossain, Md. A., Tanvir, N.I., Rahman, Md. A., Tanaka, T., & Guo, Q. (2021). Influence of the substrate, process conditions, and post annealing temperature on the properties of ZnO thin films grown by the successive ionic layer adsorption and reaction method. *ACS Omega*, 6, 2665–2674.

66. Gokul, B., Matheswaran, P., & Sathyamoorthy, R. (2013). Influence of annealing on physical properties of CdO thin films prepared by SILAR method. *Journal of Materials Science & Technology*, 29(1), 17–21.

67. Daoudi, O., Qachaou, Y., Raidou, A., Nouneh, K., Lharch, M., & Fahoume, M. (2019). Study of the physical properties of CuO thin films grown by modified SILAR method for solar cells applications. *Superlattices and Microstructures*, 127, 93–99.

68. Su, Z., Sun, K., Han, Z., Liu, F., Lai, Y., Li J., & Liu, Y. (2012). Fabrication of ternary Cu–Sn–S sulfides by a modified successive ionic layer adsorption and reaction (SILAR) method. *Journal of Materials Chemistry*, 22, 16346.

69. Kalandaragh, Y.A., Muradov, M.B., Mammedov, R.K., & Khodayari, A. (2007). Growth process and investigation of some physical properties of CdS nanocrystals formed in polymer matrix by successive ionic layer adsorption and reaction (SILAR) method. *Journal of Crystal Growth*, 305, 175–180.

70. Lindroos, S., Kanniainen, T., Leskellä, M., & Rauhala, E. (1995). Deposition of manganese-doped zinc sulfide thin films by the successive ionic layer adsorption and reaction (SILAR) method. *Thin Solid Films*, 263, 79–84.

71. Patil, U.M., Gurav, K.V., Joo, O.-H., & Lokhande, C.D. (2009). Synthesis of photosensitive nanograined TiO$_2$ thin films by SILAR method. *Journal of Alloys and Compounds*, 478, 711–715.

72. Pawar, M.S., Sutar, M.A., Maddani, K.I., & Kandalkar, S.G. (2016). Synthesis and Characterization of Vanadium Oxide (V$_2$O$_5$) Thin Film Electrode for Electrochemical Capacitors: Effect of Annealing. *IOSR Journal of Applied Physics (IOSR-JAP)*, 8(1), 07–13.

73. Singu, B.S., Hong, S.E., & Yoon, K.R. (2019). Preparation and characterization of manganese oxide nanosheets for pseudocapacitor application. *Journal of Energy Storage*, 25, 100851.

74. Ubale, A.U., & Belkhedkar, M.R. (2015). Size dependent physical properties of nanostructured α-Fe$_2$O$_3$ thin films grown by successive ionic layer adsorption and reaction method for antibacterial application. *Journal of Materials Science & Technology*, 31(1), 1–9.

75. Kandalkar, S.G., Gunjakar, J.L., & Lokhande, C.D. (2008). Preparation of cobalt oxide thin films and its use in supercapacitor application. *Applied Surface Science*, 254, 5540–5544.

76. Gund, G.S., Lokhande, C.D., & Park, H.S. (2018). Controlled synthesis of hierarchical nanoflake structure of NiO thin film for supercapacitor application. *Journal of Alloys and Compounds*, 741, 549–556.

77. Ganapathi, S.K., Kaur, M., Shaheera, M., Pathak, A., Gadkari, S.C., & Debnath, A.K. (2021). Highly sensitive NO$_2$ sensor based on ZnO nanostructured thin film prepared by SILAR technique. *Sensors & Actuators: B. Chemical*, 335, 129678.

78. Suresh Kumar, P., Dhayal Raj, A., Mangalaraj, D., & Nataraj, D. (2010). Hydrophobic ZnO nanostructured thin films on glass substrate by simple successive ionic layer absorption and reaction (SILAR) method. *Thin Solid Films*, 518, e183–e186.

79. Sahin, B., & Aydin, R. (2018). SILAR derived CdO films: Effect of triethanolamine on the surface morphology and optical bandgap energy. *Physica B: Physics of Condensed Matter*, 541, 95–102.

80. Deshpande, N.G., Vyas, J.C., & Sharma, R. (2008). Preparation and characterization of nanocrystalline tin oxide thin films deposited at room temperature. *Thin Solid Films*, 516, 8587–8593.

81. Raut, S.S., Bisen, O., & Sankapal, B.R. (2017). Synthesis of interconnected needle-like Bi_2O_3 using successive ionic layer adsorption and reaction towards supercapacitor application. *Ionics*, 23(7), 1831–1837.

82. Pineda-Leon, H.A., Gutierrez-Heredia, G., De Leon, A., Ochoa-Landin, R., Ramirez-Bon, R., Flores-Acosta, M., & Castillo, S.J. (2016). Comparative study of PbS thin films growth by two different formulations using chemical bath deposition. *Chalcogenide Letters*, 13, 161–168.

83. Mane, R.S., & Lokhande, C.D. (2000). Chemical deposition method for metal chalcogenide thin films. *Materials Chemistry and Physics*, 65, 1–31.

84. Pawar, S.M., Pawar, B.S., Kim, J.H., Joo, O.-S., & Lokhande, C.D. (2011). Recent status of chemical bath deposited metal chalcogenide and metal oxide thin films. *Current Applied Physics*, 11, 117–161.

85. Jana, S., Maity, R., Das, S., Mitra, M.K., & Chattopadhyay, K.K. (2007). Synthesis, structural and optical characterization of nanocrystalline ternary $Cd_{1-x}Zn_xS$ thin films by chemical process. *Physica E*, 39, 109–114.

86. Shaikh, S.K., Inamdar, S.I., Ganbavle, V.V., & Rajpure, K.Y. (2016). Chemical bath deposited ZnO thin film based UV photoconductive detector. *Journal of Alloys and Compounds*, 664, 242–249.

87. Sharma, N.C., Kainthla, R.C., Pandya D.K., & Chopra, K.L. (1979). Electroless deposition of semiconductor films. *Thin Solid Films*, 60(1), 55–59.

88. Chaudhari, K.B., Rane, Y.N., Shende, D.A., Gosavi, N.M., & Gosavi, S.R. (2019). Effect of annealing on the photocatalytic activity of chemically prepared TiO_2 thin films under visible light. *Optik-International Journal for Light and Electron Optics*, 193, 163006.

89. Tian, Y., Liu, Z., Xue, R., & Huang, L. (2016). An efficient supercapacitor of three-dimensional MnO_2 film prepared by chemical bath method. *Journal of Alloys and Compounds*, 671, 312–317.

90. Rahman, G., Najaf, Z., Shaha, A. ul H.A., & Mian, S.A. (2020). Investigation of the structural, optical, and photoelectrochemical properties of α-Fe_2O_3 nanorods synthesized via a facile chemical bath deposition. *Optik – International Journal for Light and Electron Optics*, 200, 163454.

91. Turan, E., Zeybekğlu, E., & Kul, M. (2019). Effects of bath temperature and deposition time on Co_3O_4 thin films produced by chemical bath deposition. *Thin Solid Films*, 692, 137632.

92. Yu, J.-H., Nam, S.-H., Gil, Y.E., & Boo, J.-H. (2020). The effect of ammonia concentration on the microstructure and electrochemical properties of NiO nanoflakes array prepared by chemical bath deposition. *Applied Surface Science*, 532, 147441.

93. Sultana, J., Paul, S., Karmakar, A., Yi, R., Dalapati, G.K., & Chattopadhyay, S. (2017). Chemical bath deposited (CBD) CuO thin films on n-silicon substrate for electronic and optical applications: Impact of growth time. *Applied Surface Science*, 418, 380–387.

94. Ghazali, M.N.I., Izmi, M.A., Mustaffa, S.N.A., Abubakar, S., Husham, M., Sagadevan, S., & Paiman, S. (2021). A comparative approach on One-Dimensional ZnO nanowires for morphological and structural properties. *Journal of Crystal Growth*, 558, 125997.

95. Hone, F.G., Tegegne, N.A., Dejene, F.B., & Andoshe, D.M. (2021). Nanofiber cadmium oxide thin films prepared from ethanolamine complexing agent by solution growth method. *Optik – International Journal for Light and Electron Optics*, 243, 167402.

96. Khallaf, H., Chen, C.-T., Chang, L.-B., Lupan, O., Dutta, A., Heinrich, H., Haque, F., del Barco, E., & Chow, L. (2012). Chemical bath deposition of SnO_2 and Cd_2SnO_4 thin films. *Applied Surface Science*, 258, 6069–6074.

97. Wang, Y., Jiang, L., Tang, D., Liu, F., & Lai, Y. (2015). Characterization of porous bismuth oxide (Bi_2O_3) nanoplates prepared by chemical bath deposition and post annealing. *RSC Advances*, 5, 65591–65594.

4 Modern Techniques to Explore Applications of Metal Oxide Phosphors

D. P. Awade

G. N. Khalsa College

CONTENTS

4.1 X-RAY DIFFRACTION TECHNIQUE (XRD TECHNIQUE)

X-ray crystallography technique is the widely used and most popular technique for structure determination of synthesized metal oxide phosphor materials [1–3]. X-ray diffraction (XRD) analysis identifies the long-range order (i.e., the structure) of crystalline materials and the short-range order of non-crystalline materials. From this information, we can deduce lattice constants and phases, average grain size, degree of crystallinity, crystal defects [4], etc. Short wavelength X-rays in the range of a few angstroms to 0.1 angstrom (1–120 keV) are used for this purpose.

DOI: 10.1201/9781003366232-4

X-rays are best suited for probing and understanding the atomic and molecular arrangements of the materials as their wavelength is of atomic scale. Further, these energetic radiations can penetrate deep inside the materials; hence, structural information can be obtained by analyzing the diffraction pattern obtained.

Incident X-rays primarily interact with the electrons in the atoms of the exposed sample material. This collision leads to deflection of incident radiation. If the wavelength of these scattered X-rays do not change, this process is called elastic scattering where only momentum transfer takes place. By measuring the angles and intensities of these diffracted X-rays, information about the sample material can be extracted. Diffracted waves from different atoms can interfere with each other and the resultant intensity distribution is strongly modulated by this interaction. If the atoms are arranged in a periodic fashion as in crystals, the diffracted waves will consist of sharp interference maxima with the same symmetry as in the distribution of atoms. Interpretation of these diffraction patterns leads to information on the distribution of atoms in the sample material. When certain geometric requirements are met, X-rays scattered from a crystalline solid can constructively interfere, producing a diffracted beam. W. L. Bragg proposed the following condition for constructive interference:

$$n\lambda = 2d\sin\theta \tag{4.1}$$

The above equation is called Bragg equation [5]. Here, n is the order of diffraction, λ is the wavelength, θ is the scattering angle measured in degrees, and d is the interplanar spacing between similar atomic planes in a crystal measured in angstroms. For practical reasons, a diffractometer measures an angle twice that of the θ angle [6,7]. The schematic diagram of the XRD system is shown in Figure 4.1.

The diffraction pattern so obtained is like a fingerprint of that crystalline substance. Therefore, the crystalline phase of a material can also be identified by examining the diffraction pattern. The width of the diffraction lines is closely related to the size of the particles, the size distribution of the particles, and the defects present in the crystals.

4.1.1 DEBYE–SCHERRER METHOD

This method is deployed for the phosphor samples in powder form. The advantage of this method is that all Bragg angles can be measured simultaneously by exposing them to the X-ray

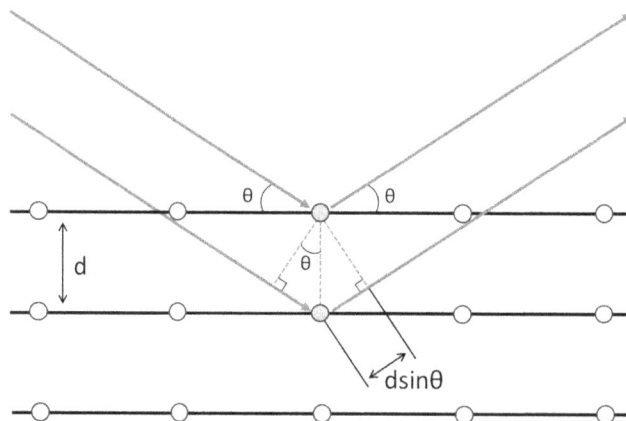

FIGURE 4.1   The schematic diagram of Bragg's diffraction [Wikimedia, Public domain].

beam only once. This method is used for sample identification and structures of powders of nanoparticles and also to study isomorphic crystals [8].

4.1.2 SCHERRER FORMULA

The Scherrer equation is the relationship between the size of sub-micrometer particles or crystallites in a solid form and the line width of XRD peaks in a diffraction pattern. It was suggested by Paul Scherrer and is used for the determination of the size of particles of crystals in a powder form. It is limited to nanoparticle size [9] and not applicable to grains larger than 0.1 μm. The Debye–Scherrer [10] equation can be used to calculate the size of particles below 60 nm. It is given as:

$$\text{Crystalline size D} = \frac{K\lambda}{\beta \cos \theta} \tag{4.2}$$

where D is the size of the particle, K is known as the Scherer's constant (K=0.94), λ is the wavelength of X-rays used (1.54178 Å), θ is the angle of diffraction, and β signifies the full width at half maximum (FWHM) of the diffraction peak.

In a diffraction pattern, the broadening of the peak is measured in the form of FWHM. Peak broadening may result due to instrumental broadening and/or sample broadening. Instrumental broadening can be found by comparing the diffraction pattern of near-perfect crystals like KCL, lanthanum hexaboride (LaB6), BaF2, etc.

Powder XRD is also used popularly for the characterization of nanoscale materials. Bulk information about the parameters such as phase, particle size, purity, morphology, etc., can be obtained from the analysis. The information obtained is then compared with the microscopic observation that covers a small region with a small number of particles. That will help in the confirmation of microscopic data representative of the bulk sample. Thus, it helps in phase identification, sample purity [11], crystallite size, and, in some cases, morphology [12].

However, it is observed that the information contained in powder XRD data for nanoscale materials is not always fully harnessed and needs to be supported by additional analyses such as TEM to be more scientifically trustworthy and correct [13].

4.1.3 ADVANTAGES AND LIMITATIONS

The X-ray powder diffraction technique is widely and popularly used for structure confirmation as it has several advantages:

- A unique diffraction pattern is associated with each material. So, the as-synthesized samples can be identified by comparing their diffraction pattern with the available database [14]. Also, thepercentage purity of the sample can be identified from the diffraction pattern obtained.
- It is a reliable, nondestructive method of sample identification in which the sample remains intact. The same sample can be used for other purposes later.
- Gives information about crystallite size and morphology as well.
- High sensitivity, user-friendliness, easy operational procedure, and fast data recording are some of the basic advantages associated with the tabletop compact model of the XRD machine available in the market (Figures 4.2 and 4.3).

However, a few disadvantages of this technique include the use of harmful X-rays and the requirement of a standard database for reference to match.

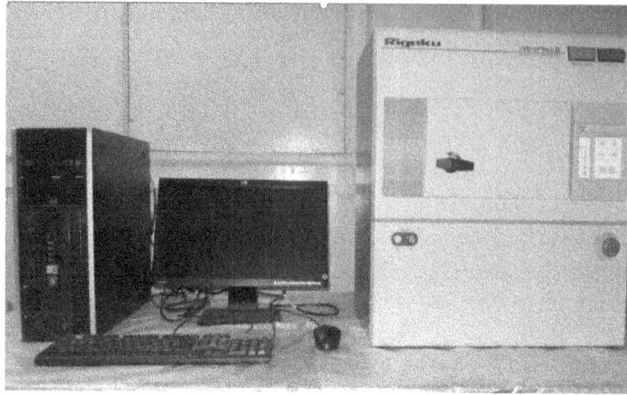

FIGURE 4.2 Actual photograph of Rigaku Miniflex II X-ray diffractometer.

FIGURE 4.3 Actual photograph of goniometer of Rigaku Miniflex II X-ray diffractometer.

4.2 SCANNING ELECTRON MICROSCOPY

Scanning electron microscopy (SEM) is popularly used to investigate the surface morphology of metal oxide phosphor materials. SEM produces images of the sample by scanning it with a focused beam of electrons.

4.2.1 SEM MECHANISM

The interaction of high-energy electrons with the sample provides different types of signals [15]. The signals produced are because of:

1. secondary electrons (SE)
2. backscattered electrons (BSE)
3. Auger electrons
4. Characteristic X-rays (EDS).

Figure 4.4 shows the schematic representation of this interaction which leads to different types of analysis techniques such as SEM, TEM, EDS, etc.

SEM works at 300,000× magnification and even more for modern equipment, and precise images can be obtained for a wide range of materials. It is possible to estimate the diameter, length, thickness, density, shape, and orientation of the phosphor materials. An SEM sample should be dry and electrically conductive to get the best results. It is required to select the appropriate accelerating

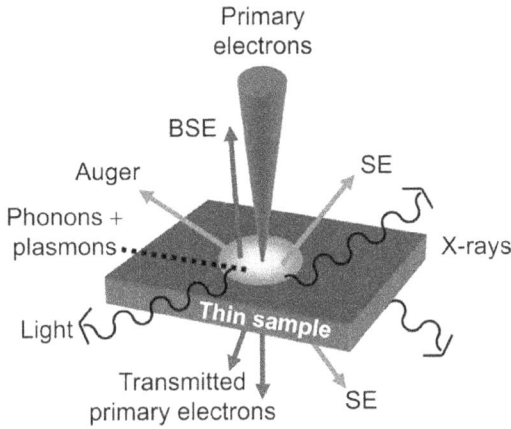

FIGURE 4.4 Interaction of electrons with the sample material producing different types of signals [16].

voltage for specific samples. For metal oxide samples, generally, 10–30 kV is preferred. An optimal aperture diameter should be selected for obtaining high-resolution SE images. The size (cross-sectional diameter) that the cone of the beam makes on the surface of the sample affects the resolution of the image and the number of electrons generated and therefore the graininess of the image. At low magnifications, a larger spot size is preferred than at higher magnifications. Sample height or working distance also plays an important role and needs to be selected for a specific sample. However, modern machines are more operator-friendly and rely on automatic adjustment (air circuit breaker buttons) along with contrast and brightness controls.

4.2.2 GENERAL MECHANISM OF SEM APPARATUS

Figure 4.5 shows the schematic diagram of the SEM apparatus. It uses electrons emitted from tungsten or lanthanum hexaboride (LaB6) thermionic emitters for visualization of the surface of the sample. The filament is heated resistively by a current to achieve a temperature between 2,000 and 2,700 K. This results in emission (EM) of thermionic electrons from the tip over an area of about $100 \times 150 \mu m$. The electron gun generates electrons and accelerates them to energy in the range of 0.1–30 keV toward the sample. A series of lenses focus the electron beam on the sample where it interacts with the sample to a depth of approximately 1 μm. When the electron beam impinges on the specimen, many types of signals are generated and any of these can be displayed as an image. The two signals most often used to generate SEM images are SE and BSE. Most of the electrons are scattered at large angles (from 0° to 180°) when they interact with the positively charged nucleus. These elastically scattered electrons usually called "backscattered electrons" (BSE) are used for SEM imaging. Some electrons are scattered elastically due to the loss in kinetic energy upon their interaction with orbital shell electrons. Incident electrons may knock off loosely bound conduction electrons out of the sample. These SEs along with BSEs are widely used for SEM topographical imaging. Both SE and BSE signals are collected when a positive voltage is applied to the collector screen in front of a detector. When a negative voltage is applied on the collector screen, only a BSE signal is captured because low-energy SEs are repelled. These captured SEs and BSEs from the sample are detected by detectors and these detectors transfer these detected electrons into an electronic signal which is sent to a computer to display the image [17].

Recently various compact machines are available in the market like Philips XL 30 SEM. SEM machines with more advanced options such as SNE-4500M Tabletop SEM are available in the market with additional features such as EDS elemental analysis, backscatter detector (BSE), cooling stage with anti-vibration table [18].

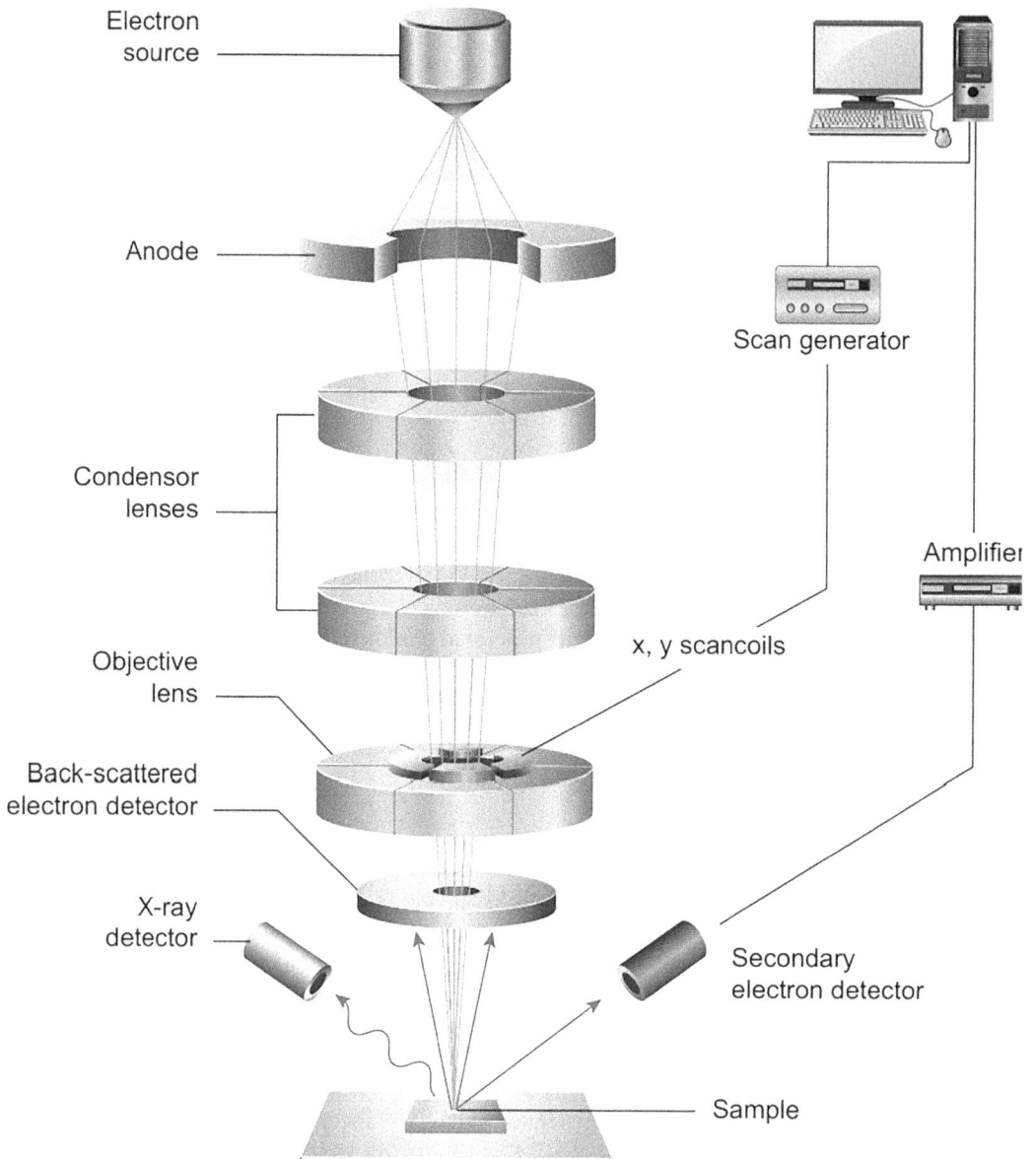

FIGURE 4.5 Schematic diagram of the scanning electron microscopy [16].

4.2.3 ADVANTAGES AND LIMITATIONS

SEM technologies have several advantages and play a key role in the morphology observation of metal oxide phosphors. A few are listed below:

- High resolution at high magnification with a good depth of focus.
- Resolving power is in the order of 1 nm.
- Detailed three-dimensional topological imaging is possible.
- Fast and easy operation with compact models available in the market.
- Sample preparation is easy compared to TEM.
- Can provide an estimate of the diameter, thickness, density, shape, and orientation of the particles in phosphor materials.

However, there are a few limitations as listed below:

- SEM cannot work for wet samples.
- Elemental analysis below a micrometer scale is not possible with this technique.
- It is not possible to record the image of an electrically nonconductive sample. Such samples need to be coated first.
- Measurements involving height (z-axis) cannot be taken directly in an SEM.

4.3 TRANSMISSION ELECTRON MICROSCOPY (TEM)

TEM images are formed by exposing the sample materials to a high-energy electron beam in a high vacuum. In this technique, transmitted electrons through the sample are detected to form an image. TEM machine consists of an electron gun at the top followed by a series of electromagnetic lenses, a sample holder, and an image detector at the bottom (Figure 4.6).

High-scale magnifications of up to 1,000,000× with resolution below 1 nm are possible with this technique. TEM images are required to be with a scale bar to calculate the actual size of structures in the image.

4.3.1 ADVANTAGES AND LIMITATIONS

This advanced characterization technique is associated with several advantages such as:

- High magnification is possible with this technique which helps in examining the structure of the specimen in detail. It is best suited for nanosized structures.
- Better resolution at high magnification is possible with this technique. It provides resolutions better than 1 nm. That helps study the defects and dislocations in the sample in detail.
- It is possible to have better microanalysis of the elemental composition with high resolution in TEM as compared to SEM.
- Diffraction patterns obtained by TEM can provide information on crystal structure, symmetry, and orientation of the crystal in periodic structured samples.
- It is an important tool in the study of material science since it is helpful to understand the morphology, crystal structure, orientations, chemical composition, and impurities in detail.

However, there are certain limitations as well, such as:

- Color imaging is not possible with the TEM technique.
- Surface morphology is not possible with only TEM imaging.
- There are limitations on sample thickness, as electrons cannot readily penetrate through the sample thicker than 200 nm.
- Sample preparation requires a specialized technique such as TEM imaging.

4.4 ENERGY-DISPERSIVE SPECTROSCOPY (EDS)

This is a microanalysis technique and is sometimes referred to as energy-dispersive X-ray spectroscopy (EDX). This microanalytical technique is deployed in qualitative as well as quantitative analysis of the chemical composition of metal oxide phosphor samples.

In this, an electron beam is focused on the sample in either SEM or TEM. The interaction of these electrons with the constituent atoms of the sample produces bremsstrahlung X-rays and characteristic X-rays. These X-rays are detected by an energy-dispersive detector that produces an analytical signal. The elements can be identified by the energies of characteristic X-rays. The intensities

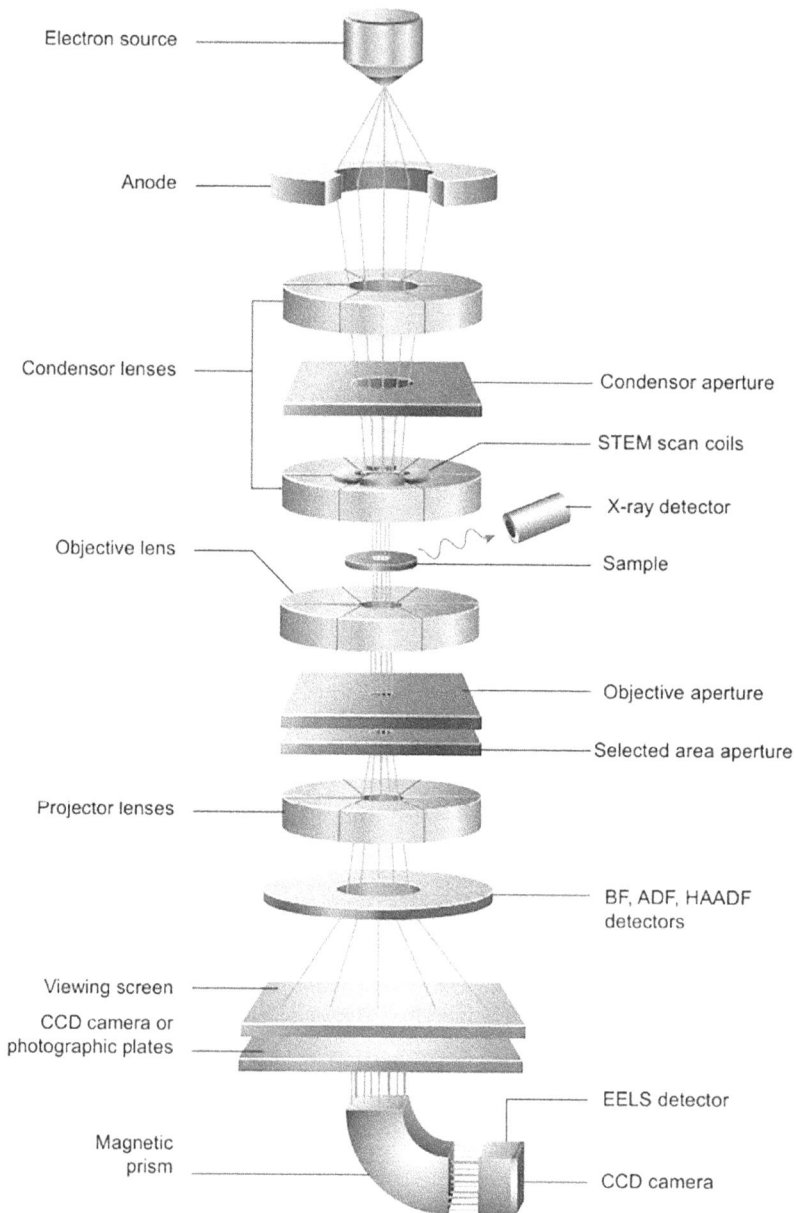

FIGURE 4.6   Schematic diagram of the transmission electron microscopy [16].

of these characteristic X-ray peaks help in judging the concentrations/quantity of these elements in the sample.

4.4.1   ADVANTAGES AND LIMITATIONS

The advantages and limitations of this technique are listed below:

- EDS along with SEM/TEM provides qualitative and qualitative results. These two techniques in combination provide information on the composition of the scanned metal oxide material sample.

TABLE 4.1

Comparison of SEM, TEM, and EDS Analysing Methods

Method	Principal and Measuring Method	Measuring Item	Energy of Incident Electron Beam
SEM (scanning electron microscopy)	By detecting the secondary electron	Surface morphology	1–30 kV
TEM (transmission electron microscopy)	By detecting the absorption and diffracted electrons	Crystal defect Crystal size Crystal structure	100–1,000 kV
EDS (energy-dispersive X-ray spectroscopy)	By detecting the emitted characteristic X-rays for different types of atoms in the sample	Atomic composition and impurity percentage in sample	10–100 keV

- The outcome of the EDS analysis is in the form of a spectrum showing a plot of the number of X-rays detected versus their energies. Characteristic X-rays and their relative intensities allow the presence of constituent elements and their relative quantity.
- A map/image is also obtained as an output showing the presence of a concentration of a particular element over the area of the sample.
- Quantitative analysis of elements present in different phases by their weight percentage can be obtained by comparing them with standard references.

Limitations
- The detection limit of EDS analysis varies with the machine with which it is clubbed, i.e., SEM/TEM.
- It is in the range of 0.1–0.5 wt.% in SEM and ~0.01–0.1 wt.% in TEM. It also lacks the sensitivity for trace element analysis.
- It is considered a nondestructive analytical technique, but practically, the materials will experience some damage under an electron beam.

A comparison of SEM, TEM, and EDS analyzing methods is enclosed in a tabular form in Table 4.1, as shown below. Each method has different requirements and application areas.

4.5 FLUORESCENCE SPECTROPHOTOMETRY FOR PHOTOLUMINESCENCE EXCITATION (PLE) AND EMISSION (PL)

Phosphors are the luminescent materials that emit radiation when suitable excitation (EX) is used. A luminescent system generally consists of a host lattice and a luminescent center termed an "activator" [19]. This activator absorbs the incoming EX radiation and goes to the excited state. The excited state returns to the ground state through a quasi-stable state by EM of radiation leading to luminescence. In some cases, the EX radiation is absorbed by another ion termed "sensitizer" which then transfers it to the activator. In many cases, the host lattice transfers its EX energy to the activator, so that the host lattice acts as a sensitizer. Luminescence is classified depending upon the type of EX source [20,21]. A prefix added to the term luminescence usually indicates EX source. In the case of photoluminescence (PL), photons are the EX source. Photoluminescent metal oxide phosphors can be used in fluorescent lamps [22,23], PL-LCD [24], plasma display [25], LASERs, LSCs, paints, inks, solar cells [26], white LEDs [27,28], etc. Hence, the measurement of EX and EM spectra is a basic and essential characterization of metal oxide phosphors. That will decide the suitability of as-synthesized phosphors for various applications.

PL spectra (EX and EM) for metal oxide phosphors are recorded with the help of a fluorescence spectrophotometer.

4.5.1 General Mechanism of Fluorescence Spectrophotometer

The block diagram (Figure 4.7) below shows a general fluorescence spectrophotometer. The fluorescence spectrophotometer consists of a xenon lamp to provide EX to the metal oxide phosphor samples. This white light enters the EX side of a spectroscope. The EX spectrum is recorded after scanning the sample for all wavelengths whereas to record the EM spectra, a particular wavelength of EX source is selected and kept fixed. Then, the EX light reaches the specimen after passing through the beam splitter. One part of the beam travels toward the monitor detector that detects the intensity of the EX source. Photosensitive devices such as a phototube, photodiode, or photomultiplier (PM) tube are preferably used. The other part of the beam is allowed to fall on the specimen. This beam excites the sample that in turn emits fluorescent light.

When the EX light reaches the specimen, the specimen is excited to emit fluorescent light. The emitted light enters the fluorescence detector after passing through the spectroscope. The detector is a PM tube that converts the fluorescent light into an analog electric signal. The A/D conversion circuit is used to convert it into a digital signal. Recently, modern compact models of fluorescence spectrophotometer are in the market. The actual photograph of a tabletop compact Hitachi-7000 Fluorescence Spectrophotometer and its inner view are shown in Figures 4.8 and 4.9, respectively.

4.5.2 Recording of Good Spectra

Various factors need to be addressed for recording noise-free EX and EM spectra. A few are discussed below:

- A proper inlet/outlet slit width needs to be selected for recording noise-free and accurate spectra. With a large slit width, the noise level is lower as the amount of light entering increases; however, minute peaks of a spectrum cannot be recorded. For a narrow slit width, minute peaks can be recorded but the light entering decreases leading to a noisy data recording.
- Optimal response should be selected for recording proper PL and PLE spectrum depending on the scanning speed.

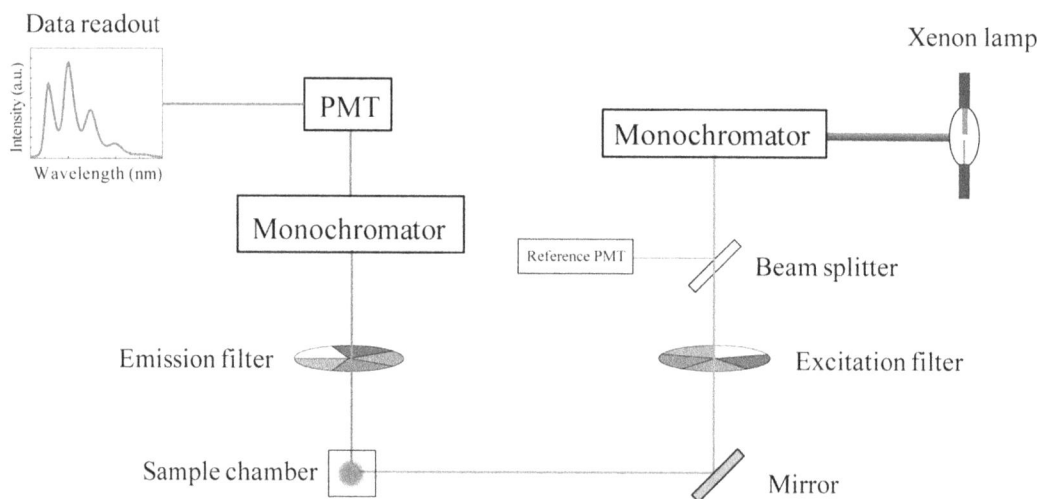

FIGURE 4.7 Optical system diagram of a fluorescence spectrophotometer.

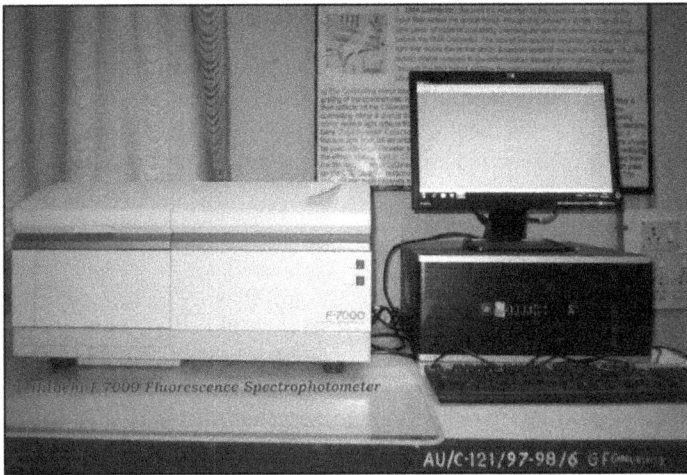

FIGURE 4.8 Typical Hitachi F-7000 fluorescence spectrophotometer.

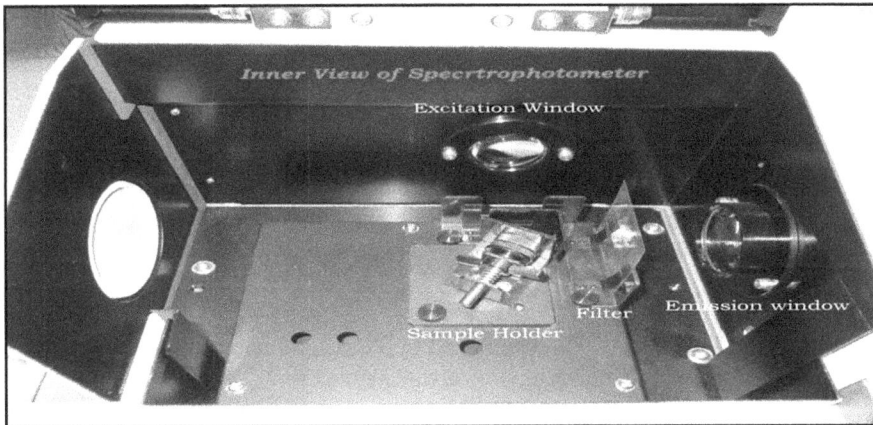

FIGURE 4.9 Inner view of typical fluorescence spectrophotometer.

- It is advisable to record PL EX spectra first by setting the EM wavelength at zero order and keeping other parameters as specified in the manual. EX bands were identified from these spectra and the EM spectra were scanned for identified EX wavelengths.

4.6 THERMOLUMINESCENCE READER

Thermoluminescence (TL) is better described as thermally stimulated luminescence. The primary agent for the induction of TL in a material is the ionizing radiation (X-rays or gamma rays) or sometimes even UV rays to which the material is exposed. When these radiations are incident on the materials, some of the deposited energy is stored in the lattice at defect sites, color centers, etc. Part of this irradiation energy is used to transfer electrons to traps. This energy, stored in the form of trapped electrons, is then released by raising the temperature of the material, and the released energy is converted into luminescence. This trapping process and the subsequent release of the stored energy find an important application in ionizing radiation dosimetry and in the operation of long persistent phosphors.

4.6.1 General Mechanism of Thermoluminescence Reader

For TL measurements, metal oxide samples need to be irradiated with different irradiation sources. Dosimetry is a fundamental part of quality programs assuring that the irradiation procedure is carried out according to standard specifications. In order to minimize errors in radiation doses, it must be accurate. The following instruments are generally used for different exposures.

Sources of Irradiation
1. Gamma chamber 900: for high dose of γ-rays.
2. Theratron 780E: for low dose of γ-rays.
3. YXLON International X-ray machine: for X-ray dose of different energies.

After irradiation with a suitable source, TL is recorded for metal oxide phosphor materials. Radiation dosimetry is mainly based on thermoluminescence dosimetry (TLD). This method offers considerable advantages because of its high precision, low cost, wide range, etc. In 1980, Feher and coworkers [29–31] developed portable battery-operated TLD readers. These devices used $CaSO_4$ bulb dosimeter, and the evaluation technique was based on analog timing circuits and analog to digital conversion of the PM current with a readout precision of micro Gy. The measured values were displayed and manually recorded. The version with an external power supply was used for space dosimetry as an on-board TLD reader [4]. With the advent of computers and the development of interfacing cards, a lot of automatic readers manufactured by Harshaw, Nucleonix, etc., are available.

The TL reader should necessarily have the following components:

- Heating system for a controlled increase of metal oxide phosphor temperature.
- A device for detecting light emission and converting it into an electrical signal.
- A device for measurement of this signal.
- A recording device.

It is observed that the TL systems are customized in the case of specialized research work and a number of commercially manufactured devices are available.

Heating of the sample is usually achieved by ohmic heating, i.e., an electrical current is passed through a metal strip onto which the sample is placed. Alternatives include the use of streams of hot gas or of an infrared heating lamp. The temperature may be monitored by means of a thermocouple, one junction of which is usually welded to the bottom of the heating strip. The signal from the thermocouple can be fed back to a temperature controller, which ensures that the temperature increases in the desired manner. Heating may take place in an air medium generally or specialized heating in oxygen-free nitrogen or argon is required in order to eliminate unauthentic sources of luminescence such as chemiluminescence or triboluminescence

Light detection is performed with a PM tube. A lens is sometimes used to focus the emitted light onto the photocathode. In order to optimize the sensitivity of the system, it is important to try to match the response of the photocathode to the wavelength of the TL emission.

Even when no light is incident on the photocathode, a small current is produced owing to thermionic emission. For sensitive work, it may be necessary to reduce this by cooling the PM tube. In many systems, a DC ammeter (electrometer) amplifies the current from the PM tube. This current is used to drive the y-axis mechanism of a chart recorder. A signal from the thermocouple is then used to drive the x-axis mechanism of the recorder. An alternative method, often used in TLD, is to digitize the PM signal using a charge-to-pulse converter and then count the resulting pulses. The total number of pulses counted is proportional to the integrated light output from the phosphor. In order to check for changes in the sensitivity of the TL reader, many devices, particularly commercial systems, incorporate a reference light source, and the signal produced by this source is measured

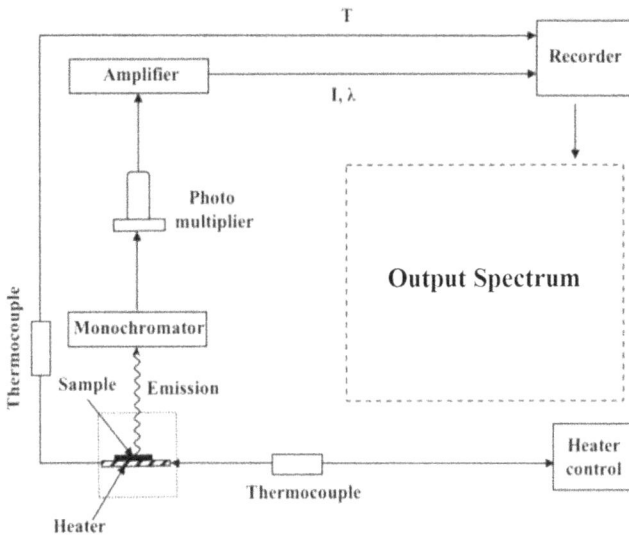

FIGURE 4.10 Schematic block diagram of TL reader system [32].

at frequent intervals. A schematic block diagram of a TL readout system incorporating the above-mentioned features is shown in Figure 4.10.

4.6.2 APPLICATIONS AND LIMITATIONS

Recently, extensive studies have been undertaken on TL and TL-related phenomena; new and sensitive TL phosphors were developed [33]. TL is a very good technique used in research in basic as well as applied sciences. It is already applied in various fields and is now extended into a whole spectrum of disciplines such as archeology, forensic sciences, geology, radiation dosimetry, radiation physics, solid-state physics, space sciences, and many more [34–36].

Information about the trapping process and the release of trapped electrons is obtained from the TL spectrum, in which, after turning off the irradiating source, the thermally stimulated luminescence is monitored under a condition of steadily increasing temperature. The shape and position of the resultant TL glow curves can be analyzed to extract information on the various parameters of the trapping process: trap depth, trapping and re-trapping rates, etc. The physical mechanisms governing the trapping and release of electrons are discussed in detail in the text by McKeever [37].

TL is the most widely used technology for evaluating personal and environmental radiation exposure. Radiation processing is a growing industrial activity that deals with radiation doses. It is used, in particular, for sterilization of medical and pharmaceutical products, water purification, sludge treatment, and delaying ripening.

In the medical field, measurements based on mailed TLD are being used for quality audits of dosimetry in the treatment of cancer patients, quality assurance checks of the performance of diagnostic and therapy X-ray machines, etc.

There are various advantages of TLD.

- A primary advantage of TLD is that it is able to measure a great range of radiation doses.
- TLD is simple to use. The doses from TLDs can be easily obtained and they can be read on-site easily.

Limitations
- The TLD dosimeter **can** be read only one time. At the end of a readout process, TLD is zero down.

- The TL sensitivity is affected by the thermal quenching of the luminescence efficiency. For a higher heating rate, TL peak shifts to higher temperatures, and luminescence efficiency will be reduced due to thermal quenching.

4.7 OPTICALLY STIMULATED LUMINESCENCE (OSL)

The use of OSL for radiation dosimetry was first proposed in 1956. The mechanism of OSL involves the illumination of an irradiated crystal with a specific wavelength of light to initiate the movement of charge trap sites to luminescent centers. Total luminescence, which depends on the amount of stimulation imparted to the crystal, is proportional to the dose. The OSL mechanism is similar to TL. The basic difference is the stimulant, light versus heat.

McKeever [38] presented an excellent review of various light stimulation modes. They include continuous wave (CWOSL), linear modulation (LMOSL), and pulsed (POSL). In brief, CWOSL involves simultaneous sample illumination with constant intensity light and monitoring of the stimulated luminescence EM. Resolution between stimulation and EM light is accomplished by appropriate filter and wavelength discrimination. A linear increase in the stimulation light intensity is used in LMOSL resulting in a peak. This peak is the result of a linear increase in OSL output followed by a nonlinear return to zero. In POSL, a pulsed stimulation source is used and discrimination between EX and EM light is accomplished by time resolution.

4.7.1 Advantages and Limitations

OSL studies find application in medical fields as discussed by Yukihara et al. in detail [39]. The OSL techniques have several advantages [40] over conventional TL techniques such as:

- The readout method is optical; hence, there is no requirement for a heating scheme for the heating of samples.
- Problems due to thermal quenching of the luminescence efficiency are removed.
- Different readout methods are available.
- OSL dosimeters can be read multiple times.
- OSL dosimeters have a simple design; they can be read at room temperature and are not affected by the environment normally.
- OSL dosimeters do not require annealing for readings.

Limitations
- OSL dosimeters are costly as compared to TL dosimeters.

4.8 FOURIER TRANSFORM INFRARED SPECTROSCOPY (FTIR)

In Fourier transform infrared spectroscopy (FTIR), IR radiations are passed through the sample. These radiations are selectively absorbed by different covalent bonds and functional groups present in the sample. The transmitted spectrum is recorded on a graph with wavenumber on X-axis and corresponding transmittance on Y-axis. These transmittance patterns are different for different molecules and act as their fingerprints. The molecular bonds and functional groups present in the sample can be identified by comparing them to the known reference compound.

4.8.1 Working of FTIR Spectrometer

An FTIR instrument relies upon the interference of various frequencies of light to collect a spectrum. The spectrometer consists of a source, beam splitter, two mirrors, laser, and a detector. The beam splitter and mirrors are collectively called interferometer. The assembled whole is shown

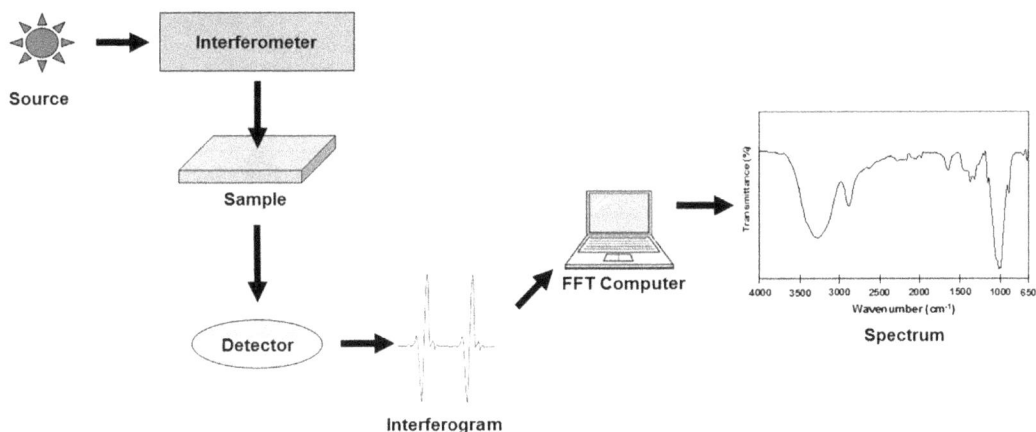

FIGURE 4.11 Schematic diagram of FTIR technique [41].

in Figure 4.11. The IR light from the source strikes the beam splitter, which produces two beams of roughly the same intensity. One beam strikes a fixed mirror and returns, while the second strikes a moving mirror. Laser parallels the IR light and also goes through the interferometer. The moving mirror oscillates at a constant velocity, timed using the laser frequency. Two beams are reflected from the mirrors and are recombined at the beam splitter. If the distance the two beams travel is the same, then they will recombine constructively. However, if the beam from the moving mirror has traveled a different distance (further or shorter) than the beam from the fixed mirror, then recombination will result in some destructive interference. The movement of the mirror thus generates an interference pattern during the motion. The IR beam next passes through the sample, where some energy is absorbed and some is transmitted. The transmitted portion reaches the detector, which records the total intensity. The raw detector response yields an interferogram. The interference pattern contains information about all wavelengths being transmitted at once, which is a function of the source, beam splitter, mirrors, and sample. This signal is digitized and processed using a computer. Untangling of the frequencies into a spectrum is done by the Fourier transform algorithm, which gives the name to the entire spectrometer. This produces a "single-beam" spectrum. A reference or "background" single beam is collected without a sample; the sample single beam is collected with the only change being the presence of the sample. The ratio of these two leads to the spectrum.

4.8.2 ADVANTAGES AND LIMITATIONS

- FTIR technology uses the combination of an interferometer with IR source and the mathematical Fourier transform wave function making it significantly faster than older techniques.
- It is a nondestructive technique, i.e., it does not destroy the sample.
- It is precise and more sensitive.
- FTIR spectrometry is a useful tool for identifying both organic and inorganic compounds and various functional groups.
- Extensive spectral libraries for compound and mixture identifications.
- This technique works in ambient conditions, i.e., normal temperature and pressure (vacuum is not required).
- It is a complementary technique to Raman spectroscopy.

Limitations
- It shows limited surface sensitivity up to a thickness of 100 nm.

FIGURE 4.12 Typical ZWL-600 lumen meter setup showing spectrometer, integrating sphere, power supply, and connected to PC.

- Standard reference is required for analysis.
- FTIR spectrum is shown by specific inorganic species.
- The water strongly absorbs infrared light. Hence, it may interfere with the analysis of dissolved, suspended, or wet samples.
- Simple cations and anions, like Na^+ and Cl^-, do not absorb FTIR light and hence cannot be detected by FTIR.
- FTIR cannot analyze metals that reflect light metals that do not absorb light in the IR range and are therefore not suitable for FTIR spectroscopy.

4.9 FABRICATED LED CHARACTERIZATION

One of the popular applications of metal oxide phosphor is for solid-state lighting (LED lighting). Various parameters of fabricated LEDs such as relative spectrum power distribution, chromaticity, dominant wavelength, peak wavelength, color purity, color temperature, color rendering index, half-wave width, radiant power, etc., can be measured using a lumen meter. Typical lumen meter (ZWL-600, ZVISION) is available in the market. It is a photometric, colorimetric, and electric test system for different kinds of LED components, mainly for material inspection, research inspection, and standard establishment in the LED enterprise.

It has the following parts (i) spectrometer: for the simulation of visual function and automatic correction for different color spectrums; measure different power LEDs in ms time. (ii) optical fiber: It is made of a high-density SiO_2 material and used for optical signal propagation between integrating sphere and spectrometer with low loss. (iii) testing instrument: It uses CNC constant current drive and LCD display; it can offline measure luminous flux, luminous intensity, leakage current, voltage, etc. The RS-232 series communication port is used as a remote control of the testing instrument. (iv) luminous intensity testing device: It measures luminous intensity and spread angle, compliant with CIEA/B standard. (v) integrating sphere: It works with a spectrometer to measure the photometric parameters (Figure 4.12).

This device is very helpful for measuring various necessary parameters of the fabricated phosphor-converted LEDs at a laboratory scale.

REFERENCES

1. Hakuta, Y, Haganuma, T, Sue, K, Adschiri, T, and Arai, K (2003). Continuous production of phosphor YAG: Tb nanoparticles by hydrothermal synthesis in supercritical waste. *Mater Res Bull*, 38(7), 1257–65. https://doi.org/10.1016/S0025–5408(03)00088–6.
2. Ekambaram, S, and Patil, KC (1997). Synthesis and properties of Eu^{2+} activated blue phosphors. *J Alloys Compds*, 248(1–2), 7–12. https://doi.org/10.1016/S0925–8388(96)02622–9.

3. Colmont, M, Boutinaud, P, Latouche, C, Massuyeau, F, Huve, M, Zadoya, A, and Jobic, S (2020). Origin of luminescence in La$_2$MoO$_6$ and La$_2$Mo$_2$O$_9$ and their Bi-doped variants. *Inorg Chem*, 59(5), 3215–20. https://doi.org/10.1021/acs.inorgchem.9b03580.

4. Khan, H, Yerramilli, AS, D'Oliveira, A, Alford, TL, Boffito, DC, and Patience, GS (2020). Experimental methods in chemical engineering: X-ray diffraction spectroscopy-XRD. *Canad J Chem Eng*, 98(6), 1255–66. https://doi.org/10.1002/cjce.23747.

5. Elton, LRB, and Jackson, DF (1966). X-ray diffraction and the Bragg law. *Am J Phys*, 34, 11, 1036–8. https://doi.org/10.1119/1.1972439.

6. Azaroff, LV (1974). *X-Ray Diffraction*. McGraw Hill Company.

7. Cullity, BD (1967). *Elements of X-ray Diffraction*. Adison–Wesley Publ. Co., London, 189.

8. Taylor, A, and Sinclair, H (1945). On the determination of lattice parameters by the Debye-Scherrer method. *Proc Phys Soc*, 57(2), 126. https://doi.org/10.1088/0959–5309/57/2/306.

9. Uvarov, VL, and Popov, I (2007). Metrological characterization of X-ray diffraction methods for determination of crystallite size in nano-scale materials. *Mater Charact*, 58(10), 883–91. https://doi.org/10.1016/j.matchar.2006.09.002.

10. Holzwarth, U, and Gibson, N (2011). The Scherrer equation versus the Debye-Scherrer equation. *Nat Nanotechnol*, 6(9), 534–534. https://doi.org/10.1038/nnano.2011.145.

11. Yu, X, Pan, H, Wan, W, Ma, C, Bai, J, Meng, Q, Ehrlich, SN, Hu, Y-S, and Yang, X-Q (2013). A size-dependent sodium storage mechanism in Li$_4$Ti$_5$O$_{12}$ investigated by a novel characterization technique combining in situ X-ray diffraction and chemical sodiation. *Nano Lett*, 13(10), 4721–7. https://doi.org/10.1021/nl402263g.

12. Chauhan, A, and Chauhan, P (2014). Powder XRD technique and its applications in science and technology. *J Anal Bioanal Tech*, 5(5), 1–5. https://doi.org/10.4172/2155–9872.1000212.

13. Holder, CF, and Schaak, RE (2019). Tutorial on powder X-ray diffraction for characterizing nanoscale materials. *Acs Nano*, 13(7), 7359–65. https://doi.org/10.1021/acsnano.9b05157.

14. Epp, J (2016). X-ray diffraction (XRD) techniques for materials characterization. In *Materials Characterization Using Nondestructive Evaluation (NDE) Methods* (pp. 81–124). Woodhead Publishing. https://doi.org/10.1016/B978-0-08-100040-3.00004–3

15. https://myscope.training/#/SEMlevel_2_25.

16. Inkson, BJ (2016). Scanning electron microscopy (SEM) and transmission electron microscopy (TEM) for materials characterization. In *Materials Characterization Using Nondestructive Evaluation (NDE) Methods* (pp. 17–43). Woodhead Publishing. https://doi.org/10.1016/B978-0-08-100040-3.00002–X.

17. Cullity, BD (1967). *Elements of X-ray Diffraction*. Adison–Wesley Publ. Co., London, 189.

18. https://www.nanoimages.com/tabletop-sem-products/sne-4500m-tabletop-sem/.

19. Blasse, G, and Grabmaier, BC (1994). A general introduction to luminescent materials. In *Luminescent Materials* (pp. 1–9). Springer, Berlin, Heidelberg. https://doi.org/10.1007/978-3-642-79017-1_1.

20. Lakshmanan, A (2008). *Luminescence and Display Phosphors: Phenomena and Applications*. Nova Publishers.

21. Valeur, B, and Berberan-Santos, MN (2011). A brief history of fluorescence and phosphorescence before the emergence of quantum theory. *J Chem Edu*, 88(6), 731–8. https://doi.org/10.1021/ed100182h.

22. Blasse, G (1966). On the Eu^{3+} fluorescence of mixed metal oxides. IV. The photoluminescent efficiency of Eu^{3+}-activated oxides. *J Chem Phys*, 45(7), 2356–60. https://doi.org/10.1063/1.1727946.

23. Wu, Y, Yin, X, Zhang, Q, Wang, W, and Mu, X (2014). The recycling of rare earths from waste tricolor phosphors in fluorescent lamps: A review of processes and technologies. *Resource Conserv Recycl*, 88, 21–31. https://doi.org/10.1016/j.resconrec.2014.04.007.

24. Liu, Z, Sinatra, L, Lutfullin, M, Ivanov, YP, Divitini, G, De Trizio, L, and Manna, L, (2022). One hundred-nanometer-sized CsPbBr3/m-SiO$_2$ composites prepared via molten-salts synthesis are optimal green phosphors for LCD display devices. *Adv Energy Mater*, 12(38), 2201948. https://doi.org/10.1002/aenm.202201948.

25. Sun, XD, Gao, C, Wang, J, and Xiang, XD (1997). Identification and optimization of advanced phosphors using combinatorial libraries. *Appl Phys Lett*, 70(25), 3353–5. https://doi.org/10.1063/1.119168.

26. Lian, H, Hou, Z, Shang, M, Geng, D, Zhang, Y, and Lin, J (2013). Rare earth ions doped phosphors for improving efficiencies of solar cells. *Energy*, 57, 270–83. https://doi.org/10.1016/j.energy.2013.05.019.

27. Silver, J, and Withnall, R (2008). Color conversion phosphors for LEDs. *Lumines Mater Appl*, 75–110.

28. Xie, RJ, and Hirosaki N (2007). Silicon-based oxynitride and nitride phosphors for white LEDs-A review. *Sci Technol Adv Mater*, 8(7–8), 588. https://doi.org/10.1016/j.stam.2007.08.00.

29. Feher, I, Deme, S, Szabo, B, Vagvolgyi, J, Szabo, PP, Csoke, A, Ranky, M, and Akatov, YA (1981). A new thermoluminescent dosimeter system for space research. *Adv Space Res*, 1(14), 61–6. https://doi.org/10.1016/0273–1177(81)90244–1.

30. Deme, S, Apathy, I, Hejja, I, Lang, E, and Feher, I (1999). Extra dose due to extravehicular activity during the NASA4 mission, measured by an on-board TLD system. *Radiat Protec Dosim*, 85(1–4), 121–4. https://doi.org/10.1093/oxfordjournals.rpd.a032815.

31. Akatov, YA, Arkhangelsky, VV, Aleksandrov, AP, Feher, I, Deme, S, Szabo, B, Vagyolgyi, J, Szabo, PP, Csoke, A, Ranky, M, and Farkas, B (1984). Thermoluminescent dose measurements on-board Salyut type orbital stations. *Adv Space Res*, 4(10), 77–81. https://doi.org/10.1016/0273–1177(84)90227–8.

32. Qiu, J, Li, Y, and Jia, Y (2020). *Persistent Phosphors: From Fundamentals to Applications*. Woodhead Publishing. https://doi.org/10.1016/B978–0–12–818637–4.00004–5.

33. Chand, S, Mehra, R, and Chopra, V (2021). Recent developments in phosphate materials for their thermoluminescence dosimeter (TLD) applications. *Luminescence*, 36(8), 1808–17. https://doi.org/10.1002/bio.3960.

34. Braunlich, P (1968). Possibility of observing negative thermally stimulated conductivity. *J Appl Phys*, 39(6), 2953–6. https://doi.org/10.1063/1.1656701.

35. DeWerd, LA (1979). Thermally stimulated relaxation in solids. *Top Appl Phys*, 37.

36. Nambi, KS (1977). *Thermoluminescence: Its Understanding and Applications*. Instituto de Energia Atomica.

37. McKeever, SW (1988). *Thermoluminescence of Solids*. Cambridge University Press.

38. McKeever, SW (2001). Optically stimulated luminescence dosimetry. *Nucl Instrum Methods Phys Res Sect B: Beam Interact Mater Atoms*, 184(1–2), 29–54. https://doi.org/10.1016/S0168-583X(01)00588–2.

39. Yukihara, EG, and McKeever, SW (2008) Optically stimulated luminescence (OSL) dosimetry in medicine. *Phys Med Biol*, 53(20), R351. https://doi.org/10.1088/0031-9155/53/20/r01.

40. Botter-Jensen, L, McKeever, SW, and Wintle, AG (2003). *Optically Stimulated Luminescence Dosimetry*. Elsevier.

41. Mohamed, MA, Jaafar, J, Ismail, AF, Othman, MH, and Rahman, MA (2017). Fourier transform infrared (FTIR) spectroscopy. In *Membrane Characterization* (pp. 3–29). Elsevier. https://doi.org/10.1016/B978–0–444–63776–5.00001–2.

5 Metal Oxides
Radiation Dosimetry Materials

R.K. Joshi
Dr. D.Y. Patil Institute of Technology

A.D. Sontakke
Ultrect University

CONTENTS

DOI: 10.1201/9781003366232-5

5.1 INTRODUCTION

5.1.1 BIOLOGICAL EFFECTS OF RADIATION

Primarily, biological effects of radiation are divided into two categories namely, non-stochastic effects and stochastic effects. Non-stochastic effects deal with high doses of radiation over a short period of time producing acute or short-term effects. They are characterized by a threshold above which the damage increases with the dose. The threshold for tissue reactions is between 0.1 and 0.5 Gy.

Stochastic effects represent exposure to low doses of radiation over an extended period of time producing chronic or long-term effects. The lower limit at which radiation can be detected should be kept as low as possible because small doses acquired over longer durations have harmful effects.

The biological effects of ionizing radiation mainly depend on radiation intensity, energy type of the radiation, exposure time, area exposed and depth of energy deposition.

5.1.2 RADIATION UNITS

The dose received by the body or organ and the biological effectiveness of that dose can be specified by different quantities such as the absorbed dose, the equivalent dose and the effective dose.

Radioactivity: It is a measure of ionizing radiation released by a radioactive material and is usually measured in Becquerel (Bq, international unit) and Curie (Ci, U.S. unit).

Absorbed dose: The absorbed dose is the amount of energy deposited by radiation in a mass (for example, water, rock, air, people, etc.). The units for the absorbed dose are gray (Gy, international unit) and rad (rad, U.S. unit).

Equivalent dose: The product of the absorbed dose (rad) in tissue and a radiation weighting factor. The equivalent dose is expressed in units of rem (or sievert) (1 rem=0.01 sievert). Radiation weighting factors are needed because different types of radiation (like alpha, beta, gamma, and neutrons) can have different effects even if the absorbed dose is the same. For example, for gamma (photon) and beta (electron) radiation, the radiation weighting factor is 1, and therefore, for example, an absorbed dose of 1 mGy in an organ equals an equivalent dose of 1 mSv to that organ.

Effective dose: It is calculated for the whole body. It is the addition of equivalent doses to all organs, each adjusted to account for the sensitivity of the organ to radiation. The units for effective dose are Sievert (Sv) and Rem (rem).

5.1.3 LUMINESCENCE DOSIMETRY

Luminescence is the emission of light in the visible or near-visible region from a material following the initial absorption of energy from an external source viz, ultraviolet or high-energy radiation. The wavelength of the light emitted is the characteristic of the material and not of the incident radiation. The materials which exhibit this property are called luminescent materials. Luminescence dosimetry is the technique used to estimate the absorbed dose of ionizing radiation with the help of detectors that exhibit luminescence. The luminescence intensity scales with the energy absorbed from the radiation field. Depending upon the mode of excitation, luminescence is classified into different techniques as mentioned in Figure 5.1. The most commonly used luminescent techniques in the field of radiation dosimetry are thermoluminescence (TL), optically stimulated luminescence (OSL), radioluminescence and mechanoluminescence (or triboluminescence).

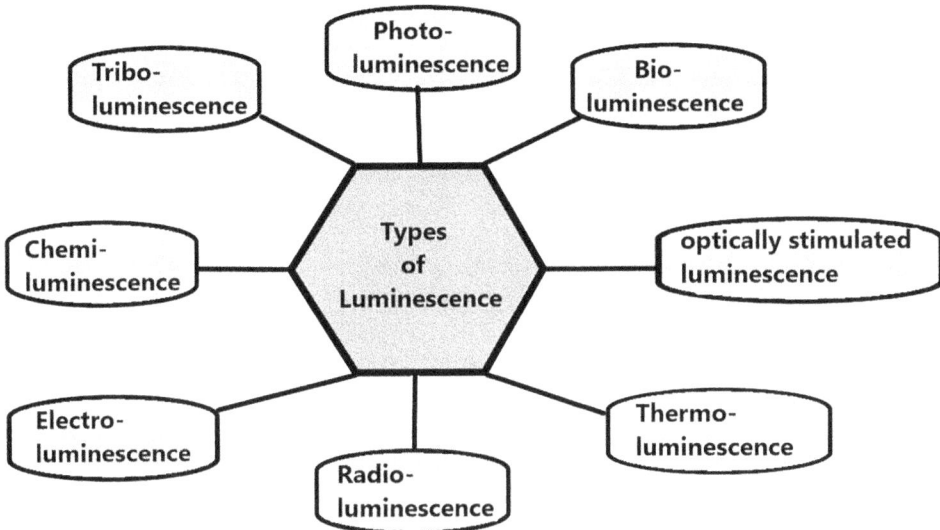

FIGURE 5.1 Various types of luminescence.

5.1.4 DEVELOPMENT OF METAL OXIDE–BASED PHOSPHORS FOR DOSIMETRIC APPLICATIONS

Alkaline earth borates are considered important luminescent materials because of their excellent chemical and thermal stability, simple synthesis and cheap raw material. Borates are one of the earliest identified materials for radiation dosimetry by the TL process. Borate phosphors are important because of the ease of synthesis in bulk quantities, simple glow curve structure, increased neutron and gamma sensitivity, near-tissue equivalence in some cases and simple thermal treatment procedure for reuse. Tissue equivalent TL/OSL materials enriched in lithium (^{7}Li) and/or boron (^{10}B) with appreciable neutron and gamma sensitivity have gained considerable interest due to their suitability for dosimetric applications. In this regard, metal oxide–based TL/OSL phosphors play a key role because of their near-tissue-equivalent absorption coefficient and neutron sensitivity due to the high thermal neutron absorption cross-section of ^{10}B [1,2]. Some important TL materials that have been developed over the years include (i) borates: $Li_2B_4O_7$:Mn [3], $Li_2B_4O_7$:Cu [4], LiB_3O_5 [5], SrB_4O_7 [6], $LiSrBO_3$:RE [7], $LiCaBO_3$:RE [8], $NaSr_4(BO_3)_3$:RE [9], $KSr_4(BO_3)_3$:RE [9], $LiKB_4O_7$ [10], $LiBa_2B_5O_{10}$:RE [11], $NaSrBO_3$ [12], MgB_4O_7:Dy/Tm [13] and SrB_6O_{10}:Tb [14], (ii) sulphates: $CaSO_4$:Dy/Tm [15], $SrSO_4$:Tb [16] and $BaSO_4$:Eu [17], (iii) oxides: BeO [18], Al_2O_3:Si, Ti [19], and Al_2O_3:C [20] and (iv) silicates: Mg_2SiO_4:Tb [21]. Since these TL phosphors have found several applications in various fields, the development of many such phosphors or improvement of the characteristics of the existing phosphors has become a topic of interest now.

The phenomenon of "thermoluminescence" was initially observed by a common man in a fluorite mineral. When heated in darkness, they exhibit a transient glow [22,23]. The scientific report of "thermoluminescence" was submitted by Sir Robert Boyle in 1663. As early as 1895, the physical process for the thermal release of stored radiation-induced luminescence (TL) was used for the detection of ionizing radiation by Wiedemann and Schmidt [24] in Erlangen. They observed that natural fluorite and manganese-doped calcium fluorite exhibit a very intense luminescence when they are heated in darkness with negligible fading even after storage for a few weeks, hence widely used in solid-state dosimetry. Wick [25] reported that certain materials exhibit TL only after irradiation by X-rays, not before irradiation.

First, the TL glow curve was recorded at the Przibram Institute in Vienna by Urbach and Frisch [26]. Continuous monitoring of radiation released to the environment has become a major concern for industrialized nations and thermoluminescent dosimeter (TLD) usage is an important factor in this type of activity. If the exposure levels are low, long exposure times are required; thus, long-term stability of TLD becomes vitally important along with extreme sensitivity. Studies of doped lithium fluoride were carried out at the University of Wisconsin under the guidance of Daniels. Later, Cameron and his coworkers in collaboration with the Harshaw Company developed magnesium- and titanium-activated phosphor which is now widely distributed under the name TLD-100. For further development in their dosimetric properties, vast research was carried out by several researchers like Nakajima et al. [27], Bilski et al. [28], Chen and Stoebe [29], Tang et al. [30], Binder et al. [31], Yamashita et al. [32,33], Alig et al. [34], Kandarakis et al. [35], Necmeddin Yazici et al. [36], Morales et al. [37], Kalchgruber et al. [38], Ishii et al. [39], Kitis et al. [40] and Furetta et al. [41]. The first thermoluminescent material based on lithium borate activated by Mn is commercialized by Harshaw under the name TLD-800 [42,43]. But the material was less sensitive to TL with the emission at 600 nm. It is observed that co-doping of copper into the host lattice enhances the TL sensitivity of the phosphor significantly [44]. Due to the appreciable dose linear response for beta radiation and lower band gap, ZnS:Cu nanophosphors can be considered a promising material for the TLD application for ionizing as well as non-ionizing (UV) radiation [36].

Farrington Daniels suggested the utilization of TL for dosimetry applications for the first time. He also suggested its utilization for geological and archaeological age determination. As a TL dating tool, natural quartz was found to be the principal dosimeter material [46,45] which has been primarily studied for archaeological dating applications [46]. However, because of a complex TL glow curve, the progress in the application of quartz to general radiation dosimetry was very slow. Also, TLD has its inherent disadvantages particularly in detecting very low doses due to problems of infrared contribution from the heater and thermal quenching of TL sensitivity in the phosphor materials.

In recent years, OSL has gained a lot of attention during the last decade as an alternative to the thermally stimulated luminescence (TSL) readout technique applicable for passive dosimetry of ionizing radiation [47,48]. The OSL technique for radiation dosimetry has a number of advantages over the TSL technique that leads to the gradual replacement of conventional TSL dosimetry with OSL dosimetry. The specific requirements of OSL materials for dosimetry applications stimulate the search and development of new materials applicable to this purpose. This technique is finding widespread application in a variety of radiation dosimetry fields including personal monitoring, environmental monitoring, retrospective dosimetry, space dosimetry, etc.

The use of OSL as a personnel dosimetry technique, however, is not yet so widespread. First Antonov-Romanovskii et al. [49] and later Brauenlich et al. [50] and Sanborn and Beard [51] reported the use of OSL as a tool for personnel monitoring. However, the use of OSL in radiation dosimetry has not been extensively reported mainly because of the lack of good luminescent material which must be highly sensitive to radiation and must have high optical stimulation efficiency, a low effective atomic number and good fading characteristics. MgS, CaS, SrS and SrSe doped with different rare earth (RE) elements such as Ce, Eu and Sm were the first phosphors suggested for OSL dosimetry [9,51]. However, Al_2O_3:C is routinely used as an OSL dosimetric material for radiation dosimetry due to its excellent dosimetric characteristics. In addition to it, extensive research is carried out towards the development of oxide-based phosphors for various dosimetry applications. Some of these phosphors are discussed in detail in further sections.

5.2 THERMOLUMINESCENCE DOSIMETRY

5.2.1 TL MECHANISM

In TL, the luminescence emission is triggered by heating the irradiated phosphor material (Figure 5.2). During irradiation, the material absorbs the energy from ionizing radiation and the energy absorbed during excitation is stored by electron–hole pair production, exciton creation and/or

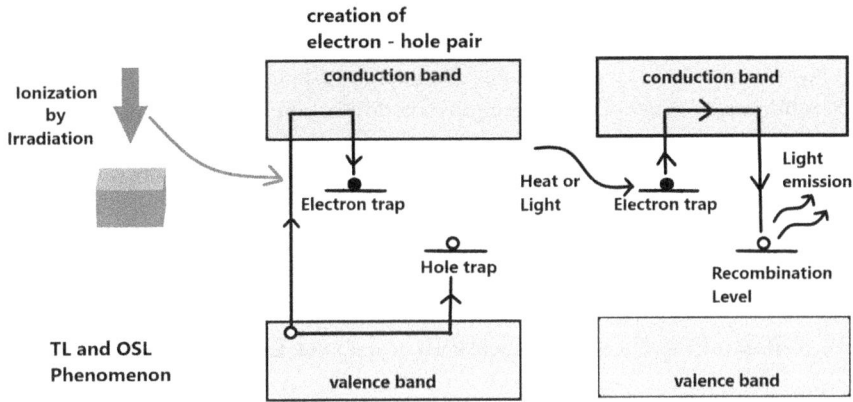

FIGURE 5.2 Mechanism of thermoluminescence and optically stimulated luminescence.

trapping of charges at defects present in the host lattice. On heating the material, the stored energy is released in the form of luminescence as the sample temperature is increased. This luminescence originates from electron–hole recombination or vacancy–interstitial recombination. In both cases, electrons undergo de-excitation from metastable excited states to the ground state, thereby restoring equilibrium. The recombination process may result in the emission of phonons (non-radiative recombination) or photons (radiative recombination). It is the latter that is monitored when recording TL emissions. Thus, the system returns to equilibrium with a portion of the excess energy emitted as light.

The intensity of emitted light, referred to as TL intensity, is a function of temperature (time). The resulting curve is termed a glow curve which consists of one or more isolated/overlapping peaks. The glow curve also depends on the light-sensitive detector used to record the spectrum and its spectral characteristics. A peak in the glow curve, preferably isolated or well resolved, is considered a dosimetric peak. In TLD, the TL intensity of the dosimetric peak is directly proportional to the absorbed dose. Thus, by a suitable calibration of the dosimeter, unknown doses absorbed by the material can be estimated by measuring TL.

5.2.2 CHARACTERISTICS OF TLD PHOSPHORS

In order to use a phosphor for dosimetric applications, it should satisfy the following important properties:

- Phosphor should be tissue equivalent.
- The phosphor should exhibit a linear relationship between TL output and the absorbed dose (D).
- Sensitivity of any TL phosphor is nothing but TL signal strength per unit of absorbed dose. The relative sensitivity of any TL phosphor can be determined by comparing the TL signal of the phosphor of interest with that of the standard phosphor (normally, TLD-100 or $CaSO_4$:Dy is used as standard).
- Fading is a loss of TL signal with time which occurs due to thermally and/or optically stimulated release of electrons. High thermal fading is observed in the phosphor exhibiting TL peak at low temperatures below 180°C. For personnel monitoring purposes, the phosphor should exhibit a TL peak in the range of 200°C–250°C.
- Phosphors must exhibit a simple glow curve structure with a single or a well-resolved dosimetric peak.
- Synthesized phosphors should be physically and chemically stable.

- The characteristic emission of a phosphor must lie in the high-efficiency region (300–500 nm) of a standard photomultiplier tube.
- The phosphor should have a reasonably flat energy response that is the TL output should be the same over a range of energies of absorbed radiation.
- Negligible or low non-radiative-induced signal from the phosphor material.
- Should have a very good shelf life. Its sensitivity should not show any decrease if the phosphor is stored without use for a long duration.
- It should be possible to make a dosimeter of any form, size and shape from the phosphor without any change in its sensitivity or other dosimetric characteristics.

5.2.3 OXIDE-BASED PHOSPHORS FOR THERMOLUMINESCENCE DOSIMETRY

5.2.3.1 Lithium Tetraborate

Phosphors like lithium tetraborate and MTB are nearly tissue equivalent (Z_{eff} = 7.25). Lithium tetraborate doped with manganese (TLD-800) was reported for the first time as TLD [52]. However, it was found to exhibit poor TL intensity, limited dose linearity, fading and energy dependence, hence limiting its use as a good dosimetric phosphor. Thus, several attempts were made by researchers to improve the dosimetric properties either by changing the method of synthesis and/or by adding different co-dopants or activators in the host lattice. Studies involve the preparation of phosphor materials in various forms, like single crystals [53], polycrystalline powders [54] and glasses [55]. Modifiers like alkali/alkaline metals were used to strengthen the relative stability of borate glass [56,57], and activators like dopants and co-dopants of either transition metals [56,58] or REs [59,60] were added to the host in order to enhance luminescence. It was observed that with the addition of copper, the emission wavelength shifted from the red region to the ultraviolet region (360 nm) resulting in a ten-fold increase in the sensitivity as compared to TLD-800 [61]. This enhancement is attributed to the amendment of electron traps and/or trap centres due to their attractive optical-luminescent properties, high resistance to radiation damage as well as high thermal and chemical stability.

Dosimetric study of $Li_2B_4O_7$:Cu; In; Ag and $Li_2B_4O_7$:Cu was first carried out by Prokic et al. [62] and Furetta et al. [41]. Lithium-based borate phosphors are promising candidates in the field of radiation dosimetry due to their low-effective atomic number (Z_{eff} = 7.25) [63–67]. TL measurements on lithium borate were first carried out by Schulman et al. [68]. Researchers took several efforts in enhancing the TL sensitivity of these glass materials by introducing different dopants in the form of transition and RE or lanthanide [Ln] metal ions to the host lattice [54,69–73]. This improves the dose linearity but not the TL sensitivity of the phosphor [74]. Takenaga et al. [52] reported an increase in the sensitivity of $Li_2B_4O_7$:Ag with the addition of Cu as a co-dopant. The largest enhancement in the sensitivity is observed in the material doped with Cu alone. However, doping Cu into the host lattice exhibits severe fading when illuminated by fluorescent or tungsten light.

5.2.3.2 Rare Earth-Doped Magnesium Borate

Magnesium tetraborate (MTB) is a near-tissue-equivalent material with an effective atomic number for photoelectron absorption equal to 8.4. Attempts were made for synthesizing RE-doped magnesium borates through different synthesis routes to optimize the TL properties [8,75–81,]. MgB_4O_7:Dy, Tm [76] exhibits TL peaks around 190°C and 210°C, respectively. Kawashima et al. [82] reported that Tb-doped MgB_4O_7 exhibits the highest TL sensitivity. The TL glow curve showed an intense and well-defined TL peak having a maximum at 220°C with a shoulder around 330°C. Annalakshmi et al. [83] reported the presence of a single dosimetric TL peak of 250°C in Gd-doped MgB_4O_7 phosphor. The incorporation of monovalent dopants such as Na, Li and Ag results in an enhancement of the TL sensitivity without affecting the position of a TL peak. Although TL dosimeters may give consistent and efficient TL output, interpreting the role of defects in the TL mechanism and understanding the underlying physics of these complex problems are still a big challenge to the fraternity. Therefore, research should be undertaken to determine the trapping parameters and TL peak structure, which may give one the ability to control the TL properties of the material. One can use the knowledge obtained from these studies to grow the materials in such a way that the desirable dosimetric properties can be achieved.

5.2.3.3 Mg_2SiO_4:Tb

Dosimetric applications of Mg_2SiO_4:Tb were first studied in 1971 [84,85], and since then, it has been widely used as it is highly sensitive to gamma radiation. It was first introduced by a Japanese group [86]. The phosphor was subsequently produced on a commercial scale and marketed by Kasei Optonix Ltd., Japan. Lakshmanan and coworkers [87–89] have done a detailed characterization of Mg_2SiO_4:Tb TL phosphor powder marketed by Kasei Optonix Ltd., Japan. The phosphor was reported to give 1,100 times higher TL sensitivity than that of $CaSO_4$:Dy [88]. Recently, Prokic and Yukihara [90] reported the development of ultra-high TL-sensitive Tb-doped Mg_2SiO_4 sintered under the reaction between the stoichiometric amounts of MgO and SiO_2 with the addition of optimal Tb_4O_7 activator concentration of about 12 mg atoms Tb per mole of Mg_2SiO_4. Its TL sensitivity is reported to be 110 times than that of TLD-100 with a detection threshold of 0.5 μGy.

5.2.3.4 Rare Earth-Doped Calcium Sulphate

Dy-doped calcium sulphate phosphor is another material of choice for monitoring the radiation (gamma or X-ray) dose received by personnel working with radiation. Yamashita et al. [16,91] reported several RE-doped $CaSO_4$ phosphors synthesized through different routes. Among them, $CaSO_4$:Dy and $CaSO_4$:Tm showed the highest sensitivity with a single dosimetric peak around 220°C. A long way back, there was an investigation of $CaSO_4$:Dy and $CaSO_4$:Dy, Cu material for high-dose dosimetry applications [45,92]. Co-doping of Cu in $CaSO_4$:Dy phosphor shows no growth of a high-temperature peak structure (350°C) even for large doses [93].

For large-scale radiation protection dosimetry service, it is embedded in a Teflon matrix with varying thicknesses. In the current era, $CaSO_4$:Dy Teflon-embedded TL discs having a thickness of 0.8 mm are routinely used for personnel monitoring applications. Guckan et al. [94] investigated the OSL dosimetric properties of $CaSO_4$:Eu and Junot et al. [95] studied the potential of new $CaSO_4$-based detectors with different combinations of impurities, using TL and OSL techniques. TL properties of $CaSO_4$ with different impurities have been extensively studied [16,91–97] and $CaSO_4$ doped with dysprosium ($CaSO_4$:Dy) has been used for personal dosimetry with TL technique for decades and its TL properties are well-known [98].

5.2.3.5 Quartz (SiO_2)

Quartz is one of the polymorphs of silica (SiO_2) and is a very abundant material in ancient ceramic fragments and geological sediments. It is made up of a continuous framework of SiO_4 (silicon–oxygen tetrahedral), with each oxygen being shared between two tetrahedra. It is the most stable form at ambient temperature amongst all others. Many natural crystals viz. aluminates, aluminosilicates, feldspars and silicates exhibit TL and that can be utilized for various applications. Amongst these classes of materials, natural quartz has been widely used in retrospective dosimetry and archaeological dating. The use of TL as a dating tool frequently relies on quartz as the principal dosimeter material [99]. The TL properties of quartz have been studied primarily for archaeological dating applications [46]. However, there is a slow progress in the application of quartz to general radiation dosimetry due to its complex TL glow curve structure as well as the wide range of different chemical forms in which it occurs. Also, TLD has its inherent disadvantages particularly in detecting very low doses due to the problems of infrared contribution from the heater and thermal quenching of TL sensitivity in the phosphor materials. Huntley et al. [100] proposed a new technique in which trapped electrons are stimulated optically rather than thermally and such a technique was called optically stimulated luminescence. OSL was also studied in natural and synthetic quartz and feldspars [101,102]. Annealing of quartz undergoes phase inversion, namely 573°C and 870°C leads to a significant change in OSL signals. Recently, Luis A. Benevides et al. [103] reported the potential use of Cu-doped fused quartz fiberoptic coupled dosimeter for clinical applications. Intrusion for the use of OSL for real-time/online dosimetry appears to stem from the studies of plastic scintillators which have been found useful for real-time measurements of small radiation fields. The availability of OSL dosimeter of a size similar to a plastic scintillator has opened up the possibility of replacing these scintillators with OSL dosimeter. Recent advances in the real-time dosimetry are the use of OSL for in vivo measurements which involves remotely placed light sources and optical fibers to simultaneously stimulate the OSL dosimeter and detect the resulting luminescence [104,105]. For this, a new material consisting of Cu-doped fused quartz has been recently developed at the naval research laboratory by Justus et al. [106]. It exhibits excellent optical transparency and characteristics which makes it ideal for dosimetry applications including excellent TL and scintillation properties in addition to OSL [107–111]. The fused quartz developed by Justus and coworkers prepared at 1,100°C using a patented method [112], shows prominent emission of Cu^+ around 550 nm with shoulder around 400 nm. Therefore, the sample is only suitable for infrared-stimulated luminescence, not for blue-stimulated luminescence [101]. It has been observed earlier that natural quartz doped with Cu using electro diffusion technique shows emission around 360 nm with a shoulder around 550 nm whereas the synthetic quartz doped with thermodiffusion at 1,700°C shows prominent emission around 515 nm with a shoulder around 395 nm. Crystalline SiO_2 has been successfully synthesized through a chemical route. Monovalent impurities like Cu^+ and Ag^+ are doped successfully.

5.3　OPTICALLY STIMULATED LUMINESCENCE (OSL) DOSIMETRY

5.3.1　OSL MECHANISM

OSL has become the technique of choice for many areas of radiation dosimetry fields, including personal monitoring, environmental monitoring, retrospective dosimetry (including geological dating and accident dosimetry), space dosimetry and many more. OSL is the luminescence emitted from an irradiated insulator or semiconductor during exposure to light (Figure 5.2). This process begins with irradiation causing ionization of valence electrons and creation of electron–hole pairs. Pre-existing defects within the material then localize the free electrons and holes through non-radiative trapping transitions. Subsequent illumination of the irradiated sample leads to the absorption of energy by the trapped electrons and transition from the localized trap into the delocalized conduction band. Recombination of the freed electrons with the localized holes results in radiative emission and OSL, whose intensity is the function of the dose of radiation absorbed by the sample. Because the OSL signal is sensitive to light, the irradiation of samples must be done under strictly controlled no-light conditions. OSL is sometimes confused with the related phenomenon of photoluminescence (PL) that can be stimulated from similar materials. However, PL is generally not dependent upon the irradiation of the sample. PL is the excitation of an electron (via the absorption of light) in a crystal defect within the material, resulting in the excitation of the electron from the defect's ground state to an excited state. Relaxation back to the ground state results in emission of luminescence, the intensity of which is proportional to the concentration of excited defects.

The basis of OSL measurement is to stimulate an irradiated sample with light of a selected wavelength and monitor the emission from the sample at a different wavelength. Different modes of stimulation which are widely used are continuous wave-OSL (CW-OSL), linear-modulation OSL (LM-OSL) and pulsed OSL (POSL).

CW-OSL: In this technique, one uses a laser or a broadband source and monochromator (or filter) to select a particular stimulation wavelength. The luminescence emission is monitored continuously while the stimulation beam is on and narrowband filters are used to discriminate between the excitation light and emission light and prevent scattered stimulation light from entering the detector. OSL is monitored from the instant that the stimulation source is switched on and is usually in the form of an exponential-like decay until all the traps are emptied and the luminescence ceases. The integrated emission (i.e. the area under the decay curve, less background) is recorded and used to determine the dose of absorbed radiation.

LM-OSL: In this technique, the OSL output is observed to increase, initially linearly as the stimulation power increases until the traps become depleted after which the OSL intensity decreases non-linearly to zero. Thus, the OSL signal is in the shape of a peak, the position of which depends upon the rate of a linear increase in the intensity of the stimulating light and the photoionization cross-section of the trap being emptied. For a given ramp rate and stimulation wavelength, traps with different photoionization cross-section values thus appear at different times and the method is thus able to distinguish between OSL originating from different traps.

POSL: Pulsed OSL results when the stimulation source is pulsed at a particular modulation frequency and with a particular pulse width appropriate to the lifetime of the luminescence being observed. In this mode of excitation, one measures the OSL emission only between pulses. In this way, discrimination between the excitation light and the emission light is achieved by time resolution, rather than wavelength resolution.

5.3.2 CHARACTERISTICS OF OSL PHOSPHOR

In general, the more desirable properties of an OSL dosimetry (OSLD) phosphor are as follows:

- Tissue equivalent
- Chemically inert
- High sensitivity
- Trapping centres should be thermally stable and optically sensitive
- Good separation between emission and stimulation bands
- Dose erasure by optical bleaching
- Must be able to detect the minimum possible dose
- Good linearity over the wide range
- Easy to reset
- Negligible fading

5.3.3 OXIDE-BASED PHOSPHORS FOR OSL DOSIMETRY

5.3.3.1 Beryllium Oxide

Beryllium oxide (BeO) is a hexagonal crystalline insulator and one of the most popular luminescent materials in TL and OSL applications [113]. BeO was initially suggested as a TL dosimeter. Taking into account the effect of light-induced fading, Rhyner and Miller [114] reported the use of BeO as an OSL dosimeter. Later on, its properties have been extensively improved due to the efforts taken by Bulur and Göksu [115] and hence find applications in different fields like medical dosimetry, environmental dosimetry [116], 2D radiation fields, [117] etc. The material has many advantageous dosimetric characteristics. BeO was first discovered in 1881 by the bright radioluminescence of BeO powder [118]. After the material was originally proposed as a TL dosimeter [119], the delayed OSL signals have been investigated for dosimetric applications [120]. Thermalox 995 chips were first investigated using the CW-OSL technique and recommended as a promising OSL material [115]. Single Li- and Na-doped BeO samples were reported to be promising for TLD with the linearity of dose response and low fading properties [121]. Recently, low-temperature luminescence, thermally stimulated luminescence and spectroscopic characteristics of BeO:Mg and BeO:Zn single crystals were also reported in detail [122,123]. In our recent study, it was found that the dosimetric properties of BeO pellets can be improved by doping Al and Ca separately as well as doping them together [124]. In another study, the effect of metal dopant on TL and OSL signals of BeO was investigated, and it was reported that Na, Li, K and Cu dopant elements showed promise for OSL dosimetry [125]. In particular, it has been reported that a passive admixture of Na^+ ion, just like oxygen vacancies, acts as a charge compensator between the lanthanide (Ln) elements and helps transfer energy between them, thus remarkably enhancing luminescence efficiency [126,127]. RE elements are mostly used as dopants and play an important role in the production of highly sensitive TL and OSL phosphors. Specifically, luminescence of BeO-doped Ln^{3+} ions has been studied frequently in recent years, and the behaviour of almost all lanthanides in BeO has been investigated separately [127,128]. This study aims to enhance the luminescence of BeO by doping it with alkali metal Na^+ (a passive admixture), RE elements, Yb3+ (as a sensitizer) and Dy3+ (as an activator). Despite numerous encouraging properties, the material is highly toxic in powder form and often more costly than other luminescent materials such as Al_2O_3, leading to limitations in its commercial production [128,129].

5.3.3.2 Carbon-Doped Aluminium Oxide

A research group of Urals Polytechnical Institute, Russia [130], introduced carbon-doped Al_2O_3 for the first time which has opened the possibility of several promising applications for ultra-high sensitivity dosimetry, particularly for short-term exposure in personal, environmental and extremity dosimetry. It is one of the earliest materials studied for the application of radiation dosimetry owing to its superior thermal and chemical stability and low-effective atomic number. A lot of work has been carried out by researchers for the development of phosphor, but this phosphor was found to be optically sensitive and also shows thermal quenching, limiting its use as a TL phosphor. Taking into consideration the drawbacks of using Al_2O_3:C as TL phosphors, Markey et al. [131] developed Al_2O_3:C as an OSL material which has solved the problem of thermal quenching. As a consequence, Al_2O_3:C has now been emerged as a material of choice for radiation dosimetry using the OSL technique [132,133].

   Al_2O_3:C was introduced in the 1990s as a high-sensitivity TL material with a reported sensitivity 40–60 times that of LiF:Mg, Ti TL detectors [118]. The crystal is grown in a reducing atmosphere in the presence of carbon to increase the concentration of oxygen vacancies responsible for the main emission band centred at 415 nm. The material was then developed for OSL applications [134–137]. The detector response is linear for several orders of magnitude, up to ~10 Gy.

   Al_2O_3:C is also used as a probe in optical fiber dosimetry systems for medical and environmental applications [138]. In addition, Al_2O_3:C-based dosimeters were recently made sensitive to neutrons with the addition of neutron converters in the dosimeter formulation. Al_2O_3 doped with carbon and magnesium (Al_2O_3:C, Mg) has been recently developed for volumetric optical memory storage and nuclear track detection [139,140] and is currently being investigated for OSL applications.

5.3.3.3 Lithium Aluminate

Lithium aluminate ($LiAlO_2$) is one of the best alternatives to aluminium oxide. It has a structure congruent to Al_2O_3, so that it can be expected to exhibit similar OSL properties. The advantage of this choice will be a small reduction of effective atomic number (10.7 for lithium aluminate) and the possibility of application in neutron fields. This last feature is possible due to the fact that $LiAlO_2$ contains a Li-6 isotope, which shows a very high cross-section for reaction with thermal neutrons. Preliminary results of attempts to develop a new OSL material based on LiAlO2 were demonstrated in several works. Mittani [141] proposed doping $LiAlO_2$ with terbium while Dhabekar [142] and Teng [143] suggested activation with manganese. Lee et al. [144] showed that undoped $LiAlO_2$ may exhibit a very high OSL signal. The purpose of this study is to characterize the OSL properties of $LiAlO_2$ samples and to determine the influence of doping on these properties. $LiAlO_2$ shows also a significant TL signal [144], which however was less thoroughly investigated so far.

   $LiAlO_2$ diffused with copper has been found to have potential use as an OSL dosimeter. The OSL effect arises from electron–hole recombination occurring at Cu^{2+} ions substituting for Li^+ ions [145]. The major intrinsic defect after neutron irradiation is an F^+ centre; an oxygen vacancy with one trapped electron. This defect has two states, a stable state that survives up to 500°C and a metastable state that survives up to 200°C. $LiAlO_2$ is an insulator that can be grown into three distinct crystal structures which are referred to in the literature as α-phase, β-phase and γ-phase [146]. α-$LiAlO_2$ or β-$LiAlO_2$ transforms into γ-$LiAlO_2$ at a high temperature [147]. γ-$LiAlO_2$ is the most commonly occurring phase and the one used in various applications.

5.4 APPLICATIONS OF LUMINESCENCE DOSIMETRY

5.4.1 MEDICAL DOSIMETRY

Various techniques in the medical field which involve the use of such solid-state dosimeters are radiotherapy, teletherapy, mammography, computed tomography, brachytherapy and radiography. Radiotherapy is the use of ionizing radiation to treat a disease. Its purpose is to deliver a very high dose to the target volume with a minimal dose to normal tissues. The most common application is the treatment of cancer using X-ray and gamma rays to prevent the proliferation of malignant cells by reducing the rate of mitosis or preventing the synthesis of DNA. If the radiation beam is produced by an X-ray tube or ^{60}Co radioactive source located at a certain distance from the body of the patient, it is called external radiotherapy or teletherapy. When a sealed source is implanted within the patient, the technique is called brachytherapy. $Li_2B_4O_7$ TL dosimeters are particularly suited for applications in radiation dosimetry, especially radiation therapy and clinical applications, since they are tissue equivalent with an effective atomic number of 7.3, which is very close to that of soft biological tissue (Z_{eff}=7.4), making them convenient for measurements independent of the photon energy used in radiation treatments.

5.4.2 PERSONNEL DOSIMETRY

Personnel dosimetry is of fundamental importance in the disciplines of radiation dosimetry and radiation health physics and is primarily used to estimate the radiation dose deposited in an individual wearing the device. A major requirement of OSLDs in these applications is that they are approximately tissue equivalent. Thus, materials with effective atomic numbers (Z_{eff}) near that of human tissue (Z_{eff}=7.6) are desired. The dose equivalent range of interest is from approximately 10 μSv to 1 Sv, with a required uncertainty better than approximately 10%.

5.4.3 ENVIRONMENTAL DOSIMETRY

Tissue equivalence is not an issue with dose estimation to the environment, for which the only quantity of interest is the absorbed dose D (in Gy). The primary interest in this field is the impact of "man-made" radiation on the general public. Sources of such man-made radiation include nuclear waste disposal, emissions from nuclear power and reprocessing plants and the nuclear weapon industry. Major requirements of dosimeters for these applications include high sensitivity (to enable short- or long-term monitoring) and stability with respect to adverse weather and changing light levels and temperatures. With natural dose rates of a few mGy (or mSv) per year, the dose range of interest depends on the exposure time, but a minimum measurable dose of approximately 1 μGy is desirable for short-term measurements. OSL techniques and materials hold significant promise for these applications due to their high sensitivity and ease of use.

5.4.4 ARCHAEOLOGICAL OR GEOLOGICAL DATING

Retrospective dosimetry is the term used when determining the dose of absorbed radiation to environmental or locally available materials in situations where conventional, synthetic dosimeters were not in place at the time of radiation exposure. Two major categories exist, namely (i) luminescence dating, in which one wishes to establish the dose delivered to natural materials during the exposure of materials to natural radiation over the lifetime of the sample in a certain environment and (ii) accident dosimetry, in which one uses OSL from natural or locally available materials to reconstruct the dose imparted to an area during a radiation accident. CW-OSL is the technique most often used in these applications.

The use of luminescence as a tool for retrospective dosimetry, archaeological and geological dating, detection of food irradiation, etc. has always relied on quartz inclusions as the main dosimetric material. Nevertheless, there are some other minerals, mainly potassium-rich feldspars, whose luminescence emissions continue to grow at far higher doses than quartz. These potassium-rich minerals are also more-efficient phosphors. They are used for dosimetric purposes as they are highly sensitive and reproducible and possess low fading and good dose linearity in the ranges of interest. Moreover, these materials possess the added value of their ubiquitous nature which makes them undoubtedly useful for radiation dose assessment in uncontrolled dosimetric areas. Farrington Daniels, who was the first to develop the use of TL for dosimetry, also suggested its utilization for geological and archaeological age determination. An essential feature of TL necessary for its utilization for dating is the stability portion of the glow curve being used. The main source of TL in the active thermoluminescent minerals used for dating archaeological samples is thorium, uranium and potassium. One can use the TLD phosphors such as calcium fluoride, calcium sulphate or aluminium oxide which are so sensitive that exposure to such samples for a few weeks induces a measurable level of TL. By comparing the obtained TL with that of the standard samples containing a proper amount of thorium, uranium and potassium, the amount of dose given can be estimated.

5.5   CONCLUSION

This chapter discusses borate-based TL and OSL phosphors for their possible applications in the field of radiation dosimetry. It highlights the short review of some of the well-known TL and OSL phosphors used for personnel monitoring of radiation doses and many other applications. It can be observed that a good number of highly sensitive and near-tissue-equivalent TL phosphor materials have been developed, characterized and recommended for their use in various fields such as radiation dosimetry by researchers. However, these developed phosphors could be useful for their practical application in radiation dosimetry provided they fulfil the requirements of regulatory bodies for "passive integrating dosimetry systems for personal and environmental monitoring". As per the current scenario, $CaSO_4$:Dy and LiF:Mg, Cu, P (in TL mode) and Al_2O_3:C (in OSL mode) are the leading materials widely used for the luminescence dosimetry programmes. Considering the current tremendous growth in the development of TL and OSL materials, it is hoped that new and more sensitive luminescent materials will be available in the coming years for use in the field of radiation dosimetry (Figure 5.3).

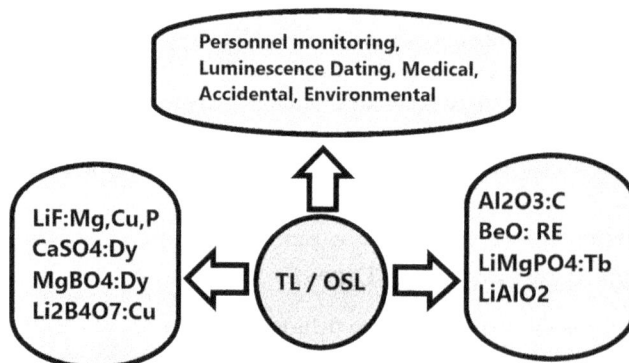

FIGURE 5.3   Some of the well-known TL/OSL materials and applications of TL and OSL dosimetry.

REFERENCES

1. Budzanowski, M., P. Bilski, P. Olko, T. Niewiadomski, B. Burgkhardt, and E. Piesch. New TL detectors for personal neutron dosimetry. *Radiation Protection Dosimetry* 47, no. 1–4 (1993): 419–423.
2. Lakshmanan, A. R. A review on the rŏle of thermoluminescent dosimeters in fast-neutron personnel dosimetry. *Nuclear Tracks and Radiation Measurements* 6, no. 2–3 (1982): 59–78.
3. Schulman, J. H., R.D. Kirk, and E.J. West. USAEC symposium series 650637, luminescence dosimetry (1967): 113.
4. Mutsuo, T., Y. Osamu, and Y. Tadaoki. *Nuclear Instruments and Methods* 175 (1980): 77.
5. Özdemir, Z., G. Zbayoğlu, and A. Yilmaz. Investigation of thermoluminescence properties of metal xide doped lithium triborate. *Journal of Materials Science* 42, no. 20 (2007): 8501–8508.
6. Santiago, M., C. Grasseli, E. Caselli, M. Lester, A. Lavat, and F. Spano. *Physica Status Solidi.*
7. Jiang, L. H., Y. L. Zhang, C. Y. Li, J. Q. Hao, and Q. Su. Thermoluminescence studies of LiSrBO3: RE3+ (RE= Dy, Tb, Tm and Ce). *Applied Radiation and Isotopes* 68, no. 1 (2010): 196–200.
8. Anishia, S. R., M. T. Jose, O. Annalakshmi, V. Ponnusamy, and V. Ramasamy. Dosimetric properties of rare earth doped LiCaBO3 thermoluminescence phosphors. *Journal of Luminescence* 130, no. 10 (2010): 1834–1840.
9. Lihong, J., Z. Yanli, L. Chengyu, P. Ran, S. Lili, Z. Su, H. Jingquan, and S. Qiang. Thermoluminescence characteristics of NaSr4 (BO3) 3: Ce3+ under β-ray irradiation. *Journal of Rare Earths* 27, no. 2 (2009): 320–322.
10. Annalakshmi, O. Dosimetric character thermolumines. Phd diss., Homi Bhabha National Institute Mumbai (2013).
11. Liu, L., Y. Zhang, J. Hao, C. Li, S. Wang, and Q. Su. Thermoluminescence studies of LiBa2B5O10: RE3+ (RE= Dy, Tb and Tm). *Journal of Physics and Chemistry of Solids* 68, no. 9 (2007): 1745–1748.
12. Wu, L., X. L. Chen, Y. Zhang, Y. F. Kong, J. J. Xu, and Y. P. Xu. Ab initio structure determination of novel borate NaSrBO3. *Journal of Solid State Chemistry* 179, no. 4 (2006): 1219–1224.
13. Prokić, M. J. N. I. Development of highly sensitive CaSO4: Dy/Tm and MgB4O7: Dy/Tm sintered thermoluminescent dosimeters. *Nuclear Instruments and Methods* 175, no. 1 (1980): 83–86.
14. NecmeddinYazici, A., M. Dogan, and V. Emifjadafar. HuseyinToktamis. *Nuclear Instruments and Methods in Physics Research* 246 (2006): 402.
15. Liu, L., J. Hao, C. Li, Q. Tang, C. Zhang, Q. Su, and S. Wang. Thermoluminescence characteristics of SrB6O10: Tb. *Journal of Rare Earths* 24, no. 3 (2006): 276–280.
16. Yamashita, T., N. Nada, H. Onishi, and S. Kitamura. Calcium sulfate activated by thulium or dysprosium for thermoluminescence dosimetry. *Health Physics* 21, no. 2 (1971): 295–300.
17. Dixon, R. L., and K. E. Ekstrand. The systematics of TL in rare earth activated sulfate lattices. In *Proc. 4th Int. Conf. Luminescence Dosimetry, Institute of Nuclear Physics,* vol. 481, 1974, Krakow.
18. Tochilin, E., N. Goldstein, and W. G. Miller. Beryllium oxide as a thermoluminescent dosimeter. *Health Physics* 16, no. 1 (1969): 1–7.
19. Altunal, V., Z. Yegingil, T. Tuken, T. Depci, A. Ozdemir, V. Guckan, N. Nur, K. Kurt, and E. Bulur. Optically stimulated luminescence characteristics of BeO nanoparticles synthesized by sol-gel method. *Radiation Measurements* 118 (2018): 54–66.
20. Mehta, S. K., and S. Sengupta. Al2O3 phosphor for thermoluminescence dosimetry. *Health Physics* 31, no. 2 (1976): 176–177.
21. Akselrod, M. S., V. S. Kortov, D. J. Kravetsky, and V. I. Gotlib. Highly sensitive thermoluminescent anion-defect alpha-Al2O3: C single crystal detectors. *Radiation Protection Dosimetry* 33, no. 1–4 (1990): 119–122.
22. Becker, K., and A. Scharmann. Einfiihrung in die Festkorperdosimetrie. *Verlag Karl Thiemig* 56 (1975).
23. Murthy, K. V. R. Thermoluminescence and its applications: A review. In *Defect and Diffusion Forum,* vol. 347, pp. 35–73. Trans Tech Publications Ltd, 2014.
24. Sahare, P. D., and S. K. Srivastava. Lyoluminescence dosimetry of high-energy γ radiation using MgB4O7: Mn2+. *Journal of Radioanalytical and Nuclear Chemistry* 307, no. 1 (2016): 31–36.
25. Wick, G. C., and B. Zumino. *Physical Review* 80 (1950): Z68.
26. Scharmann, A., M. Bohm, G. Born, R. Grasser, and A May. *Einfuhrung in die umineszenz.* VerlagKarl Thiemig, Miinchen, 1971.
27. Bhatt, B. C., and M. S. Kulkarni. Thermoluminescent phosphors for radiation dosimetry. In *Defect and Diffusion Forum,* vol. 347, pp. 179–227. Trans Tech Publications Ltd, 2014.
28. Bilski, P., M. Budzanowski, and P. Olko. Dependence of LiF: Mg, Cu, P (MCP-N) glow-curve structure on dopant composition and thermal treatment. *Radiation Protection Dosimetry* 69, no. 3 (1997): 187–198.

29. Chen, T-C., and T. G. Stoebe. Role of copper in LiF: Mg, Cu, P thermoluminescent phosphors. *Radiation Protection Dosimetry* 78, no. 2 (1998): 101–106.

30. Tang, K., H. Zhu, W. Shen, and B. Liu. A new high sensitivity thermoluminescent phosphor with low residual signal and good stability to heat treatment: LiF: Mg, Cu, Na, Si. *Radiation Protection Dosimetry* 100, no. 1–4 (2002): 239–242.

31. Binder, W., S. Dialerhoff, and J. R. Cameron, *Proc. Int. Conf. Lumin. Dosim.*, p. 43, Gatinberg, 1968.

32. Taikar, D. R., P. D. Belsare, and S. V. Moharil. Study of thermoluminescence in Cu doped CaO phosphor. In *AIP Conference Proceedings*, vol. 2104, no. 1, p. 030034. AIP Publishing LLC, 2019.

33. Yamashita, T., N. Nada, H. Onishi, and S. Kitamura, *Health Physics*. 21 (1971): 295.

34. Alig, R. C., and S. Bloom. Cathodoluminescent efficiency. *Journal of the Electrochemical Society* 124, no. 7 (1977): 1136.

35. Kandarakis, I., D. Cavouras, D. Nikolopoulos, A. Anastasiou, N. Dimitropoulos, N. Kalivas, E. Ventouras, I. Kalatzis, C. Nomicos, and G. Panayiotakis, *Radiation Measurements* 39 (2005): 263.

36. Yazici, A. N, M. Öztaş, and M. Bedir. The thermoluminescence properties of copper doped ZnS nanophosphor. *Optical Materials* 29, no. 8 (2007): 1091–1096.

37. Zarate-Morales, A., and A. E. Buenfil. Environmental gamma dose measurements in Mexico City using TLD. *Health Physics* 71, no. 3 (1996): 358–361.

38. Kalchgruber, R., H. Y. Göksu, E. Hochhäuser, and G. A. Wagner. *Radiation Measurements* 35 (2002) 585.

39. Ishii, M., Y. Kuwano, S. Asaba, T. Asai, M. Kawamura, N. Senguttuvan, T. Hayashi et al. Luminescence of doped lithium tetraborate single crystals and glass. *Radiation Measurements* 38, no. 4–6 (2004): 571–574.

40. Kitis, G., C. Furetta, M. Prokic, and V. Prokic. Kinetic parameters of some tissue equivalent thermoluminescence materials. *Journal of Physics D: Applied Physics* 33, no. 11 (2000): 1252.

41. Furetta, C., M. Prokic, R. Salamon, V. Prokic, and G. Kitis. Dosimetric characteristics of tissue equivalent thermoluminescent solid TL detectors based on lithium borate. *Nuclear Instruments and Methods in Physics Research Section A: Accelerators, Spectrometers, Detectors and Associated Equipment* 456, no. 3 (2001): 411–417.

42. Mahesh, K., P-s. Weng, and C. Furetta. *Thermoluminescence in Solids and Its Applications*. Nuclear Technology Pub., 1989.

43. Azorin, J., C. Furetta, and A. Scacco. Preparation and properties of thermoluminescent materials. *Physica Status Solidi (a)* 138, no. 1 (1993): 9–46.

44. Takenaga, M., O. Yamamoto, and T. Yamashita, *Nuclear Instruments and Methods in Physics Research Section B* 175 (1980): 77.

45. McKeever, S. W. S. Thermoluminescence in quartz and silica. *Radiation Protection Dosimetry* 8, no. 1–2 (1984): 81–98.

46. Aitken, M. J. *Thermoluminescence Dating*. London, Orlando, San Diego, New York, Austin, Montreal. (1985).

47. Yukihara, E. G., and S. W. S. McKeever. *Optically Stimulated Luminescence: Fundamentals and Applications*. John Wiley & Sons, 2011.

48. Yukihara, E. G., S. W. S. McKeever, and M. S. Akselrod. State of art: Optically stimulated luminescence dosimetry–Frontiers of future research. *Radiation Measurements* 71 (2014): 15–24.

49. Antonov-Romanovskii, V.V., I. F. Keirum-Marcus, M. S. Poroshina, and Z. A. Trapeznikova, Conference of academy of sciences of USSR (1995), USAEC report AEC-tr-2435, 239 (1996).

50. Braeunlich, P., D. Shafer, and A. Sharmann, *Proc. Int. Conf. Lumin. Dosim.*, 1965, USAEC, vol. 57 (1967)

51. Sanborn, E. N., and E. L. Beard. *Proc. Int. Conf. Lumin. Dosim.*, 1965, USAEC, vol. 183 (1967).

52. Takenaga, M., O. Yamamoto, and T. Yamashita. Preparation and characteristics of Li2B4O7: Cu phosphor. *Nuclear Instruments and Methods* 175, no. 1 (1980): 77–78.

53. Rao, R. P., M. DeMurcia, and J. Gasiot. Optically stimulated luminescence dosimetry. *Radiation Protection Dosimetry* 6, no. 1–4 (1983): 64–66.

54. Pekpak, E., A. Yilmaz, and G. Ozbayoglu. An overview on preparation and TL characterization of lithium borates for dosimetric use. *The Open Mineral Processing Journal* 3, no. 1 (2010).

55. Schulman, J. H., R. D. Kirk, and E. J. West. For thermoluminescence dosimetry. *Luminescence Dosimetry: Proceedings* 8 (1967): 113.

56. Xiong, Z., P. Ding, Q. Tang, J. Chen, and W. Shi. *Advanced Materials Research* 160 (2011): 252.

57. Sabharwal, S. C. Kinetics of thermally stimulated luminescence from alkaline earth borates. *Journal of Luminescence* 109, no. 2 (2004): 69–74.

58. Rao, G. Venkateswara, P. Yadagiri Reddy, and N. Veeraiah. Thermoluminescence studies on Li2O–CaF2–B2O3 glasses doped with manganese ions. *Materials Letters* 57, no. 2 (2002): 403–408.
59. Manam, J., and S. K. Sharma. Thermally stimulated luminescence studies of undoped and doped K2B4O7 compounds. *Nuclear Instruments and Methods in Physics Research Section B: Beam Interactions with Materials and Atoms* 217, no. 2 (2004): 314–320.
60. Takenaga, M., O. Yamamoto, and T. Yamashita. A new phosphor Li2B4O7: Cu for TLD. *Health Physics* 44, no. 4 (1983): 387–393.
61. Proki, M. Dosimetric characteristics of Li_2B4O_7: Cu, Ag, P solid TL detectors. *Radiation Protection Dosimetry* 100, no. 1–4 (2002): 265–268.
62. Prokic, M. Lithium borate solid TL detectors. *Radiation Measurements* 33, no. 4 (2001): 393–396.
63. Rodriguez Chialanza, M., J. Castiglioni, and L. Fornaro. Crystallization as a way for inducing thermoluminescence in a lead borate glass. *Journal of Materials Science* 47, no. 5 (2012): 2339–2344.
64. Kipke, A., and H. Hofmeister. Formation of silver nanoparticles in low-alkali borosilicate glass via silver oxide intermediates. *Materials Chemistry and Physics* 111, no. 2–3 (2008): 254–259.
65. Ignatovych, M., M. Fasoli, and A. Kelemen. Thermoluminescence study of Cu, Ag and Mn doped lithium tetraborate single crystals and glasses. *Radiation Physics and Chemistry* 81, no. 9 (2012): 1528–1532.
66. Swamy, B. J. R., B. Sanyal, Y. Gandhi, R. M. Kadam, V. Nata Rajan, P. Raghava Rao, et al. Thermoluminescence study of MnO doped borophosphate glass samples for radiation dosimetry. *Journal of Non-Crystalline Solids* 368 (2013): 40–44.
67. Szumera, M., and I. Wacławska. Thermal study of Mn-containing silicate–phosphate glasses. *Journal of Thermal Analysis and Calorimetry* 108, no. 2 (2012): 583–588.
68. Schulman, J. H., R. D. Kirk, and E. J. West. Use of lithium borate for thermoluminescence dosimetry. In *Proceedings of the International Conference on Luminescence Dosimetry*, 1965.
69. Ivascu, C., I. B. Cozar, L. Daraban, and G. Damian. Spectroscopic investigation of P2O5–CdO–Li2O glass system. *Journal of Non-crystalline Solids* 359 (2013): 60–64.
70. Kutub, A. A., M. S. Elmanhawaawy, and M. O. Babateen. Studies on gamma-irradiated sodium tetraborate glasses containing ytterbium. *Solid State Science and Technology* 15 (2007): 191–202.
71. Mady, F., M. Benabdesselam, and W. Blanc. Thermoluminescence characterization of traps involved in the photo darkening of ytterbium-doped silica fibers. *Optical Letters* 3521 (2010): 3541–3543.
72. Rojas, S. S., K. Yukimitu, and A. C. Hernandes. Dosimetric properties of UV irradiated calcium co-doped borate glass–ceramic. *Nuclear Instruments and Methods in Physics Research Section B: Beam Interactions with Materials and Atoms* 266, no. 4 (2008): 653–657.
73. Takam, R., E. Bezak, G. Liu, and L. Marcu. The use of enriched 6 Li and 7 Li LiF, Mg, Cu, P glass-rod thermoluminescent dosimeters for linear accelerator for out of-field radiation dose measurements. *Radiation Protection Dosimetry* 150(1) (2012): 22–33.
74. Rzyski, B. M., and S. P. Morato. Luminescence studies of rare earth doped lithium tetraborate. *Nuclear Instruments and Methods* 175, no. 1 (1980): 62–64.
75. Huy, B. T., V. X. Quang, and H. T. B. Chau. Effect of doping on the luminescence properties of Li2B4O7. *Journal of Luminescence* 128, no. 10 (2008): 1601–1605.
76. Prokic, M. Dosimetric properties of $Li_2B_4O_7$:Cu, Ag, P solid TL detectors. *Radiation Measurements* 33 (2001): 393–396.
77. Mirjana, P. Effect of lithium co-dopant on the thermoluminescence response of some phosphors. *Applied Radiation and Isotopes* 52 (2000): 97–103
78. Annalakshmi, O., M. T. Jose, and G. Amarendra. Dosimetric characteristics of manganese doped lithium tetraborate–An improved TL phosphor. *Radiation Measurements* 46, no. 8 (2011): 669–675.
79. Annalakshmi, O., M. T. Jose, U. Madhusoodanan, B. Venkatraman, and G. Amarendra. Kinetic parameters of lithium tetraborate based TL materials. *Journal of Luminescence* 141 (2013): 60–66.
80. Lochab, S. P., A. Pandey, P. D. Sahare, R. S. Chauhan, N. Salah, and R. Ranjan. Nanocrystalline MgB4O7: Dy for high dose measurement of gamma radiation. *Physica Status Solidi (a)* 204, no. 7 (2007): 2416–2425.
81. Abtahi, A., T. Haugan, and P. Kelly. Investigation of the dosimetric properties of MgB_4O_7: Dy under laser heating. *Radiation Protection Dosimetry* 21, no. 4 (1987): 211–217.
82. Kawashima, Y. S., C. F. Gugliotti, M. Yee, S. H. Tatumi, and J. C. R. Mittani. Thermoluminescence features of MgB4O7: Tb phosphor. *Radiation Physics and Chemistry* 95 (2014): 91–93.
83. Annalakshmi, O., M. T. Jose, U. Madhusoodanan, B. Venkatraman, and G. Amarendra. Synthesis and thermoluminescence characterization of MgB4O7: Gd, Li. *Radiation Measurements* 59 (2013): 15–22.

84. McGovern, P. E. Current scientific techniques in archaeology. *American Journal of Archaeology* 92, no. 2 (1988): 285–286.

85. Debenham, N. C., and M. J. Aitken. Thermoluminescence dating of stalagmitic calcite. *Archaeometry* 26, no. 2 (1984): 155–170.

86. Hashizume, T., Y. Kato, T. Nakajima, T. Toryu, H. Sakamoto, N. Kotera, and S. Eguchi. A new thermoluminescence dosimeter of high sensitivity using a magnesium silicate phosphor. IAEA-SM-143/11, Vienna (1971): 91–98.

87. Lakshmanan, A. R., and K. G. Vohra. Gamma radiation induced sensitization and photo-transfer in Mg2SiO4: Tb TLD phosphor. *Nuclear Instruments and Methods* 159, no. 2–3 (1979): 585–592.

88. Lakshmanan, A. R., S. S. Shinde, and R. C. Bhatt. Ultraviolet-induced thermoluminescence and phosphorescence in Mg2SiO4: Tb. *Physics in Medicine & Biology* 23, no. 5 (1978): 952–960

89. Lakshmanan, A. R., and R. C. Bhatt. Photon energy dependence of sensitized TLD phosphors. *Nuclear Instruments and Methods* 171, no. 2 (1980): 259–263.

90. Prokić, M., and E. G. Yukihara. Dosimetric characteristics of high sensitive Mg2SiO4: Tb solid TL detector. *Radiation Measurements* 43, no. 2–6 (2008): 463–466.

91. Chagas, M. A. P., M. G. Nunes, L. L. Campos, and D. N. Souza. TL properties of anhydrous CaSO4: Tm improvement. *Radiation Measurements* 45, no. 3–6 (2010): 550–552.

92. Piesch, E., B. Burgkhardt, and D. Singh, *Proc. 5th Int. Conf. on Luminescence Dosimetry*, Sao Paulo, 1977.

93. Srivastava, J. K., B. C. Bhatt, and S. J. Supe. Thermoluminescence characteristics of CaSO4 doped with Dy and Cu. *Radiation Protection Dosimetry* 40, no. 4 (1992): 271–274.

94. Guckan, V., V. Altunal, N. Nur, T. Depci, A. Ozdemir, K. Kurt, Y. Yu, I. Yegingil, and Z. Yegingil. Studying CaSO4: Eu as an OSL phosphor. *Nuclear Instruments and Methods in Physics Research Section B: Beam Interactions with Materials and Atoms* 407 (2017): 145–154.

95. Junot, D. O., M. A. Couto dos Santos, P. L. Antonio, L. V. E. Caldas, and D. N. Souza. Feasibility study of CaSO4: Eu, CaSO4: Eu, Ag and CaSO4: Eu, Ag (NP) as thermoluminescent dosimeters. *Radiation Measurements* 71 (2014): 99–103.

96. Bahl, S., S. P. Lochab, and P. Kumar. CaSO4: Dy, Mn: A new and highly sensitive thermoluminescence phosphor for versatile dosimetry. *Radiation Physics and Chemistry* 119 (2016): 136–141.

97. Nunes, M. G., and L. L. C. Rodrigues. Dose profile of electron beams obtained with CaSO4: Ce, Eu thermoluminescent dosimeters. *Brazilian Journal of Medical Physics* 6, no. 3 (2012): 156–162.

98. Campos, L. L., and M. F. Lima. Dosimetric properties of $CaSO_4$: Dy teflon pellets produced at IPEN. *Radiation Protection Dosimetry* 14, no. 4 (1986): 333–335.

99. McKeever, S. W. S. Thermoluminescence in quartz and silica. *Radiation Protection Dosimetry* 8, no. 1–2 (1984): 81–98.

100. Huntley, D. J., D. I. Godfrey-Smith, and M. L. W. Thewalt. Optical dating of sediments. *Nature* 313, no. 5998 (1985): 105–107.

101. Justus, B. L., S. Rychnovsky, M. A. Miller, K. J. Pawlovich, and A. L. Huston. Optically stimulated luminescence radiation dosimetry using doped silica glass. *Radiation Protection Dosimetry* 74, no. 3 (1997): 151–154.

102. Bøtter-Jensen, L., S. W. S. McKeever, and A. G. Wintle. *Optically Stimulated Luminescence Dosimetry*. Elsevier, 2003.

103. Benevides, L. A., A. L. Huston, B. L. Justus, P. Falkenstein, L. F. Brateman, and D. E. Hintenlang. Characterization of a fiber-optic-coupled radioluminescent detector for application in the mammography energy range. *Medical Physics* 34, no. 6 Part 1 (2007): 2220–2227.

104. Polf, C., S. W. S. McKeever, M. S. Akselrod, and S. Holmstrom. A real-time, fibre optic dosimetry system using Al2O3 fibres. *Radiation Protection Dosimetry* 100, no. 1–4 (2002): 301–304.

105. Ranchoux, G., S. Magne, J. P. Bouvet, and P. Ferdinand. Fibre remote optoelectronic gamma dosimetry based on optically stimulated luminescence of Al2O3: C. *Radiation Protection Dosimetry* 100, no. 1–4 (2002): 255–260.

106. Justus, B. L., C. D. Merritt, K. J. Pawlovich, A. L. Huston, and S. Rychnovsky. Optically stimulated luminescence dosimetry using doped fused quartz. *Radiation Protection Dosimetry* 84, no. 1–4 (1999): 189–192.

107. Huston, A. L., and B. L. Justus. US Patent 5,585,640 (1996).

108. Huston, A.L., and B. L. Justus. US Patent 5,606,163 (1997).

109. Huston, A. L., and B. L. Justus. US Patent 5,656,815 (1997).

110. Huston, A. L., and B. L. Justus. US Patent 5,811,822 (1998).

111. Huston, A. L., and B. L. Justus. US Patent 6087666 (2000).

112. Justus, B. L., K. J. Pawlovich, C. D. Merritt, and A. L. Huston. Optically and thermally stimulated luminescence characteristics of Cu1+-doped fused quartz. *Radiation Protection Dosimetry* 81, no. 1 (1999): 5–10.
113. Ogorodnikov, I. N., V. Yu Ivanov, and A. V. Kruzhalov. Short-wavelenght luminescence and thermostimulated processes in single crystals of BeO. *Radiation Measurements* 24, no. 4 (1995): 417–421.
114. Rhyner, C. R., and W. G. Miller. Radiation dosimetry by optically-stimulated luminescence of BeO. *Health Physics* 18, no. 6 (1970): 681–684. (b) E. Bulur and H. Y. Goeksu, *Radiation Measurements* 29 (1998): 639.
115. Bulur, E., and H. Y. Goeksu. OSL from BeO ceramics: New observations from an old material. *Radiation Measurements* 29, no. 6 (1998): 639–650.
116. Sommer, M., R. Freudenberg, and J. Henniger. New aspects of a BeO-based optically stimulated luminescence dosimeter. *Radiation Measurements* 42, no. 4–5 (2007): 617–620.
117. Jahn, A., M. Sommer, and J. Henniger. 2D-OSL-dosimetry with beryllium oxide. *Radiation Measurements* 45, no. 3–6 (2010): 674–676.
118. Akselrod, M. S., V. S. Kortov, D. J. Kravetsky, and V. I. Gotlib. Highly sensitive thermoluminescent anion-defective alpha-Al2O3: C single crystal detectors. *Radiation Protection Dosimetry* 32, no. 1 (1990): 15–20.
119. Scarpa, G. The dosimetric use of beryllium oxide as a thermoluminescent material: A preliminary study. *Physics in Medicine & Biology* 15, no. 4 (1970): 667.
120. Rhyner, C. R., and W. G. Miller. Radiation dosimetry by optically-stimulated luminescence of BeO. *Health Physics* 18, no. 6 (1970): 681–684.
121. Yamashita, T., Y. Yasuno, and M. Ikedo. Beryllium oxide doped with lithium or sodium for thermoluminescence dosimetry. *Health Physics* 27, no. 2 (1974): 201–206.
122. Ogorodnikov, I. N., M. D. Petrenko, and V. Yu Ivanov. Low-temperature luminescence and thermoluminescence from BeO: Zn single crystals. *Optical Materials* 62 (2016): 219–226
123. Ogorodnikov, I. N., M. D. Petrenko, and V. Yu Ivanov. Low-temperature luminescence and thermally stimulated luminescence of BeO: Mg single crystals. *Physics of the Solid State* 60, no. 1 (2018): 134–146
124. Altunal, V., V. Guckan, A. Ozdemir, N. Can, and Z. Yegingil. Luminescence characteristics of Al-and Ca-doped BeO obtained via a sol-gel method. *Journal of Physics and Chemistry of Solids* 131 (2019): 230–242
125. Altunal, V. O. L. K. A. N., V. Guckan, A. Ozdemir, K. Kurt, A. H. M. E. T. Ekicibil, and Z. Yegingil. Investigation of luminescence properties of BeO ceramics doped with metals for medical dosimetry. *Optical Materials* 108 (2020): 110436.
126. Altunal, V., V. Guckan, A. Ozdemir, A. Ekicibil, F. Karadag, I. Yegingil, Y. Zydhachevskyy, and Z. Yegingil. A systematic study on luminescence characterization of lanthanide-doped BeO ceramic dosimeters. *Journal of Alloys and Compounds* 876 (2021): 160105.
127. Altunal, V., V. Guckan, A. Ozdemir, Y. Zydhachevskyy, Y. Lawrence, Y. Yu, and Z. Yegingil. Three newly developed BeO-based OSL dosimeters. *Journal of Luminescence* 241 (2022): 118528.
128. Hazen, R. M., and L. W. Finger. High-pressure and high-temperature crystal chemistry of beryllium oxide. *Journal of Applied Physics* 59, no. 11 (1986): 3728–3733.
129. Watanabe, S., T. K. Gundu Rao, P. S. Page, and B. C. Bhatt. TL, OSL and ESR studies on beryllium oxide. *Journal of Luminescence* 130, no. 11 (2010): 2146–2152.
130. Markey, B. G., L. E. Colyott, and S. W. S. McKeever. Time-resolved optically stimulated luminescence from α-Al2O3: C. *Radiation Measurements* 24, no. 4 (1995): 457–463.
131. McKeever, S. W. S., M. W. Blair, E. Bulur, R. Gaza, R. Gaza, R. Kalchgruber, D. M. Klein, and E. G. Yukihara. Recent advances in dosimetry using the optically stimulated luminescence of Al2O3: C. *Radiation Protection Dosimetry* 109, no. 4 (2004): 269–276.
132. De Azevedo, W. M., G. B. De Oliveira, E. F. da Silva Jr, H. J. Khoury, and E. F. Oliveira de Jesus. Highly sensitive thermoluminescent carbon doped nanoporous aluminium oxide detectors. *Radiation Protection Dosimetry* 119, no. 1–4 (2006): 201–205.
133. Akselrod, M. S., A. C. Lucas, J. C. Polf, and S. W. S. McKeever. Optically stimulated luminescence of Al2O3. *Radiation Measurements* 29, no. 3–4 (1998): 391–399.
134. Akselrod, M. S., and S. W. S. McKeever. A radiation dosimetry method using pulsed optically stimulated luminescence. *Radiation Protection Dosimetry* 81, no. 3 (1999): 167–175.
135. Colyott, L. E., M. S. Akselrod, and S. W. S. McKeever. Phototransferred thermoluminescence in alpha-Al2O3: C. *Radiation Protection Dosimetry* 65, no. 1–4 (1996): 263–266.
136. Markey, B. G., S. W. S. McKeever, M. S. Akselrod, L. Bøtter-Jensen, N. Angersnap Larsen, and L. E. Colyott. The temperature dependence of optically stimulated luminescence from alpha-Al2O3: C. *Radiation Protection Dosimetry* 65, no. 1–4 (1996): 185–189.

137. Andersen, C. E., L. Bøtter-Jensen, and S. Mattsson. Medical applications of luminescence dosimetry using fibre-coupled Al_2O_3: C crystals. *Radiological Protection in Transition* (2005): 143.

138. Akselrod, M. S., A. E. Akselrod, S. S. Orlov, S. Sanyal, and T. H. Underwood. Fluorescent aluminum oxide crystals for volumetric optical data storage and imaging applications. *Journal of Fluorescence* 13, no. 6 (2003): 503–511.

139. Sykora, G. J., and M. S. Akselrod. Photoluminescence study of photochromically and radiochromically transformed Al2O3: C, Mg crystals used for fluorescent nuclear track detectors. *Radiation Measurements* 45, no. 3–6 (2010): 631–634.

140. Mittani, J. C., M. Prokić, and E. G. Yukihara. Optically stimulated luminescence and thermoluminescence of terbium-activated silicates and aluminates. *Radiation Measurements* 43, no. 2–6 (2008): 323–326.

141. Dhabekar, B., E. A. Raja, S. Menon, T. K. Gundu Rao, R. K. Kher, and B. C. Bhatt. ESR, PL and TL studies of LiAlO2: Mn/Ce phosphor. *Radiation Measurements* 43, no. 2–6 (2008): 291–294.

142. Teng, H., S. Zhou, H. Lin, T. Jia, X. Hou, and J. Wang. Growth and characterization of high-quality Mn-doped LiAlO 2 single crystal. *Chinese Optics Letters* 8, no. 4 (2010): 414–417.

143. Lee, J. I., A. S. Pradhan, J. L. Kim, I. Chang, B. H. Kim, and K. S. Chung. Preliminary study on development and characterization of high sensitivity LiAlO2 optically stimulated luminescence material. *Radiation Measurements* 47, no. 9 (2012): 837–840.

144. Kananen, B. E., E. S. Maniego, E. M. Golden, N. C. Giles, J. W. McClory, V. T. Adamiv, Y. V. Burak, and L. E. Halliburton. Optically stimulated luminescence (OSL) from Ag-doped $Li_2B_4O_7$ crystals. *Journal of Luminescence* 177 (2016): 190–196.

145. Chou, M. M. C., H. C. Huang, D-S. Gan, and C. W. C. Hsu. Defect characterizations of γ-LiAlO2 single crystals. *Journal of Crystal Growth* 291, no. 2 (2006): 485–490.

146. Kwon, S. W., and S. B. Park. Effect of precursors on the preparation of lithium aluminate. *Journal of Nuclear Materials* 246, no. 2–3 (1997): 131–138.

147. Chou, M. M. C., S. J. Huang, and C. W. C. Hsu. Crystal growth and polishing method of lithium aluminum oxide crystal. *Journal of Crystal Growth* 303, no. 2 (2007): 585–589.

6 Metal Oxides
Solid-State Lighting

P. P. Lohe,
Jhulelal Institute of Technology

S. J. Tamboli
Faculty of Natural and Agricultural Sciences

CONTENTS

6.1 INTRODUCTION

6.1.1 Brief History of Lighting

Light occupies central position in the evolution of mankind. Since the beginning of civilization, man has always been fascinated by Light. This is evident from quotations in scripture. "तमसोमाज् योतरि गमय" in Sanskrit or "fiat lux" in the book of Genesis that are most celebrated examples. Lighting can be considered as an index of civilization. Earth by night, as viewed from the satellites, shows illuminated spots and dark areas (Figure 6.1). Intensely illuminated areas correspond to developed countries while dark areas to underdeveloped countries. A similar figure for India enables us to identify metropolis, urban and rural areas easily in terms of bright spots and dark regions (Figure 6.2).

DOI: 10.1201/9781003366232-6

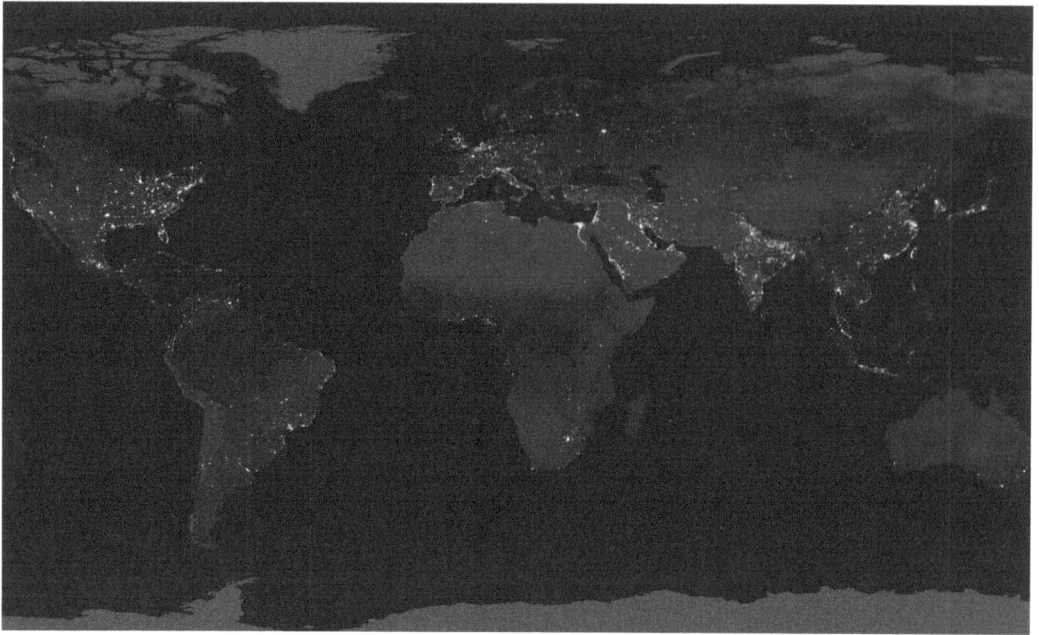

FIGURE 6.1 Earth by night, as viewed from the satellite.

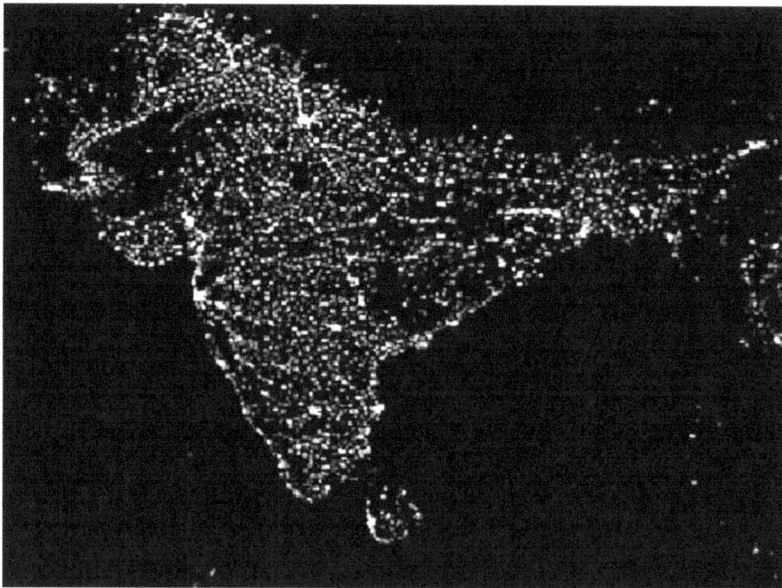

FIGURE 6.2 India by night, as viewed from the satellite.

Sun is the natural source of light. We get natural light from sunrise to sunset. Outside these timings, we must find out artificial source of light. Ancient artificial lighting was achieved by burning fossil fuels. In the primitive form, this source does not have constant intensity and needs continuous attention to keep it glowing. More durable sources were the torches made from slow-burning materials. Somewhere down the line, oils were used to provide light sources with more constant intensity. Oil vapours constitute a still better source of light. Higher intensities could be achieved by using

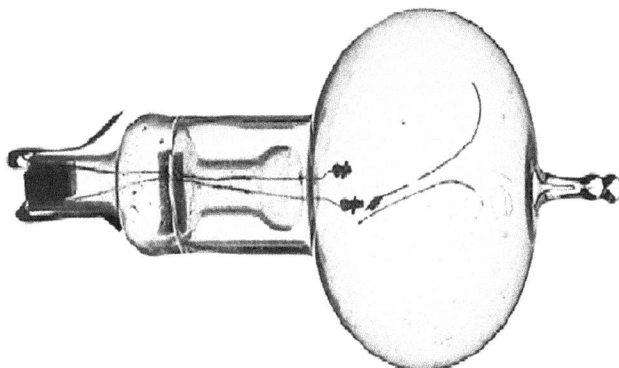

FIGURE 6.3 Original carbon-filament bulb from Thomas Edison.

vapours at elevated pressures with the help of gas stoves. In all these sources, the direct use of fossil fuels leads to fire hazard. After suffering a lot from ensuing fires, a safety lamp was invented by using electricity for producing light.

In 1802, pioneer experiment on producing light using electricity was conducted by Humphry Davy [1]. He termed this invention as the "Electric Arc lamp". This lamp was not robust, and the light it produced was dazzling. A first practicable electric lamp was patented by Thomas Edison on October 14, 1878, under the title "Improvement In Electric Lights". This was followed by an improved device which used "a carbon filament or strip coiled and connected to platina contact wires". A photograph of this antique lamp is shown in Figure 6.3. Edison soon established a company for manufacturing and marketing electric bulbs [2].

The electric bulb is based on black body radiation. The filament is heated to very high temperature by passing current. The black body radiation is distributed over a very wide spectral range from ultraviolet to infrared. Radiation in the visible window of 400–700 nm only is useful for lighting. Thus, this is not the most efficient way of converting electricity into light.

A more efficient way of converting the electricity into light is gas discharge. The most efficient and successful lamp based on the phenomenon of gas discharge uses mercury vapour. In 1906, Kuch and Retschinky [3] reported the production of light by mercury (Hg) discharge. Efficient emission was obtained much latter when a compact lamp could be developed using tungsten wires sealed in silica [4]. In a mercury vapour discharge, accelerated electrons collide with mercury atoms and take them to an excited state. Radiative de-excitation to ground state produces light. However, more than 90% light is in the ultraviolet region (253.7 nm). This has to be converted to the visible region. This is accomplished by using luminescence phenomenon. Lighting based on mercury discharge is thus popularly called as fluorescent lighting.

During World War 2, it was necessary to run the factories round the clock. Efficient lighting was needed. "Fluorescent lighting" was the solution. Fluorescence lighting technology developed in Great Britain was freely given to India (Figure 6.4). In the fluorescent lamp, $Ca_{4.7}Sb_{0.03}Mn_{0.15}(PO_4)_3$ $(F, Cl)_{0.95}$ is used as the phosphor for converting ultraviolet to visible light. Sb gives blue emission and Mn Yellow. The combination gives the illusion of white light.

Colour rendition of "tube light" is poor. Red colour is not properly seen as the lamp is deficient in this colour. Adding red phosphor decreases the effective lamp output. Tri-colour lamp was proposed by Koedam and Opstelten in 1971 [5] and developed by Thornton in 1972 [6]. It is based on "three-colour human vision". Any colour can be generated by mixing the primary red, green and blue colours in proper proportions. Thus, a white lamp can be fabricated using the blend of three phosphors with emissions in form of narrow bands around 450 (blue), 540 (green) and 610 nm (red). New phosphors replacing halophosphate were developed. These are BAM, $BaMgAl_{10}O_{17}:Eu^{2+}$ (450 nm), CAT, $Ce_{0.67}Tb_{0.33}MgAl_{11}O_{19}$ (543 nm) and $Y_2O_3:Eu^{3+}$ (611 nm). The use of efficient phosphors based

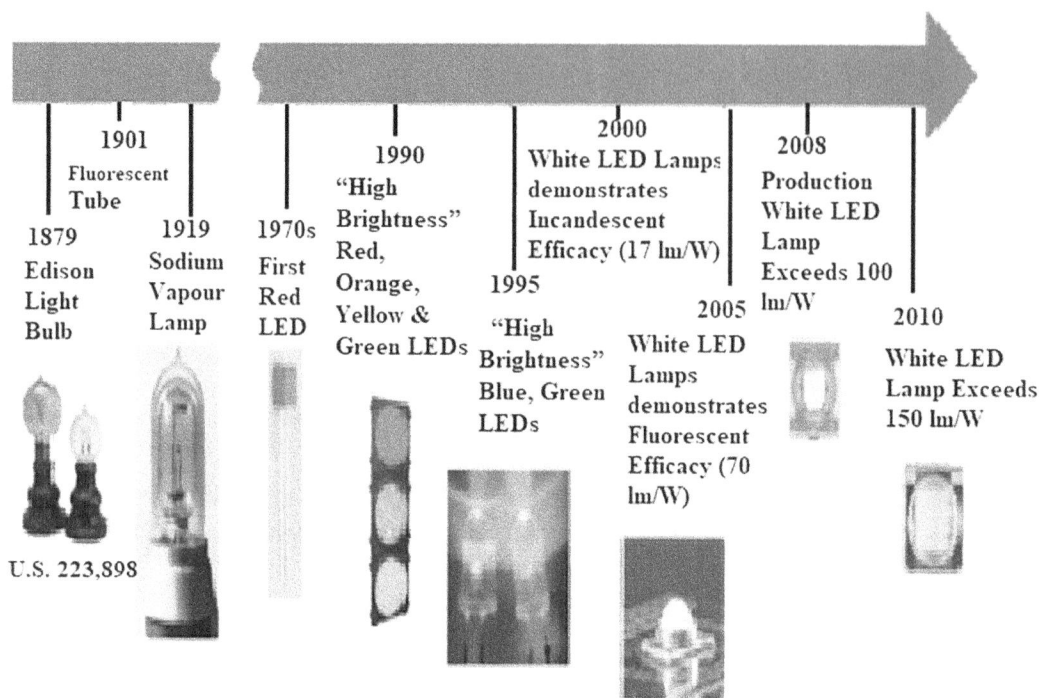

FIGURE 6.4 Evolution of lighting.

on rare earth activators enabled handling of much higher wall load. Size of tri-colour lamps could be reduced a great deal as compared to that of tube light. These lamps were thus named as compact fluorescent lamps (CFL).

Compact fluorescent lamps are efficient and aesthetic, and also have long life. However, they use mercury. Mercury can pose serious threat to environment. Next revolution in the lighting technology began in 1995 when efficient blue LEDs based on InGaN nitride semiconductors were invented. When these LEDs are coated with yellow emitting phosphors, a white LED lamp is produced. White LEDs, based on blue LED chips coated with a yellow emitting phosphor (YAG:Ce), were first reported in 1997. With the introduction of the blue InGaN technology, OSRAM OS (former Siemens Opto Semiconductors) introduced the first white LED on the market in 1998. Between 1996 and 2000, in US alone, about 100 companies started the manufacture of white LED lamps. Blue emission is provided by InGaN LED. YAG:Ce^{3+} phosphor coated on the LED absorbs part of blue light and converts it into yellow light. A combination of blue and yellow creates illusion of white. As LEDs are solid-state devices, the new technology is called as solid-state lighting. Initial LED lamps used only a single yellow-emitting phosphor. It does not have good colour rendition. However, it is compact, more efficient and more robust compared to CFL. The next generation of LED lamps is based on two phosphors, red and green. In combination with blue chip, these produce full colour gamut.

6.1.2 BLUE LED INVENTION

Progress in GaN synthesis and p-type doping led to the much awaited blue LED invention. This invention brought revolution in lighting as well as several other applications. "The first high-brightness blue LED was demonstrated by Shuji Nakamura of Nichia Corporation in 1994 and was based on InGaN [7]". This could be realized due to success in depositing GaN on sapphire substrate as well as p-type doping, largely through efforts of "Isamu Akasaki and H. Amano in Nagoya [8]".

TABLE 6.1

Emission Wavelengths of LED Semiconductor Materials

Semiconductor Material	Wavelength (nm)	Emission Colour
GaAs	840	Infrared
AlGaAs	780, 880	Red, infrared
GaP	55, 565, 700	Green, yellow, red
GaAsP	590–620	Yellow, orange, red
AlGaInP	590–620	Amber, yellow, orange
InGaN	390, 420, 515	Ultraviolet, blue, green
AlGaInN	450, 570	Blue, green
AlGaN	220–360	Ultraviolet
AlN	210	Ultraviolet

Further improvements were achieved by using transparent contact made from indium tin oxide (ITO) [9]. Alloying GaN with InN and AlN widened the range of wavelength over which LEDs could be designed. These LEDs could be used as viable light source for illumination applications, such as automotive headlights, interior/exterior lighting and full colour displays (Table 6.1).

6.1.3   LED Lamps

Though blue LED has several applications, lighting has received most wide attention. This is evident from the Nobel citation. On 7 October 2014, Professor Staffan Normark, Permanent Secretary of the Royal Swedish Academy of Sciences announced on behalf of the Nobel committee its decision to award Nobel Prize of 2014 to Shuji Nakamura, Isamu Akasaki and Hiroshi Amano "for the invention of efficient blue light-emitting diodes which has enabled bright and energy-saving light sources".

White light can be generated by combining blue LED with red and green. A more popular method which is developed following the invention of blue LED is partially convert the blue light into yellow using a phosphor like YAG:Ce. A combination of blue and yellow creates illusion of white. Since 1995, rapid developments have taken place in the field of white light sources using blue LED chips. LED lamps are now competing with traditional sources like tube lights, CFL, sodium vapour pressure lamps and mercury lamps.

6.2   PHOSPHORS FOR LED LAMPS

LED emission is in the form of a narrow band. However, for lighting purpose, white emission is required. This can be achieved by using suitable phosphors. For example, if LED emission is in the n-UV region, then a blend of three phosphors which can absorb this light and emit in the primary colours can be used to obtain white LED lamps. On the other hand, most efficient LEDs are available in the blue region. Bright energy-efficient white LED lamps can be obtained in two ways, one is two phosphor-converted LED in which a blend of green and red emitting phosphors excitable by blue light is used. A simpler way is to use a single yellow-emitting phosphor which can absorb part of LED emission and convert to yellow light. A combination of yellow and blue produces illusion of white.

6.2.1   One Phosphor-Converted LED Lamps

Immediately after discovery of blue LED, its application for solid-state lighting was pursued with great vigour. Partial conversion of blue emission to yellow was the most convenient and low-cost method. These efforts revolutionized the domestic lighting. Even today, this approach dominates the LED lamp market. Various phosphors used for this purpose are reviewed in the following.

6.2.1.1 YAG:Ce

The YAG:Ce phosphor developed in 1967 by Blasse and Brill has been practically used as cathode ray tube phosphors (P 46 and P 48) [10]. Blasse and Brill [11] also showed that this phosphor emitted in the yellow region when excited by blue light. Emission colour could be tuned by substituting gadolinium at yttrium site for the red shift, and galium at aluminium site for blue shift. Modifications brought in the PL characteristics by substitutions continue to interest researcher to date [12–14]. The use of the long wavelength excitation band centred on 460 nm was reported by Wyner and Daigneault [15] and Van Kemenade et al. [16]. They described the application of YAG:Ce in low-pressure mercury vapour discharge lamps to absorb the Hg-plasma lines in the blue/violet part of the spectrum, viz. 405 and 436 nm. The conversion of the violet and blue emission lines into yellow light adds to the white light emitted by the halophosphate phosphor (the commonly used phosphor in those days) to create a warmer white light. YAG:Ce is also a commercial phosphor (Sylvania 251) used for improving CRI in HPMV lamps. In the 1990s, YAG:Ce and the higher density analogue LuAG:Ce were proposed as fast and efficient scintillator materials, and interest in this topic is continuing to date [17,18]. YAG:Ce scintillators are used in a variety of applications that include a calorimeter for a proton computed radiography [19]. A recent application of Ce^{3+}-doped YAG luminescence is the use under single-crystalline disks to monitor and control the dose of vacuum/ultraviolet (VUV) radiation in wafer steppers. The dose applied for photoetching is measured by converting a small part of the VUV radiation from an excimer laser pulse into visible light [20]. Active-tip based, near-field optics are realized by coating a standard tip with Ce^{3+}-doped YAG particles [21]. YAG: Ce^{3+} nanoparticles conjugated with biomolecules work as biomarkers for fluorescent bioimaging [22,23]. Recently, Ce^{3+}–Yb^{3+}-co-doped YAG has been suggested as a downconversion material for converting the visible and near UV solar radiation, with quantum efficiencies exceeding 100%, to IR region where c-Si solar cells are most efficient [24,25]. Ce^{3+}-Nd^{3+} energy transfer in YAG host is used to obtain efficient IR-emitting phosphors [26]. Ce^{3+}-doped YAG is the most widely applied phosphor in white light LEDs [27–29]. Part of the blue light from the (In, Ga)N LED is absorbed by a thin layer of Ce^{3+}-doped YAG and then it is converted into yellow light. The combination of blue and yellow gives a bright white light source with an overall energy efficiency that is exceeding that of the compact fluorescent lamp.

Research on YAG:Ce has been conducted on single crystals, ceramics, single crystal films (SCF), bulk and nano-powders. Due to high melting point, it is difficult to grow transparent single crystals needed for scintillation studies. Moreover, the single crystals contain lattice defects that lead to UV emission with longer decay times [30]. Transparent ceramics obtained by sintering powders are more suited. Alpha and beta particles have small range and thus SCF are useful in their detection. At present, the fields of applications of SCF as luminescent detectors of ionising radiation include scintillators for monitoring of alpha and beta and "phosvich"-type scintillators/detectors for registration of the components of mixed ionizing fluxes, screens for visualization of X-ray images with high spatial resolution and screens for electron-beam tubes [31]. Mechanical robustness facilitates the preparation of SCF of YAG:Ce down to about 1 μm thickness by liquid-phase epitaxy (LPE) [32]. Nanoparticles can be easily attached to biomolecules due to comparable size, and hence, they are important in bioimaging and similar applications. They also offer ease of coating needed in an application like phosphor-sensitized solar cells. For white LED applications, it is reported that YAG:Ce in nano-sized order performs better than it does in larger particles [33]. This is because nano-sized YAG:Ce can reduce internal light scattering when coated onto a blue LED surface.

A good overview of the different synthesis methods for crystalline powders is provided by Pan et al. [34]. There are several methods to prepare Ce:YAG phosphor, but to achieve high-emission efficiency, a fine non-agglomerated powder with good purity, narrow particle size distribution and high crystallinity are prerequisites. YAG phosphors doped with activators are mainly synthesized by solid-state reaction techniques that require high sintering temperatures (above 1,800°C) to eliminate YAM and YAP phases [35]. Fluxes such as BaF_2 or YF_3 can lower the sintering temperature down to 1,500°C [36]. Several novel methods have been used to obtain phase pure YAG:Ce^{3+} at lower temperatures. Thus, YAG:Ce phosphor has been prepared using novel methods such as precipitation [37–39], heterogeneous precipitation [40], spray pyrolysis from ethylenediaminetetraacetic acid solution [41], spray pyrolysis of polymeric precursors [42], flame spray pyrolysis [43,44], sol-gel pyrolysis [45], sol-gel [46–48], sol-gel combustion [49], glycothermal [50], combustion synthesis using urea [51,52] or carbohydrazide [53] fuels, in situ esterification [54], microwave homogeneous synthesis, which is a combination of the traditional homogeneous co-precipitation with microwave irradiation [55], etc.

Though several soft chemical routes have been explored for the synthesis of YAG, most of these methods are complex and high cost for the industrialization. Moreover, phase pure materials are not obtained in one step, but prolonged annealing at temperatures around 1,000°C or above becomes necessary.

6.2.1.2 Substituted YAG:Ce

The blue chip/YAG:Ce system has many advantages. But the lamps fabricated in this manner give a poor colour rendering because the resulting light is typically deficient in the green and red colours. Apart from poor colour rendering, there are several other disadvantages of YAG:Ce phosphor which are known and well-documented [56].

Combining the yellow emission of YAG:Ce with blue emission of InGaN chip produces white light but with high colour temperature (CT), of the order of 6,000 K. Thus, only "cool white" emission is possible. For reducing CT, the colour coordinates need to be shifted to right. For this, red component has to be enhanced and blue reduced. This cannot be achieved by merely increasing the phosphor load.

In most commonly used lamp assembly, phosphor is deposited directly on the LED chip. Effect of temperature on phosphor characteristics becomes crucial. Relatively, large power over small chip area results in heating the chip to a temperature of 400–450 K. Efficiency of YAG:Ce and hence the LED decreases due to such temperature rise. "The thermal quenching behaviour of phosphor itself is much better [57]. The reduction is related to the concentration quenching at higher temperatures" [17].

Both excitation and emission spectra are modified due to rise in temperature. For obtaining higher light output, the diode has to be operated at higher current. However, at higher current density, non-radiative losses increase. This adds to the heating effect. This leads to red shift in diode emission. This shift can detune the diode with respect to the excitation spectrum of YAG:Ce. To avoid the detuning, the excitation band of the phosphor has to be as broad as possible. Another problem associated with the use of YAG:Ce is the "halo effect" (blue/yellow colour separation). YAG:Ce is not properly mixed in resin due to high density, resulting in colour unevenness [58].

Various solutions were sought to overcome some of these problems. Two types of approaches will be discussed as

  a. finding replacement for YAG:Ce and
  b. improving performance of YAG:Ce.

For increasing CRI and reducing CT, the phosphor emission has to be red-shifted. The red shift can be achieved by partially replacing Y^{3+} with Gd^{3+} [59–61], Tb^{3+} [62–64]. In contrast, emission becomes blue-shifted when Al^{3+} is partly replaced by Ga^{3+} [65–67]. The substitutions not only modify emission and CRI but alter intensities as well [68]. Also, by and large, these modified YAG:Ce phosphors have inferior thermal stability [14].

Ce^{3+}-Pr^{3+} [69,70] or Ce^{3+}-Eu^{3+} [71] transfer also results in some red emission which improves CRI [72]. However, light output decreases severely with increase in CRI. Similar effect has been noticed by Gd^{3+} [60] addition. Recently, addition of AlN [73], Si_3N_4 [74,75] or Cu^+ [76], Mn [77], Mn with Si [78–81] and other charge compensators [82], Sb^{3+} [83], Tb^{3+} [68] has been tried to achieve the same purpose, and these efforts were claimed to be more successful. Co-doping with Li is claimed to broaden the excitation band [84].

By way of finding replacement for YAG:Ce, in recent years, TbAG:Ce^{3+} [73] CaO:Ce, Na [85], (La, Gd)Sr$_2$(Al, B)O$_5$:Ce [86], Ca$_2$SiO$_4$:Ce, Li$^+$ [87], La$_4$Ca(SiO$_4$)$_3$O:Ce [88], Ca$_3$Sc$_2$Si$_3$O$_{12}$: Ce [89,90], Ca$_3$Sc$_2$Si$_3$O$_{12}$:Ce, Mn [91,92], Ca$_{1.5}$Y$_{1.5}$Al$_{3.5}$Si$_{1.5}$O$_{12}$:Ce^{3+} [93], CaY$_2$Al$_4$SiO$_{12}$:Ce^{3+} [94], (Ca, Lu)$_3$(Sc, Mg)$_2$Si$_3$O$_{12}$:Ce [95,96], Y$_{(3-x)}$Ce$_x$Mg$_2$AlSi$_2$O$_{12}$ [80], (Sr, Ba)$_3$SiO$_5$:Eu^{2+} [97,98], Sr$_3$AlO$_4$F: Ce [99], Sr$_3$LuAl$_2$O$_{7.5}$:Ce^{3+} [100], Y$_2$(Ca, Sr)F$_4$S$_2$:Ce^{3+} [101], BaS:Yb^{2+} [102] CaZnOS:Ce^{3+} [103] AE(Al, Si)N2:Ce [104], CaSi$_2$O$_2$N$_2$:Ce^{3+}, Eu^{2+} [105], Y-sialon:Ce^{3+} [106], Ca-alpha-SiAlON:Eu^{2+} [107–109], Ca$_{1.5}$Ba$_{0.5}$Si$_5$N$_6$O$_3$:Eu^{2+} [110], La$_{4-x}$Ca$_x$Si$_{12}$O$_{3+x}$N$_{18-x}$:Eu^{2+} (x = 1.456) [111], Y$_3$Si$_6$N$_{11}$:Ce^{3+} [112] SrAlSi$_4$N$_7$:Ce^{3+} [113] Ca$_2$BO$_3$Cl:Eu^{2+} [114–117], Ca$_3$Si$_2$O$_7$:Eu^{2+} [118], Sr$_3$SiO$_5$:Eu^{2+} [119], Sr$_2$ZnS$_3$:Eu^{2+} [120], Sr1-xZnxSe:Eu^{2+} [121], Sr1-xCaxSe:Eu^{2+} [122], Sr$_8$Al$_{12}$O$_{24}$S$_2$:Eu^{2+} [123], Ba$_2$Mg(BO$_3$)$_2$:Eu^{2+} [124], Sr$_8$MgSc(PO$_4$)$_7$:Eu^{2+} [125], NaCaPO$_4$:Mn [126,127], CaO:Ce, Li [128] and carbogenic dots [129] phosphors have been investigated with more or less success. Blending YAG:Ce with red emitting phosphor like SrS:Eu, LiSrBO$_3$:Eu [130] has also been attempted to improve CRI [131].

6.2.1.3 Double-Substituted Garnets

During early stages, simple substitutions at a single site such as Tb, Gd, at Yttrium site, Ga, at Al site were experimented with. These experiments resulted in improvements in colour rendition, thermal quenching, correlated colour temperature (CCT), etc. However, the resulting phosphors could not be treated as a final solution. The next step was to explore substitution using pair of ions at different sites, again maintaining the garnet structure. The pairs are constituted of AE-M(IV). AE stands for a divalent ion like one of the alkaline earths Mg, Ca, Sr or Ba, and M(IV) is a tetravalent element like Si, Ge or Zr [132].

Such an approach led to many phosphors which were superior to YAG:Ce in some respect or other. For example, CaY$_2$Al$_4$SiO$_{12}$:Ce^{3+} [94,133], CaLu$_2$Al$_4$SiO$_{12}$:Ce^{3+} [134], MgY$_2$Al$_4$SiO$_{12}$:Ce^{3+} [135], BaY$_2$Al$_4$SiO$_{12}$:Ce^{3+}[136,137], Y2BaAl4SiO12:Tb3+, Eu^{3+} [138], Ba$_2$YAl$_3$Si$_2$O$_{12}$:Ce^{3+} [139], BaLu$_2$Al$_2$Ga$_2$SiO$_{12}$:Ce [140], Y$_3$Mg$_2$AlSi$_2$O$_{12}$:Ce^{3+} [80], Y$_2$Mg$_3$Ge$_3$O$_{12}$:Dy^{3+}, Eu^{3+} [141], Y$_2$Mg$_2$Al$_2$Si$_2$O$_{12}$:Eu^{2+}, Ce^{3+} [142], Lu$_3$(Al, Mg)$_2$(Al, Si)$_3$O$_{12}$:Ce^{3+} [143], Lu$_2$Mg$_2$Al$_2$Si$_2$O$_{12}$:Eu^{2+} [144], Y$_3$MgSiAl$_3$O$_{12}$:Ce^{3+} [145,146], Lu$_3$MgAl$_3$SiO$_{12}$ [147], (Lu$_2$M)(Al$_4$Si)O$_{12}$:Ce^{3+} [148], Lu$_2$SrAl$_4$SiO$_{12}$:Ce^{3+} [149], Ca$_2$YHf$_2$Al$_3$O$_{12}$:Ce [150,151], BaLu$_2$Al$_2$Ga$_2$SiO$_{12}$:Pr^{3+} [152], Ca$_2$LuZr$_2$(AlO$_4$)$_3$:Ce^{3+} [153], Gd$_3$Mg$_x$Ge$_x$Al$_{5-2x}$O$_{12}$:Ce [154].

In most of these cases, Ce^{3+} had been used as an activator. The emission is in the yellow and yellow-red region. These phosphors are thus suitable for one-phosphor-converted LED lamps based on blue + yellow = white logic. Further improvements can be achieved by using two phosphors, green and red, thus making wide colour gamut possible in combination with blue chip.

6.2.2 Two-Phosphor-Converted White LED

For generating all possible colours, all three primary colours red, green and blue need to be combined. Blue colour is already present in the chip emission. A blend of two phosphors which can partially convert chip emission to green and red is required.

6.2.2.1 Green-Emitting Phosphors

6.2.2.1.1 LuAG:Ce

Use of Lu in place of Y shifts Ce emission to shorter wavelength. LuAG:Ce^{3+} is thus a green-emitting phosphor that can be excited by blue LED. Though it is commercially available, it has been investigated to a lesser extent. It was primarily developed for obtaining improved scintillation performance. Its predecessor, YAG has lower density (4.56 g/cm^3). LuAG has a high density (LuAG = 6.73 g/cm^3) and a high effective atomic number (Z$_{eff}$ = 60) and therefore high stopping power for ionizing radiation. Thus, LuAG:Ce^{3+} is regarded as a particularly promising scintillator due to its allowed electric dipole 5d–4f transition (short emission lifetime) and high theoretical light yield.

6.2.2.1.2 BaSrSiO$_4$:Eu^{2+}

Sr$_2$SiO$_4$:Eu^{2+} is an efficient yellow emitting phosphor. Partial replacement of Sr by Ba shifts the emission to green.

> Green barium strontium silicate phosphor (BaSrSiO$_4$:Eu^{3+}, Eu^{2+}) was synthesized using a solid-state reaction method in air and reducing atmosphere. Investigation of the firing temperature indicates that a single phase of BaSrSiO$_4$ is formed when the firing temperature is higher than 1400°C. The effect of firing temperature and doping concentration on luminescent properties are investigated. The light-emitting property was the best when the molar content of Eu$_2$O$_3$ was 0.025 mol. Also, the luminescent brightness of the BaSrSiO$_4$ fluorescent substance was the best when the particle size of the barium was 0.5 μm. BaSrSiO$_4$ phosphors exhibit the typical green luminescent properties of Eu^{3+} and Eu^{2+}. The maximum emission band of the BaSrSiO$_4$: Eu^{3+}, Eu^{2+} was 520 nm [155].

Detailed synthesis has been described recently by Thi et al. [156,157]. (BaSr)SiO$_4$ belongs to a series of host materials for phosphor when doped with Eu^{2+}, which has short decay time and high luminescence characteristics under long wavelength UV [158].

6.2.2.1.3 CaSc$_2$O$_4$:Ce

Luminescence of various activators, in CaSc$_2$O$_4$ host, has been thoroughly studied [159]. The phosphors have been mostly prepared by conventional solid-state reaction. CaSc$_2$O$_4$ belongs to orthorhombic system (space group – Pnam). The unit cell contains four formula units. Ca is 8 co-ordinated. Scandium has two types of sites; both are 6 co-ordinated. The structure is made up of "double octahedral Sc$_2$O$_4$-framework and Ca atoms residing within the framework with the bicapped trigonal prismatic site" [160].

Calcium scandate became relevant for LED applications due to its green emission that can be excited by blue light [161]. CaSc$_2$O$_4$:Ce is a well-known green-emitting phosphor which has remarkable thermal stability and high quantum efficiency. The excitation spectrum overlapping with blue LED emission makes it almost perfect green-emitting phosphor for two-phosphor-converted LED applications [162].

CaSc$_2$O$_4$ is also a suitable host for upconversion owing to low phonon frequency (cut-off frequency 540 cm^{-1}). In fact, even Y$_2$O$_3$, a commonly explored host has higher phonon frequency. Lanthanides suitable for upconversion such as Er^{3+}, Ho^{3+} or Tm^{3+}, and co-doped with Yb^{3+}, exhibit significant upconversion efficiencies.

> The spectroscopic properties (radiative lifetimes, quantum efficiency, branching ratios) of CaSc$_2$O$_4$ doped with Er^{3+}, Tm^{3+}, or Ho^{3+} were estimated from Judd-Ofelt analyses on ceramic samples. The upconversion efficiency of a CaSc$_2$O$_4$ ceramic sample doped with 1 at.% Er^{3+} and 5% Yb^{3+} was measured. The color of the upconversion-emitted light of the CaSc$_2$O$_4$:Er:Yb changes with ytterbium concentration from green to reddish [163].

In Eu^{3+}-activated CaSc$_2$O$_4$, emission from states higher than ^{5}D$_0$ can be observed due to the low-energy phonons (540 cm^{-1}). The emission colour can thus be tuned over a wide range by just controlling Eu^{3+} concentration.

6.2.2.1.4 Ca$_3$Sc$_2$Si$_3$O$_{12}$:Ce

Synthesis and crystal structures of garnet compounds of scandium, viz., Ca$_3$Sc$_2$Si$_3$O$_{12}$, Ca$_3$Sc$_2$Ge$_3$O$_{12}$ and Cd$_3$Sc$_2$Ge$_3$O$_{12}$, have been described as early as 1977 [164]. However, luminescence in Ca$_3$Sc$_2$Si$_3$O$_{12}$ was studied only during the last decade. Shimomura et al. obtained intense luminescence in Ca$_3$Sc$_2$Si$_3$O$_{12}$ activated by Ce^{3+} in 2007 [161] and patented it [165] as a wavelength converter for LED based on blue chip. Ca$_3$Sc$_2$Si$_3$O$_{12}$:Ce^{3+} is a green-emitting phosphor with thermal stability much better than YAG:Ce. For obtaining white LED with a single phosphor, yellow emission is desired which could be achieved by Mg co-doping [166,167]. Mg served a dual purpose. Though Ce^{3+} is trivalent, it occupies Ca^{2+} site as these ions are of nearly same size. The substitution, however, necessitates charge compensation which is achieved by placing Mg^{2+} on Sc^{3+}. Hence, red shift and charge compensation are achieved by Mg co-doping. Various other co-dopants were explored for improving the properties of $Ca_3Sc_2Si_3O_{12}$:Ce^{3+} phosphor. Charge compensation can also be achieved by Na^+ doping at Ca^{2+} site. However, emission colour of $Ca_3Sc_2Si_3O_{12}$:Ce^{3+} with Na^+ as a charge compensator is green. Doping Al^{3+} at Sc^{3+} site induces red shift of the emission maximum. Al^{3+} doping results into lattice contraction, lower cell volume and increased crystal field strength. The inclusion of Al^{3+} was also found to prevent the undesired phases of ingredient oxides. As a result, crystallinity is improved, which in turn enhances the luminescence efficiency and intensity. Mn^{2+} could be incorporated at both Ca^{2+} and Sc^{3+} sites. It acts as a charge compensator at the Sc^{3+} site and also exhibits deep red emission (680 nm). On the other hand, at Ca^{2+} substitutional site, the emission is in the yellow-orange region (574 nm) [91]. Partial substitution of oxygen ion by N^{3-} produced huge shift resulting in a red emitting phosphor [90,168]. Apart from Ce^{3+}, lanthanide activators like Pr^{3+} [169–171], Eu^{3+} [172,173], Tb^{3+} [174,175], Dy^{3+} [176,177], Er^{3+} [89] and Yb^{3+} [178] have also been investigated in subsequent years. Divalent Eu^{2+} also exhibits characteristic emission in $Ca_3Sc_2Si_3O_{12}$ host. Eu^{2+} emission is at unusually long wavelengths [179].

Energy transfer studies involving pair of dopants have also been carried out. Interesting results involving energy transfer have been reported for pairs Ce^{3+}-Er^{3+} [180], Ce^{3+}-Pr^{3+} [181], Tb^{3+}-Ce^{3+} [182], Ce^{3+}-Eu^{2+} [183], Eu^{2+}-Yb^{3+} [184] and Ce^{3+}-Mn^{2+} [92,185,186].

$Ca_3Sc_2Si_3O_{12}$:Ce^{3+} phosphor was initially studied for application in solid-state lighting. Later, cathodoluminescence in this phosphor was studied and found to be useful for field-emission display applications [187]. Phosphors incorporating other activators were suggested for various other applications. $Ca_3Sc_2Si_3O_{12}$:Pr^{3+} was found suitable as a scintillator. Broad excitation covering entire visible region and emission around 873 nm makes $Ca_3Sc_2Si_3O_{12}$:Ce^{3+}, Eu^{2+} phosphor useful as spectral converter for c-Si solar cell. $Ca_3Sc_2Si_3O_{12}$:Eu^{2+}, Yb^{3+} gives emission at longer wavelengths and can also be used for such an application. A similar application for germanium solar cell was envisaged for $Ca_3Sc_2Si_3O_{12}$:Ce^{3+}, Er^{3+}. $Ca_3Sc_2Si_3O_{12}$:Dy, $Ca_3Sc_2Si_3O_{12}$:Dy, Ce can be used in ratiometric luminescence thermometry. Cooperative energy transfer from Yb^{3+} to Tb^{3+} in $Ca_3Sc_2Si_3O_{12}$ host leads to intense upconversion.

A variety of methods have been employed in the synthesis of $Ca_3Sc_2Si_3O_{12}$. Conventional solid-state reaction is sometimes inadequate to give phase pure compound. Different fluxes such as CaF_2 [188] and LiF [189] have been used to achieve the phase purity and lower the reaction temperature. Several novel synthesis techniques such as sol–gel, gel-combustion [190], freeze-drying emulsion evaporation [191], liquid-phase epitaxy [192], etc., have been successfully adapted for preparing $Ca_3Sc_2Si_3O_{12}$-based phosphors.

Most of the substitutions in $Ca_3Sc_2Si_3O_{12}$:Ce^{3+} resulted in red shift of the emission. Luminescence at wavelengths shorter than 500 nm has seldom been achieved. $Ca_3Sc_2Ge_3O_{12}$:Ce^{3+} exhibits such emission; however, intensity is very low. It might be possible to get reasonable intensity and blue-shifted emission by partial substitution if Si by Ge. We have indeed observed such emission in $Ca_3Sc_2Si_{1.5}Ge_{1.5}O_{12}$:$Ce^{3+}$ [193].

6.2.2.2 Red-Emitting Phosphors

6.2.2.2.1 CaZnOS:Mn^{2+}

CaZnOS was discovered in 2003 and received attention of researchers immediately [194]. Clarke et al. [195] presented a detailed report of its preparation, electrical properties and structure. The synthesis had been particularly tricky due to possibility of decomposition at temperatures higher than 1,370 K [196]. CaZnOS is a semiconductor with a band gap of about 3.71 eV. CaZnOS crystallizes in a hexagonal space group $P63mc$ with a = 3.75726 Å, c = 11.4013 Å and Z = 2. In CaZnOS, each Zn is tetrahedrally coordinated by three S atoms and an O atom (ZnS3O) with

$C3V$ point symmetry. The non-centro-symmetric structure of CaZnOS, which has only few analogues, is composed of isotypic puckered hexagonal ZnS and CaO layers arranged so that [ZnS$_3$O] tetrahedrons are all aligned parallel, resulting in a polar structure [197]. Luminescence of various activators in this host has been studied. Red emission of Mn^{2+} in this host has been remarkable. There is prominent excitation in the blue region which makes it suitable as a colour conversion phosphor for LED lamps. The emission colour of PL, CL and ML in CaZnOS:Mn^{2+} can be tuned from yellow to red through varying Mn^{2+} concentration [198]. The origin of multi-luminescent colour manipulation has also been explained according to the investigation of crystal filed, quenching concentration and Mn^{2+} PL lifetimes. Rare earth ions Pr^{3+}, Ce^{3+} and Nd^{3+} have been used as co-dopants to obtain a brighter red after glow and longer persistence time in CaZnOS:Mn^{2+}.

Li's group found that CaZnOS:Nd^{3+} can effectively convert mechanical stress into near-infrared light penetrating deep into biological tissues. Significantly, CaZnOS has been specifically identified as an excellent host lattice for Mn^{2+} due to its chemical and thermal stability properties, and thus, has attracted the greatest attention in recent years. In particular, Zhang and his colleagues have reported that CaZnOS:Mn^{2+} can emit strong red light under a variety of mechanical stimuli (ultrasonic vibration, impact, friction and compression), and its EML intensity increases linearly with the applied compression load and has recoverability, indicating the practical prospect on stress sensors [199].

Up till now, LED lighting for domestic and office use with emphasis on lumen efficiency and CRI was considered. There are also other types of lighting where phosphor-converted LED lamps are replacing conventional sources. Some of these are discussed in the following sections.

6.3 PHOSPHORS FOR HORTICULTURE LED

Vegetable crops need specific weather conditions which include soil, weather and light. That is why these can be cultivated only in specific seasons and localities. Greenhouses providing these conditions are built to overcome the limitations. Traditionally, light sources with broad spectral emissions have been used for greenhouse lighting. Availability of narrow-emitting LEDs has changed the scenario.

Photosynthesis is the most important biochemical processes on Earth. It enables conversion of the incident solar energy into chemical energy used for the plant growth.

The generic formula of photosynthesis is

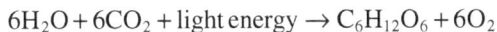

$$6H_2O + 6CO_2 + light\,energy \rightarrow C_6H_{12}O_6 + 6O_2$$

The spectral influence of light on plant development was first reported around a century ago [200], although earlier experiments had indicated the effects of specific wavelengths. The activities of photosynthetic pigments, such as chlorophylls and carotenoids, are mostly related to light-harvesting and energy transduction during photosynthesis. Chlorophylls have maximum sensitivities in the blue and red regions, around 300–400 nm and 600–700 nm, respectively (Figure 6.5). Carotenoids such as xanthophylls and carotenes absorb mainly blue-light and are also known as auxiliary photoreceptors of chlorophyll [201].

6.3.1 CONVENTIONAL LIGHT SOURCES IN HORTICULTURE

Experiments on plant growth using man-made illumination date back to 1861, when the tungsten filament lamp was not even invented [204]. Availability of such lamps provided impetus to these efforts. However, using incandescent lamps has several drawbacks. Notably, such sources have low conversion efficiency, emission lying outside the action spectrum for photosynthesis, etc.

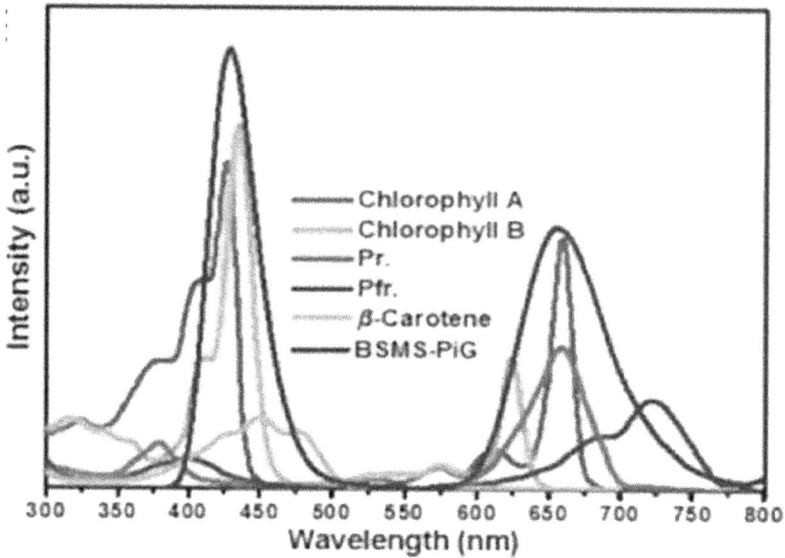

FIGURE 6.5 Absorption spectra of the most common photosynthetic and photomorphogenetic receptors in green plants: chlorophyll a (chl a), chlorophyll b (ch b), beta-carotene, phytochromes (Pfr and Pr) [202,203].

6.3.2 Horticultural LED Lighting

With incandescent lamps, only way to avoid undesired wavelengths is to use optical filters. This is an expensive method. The features like pulsed operation, dimming, etc., are also missing. These lacunae can be removed by using LED lighting. LED technology is now successfully and seamlessly integrated with horticultural lighting. The new technology presents techniques for controlling intensity, time cycle and spectral distribution of the light which can be exploited for optimization of the photosynthesis.

Light-emitting diodes (LEDs) have tremendous potential as supplemental or sole-source lighting systems for crop production both on and off Earth. Their small size, durability, long operating lifetime, wavelength specificity, relatively cool emitting surfaces and linear photon output with electrical input current make these solid-state light sources ideal for use in plant lighting designs. Because the output waveband of LEDs (single colour, non-phosphor-coated) is much narrower than that of traditional sources of electric lighting used for plant growth, one challenge in designing an optimum plant lighting system is to determine wavelengths essential for specific crops. Work at NASA's Kennedy Space Center has focused on the proportion of blue light required for normal plant growth as well as the optimum wavelength of red and the red/far-red ratio. The addition of green wavelengths for improved plant growth as well as for visual monitoring of plant status has been addressed. Like with other light sources, spectral quality of LEDs can have dramatic effects on crop anatomy and morphology as well as nutrient uptake and pathogen development. Work at Purdue University has focused on the geometry of light delivery to improve energy use efficiency of a crop lighting system. Additionally, foliar intumescence developing in the absence of ultraviolet light or other less understood stimuli could become a serious limitation for some crops lighted solely by narrow-band LEDs. Ways to prevent this condition are being investigated. Potential LED benefits to the controlled environment agriculture industry are numerous, and more work needs to be done to position horticulture at the forefront of this promising technology.

Earlier studies on using LEDs for plant growth were focussed on red- and deep blue-emitting chips [205,206]. There were two main reasons for this. The first and the foremost was the availability of such LEDs with adequate light output. Second, the action spectra of photoreceptors relevant to photosynthesis lie in this region as shown in Figure 6.5 Matching the LED emission with action spectrum is a critical parameter of horticulture LED lighting technology. LED lighting scores over the conventional greenhouse lighting by improving the energy efficiency several folds.

LEDs are the ideal light sources to be integrated into multi-layer cultivation due to their small size, controllability, electrical efficiency and absence of heat radiation which allows for mounting distances closer to the plants. LEDs of different wavelengths 410–430 nm, 630–650 nm and 690–730 nm are available and can be integrated to make lamps for horticulture. However, blue LEDs are most efficient and low cost. It is more economical to obtain deep red emission using phosphors. LEDs emitting in 410–430 nm can be coated with phosphors such as $CaS:Eu^{2+}$ (630–650 nm) (matching absorption of chi b, beta karotene, phytochrome Pr), $Al_2O_3:Cr^{3+}$ or $LiAl_5O_8:Cr^{3+}$ (matching absorption of phytochrome Pr). These LEDs can be thoroughly characterized, and suitability for horticulture can be tested. Various phosphors which can be used for preparing horticultural LED lamps have recently been reviewed by Fang et al. [207].

6.3.3 Cr³⁺-Activated Phosphors

Cr^{3+} with unfilled ($3d^3$) electron configuration is one promising candidate for red-emitting phosphors. The outstanding feature of Cr^{3+} is that it can absorb two bands around ~410 and ~560 nm and has a strong red/far-red emission, promising to be applied in far-red growth LEDs. It is noticeable that the emission spectra of Cr^{3+} ions could be a broad or narrow band, depending on the interaction of crystal field. The broadband could be obtained when Cr^{3+} ions are affected by weak crystal field (Dq/B < 2.3); in contrast, the narrowband is generated under the effects of strong crystal field (Dq/B > 2.3) [208,209]. The influence of crystal field on luminescent properties of Cr^{3+}-doped phosphors has been a growing concern in the science community recently.

Cr^{3+} exhibits efficient emission in several oxides. There is thus wide choice to choose an oxide phosphor suitable for specific horticultural lighting. Table 6.2 shows spectral data for various Cr^{3+}-doped oxide phosphors.

TABLE 6.2
Spectral Data for Various Cr3+ Doped Oxide Phosphors.

Host	Emission (nm)	Excitation 1 (nm)	Excitation 2 (nm)
$Al_{10}Ge_2O_{19}$	699	614	437
Al_5NO_6	693	614	435
$BaAl_{12}O_{19}$	702	611	435
$BaMgAl_{10}O_{17}$	691	646	453
$Ba_2Mg(BO_3)_2$	698	674	470
$CaAl_{12}O_{19}$	685	642	451
	685	656	458
$CaAl_6Ga_6O_{19}$	693	656	459
$Ca_3Al_2Ge_2O_{10}$	697	646	454
$CaGa_2O_4$	693	681	473
$CaGdAlO_4$	745	681	481
Ca_2NbAlO_6	746	713	498
Ca_2NbGaO_6	740	701	490
$CaTiO_3$	766	685	484
$CaYAlO_4$	743	693	486
$Ca14Zn6Al10O35$	712	663	466
$CdAl_2O_4$	691	653	456
$(CH_3)_2NH_2Al(SO_4)_2 \cdot 6H_2O$	704	689	479
Cr_2O_3	728	701	488

(Continued)

TABLE 6.2 (*Continued*)
Spectral Data for Various Cr3+ Doped Oxide Phosphors.

Host	Emission (nm)	Excitation 1 (nm)	Excitation 2 (nm)
$GdAlO_3$	727	611	438
$GdAl_3(BO_3)_4$	686	681	471
$GdY_2Al_3Ga_2O_{12}$	691	681	471
$LaAlO_3$	735	633	453
$LaGaO_3$	730	693	484
$La_3GaGe_5O_{16}$	701	701	484
$LaMgAl_{11}O_{19}$	693	667	464
$LaMgGa_{11}O_{19}$	738	709	494
La_2MgGeO_6	695	674	470
La_2ZnTiO_6	742	544	403
$Li_5Zn_8Al_5Ge_9O_{36}$	700	653	458
$Li_5Zn_8Ga_5Ge_9O_{36}$	700	678	471
$Li_2ZnGe_3O_8$	716	697	484
$Lu_3Al_5O_{12}$	690	626	443
$Lu_3Ga_5O_{12}$	688	674	468
$MgAl_2O_4$	689	660	459
$MgAlGaO_4$	707	605	432

6.3.4 Mn^{4+}-Activated Phosphors

Mn^{4+} is another activator belonging to ^3d_n group which can give efficient emission in the red or near-infrared region using near UV/violet excitation. Mn^{4+} has $^3d_3$ electronic configurations, preferring to enter octahedral sites in lattices, and the $^2E_g \rightarrow ^4A_{2g}$ transitions cause adjustable emitting spectral range from red to far-red in 600–750 nm. The excitation spectra are consisted of spin-allowed $^4A_{2g} \rightarrow ^4T_{2g}$ and $^4A_{2g} \rightarrow ^4T_{1g}$ transitions, which can be excited by near-ultraviolet (NUV) or blue light [210].

Mn^{4+} exhibits efficient emission in several oxides. There is thus wide choice to choose an oxide phosphor suitable for specific horticultural lighting. Table 6.3 shows spectral data for various Mn^{4+}-doped oxide phosphors.

6.4 HUMAN-CENTRIC LIGHTING

Solid-state lighting types are increasing in the general lighting market due to their energy savings, extended longevity and eco-friendly characteristics [211,212]. Two major performance metrics, the colour quality (colour rendering index, CRI or *Ra*) and the visual energy efficiency (luminous efficacy of radiation, LER), are the critical criteria to be considered when developing luminescent materials for down-converted white-light-emitting diodes (DC-WLEDs) [213]. A high LER and a high *Ra* serve as target figures of merit in the early stages of the development of WLEDs between the first and third generation of general WLED lightings. At present, the CRI is well known as a colour metric, though it is limited when used to reproduce good saturated colours of illuminated objects under white conditions. Fairly recently, the Illuminating Engineering Society of North America developed and adopted a two-measure system with a colour fidelity index (CFI, *Rf*) and a colour gamut index (CGI, *Rg*) as technical memorandum TM-30–2015 in the fourth generation of the development of WLEDs for accurate evaluations of the colour rendition and the colour performances of DC-WLED lighting sources.

However, in recent years, several other factors are considered while designing white LED for domestic lighting. In addition to both a high colour-quality level for colour perception/reproduction

TABLE 6.3

Spectral Data for Various Mn4+ Doped Oxide Phosphors.

Host	Excitation(nm)	Emission(nm)
$Ca_{14}Zn_6Ga_{10}O_{35}$	313	712
$Ca_{14}Zn_6Al_{10}O_{35}$	460	710
$Ca_3ZnAl_4O_{10}$	467	715
Gd_2ZnTiO_6	365	704
La_2LiTaO_6	495	709
$NaMgLaTeO_6$	365	705
La_2MgTiO_6	355	710
Ba_2LaNbO_6	352	677
$CaAl_{12}O_{19}$	400	654
Mg_2TiO_4	475	657
La_2ZnTiO_6	345	710
$MgAl_2Si_2O_8$	258	710
$YAlO_3$	414	714
$La(MgTi)_{1/2}O_3$	324,360	708
Ca_2LaTaO_6	325	696
Li_2MgZrO_4	335	670
$NaLaMgWO6$	342	700
Ba_2LaSbO_6	325,380	654,678
$LiLaMgTeO_6$	465	708
Ca_2LuTaO_6	351	682
Sr_2GdNbO_6	312	687
$CaYMgSbO_6$	273,294	688
Ba_2GdTaO_6	354	688
$SrMg_2La_2W_2O_{12}$	344,469	708
$CaLaMgSbO_6$	371	695,708
Sr_2GdNbO_6	312	671,687
Sr_3LiSbO_6	267,302,343,493	698
$Li_5La_3Nb_2O_{12}$	318	715
Ca_2LaNbO_6	288,311,352,506	684

and high visual efficiency to reduce energy use, another important figure of merit for smart WLEDs is the tunable capability of the circadian energy efficiency (circadian efficacy of radiation, CER) for entraining the central and local clocks of the human body with the circadian rhythm [214]. The tunability of the circadian effect has obtained greater prominence in recent years for controlling melatonin suppression/secretion, resetting the central/local clocks of individuals, and improving human health by matching the spectrum of QD-containing WLEDs with daily variations of sunlight under the natural circadian rhythm [215,216]. Light is known to influence human physiology and behaviour, as it directly stimulates the human biological clock. This endogenous circadian (24 hours) rhythm generates the daily rhythms of rest and activity and is responsible for modulating the daily rhythms in several physiological phenomena, including the sleep/wake cycle, hormone secretion, and subjective alertness and performance in humans [217]. The timing of sleep is not only regulated by circadian phenomena but also induced by a combination of circadian and homeostatic mechanisms – in other words, sleep onset results from the combination of the circadian timing and the physiological need for sleep [218]. Adjustment of physiological systems within the body is called homeostatic regulation, which involves three parts or mechanisms: **(i) the receptor, (ii) the control centre and (iii) the effector**. The receptor receives information that something in the environment is changing. The control centre or integration centre receives and processes information from the receptor. The effector responds to the commands of the control centre by either opposing or enhanc-ing the stimulus. This ongoing process continually works to restore and maintain homeostasis. For example, during body temperature regulation, temperature receptors in the skin communicate information to the brain (the control centre) which signals the effectors: blood vessels and sweat glands in the skin. As the internal and external environment of the body are constantly changing, adjustments must be made continuously to stay at or near a specific value: the set point. For millen-nia, our daily habits and routines were determined by the sun. We have been using the daily pattern of light and dark to set the timing of the biological clock in synchrony with the 24 hours Earth rota-tion [219]. However, this pattern is disrupted by the modern lifestyle. Currently, the average person spends most of their time indoors (on average, 87% [220]) with a large part of the day spent at work. During the work day, the light levels indoors in offices are commonly much lower than outdoors, while evening light levels, not only from artificial illumination sources but also from smartphones or TVs, are relatively high. Additionally, most people typically delay their bedtimes and use an alarm to comply with work schedules and other social constraints. This modern lifestyle, particu-larly its socially enforced shifts in sleep schedules, is often seen as a cause of the mismatch with our endogenous circadian rhythmicity. In fact, the majority the world's population is estimated to have social jet-lag [221], that is, a circadian rhythm that is out of phase with people's daily schedules. Social jet-lag indirectly disturbs the natural cycle of our biological clock and can have a profound effect on our mental and physical health [222].

6.4.1 Biological Clock and Sleep

In humans, sleep is regulated by two interacting, coupled mechanisms: (i) the biological clock, which generates a circadian rhythm in sleep-wake propensity, which we refer to as the circadian mechanism C; (ii) a homeostatic process, H, that represents the sleep depth which builds up during wakefulness and dissipates during sleep. The circadian factor is evident by the clock-regulated tim-ing of physiological processes; for example, the secretion of hormones like melatonin or the body temperature rhythm; while the homeostatic component is evident in sleep deprivation, i.e. the rising need for sleep after a sustained period of wakefulness. Light affects sleep through the circadian component. Light exposure throughout the day determines the circadian phase relative to the local clock time, which in turn influences the timing of key physiological properties, like the nadir of the core body temperature or the start of melatonin secretion. On the other hand, the timing of sleep influences the light input signal to the pacemaker, as, during sleep, no light reaches the eye, while during wakefulness, the amount of light that reaches the eye depends on the clock time.

In fact, there are more than 100 studies that relate circadian disruption to a wide variety of health risks and diseases, including mood disorders, depression, diabetes, obesity, cardiovascular disease and cancer [223–226]. Despite this growing scientific understanding of the impact of light on bio-logical mechanisms, supported by field studies on how light induces vitality and alertness even during office hours [227], the benefits of this understanding are not (yet) harvested by practical sys-tems. Office lighting is presently designed only to meet visual requirements, providing the amount of light that is "suitable and sufficient" to perform certain tasks and reduce visual discomfort. Less attention has been given to the biological effects of light, especially how it could be used to promote occupants' health and well-being through the circadian functions that regulate sleep, mood and alertness. Artificial light can be exploited as a means to modulate sleep-wake schedules and re-align the internal clock with the environment by phase-shifting the biological clock. Light is often men-tioned to be the main synchronizer of the human biological clock. In fact, depending on the timing and magnitude of light exposure, light either accelerates (advances) or slows down (delays) the phase of the clock. As a result, a lighting system that can shift our body's internal perception of time can potentially enhance well-being and can mitigate sleep problems in an attractive, unobtrusive way. In fact, a recent study in office buildings [228] showed that office workers who received high circadian stimulation in the morning reported better sleep and fewer depressive symptoms than those who received low circadian stimulation in the morning. While the importance of reducing evening light exposures to maintain a regular sleep–wake cycle is already widely known, this study also demonstrated the benefit of providing circadian stimulation throughout the entire workday.

In addition to light being an effective regulator of visual perception, it has now been established that it also regulates the biological clock or the circadian rhythm of humans [229]. For example, morning light is characterized by high blue content and is responsible for the onset of the biological clock in humans. Also known as the circadian clock, the biological clock which regulates the human sleep-wake cycle is an outcome of melatonin secretions in the body, directly impacted by the spectral properties of light incident on the eye over a period of time [230]. The need for a luminaire which allows for "circadian tenability" was envisaged in 2002, when Berson et al. identified a new photoreceptor, the intrinsically photosensitive retinal ganglion cells (ipRGCs) that transmit light stimuli to the suprachiasmatic nucleus (SCN) which governs the regulation of the circadian, hormonal and behavioural systems of an organism [231]. It was also discovered that the ipRGCs are maximally sensitive to blue light between 446 and 477 nm [232,233]. This brings the possibility of artificial lighting being both beneficial and detrimental to human health. Several human circadian photo-transduction models have been proposed by several researchers. Extensive studies have shown that exposure of photo-biologically active light at inappropriate timing and intensity affects human health eliciting retinal damage [234], glucose tolerance impairments [235] and negative impact on psychological and physiological health [236]. There are also past works on the effects of melatonin suppression due to light at night (LAN) on certain types of cancers such as breast cancer, colorectal cancer and endometrial cancer, and the development of tumours. On the other hand, studies have also suggested that light exposure scheduling can be implemented to treat jet-lag disorder and promote circadian entrainment in night-shift workers as a countermeasure to circadian disruption by exposing them to temporary bright light for higher productivity and restricting bright light in the latter shift period [237]. These aspects are nicely reviewed by Saw et al. [238].

Two newly proposed figures of merit, the two-measure system of colours (Rf and Rg) and the circadian tenability of the CER, should also be analyzed and optimized to develop appropriate phosphor materials for highly efficient, widely colour-reproducible and healthful DC-WLED lightings. high luminance (first generation), high thermal stability (second generation) and high CRI (third generation) were sequentially developed by DC materials and LED-related scientists and engineers to produce illumination-grade LED lightings before the two-measure system of colour metrics and the circadian-tunable system to control circadian luminance levels were considered. Therefore, a high quantum yield (QY), pure colour coordinates and tunable PL peaks must initially be considered as prerequisites in the development of DC-QD materials for energy-saving, human-centric and sun-like colour-reproducible WLED lamps.

A light source compatible with the human centric concept has to match the natural day light as closely as possible. However, the daylight colour temperature in fact changes from dawn to dusk, and our circadian cycle follows this change closely. Recent report on lighting for health and well-being by European Union has shown clearly that light has strong effect on mood, attention and alertness through the light non-visual effect on human body. Thus, attention is now being paid on dynamic change of spectral content, intensity, luminance, duration and timing over the circadian cycle of 24 hours. To mimic this circadian cycle-induced mood, broad colour tunability is a prerequisite, and it can be applied in advanced mood lighting applications (i.e. aircraft cabins). The key is to consistently maintain high CRI when changing the colour [239].

At present, phosphor-converted white light-emitting diodes (pc-WLEDs) is the mainstream approach to realize the WLEDs due to their simple process and high luminous efficiency. The WLEDs with a high colour rendering index (CRI) can be accomplished by combining the blue LED chips with green and red phosphors or the NUV LED chips with trichromatic (blue, green and red) phosphors. However, the WLEDs produced by these approaches exist a cyan cavity in the range from 480 to 520 nm, which places restrictions on the vividness of the WLEDs. Meanwhile, compared with the trichromatic phosphors, only a few cyan light-emitting phosphors, for example,

$Y_{10}Si_3Al_2O_{18}N_4$:Ce^{3+} [240], $Ca_2LuZr_2(AlO_4)_3$:Ce^{3+} [153], $Ca_2YHf_2Al_3O_{12}$:Ce^{3+} [151], are available. Recently, some Ce^{3+}-activated garnet-type cyan phosphors, such as $Ca_2YZr_2Al_3O_{12}$:Ce^{3+} and $Ca_2LuHf_2Al_3O_{12}$:Ce^{3+}, reported by Wang et al. and $Ca_2GdZr_2Al_3O_{12}$:Ce^{3+} reported by Gong et al. [132,241,242] have been designed. These three phosphors showed efficient excitation wavelengths from 410 to 420 nm and exhibited a broad cyan emission band and high IQE, meeting the conditions of cyan phosphors for NUV-based w-LEDs.

Prerequisites for such highly demanded cyan-emitting phosphors include the following: (i) it can give rise to emission in the blue-green wavelength range of 480–520 nm with a peak around 490 nm; (ii) a strong absorption in the NUV spectral range of 380–420 nm; (iii) high photoluminescence (PL) quantum efficiency (QE); (iv) a broad emission band with a large bandwidth value; (v) good chemical and thermal stability; (vi) cost-effective and process consistency for mass production; and (vii) non-pollution and environment-friendly. These exacting requirements, however, make it extremely difficult to develop such a desirable cyan phosphor for LED solid-state lighting.

This motivation has resulted in great efforts to identify cyan emitting phosphors using various inorganic hosts, such as silicates, aluminates, nitrides, sulphides, phosphates, vanadates and borates [243].

Human-centric lighting is not just about obtaining a source with good CRI, CCT, colour gamut, etc. Ideally, such lighting should give natural daylight effect. The daylight itself changes with every hour of the day. It is important for human health and psychology to feel these changes even indoor under artificial lighting. The recent trend thus is to design a lighting source with wide tunability of the above-mentioned factors. Apart from this, there are some special requirements for specific cases. For example, lighting in a hospital ward for vision impaired patients cannot be same as that for old age homes.

With the emergence of multi-channel LED-based lighting systems, spectral tuning is now realizable as opposed to traditional lights with a fixed spectrum such as fluorescent lamps and single-channel white LED luminaires. Some literature on the implementation of these multi-channel luminaires are from Chew et al. [244] illustrating the design of an eight-channel spectrally tunable luminaire integrated with a novel closed-loop algorithm and its capacity to replicate target spectra and Tang et al. [245] presenting a colour control methodology for multi-channel luminaire systems using the camera on Android smartphones as the feedback sensor.

Summarizing, science is ready with its solutions for providing human centric lighting. Thanks to the availability of compact sources like LED, diurnal and annual changes in the lighting can be incorporated and programmed within the single envelope. Future developments and technology evolution depend on other factors such as social awareness, demand, marketing, pricing, etc.

6.5   CONCLUSIONS AND OUTLOOK

With a humble beginning, in 1996, solid-state lighting has made a rapid progress. Seemingly impossible task of producing GaN-based chips has now become so routine that it has become a cottage industry in China and some Asian countries.

Initially, YAG:Ce was the only choice as the colour conversion phosphor for LED lamp. With phenomenal growth in research on LED phosphors, which was spurred by the Nobel Prize in 2014, a large number of phosphors are now available; several of them commercially. Due to this situation, LED lamp manufacturers can meet more and more stringent specifications imposed by regulating authorities. It became thus possible not only to achieve lumen efficiency and CRI but also various other criteria like colour purity, colour gamut, colour consistency, full spectrum emission, etc., can also be met. Next goal is to achieve cost reduction which may be achieved through the use of rare earth-free phosphors.

Even today, InGaN chips lead to most efficient electroluminescence (EL). These chips emit in the blue region. Efficiencies of chips in violet and nUV have also improved a great deal. With these improvements, LED lamps for plant growth can be produced which is a step forward towards reducing recurring expenditure on greenhouses. Availability of Cr^{3+}- and Mn^{4+}-activated oxide phosphors has now made it possible to produce emission that can match any type of photoreceptor in the plants. It has thus become economically viable to cultivate any type of crop in any part of the world.

A mature and rich lighting technology has made it possible to go beyond primary needs of lighting such as intensity and CRI. LED lighting can now be designed to take care of the working environment and personnel health. Summarizing, science is ready with its solutions for providing human centric lighting. Thanks to the availability of compact sources like LED, diurnal and annual changes in the lighting can be incorporated and programmed within the single envelope. Future developments and technology evolution depend on other factors such as social awareness, demand, marketing, pricing, etc.

REFERENCES

1. https://en.wikipedia.org/wiki/Arc_lamp.
2. https://www.bulbs.com/learning/history.aspx.
3. Kuch, R. and Retschinsky, T., (1906). Photometrische und spektralphotometrische Messungen am Quecksilberlichtbogen bei hohem Dampfdruck. *Annalen der Physik*, 325(8), 563–583.
4. Jonas, B., (1938). Sealing of metal leads through hard glass and silica. *Philips Tech. Rev.*, 3, 119–124.
5. Koedam, M. and Opstelten, J. J., (1971). Measurement and computer-aided optimization of spectral power distributions. *Light Res. Tech.*, 3, 205–210. https://doi.org/10.1177/096032717100300303.
6. Thornton, W., (2013). Fluorescent lamps with high color-discrimination capability. *J. Illum. Eng. Soc.*, 3, 61–64. Doi: 10.1080/00994480.1973.10732226.
7. Nakamura, S., Mukai, T. and Senoh, M., (1994). Candela-class high-brightness InGaN/AlGaN double-heterostructure blue-light-emitting diodes. *Appl. Phys. Lett.*, 64(13), 1687–1689. https://doi.org/10.1063/1.111832.
8. Akasaki, I., Amano, H., Kito, M. and Hiramatsu, K., (1991). Photoluminescence of Mg doped p-type GaN and electroluminescence of GaN p-n junction LED. *J. Lumin.*, 48 & 49, 666–670. https://doi.org/10.1016/0022-2313(91)90215-H.
9. https://andromedalighting.com/led-history/.
10. Blasse, G. and Bril, A., (1967). A new phosphor for flying-spot cathode-ray tubes for color television: Yellow-emitting Y3Al5O12-Ce3+. *Appl. Phys. Lett.*, 11(2), 53–55. https://dx.doi.org/10.1063/1.1755025.
11. Blasse, G. and Bril, A., (1967). Investigation of some Ce3+-activated phosphors. *Chem. Phys.*, 47(12), 5139–5145. https://doi.org/10.1063/1.1701771
12. Kottaisamy, M., Thiyagarajan, P., Mishra, J. and Rao, M. R., (2008). Color tuning of Y3Al5O12: Ce phosphor and their blend for white LEDs. *Mater. Res. Bull.*, 43(7), 1657–1663. https://dx.doi.org/10.1016/j.materresbull.2007.09.005.
13. Ming-Hui, C., Xue-Yan, L., Qing, W., Jian, W., Da-Peng, J., De-Zhen, S., Cheng-Jiu, Z. and Feng-Qin, H., (2010). Luminescence reduction and thermally induced luminescence quenching of Ce, Gd highly doped (Y3-x-yCexGdy) Al5O12 phosphors for white LEDs. *Chin. J. Inorg. Chem*, 26(2), 183–189.
14. Shao, Q., Li, H., Dong, Y., Jiang, J., Liang, C. and He, J., (2010). Temperature-dependent photoluminescence studies on Y2.93– xLnxAl5O12: Ce0. 07 (Ln= Gd, La) phosphors for white LEDs application. *J. Alloys Compd.*, 498(2), 199–202. https://doi.org/10.1016/j.jallcom.2010.03.159
15. Wyner, E. F. and Daigneault, A. J., (1980). Improved mercury lamp for low color temperature applications. *J. Illum. Eng. Soc.*, 9(2), 109–114. https://doi.org/10.1080/00994480.1980.10747886
16. Van Kemenade, J. T. C., Siebers, G. H. M., Heuvelmans, J. J., Hair, De J. T. W. and TerVrugt, J. W., (1987). European patent 209942 A1.
17. Bachmann, V., Ronda, C. and Meijerink, A., (2009). Temperature quenching of yellow Ce^{3+} luminescence in YAG: Ce. *Chem. Mater.*, 21(10), 2077–2084. https://doi.org/10.1021/cm8030768.
18. Kucera, M., Nitsch, K., Nikl, M., Hanuš, M. and Daniš, S., (2010). Growth and characterization of YAG and LuAG epitaxial films for scintillation applications. *J. Cryst. Growth*, 312(9), 1538–1545. https://doi.org/10.1016/j.jcrysgro.2010.01.023.
19. Valais, I. G., Michail, C. M., David, S. L., Liaparinos, P. F., Fountos, G. P., Paschalis, T. V., Kandarakis, I. S. and Panayiotakis, G. S., (2010). Comparative investigation of Ce^{3+} doped scintillators in a wide range of photon energies covering X-ray CT, nuclear medicine and megavoltage radiation therapy portal imaging applications. *IEEE Trans. Nucl. Sci.*, 57(1), 3–7. Doi: 10.1109/TNS.2009.2038273

20. Zhao, W., Mancini, C., Amans, D., Boulon, G., Epicier, T., Min, Y., Yagi, H., Yanagitani, T., Yanagida, T. and Yoshikawa, A., (2010). Evidence of the inhomogeneous Ce^{3+} distribution across grain boundaries in transparent polycrystalline Ce^{3+}-doped (Gd, Y)$_3$Al$_5$O$_{12}$ garnet optical ceramics. *Jpn. J. Appl. Phys.*, 49(2R), 022602. Doi: 10.1143/JJAP.49.022602/meta

21. Jacinto, C., Benayas, A., Catunda, T., García-Solé, J., Kaminskii, A. A. and Jaque, D., (2008). Microstructuration induced differences in the thermo-optical and luminescence properties of Nd: YAG fine grain ceramics and crystals. *Chem. Phys.*, 129(10), 104705. https://doi.org/10.1063/1.2975335

22. Asakura, R., Kusayama, I., Saito, D., Isobe, T., Kurokawa, K., Hirayama, Y., Aizawa, H., Takagi, T. and Ohkubo, M. (2007). Preparation of fluorescent poly (methyl methacrylate) beads hybridized with Y$_3$Al$_5$O$_{12}$: Ce^{3+} Nanophosphor for Biological Application. *Jpn. J. Appl. Phys.*, 46(8R), 5193. Doi: 10.1143/JJAP.46.5193/meta

23. Tsukamoto, A. and Isobe, T. (2009). Characterization and biological application of YAG:Ce^{3+} nanophosphor modified with mercaptopropyl trimethoxy silane. *J. Mater. Sci.*, 44(5), 1344–1350. https://doi.org/10.1007/s10853-008-3012-4

24. Ueda, J. and Tanabe, S., (2009). Visible to near infrared conversion in Ce3+-Yb3+co-doped YAG ceramics. *J. Appl. Phys.*, 106(4), 043101. https://doi.org/10.1063/1.3194310

25. Lin, H., Zhou, S., Teng, H., Li, Y., Li, W., Hou, X. and Jia, T. (2010). Near infrared quantum cutting in heavy Yb doped Ce$_{0.03}$Yb$_{3x}$ Y $_{(2.97-3x)}$ Al$_5$O$_{12}$ transparent ceramics for crystalline silicon solar cells. *J. Appl. Phys.*, 107(4), 043107. https://doi.org/10.1063/1.3298907

26. Li, Y., Zhou, S., Lin, H., Hou, X. and Li, W., (2010). Intense 1064 nm emission by the efficient energy transfer from Ce^{3+} to Nd^{3+} in Ce/Nd co-doped YAG transparent ceramics. *Opt. Mater.*, 32(9), 1223–1226. https://doi.org/10.1016/j.optmat.2010.04.003

27. Cuche, A., Masenelli, B., Ledoux, G., Amans, D., Dujardin, C., Sonnefraud, Y., Melinon, P. and Huant, S. (2008). Fluorescent oxide nanoparticles adapted to active tips for near-field optics. *Nanotechnology*, 20(1), 015603. Doi: 10.1088/0957-4484/20/1/015603/meta

28. Liu, Z., Liu, S., Wang, K. and Luo, X., (2010). Measurement and numerical studies of optical properties of YAG: Ce phosphor for white light-emitting diode packaging. *Appl. Opt.*, 49(2), 247–257. https://doi.org/10.1364/AO.49.000247

29. Chao, W. H., Wu, R. J., Tsai, C. S. and Wu, T. B., (2010). Surface plasmon-enhanced emission from Ag-coated Ce doped Y3Al5O12 thin films phosphor capped with a dielectric layer of SiO 2. *J. Appl. Phys.*, 107(1), 013101. https://doi.org/10.1063/1.3277015

30. Zorenko, Y., Mares, J. A., Prusa, P., Nikl, M., Gorbenko, V., Savchyn, V., Kucerkova, R. and Nejezchleb, K. (2010). Luminescence and scintillation characteristics of YAG: Ce single crystalline films and single crystals. *Radiat. Meas.*, 45(3–6), 389–391. https://doi.org/10.1016/j.radmeas.2009.09.009

31. Zorenko, Y., Gorbenko, V., Mihokova, E., Nikl, M., Nejezchleb, K., Vedda, A., Kolobanov, V. and Spassky, D., (2007). Single crystalline film scintillators based on Ce-and Pr-doped aluminium garnets. *Radiat. Meas.*, 42(4–5), 521–527. https://doi.org/10.1016/j.radmeas.2007.01.045

32. Kucera, M., Nitsch, K., Nikl, M. and Hanus, M., (2010). Defects in Ce-doped LuAG and YAG scintillation layers grown by liquid phase epitaxy. *Radiat. Meas.*, 45(3–6), 449–452. https://doi.org/10.1016/j.radmeas.2009.12.031

33. Jia, D., Wang, Y., Guo, X., Li, K., Zou, Y. K. and Jia, W., (2007) *Electrochem. Soc.*, 154, J1. Doi: 10.1149/1.2372589

34. Pan, Y., Wu, M. and Su, Q., (2004). Comparative investigation on synthesis and photoluminescence of YAG: Ce phosphor. *Mater. Sci. Eng. B.*, 106(3), 251–256. https://doi.org/10.1016/j.mseb.2003.09.031

35. Ohno, K. and Abe, T., (1994). The synthesis and particle growth mechanism of bright green phosphor YAG: Tb. *J. Electrochem. Soc.*, 141(5), 1252. Doi: 10.1149/1.2054905

36. Ohno, K. and Abe, T., (1986). Effect of BaF2 on the synthesis of the single-phase cubic Y3Al5O12: Tb. *J. Electrochem. Soc.*, 133(3), 638. Doi: 10.1149/1.2108635

37. Chiang, C. C., Tsai, M. S., Hsiao, C. S. and Hon, M. H., (2006). Synthesis of YAG: Ce phosphor via different aluminum sources and precipitation processes. *J. Alloys Compd.*, 416(1–2), 265–269. https://doi.org/10.1016/j.jallcom.2005.08.041

38. Fadlalla, H. M. H. and Tang, C. C., (2009). YAG: Ce^{3+} nano-sized particles prepared by precipitation technique. *Mater. Chem. Phys.*, 114(1), 99–102. https://doi.org/10.1016/j.matchemphys.2008.08.049

39. Jung, K. Y. and Kang, Y. C., (2010). Luminescence comparison of YAG: Ce phosphors prepared by microwave heating and precipitation methods. *Phys. B: Condens. Matter.*, 405(6), 1615–1618. https://doi.org/10.1016/j.physb.2009.12.052

40. Wang, S. F., Rao, K. K., Wu, Y. C., Wang, Y. R., Hsu, Y. F. and Huang, C. Y., (2009). Synthesis and characterization of Ce^{3+}: YAG phosphors by heterogeneous precipitation using different alumina sources. *Int. J. Appl. Ceram.* 6(4), 470–478. https://doi.org/10.1111/j.1744-7402.2008.02275.x

41. Lee, S. H., Koo, H. Y., Lee, S. M. and Kang, Y. C., (2010). Characteristics of $Y_3Al_5O_{12}$: Ce phosphor powders prepared by spray pyrolysis from ethylenediaminetetraacetic acid solution. *Ceram. Int.*, 36(2), 611–615. https://doi.org/10.1016/j.ceramint.2009.09.041

42. Mancic, L., Marinkovic, K., Marinkovic, B. A., Dramicanin, M. and Milosevic, O., (2010). YAG: Ce^{3+} nanostructured particles obtained via spray pyrolysis of polymeric precursor solution. *J. Eur. Ceram. Soc.*, 30(2), 577–582. https://doi.org/10.1016/j.jeurceramsoc.2009.05.037

43. Lee, J. S., Kumar, P., Gupta, S., Oh, M. H., Ranade, M. B. and Singh, R. K., (2009). Enhanced luminescence properties of YAG: Ce^{3+} nanophosphor prepared by flame spray pyrolysis. *J. Electrochem. Soc.*, 157(2), K25. Doi: 10.1149/1.3262609

44. Purwanto, A., Wang, W. N., Ogi, T., Lenggoro, I. W., Tanabe, E. and Okuyama, K., (2008). High luminance YAG: Ce nanoparticles fabricated from urea added aqueous precursor by flame process. *J. Alloys Compd.*, 463(1–2), 350–357. https://doi.org/10.1016/j.jallcom.2007.09.023

45. Lu, C. H. and Jagannathan, R., (2002). Cerium-ion-doped yttrium aluminum garnet nanophosphors prepared through sol-gel pyrolysis for luminescent lighting. *Appl. Phys. Lett.*, 80(19), 3608–3610. https://doi.org/10.1063/1.1475772

46. Yang, L., Lu, T., Xu, H., Zhang, W. and Ma, B. (2010). A study on the effect factors of sol-gel synthesis of yttrium aluminum garnet nanopowders. *J. Appl. Phys.*, 107(6), 064903. https://doi.org/10.1063/1.3341012

47. Lu, C. H., Hong, H. C. and Jagannathan, R., (2002). Sol-gel synthesis and photoluminescent properties of cerium-ion doped yttrium aluminium garnet powders. *J. Mater. Chem.*, 12(8), 2525–2530. https://doi.org/10.1039/B200776M

48. Ovalle, R., Arredondo, A., Diaz-Torres, L. A., Salas, P., Angeles, C., Rodriguez, R. A., Meneses, M. A. and De la Rosa, E., (2004). Concentration and crystallite size dependence of the photoluminescence in YAG: Ce3+ nanophosphor. In *Fourth International Conference on Solid State Lighting* Vol. 5530, 274–283. SPIE. https://doi.org/10.1117/12.566420

49. Jiao, H., Ma, Q., He, L., Liu, Z. and Wu, Q., (2010). Low temperature synthesis of YAG: Ce phosphors by LiF assisted sol-gel combustion method. *Powder Technol.*, 198(2), 229–232. https://doi.org/10.1016/j.powtec.2009.11.011

50. Kamiyama, Y., Hiroshima, T., Isobe, T., Koizuka, T. and Takashima, S., (2010). Photostability of YAG: Ce^{3+} nanophosphors synthesized by glycothermal method. *J. Electrochem. Soc.*, 157(5), J149. Doi: 10.1149/1.3327907/meta

51. Pan, Y., Wu, M. and Su, Q., (2004). Tailored photoluminescence of YAG: Ce phosphor through various methods. *J. Phys. Chem. Solids*, 65(5), 845–850. https://doi.org/10.1016/j.jpcs.2003.08.018

52. Yang, Z., Li, X., Yang, Y. and Li, X., (2007). The influence of different conditions on the luminescent properties of YAG: Ce phosphor formed by combustion. *J. Lumin.*, 122, 707–709. https://doi.org/10.1016/j.jlumin.2006.01.266

53. Fu, Y. P., Wen, S. B. and Hsu, C. S., (2008). Preparation and characterization of Y3Al5O12: Ce and Y_2O_3: Eu phosphors powders by combustion process. *J. Alloys Compd.*, 458(1–2), 318–322. https://doi.org/10.1016/j.jallcom.2007.03.147

54. Zhang, K., Hu, W., Wu, Y. and Liu, H., (2008). Photoluminescence investigations of $(Y_{1-x}Ln_x)_3Al_5O_{12}$: Ce (Ln3+= Gd3+, La3+) nanophosphors. *Phys. B: Condens. Matter.*, 403(10–11), 1678–1681. https://doi.org/10.1016/j.physb.2007.09.084

55. Wang, J., Zheng, S., Zeng, R., Dou, S. and Sun, X., (2009). Microwave synthesis of homogeneous YAG nanopowder leading to a transparent ceramic. *J. Am. Ceram. Soc.*, 92(6), 1217–1223. https://doi.org/10.1111/j.1551-2916.2009.03086.x

56. Smet, P. F., Parmentier, A. B. and Poelman, D., (2011). Selecting conversion phosphors for white light-emitting diodes. *J. Electrochem. Soc.*, 158(6), R37. Doi: 10.1149/1.3568524/meta

57. Ueda, J., Tanabe, S. and Nakanishi, T., (2011). Analysis of Ce3+ luminescence quenching in solid solutions between $Y_3Al_5O_{12}$ and $Y_3Ga_5O_{12}$ by temperature dependence of photoconductivity measurement. *J. Appl. Phys.*, 110(5), 053102. https://doi.org/10.1063/1.3632069

58. Maeda, T., Oshio, S., Iwama, K., Kitahara, H., Ikeda, T., Kamei, H., Hanada, Y. and Sakanoue, K., (2008). Patent US7,422,504

59. Tien, T. Y., Gibbons, E. F., DeLosh, R. G., Zacmanidis, P. J., Smith, D. E. and Stadler, H. L., (1973). Ce3+ activated Y3Al5O12 and some of its solid solutions. *J. Electrochem. Soc.*, 120(2), 278. Doi: 10.1149/1.2403436/meta

60. Satilmis, S. U., Ege, A., Ayvacikli, M., Khatab, A., Ekdal, E., Popovici, E. J., Henini, M. and Can, N., (2012). Luminescence characterization of cerium doped yttrium gadolinium aluminate phosphors. *Opt. Mater.*, 34(11), 1921–1925. https://doi.org/10.1016/j.optmat.2012.06.002

61. Latynina, A., Watanabe, M., Inomata, D., Aoki, K., Sugahara, Y., Víllora, E. G. and Shimamura, K. (2013). Properties of Czochralski grown Ce, Gd: $Y_3Al_5O_{12}$ single crystal for white light-emitting diode. *J. Alloys Compd.*, 553, 89–92. https://doi.org/10.1016/j.jallcom.2012.11.096

62. Lin, Y. S., Liu, R. S. and Cheng, B. M., (2005). Investigation of the luminescent properties of Tb^{3+}-substituted YAG: Ce, Gd phosphors. *J. Electrochem. Soc.*, 152(6), J41. Doi: 10.1149/1.1896307/meta

63. Fujita, S., Sakamoto, A. and Tanabe, S., (2008). Luminescence characteristics of YAG glass-ceramic phosphor for white LED. *IEEE J. Select. Top. Quant. Electron.*, 14(5), 1387–1391. Doi: 10.1109/JSTQE.2008.920285

64. Dotsenko, V. P., Berezovskaya, I. V., Zubar, E. V., Efryushina, N. P., Poletaev, N. I., Doroshenko, Y. A., Stryganyuk, G. B. and Voloshinovskii, A. S., (2013). Synthesis and luminescent study of Ce3+-doped terbium-yttrium aluminum garnet. *J. Alloys Compd.*, 550, 159–163. Doi: 10.1016/j.jallcom.2012.09.053

65. Hansel, R., Allison, S. and Walker, G., (2010). Temperature-dependent luminescence of gallium-substituted YAG: Ce. *J. Mater. Sci.*, 45(1), 146–150. https://doi.org/10.1007/s10853-009-3906-9

66. Kamada, K., Yanagida, T., Pejchal, J., Nikl, M., Endo, T., Tsutumi, K., Fujimoto, Y., Fukabori, A. and Yoshikawa, A., (2011). Scintillator-oriented combinatorial search in Cedoped (Y, Gd)3(Ga, Al)5O12 multicomponent garnet compounds. *J. Phys. D Appl. Phys.*, 44(50), 505104. Doi: 10.1088/0022-3727/44/50/505104/meta

67. Dorenbos, P., (2013). Electronic structure and optical properties of the lanthanide activated RE3 $(Al_{1-x}Ga_x)_5O_{12}$ (RE= Gd, Y, Lu) garnet compounds. *J. Lumin.*, 134, 310–318. https://doi.org/10.1016/j.jlumin.2012.08.028

68. Han, T., Cao, S., Peng, L., Zhu, D., Zhao, C., Tu, M. and Zhang, J., (2012). Chemical substitution effects of elements on photoluminescence properties of YAG: Ce phosphors using orthogonal experimental design. *Opt. Mater.*, 34(9), 1618–1621. https://doi.org/10.1016/j.optmat.2012.03.035

69. Wang, L., Zhang, X., Hao, Z., Luo, Y., Zhang, L., Zhong, R. and Zhang, J., (2012). Interionic energy transfer in Y3Al5O12: Ce3+, Pr3+, Cr3+ phosphor. *J. Electrochem. Soc.*, 159(4), F68. https://10.1149/2.054204jes/meta

70. Zhou, X., Zhou, K., Li, Y., Wang, Z. and Feng, Q., (2012). Luminescent properties and energy transfer of $Y_3Al_5O_{12}$: Ce^{3+}, Ln^{3+} (Ln= Tb, Pr) prepared by polymer-assisted sol-gel method. *J. Lumin.*, 132(11), 3004–3009. https://doi.org/10.1016/j.jlumin.2012.06.005

71. Yan, X., Li, W., Wang, X. and Sun, K., (2011). Facile synthesis of Ce3+, Eu3+ co-doped YAG nanophosphor for white light-emitting diodes. *J. Electrochem. Soc.*, 159(2), H195. https://10.1149/2.101202jes/meta

72. Jang, H. S., Im, W. B., Lee, D. C., Jeon, D. Y. and Kim, S. S., (2007). Enhancement of red spectral emission intensity of Y3Al5O12: Ce3+ phosphor via Pr co-doping and Tb substitution for the application to white LEDs. *J. Lumin.*, 126(2), 371–377. https://doi.org/10.1016/j.jlumin.2006.08.093

73. Song, Y. H., Choi, T. Y., Senthil, K., Masaki, T. and Yoon, D. H., (2012). Enhancement of photoluminescence properties of green to yellow emitting $Y_3Al_5O_{12}$: Ce^{3+}phosphor by AlN addition for white LED applications. *Mater. Lett.*, 67(1), 184–186. https://doi.org/10.1016/j.matlet.2011.08.114

74. Wang, X., Zhou, G., Zhang, H., Li, H., Zhang, Z. and Sun, Z., (2012). Luminescent properties of yellowish orange $Y_3Al_{5-x}Si_xO_{12-}$ xNx: Ce phosphors and their applications in warm white light-emitting diodes. *J. Alloys Compd.*, 519, 149–155. https://doi.org/10.1016/j.jallcom.2011.12.158

75. Sopicka-Lizer, M., Michalik, D., Plewa, J., Juestel, T., Winkler, H. and Pawlik, T., (2012). The effect of Al-O substitution for Si-N on the luminescence properties of YAG: Ce phosphor. *J. Eur. Ceram. Soc.*, 32(7), 1383–1387. https://doi.org/10.1016/j.jeurceramsoc.2011.04.021

76. Zhao, C., Zhu, D., Ma, M., Han, T. and Tu, M., (2012). Brownish red emitting YAG: Ce^{3+}, Cu^+ phosphors for enhancing the color rendering index of white LEDs. *J. Alloys Compd.*, 523, 151–154. https://doi.org/10.1016/j.jallcom.2012.01.131

77. Xiang, W., Zhong, J., Zhao, Y., Zhao, B., Liang, X., Dong, Y., Zhang, Z., Chen, Z. and Liu, B., (2012). Growth and characterization of air annealing Mn-doped YAG: Ce single crystal for LED. *J. Alloys Compd.*, 542, 218–221. https://doi.org/10.1016/j.jallcom.2012.07.009

78. Mu, Z., Hu, Y., Wu, H., Fu, C. and Kang, F., (2011). The structure and luminescence properties of a novel orange emitting phosphor $Y_3Mn_xAl_{5-2x}Si_xO_{12}$. *Phys. B: Condens. Matter.*, 406(4), 864–868. https://doi.org/10.1016/j.physb.2010.12.015

79. Mu, Z., Hu, Y., Wang, Y., Wu, H., Fu, C. and Kang, F., (2011). The structure and luminescence properties of long afterglow phosphor $Y_{3-x}Mn_xAl_{5-2x}Si_xO_{12}$. *J. Lumin.*, 131(4), 676–681. https://doi.org/10.1016/j.jlumin.2010.11.016

80. Katelnikovas, A., Jurkevičius, J., Kazlauskas, K., Vitta, P., Jüstel, T., Kareiva, A., Zukauskas, A. and Tamulaitis, G., (2011). Efficient cerium-based sol-gel derived phosphors in different garnet matrices for light-emitting diodes. *J. Alloys Compd.*, 509(21), 6247–6251. https://doi.org/10.1016/j. jallcom.2011.03.032.

81. Jia, Y., Huang, Y., Zheng, Y., Guo, N., Qiao, H., Zhao, Q., Lv, W. and You, H., (2012). Color point tuning of $Y_3Al_5O_{12}$: Ce^{3+} phosphor via Mn^{2+}-Si^{4+} incorporation for white light generation. *J. Mater. Chem.*, 22(30), 15146–15152. https://doi.org/10.1039/C2JM32233A.

82. Shi, Y., Wang, Y., Wen, Y., Zhao, Z., Liu, B. and Yang, Z., (2012). Tunable luminescence $Y_3Al_5O_{12}$: 0.06 Ce^{3+}, xMn^{2+} phosphors with different charge compensators for warm white light emitting diodes. *Opt. Express*, 20(19), 21656–21664. https://doi.org/10.1364/OE.20.021656.

83. Jung, H. C., Park, J. Y., Raju, G. S. R., Choi, B. C., Jeong, J. H. and Moon, B. K., (2011). Enhancement of red emission in aluminum garnet yellow phosphors by Sb^{3+} substitution for the octahedral site. *J. Am. Ceram. Soc.*, 94(2), 551–555. https://doi.org/10.1111/j.1551-2916.2010.04130.x.

84. Chen, S., Zhang, L., Kisslinger, K. and Wu, Y., (2013). Transparent $Y_3Al_5O_{12}$: Li, Ce ceramics for thermal neutron detection. *J. Am. Ceram. Soc.*, 96(4), 1067–1069. https://doi.org/10.1111/jace.12297.

85. Hao, Z., Zhang, X., Wang, X. and Zhang, J., (2012). Photoluminescence properties of CaO:Ce^{3+}, Na^+, a non-garnet yellow-emitting phosphor under blue light excitation. *Mater. Lett.*, 68, 443–445. https://doi. org/10.1016/j.matlet.2011.10.102.

86. Kim, J. H. and Jung, K. Y., (2012). Luminescence characteristics and optimization of (La, Gd) $Sr_2(Al, B)$ O_5: Ce phosphor for white light emitting diodes. *J. Lumin.*, 132(6), 1376–1381. https://doi.org/10.1016/j. jlumin.2012.01.029.

87. Jang, H. S., Kim, H. Y., Kim, Y. S., Lee, H. M. and Jeon, D. Y., (2012). Yellow-emitting γ-Ca_2SiO_4: Ce^{3+}, Li^+ phosphor for solid-state lighting: Luminescent properties, electronic structure, and white light-emitting diode application. *Opt. Express*, 20(3), 2761–2771. https://doi.org/10.1364/ OE.20.002761.

88. Li, G., Deng, D., Su, X., Wang, Q., Li, Y., Hua, Y., Huang, L., Zhao, S., Wang, H., Li, C. and Xu, S., (2011). Sol-gel synthesis of yellow-emitting $La_4Ca(SiO_4)_3O$: Ce^{3+} phosphors for white-light emitting diodes. *Mater. Lett.*, 65(13), 2019–2021. https://doi.org/10.1016/j.matlet.2011.04.026.

89. Yuanhong, L. I. U., Zhuang, W., Ronghui, L. I. U., Yunsheng, H. U., Huaqiang, H. E., Zhang, S. and Wei, G. A. O., (2012). Spectral variations of $Ca_3Sc_2Si_3O_{12}$: Ce phosphors via substitution and energy transfer. *J. Rare Earths*, 30(4), 339–341. https://doi.org/10.1016/S1002-0721(12)60049-0.

90. Liu, Y., Zhang, X., Hao, Z., Wang, X. and Zhang, J., (2011). Generation of broadband emission by incorporating N^{3-} into $Ca_3Sc_2Si_3O_{12}$: Ce^{3+} garnet for high rendering white LEDs. *J. Mater. Chem.*, 21(17), 6354–6358. https://doi.org/10.1039/C0JM04404K.

91. Liu, Y., Zhang, X., Hao, Z., Luo, Y., Wang, X. and Zhang, J., (2011). Generating yellow and red emissions by co-doping Mn^{2+} to substitute for Ca^{2+} and Sc^{3+} sites in $Ca_3Sc_2Si_3O_{12}$: Ce^{3+} green emitting phosphor for white LED applications. *J. Mater. Chem.*, 21(41), 16379–16384. https://doi.org/10.1039/C1JM11601K.

92. Liu, Y., Zhang, X., Hao, Z., Luo, Y., Wang, X., Ma, L. and Zhang, J., (2013). Luminescence and energy transfer in $Ca_3Sc_2Si_3O_{12}$: Ce^{3+}, Mn^{2+} white LED phosphors. *J. Lumin.*, 133, 21–24. https://doi. org/10.1016/j.jlumin.2011.12.052.

93. Chunyan, L. U., Huaidong, J., Liang, X., Xiang, W., Zhong, J. and Yongjun, D., (2012). A novel green-yellow emitting phosphor $Ca_{1.5}Y_{1.5}Al_{3.5}Si_{1.5}O_{12}$: Ce^{3+} and its luminescence properties. *J. Rare Earths*, 30(7), 647–650. https://doi.org/10.1016/S1002-0721(12)60106-9.

94. Katelnikovas, A., Sakirzanovas, S., Dutczak, D., Plewa, J., Enseling, D., Winkler, H., Kareiva, A. and Justel, T., (2013). Synthesis and optical properties of yellow emitting garnet phosphors for pcLEDs. *J. Lumin.*, 136, 17–25. https://doi.org/10.1016/j.jlumin.2012.11.012.

95. Liu, Y., Zhang, X., Hao, Z., Luo, Y., Wang, X. and Zhang, J., (2012). Crystal structure and luminescence properties of Lu^{3+} and Mg^{2+} incorporated silicate garnet $[Ca_{3-(x+0.06)}Lu_xCe_{0.06}](Sc_{2-y}Mg_y)$ Si_3O_{12}. *J. Lumin.*, 132(5), 1257–1260. https://doi.org/10.1016/j.jlumin.2011.12.060

96. Zhang, X., Liu, Y., Hao, Z., Luo, Y., Wang, X. and Zhang, J., (2012). Yellow-emitting $(Ca2Lu1-xCex)$ $(ScMg)$ $Si3O12$ phosphor and its application for white LEDs. *Mater. Res. Bull.*, 47(5), 1149–1152. https:// doi.org/10.1016/j.materresbull.2012.02.013.

97. Chen, L., Luo, A., Jiang, Y., Liu, F., Deng, X., Xue, S., Chen, X. and Zhang, Y., (2013). Suppressing the phase transformation and enhancing the orange luminescence of $(Sr, Ba)_3SiO_5$: Eu^{2+} for application in white LEDs. *Mater. Lett.*, 106, 428–431. https://doi.org/10.1016/j.matlet.2013.05.057.

98. Hua, Y., Ma, H., Deng, D., Zhao, S., Huang, L., Wang, H. and Xu, S., (2014). Enhanced photoluminescence properties of orange emitting $Sr_{2.96-x}Ba_xSiO_5$: Eu^{2+} phosphors synthesized with Sr_2SiO_4 as precursor. *J. Lumin.*, 148, 39–43. https://doi.org/10.1016/j.jlumin.2013.11.056.

99. Im, W. B., George, N., Kurzman, J., Brinkley, S., Mikhailovsky, A., Hu, J., Chmelka, B. F., DenBaars, S. P. and Seshadri, R., (2011). Efficient and color-tunable oxyfluoride solid solution phosphors for solid-state white lighting. *Adv. Mater.*, 23(20), 2300–2305. https://doi.org/10.1002/adma.201003640.

100. Tao, Z., Zhang, W., Qin, L., Huang, Y., Wei, D. and Seo, H. J., (2014). A yellow-emitting nanophosphor of Ce^{3+}-activated aluminate $Sr_3LuAl_2O_{7.5}$. *J. Alloys Compd.*, 588, 540–545. https://doi.org/10.1016/j.jallcom.2013.11.128.

101. Wu, Y. C., Chen, Y. C., Chen, T. M., Lee, C. S., Chen, K. J. and Kuo, H. C., (2012). Crystal structure characterization, optical and photoluminescent properties of tunable yellow-to orange-emitting $Y_2(Ca, Sr)F_4S_2$: Ce^{3+} phosphors for solid-state lighting. *J. Mater. Chem.*, 22(16), 8048–8056. https://doi.org/10.1039/C2JM16882K.

102. Yang, Y., Su, X., Li, X., Yu, F., Mi, C. and Li, G., (2014). A novel orange emitting $BaS:xYb^{2+}$ phosphor for white light LEDs. *Mater. Res. Bull.*, 51, 202–204. https://doi.org/10.1016/j.materresbull.2013.12.016.

103. Zhang, Z. J., Feng, A., Sun, X. Y., Guo, K., Man, Z. Y. and Zhao, J. T., (2014). Preparation, electronic structure and luminescence properties of Ce3+-activated CaZnOS under UV and Xray excitation. *J. Alloys Compd.*, 592, 73–79. https://doi.org/10.1016/j.jallcom.2013.12.211.

104. Le Toquin, R. and Cheetham, A. K., (2006). Red-emitting cerium-based phosphor materials for solid-state lighting applications, 423(4–6), 352–356. https://doi.org/10.1016/j.cplett.2006.03.056.

105. Hsu, C. H., Cheng, B. M. and Lu, C. H., (2011). Photoluminescent properties and energy transfer mechanism of color-tunable $CaSi_2O_2N_2$: Ce^{3+}, Eu^{2+} phosphors. *J. Am. Ceram. Soc.*, 94(9), 2878–2883. https://doi.org/10.1111/j.1551-2916.2011.04461.x

106. Park, W. B., Singh, S. P., Pyo, M. and Sohn, K. S., (2011). $Y_{6+x/3} Si_{11-x} Al_y N_{20+x-y} O_{1-x+y}$: Re^{3+}(Re= Ce^{3+}, Tb^{3+}, Sm^{3+}) phosphors identified by solid-state combinatorial chemistry. *J. Mater. Chem.*, 21(15), 5780–5785. https://doi.org/10.1039/C0JM03538F

107. Yang, Z., Wang, Y. and Zhao, Z., (2012). Synthesis, structure and photoluminescence properties of fine yellow-orange Ca-α-SiAlON: Eu^{2+} phosphors. *J. Alloys Compd.*, 541, 70–74. https://doi.org/10.1016/j.jallcom.2012.06.107

108. Yamada, S., Emoto, H., Ibukiyama, M. and Hirosaki, N., (2012). Properties of SiAlON powder phosphors for white LEDs. *J. Eur. Ceram. Soc.*, 32(7), 1355–1358. https://doi.org/10.1016/j.jeurceramsoc.2011.05.050

109. Wu, Q., Wang, Y., Yang, Z., Que, M., Li, Y., and Wang, C., (2014). Synthesis and luminescence properties of pure nitride Ca-a-sialon with the composition $Ca_{1.4}Al_{2.8}Si_{9.2}N_{16}$ by gas-pressed sintering. *J. Mater. Chem. C*, 2, 0829–0834. https://doi.org/10.1039/C3TC31471E

110. Park, W. B., Singh, S. P., Yoon, C. and Sohn, K. S., (2013). Combinatorial chemistry of oxynitride phosphors and discovery of a novel phosphor for use in light emitting diodes, $Ca_{1.5} Ba_{0.5}Si_5N_6O_3$: Eu^{2+}. *J. Mater. Chem. C*, 1(9), 1832–1839. https://doi.org/10.1039/C2TC00731B

111. Park, W. B., Shin, N., Hong, K. P., Pyo, M. and Sohn, K. S., (2012). A new paradigm for materials discovery: Heuristics-assisted combinatorial chemistry involving parameterization of material novelty. *Adv. Funct. Mater.*, 22(11), 2258–2266. https://doi.org/10.1002/adfm.201102118

112. Liu, L., Xie, R. J., Li, W., Hirosaki, N., Yamamoto, Y. and Sun, X., (2013). Yellow-emitting $Y_3Si_6N_{11}$: Ce^{3+} phosphors for white light-emitting diodes (LED s). *J. Am. Ceram. Soc.*, 96(6), 1688–1690. https://doi.org/10.1111/jace.12357

113. Ruan, J., Xie, R. J., Funahashi, S., Tanaka, Y., Takeda, T., Suehiro, T., Hirosaki, N. and Li, Y. Q., (2013). A novel yellow-emitting $SrAlSi_4N_7$: Ce^{3+} phosphor for solid state lighting: Synthesis, electronic structure and photoluminescence properties. *J. Solid State Chem.*, 208, 50–57. https://doi.org/10.1016/j.jssc.2013.09.040

114. Zhang, X., Zhang, J., Dong, Z., Shi, J. and Gong, M., (2012). Concentration quenching of Eu^{2+} in a thermal-stable yellow phosphor Ca_2BO_3Cl: Eu^{2+} for LED application. *J. Lumin.*, 132(4), 914–918. https://doi.org/10.1016/j.jlumin.2011.11.001

115. Berezovskaya, I. V., Efryushina, N. P., Voloshinovskii, A. S., Vdovenko, S. I., Kovalevskaya, I. P. and Dotsenko, V. P., (2012). Luminescent properties and stability of europium ions in Ca2BO3Cl: Eu. *Inorg. Mater.*, 48(5), 539–543. https://doi.org/10.1134/S0020168512050032

116. Liu, H., Xia, Z., Zhuang, J., Zhang, Z. and Liao, L., (2012). Surface treatment investigation and luminescence properties of SiO_2-coated Ca_2BO_3Cl: 0.02 Eu^{2+} phosphors via sol-gel process. *J. Phys. Chem. Solids*, 73(1), 104–108. https://doi.org/10.1016/j.jpcs.2011.10.011

117. Han, B., Zhang, J., Cui, Q. and Liu, Y., (2012). A potential reddish orange emitting phosphor Ca_2BO_3Cl: Eu^{3+} for white light-emitting diodes. *Phys. B: Condens. Matter.*, 407(17), 3484–3486. https://doi.org/10.1016/j.physb.2012.05.006

118. Zhang, X., Lu, Z., Meng, F., Lu, F., Hu, L., Xu, X. and Tang, C., (2012). A yellow-emitting $Ca_3Si_2O_7$: Eu^{2+} phosphor for white LEDs. *Mater. Lett.*, 66(1), 16–18. https://doi.org/10.1016/j.matlet.2011.08.054

119. Kang, E. H., Choi, S. W., Chung, S. E., Jang, J., Kwon, S. and Hong, S. H., (2011). Photoluminescence characteristics of Sr3SiO5: Eu2+ yellow phosphors synthesized by solid state method and Pechini process. *J. Electrochem. Soc.*, 158(11), J330. https://doi.org/10.1149/2.016111jes/meta

120. Petrykin, V., Okube, M., Yamane, H., Sasaki, S. and Kakihana, M., (2010). Sr2ZnS3: Crystal structure and fluorescent properties of a New Eu (II)-activated yellow emission phosphor. *Chem. Mater.*, 22(21), 5800–5802. https://doi.org/10.1021/cm1023713

121. Yu, H. J., Chung, W., Park, S. H., Kim, J. and Kim, S. H., (2011). Luminous properties of $Sr_{1-x}Zn_x$Se: Eu^{2+} phosphors for LEDs application. *J. Cryst. Growth*, 326(1), 77–80. https://doi.org/10.1016/j.jcrysgro.2011.01.056

122. Chung, W., Jung, H., Lee, C. H., Kim, J. and Kim, S. H. (2012). Spray pyrolysis prepared yellow to red color tunable $Sr_{1-x}Ca_x$Se: Eu^{2+} phosphors for white LED. *Opt. Express*, 20(12), 12885–12892. https://doi.org/10.1364/OE.20.012885

123. Dong, K., Li, Z., Xiao, S., Xiang, Z., Zhang, X., Yang, X. and Jin, X., (2012). Yellowish orange luminescence in $Sr_8Al_{12}O_{24}S_2$: Eu^{2+} phosphor. *J. Alloys Compd.*, 543, 105–108. https://doi.org/10.1016/j.jallcom.2012.07.104

124. Zhang, X., Fei, L., Shi, J. and Gong, M., (2011). Eu^{2+}-activated Ba_2Mg $(BO_3)_2$ yellow emitting phosphors for near ultraviolet-based light-emitting diodes. *Phys. B: Condens. Matter.*, 406(13), 2616–2620. https://doi.org/10.1016/j.physb.2011.03.077

125. Huang, C. H., Chen, Y. C., Chen, T. M., Chan, T. S. and Sheu, H. S., (2011). Near UV pumped yellow-emitting $Sr_8MgSc(PO_4)_7$: Eu^{2+} phosphor for white-light LEDs with excellent color rendering index. *J. Mater. Chem.*, 21(15), 5645–5649. https://doi.org/10.1039/C0JM04524A

126. Lin, C. C., Xiao, Z. R., Guo, G. Y., Chan, T. S. and Liu, R. S., (2010). Versatile phosphate phosphors $ABPO_4$ in white light-emitting diodes: Collocated characteristic analysis and theoretical calculations. *J. Am. Chem. Soc.*, 132(9), 3020–3028. https://doi.org/10.1021/ja9092456

127. Shi, L., Huang, Y. and Seo, H. J., (2010). Emission red shift and unusual band narrowing of Mn^{2+} in $NaCaPO_4$ phosphor. *J. Phys. Chem. A*, 114(26), 6927–6934. https://doi.org/10.1021/jp101772z

128. Hao, Z., Zhang, X., Luo, Y., Zhang, L., Zhao, H. and Zhang, J., (2013). Enhanced Ce^{3+} photoluminescence by Li^+ co-doping in CaO phosphor and its use in blue-pumped white LEDs. *J. Lumin.*, 140, 78–81. https://doi.org/10.1016/j.jlumin.2013.03.013

129. Chen, Q. L., Wang, C. F. and Chen, S., (2013). One-step synthesis of yellow-emitting carbogenic dots toward white light-emitting diodes. *J. Mater. Sci.*, 48(6), 2352–2357. https://doi.org/10.1007/s10853-012-7016-8

130. Zhang, J., Zhang, X., Gong, M., Shi, J., Yu, L., Rong, C. and Lian, S., (2012). $LiSrBO_3:Eu^{2+}$: A novel broad-band red phosphor under the excitation of a blue light. *Mater. Lett.*, 79, 100–102. https://doi.org/10.1016/j.matlet.2012.04.011

131. Zhang, X., Dong, Z., Shi, J. and Gong, M., (2012). Luminescence properties of colortunable zinc-codoped alikali earth sulfide phosphor for LED application. *Mater. Lett.*, 76, 113–116. https://doi.org/10.1016/j.matlet.2012.02.051

132. Wang, X. and Wang, Y., (2015). Synthesis, structure, and photoluminescence properties of Ce^{3+}-doped $Ca_2YZr_2Al_3O_{12}$: A novel garnet phosphor for white LEDs. *J. Phys. Chem. C*, 119(28), 16208–16214. https://doi.org/10.1021/acs.jpcc.5b01552

133. Zhang, Q., Li, J., Jiang, W., Lin, L., Ding, J., Brik, M. G., Molokeev, M. S., Ni, H. and Wu, M., (2021). CaY2Al4SiO12: Ce^{3+}, Mn^{2+}: A single component phosphor to produce high color rendering index WLEDs with a blue chip. *J. Mater. Chem. C*, 9(34), 11292–11298. https://doi.org/10.1039/D1TC01770E

134. Katelnikovas, A., Plewa, J., Dutczak, D., Moller, S., Enseling, D., Winkler, H., Kareiva, A. and Justel, T., (2012). Synthesis and optical properties of green emitting garnet phosphors for phosphor converted light emitting diodes. *Opt Mater*, 34, 1195–1201. https://doi.org/10.1016/j.optmat.2012.01.034

135. Pan, Z., Li, W., Xu, Y., Hu, Q. and Zheng, Y., (2016). Structure and redshift of Ce3+ emission in anisotropic expansion garnet phosphor $MgY_2Al_4SiO_{12}$:Ce. *RSC Adv.*, 6, 20458–20466. https://doi.org/10.1039/C6RA00356G

136. Ji, H., Wang, L., Cho, Y, Hirosaki, N., Molokeev, M. S., Xia, Z., Huang, Z. and Xie, R. J., (2016). New $Y_2BaAl_4SiO_{12}:Ce^{3+}$ yellow microcrystal-glass powder phosphor with high thermal emission stability. *J. Mater. Chem. C*, 4, 9872–9878. Doi: 10.1039/C6TC03422E

137. Yan, M., Seto, T. and Wang, Y., (2021). Strong energy transfer induced deep-red emission for LED plant growth phosphor $(Y, Ba)_3(Al, Si)_5O_{12}$: Ce^{3+}, Cr^{3+}. *J. Lumin.*, 239, 118352. https://doi.org/10.1016/j.jlumin.2021.118352

138. Wang, J., Peng, X., Cheng, D., Zheng, Z. and Guo, H., (2021). Tunable luminescence and energy transfer in $Y_2BaAl_4SiO_{12}$:Tb^{3+}, Eu^{3+} phosphors for solid-state lighting. *J. Rare Earths*, 39, 284–290. https://doi.org/10.1016/j.jre.2020.06.010

139. Huang, D., Liu, Z., Wang, B., Che, H., Zou, M., Zeng, Q., Lian, H. and Lin, J., (2021). Highly efficient yellow-orange emission and superior thermal stability of $Ba_2YAl_3Si_2O_{12}$: Ce^{3+} for high-power solid lighting. *J. Am. Ceram. Soc.*, 104(1), 524–534. https://doi.org/10.1111/jace.17439

140. Liang, M., Xu, J., Qiang, Y., Kang, H., Zhang, L., Chen, J., Liu, C., Luo, X., Li, Y., Zhang, J. and Ouyang, L., (2021). Ce3+ doped BaLu2Al2Ga2SiO12 – A novel blue-light excitable cyan-emitting phosphor with ultra-high quantum efficiency and excellent stability for full-spectrum white LEDs. *J. Rare Earths.*, 39(9), 1031–1039. https://doi.org/10.1016/j.jre.2020.09.015

141. Wang, Z., Cheng, L., Tang, H., Yu, X., Xie, J., Mi, X., Liu, Q. and Zhang. X., (2021). Garnet type $Y_2Mg_3Ge_3O_{12}$: Dy^{3+}/Eu^{3+} phosphors excited near ultraviolet: Luminescence properties and energy transfer mechanisms. *J. Solid State Chem.*, 301, 122295. https://doi.org/10.1016/j.jssc.2021.122295

142. Zhang, X., Zhang, D., Zheng, B., Zheng, Z., Song, Y., Zheng, K., Sheng, Y., Shi, Z. and Zou, H., (2021). Luminescence and energy transfer of color-tunable $Y_2Mg_2Al_2Si_2O_{12}$: Eu^{2+}, Ce^{3+} phosphors. *Inorg. Chem.*, 60(8), 5908–5916. https://doi.org/10.1021/acs.inorgchem.1c00317

143. Ji, H., Wang, L., Molokeev, M. S., Hirosaki, N., Xie, R., Huang, Z., Xia, Z., ten Kate, O. M., Liu, L and Atuchin, V. V., (2016). Structure evolution and photoluminescence of $Lu_3(Al, Mg)_2(Al, Si)_3O_{12}$:$Ce^{3+}$ phosphors: New yellow-color converters for blue LED-driven solid state lighting. *J. Mater. Chem. C*, 6855–6863. Doi: 10.1039/C6TC00966B

144. Zheng, Z., Zhang, D., Zheng, B., Song, Y., Zhang, X., Zheng, K., Sheng, Y. and Zou, H., (2022). Two strategies to achieve color adjustment of Eu^{2+}-doped garnet $Lu_2Mg_2Al_2Si_2O_{12}$ phosphors. *J. Lumin.*, 243, 118651. https://doi.org/10.1016/j.jlumin.2021.118651

145. He, C., Ji, H., Huang, Z., Wang, T., Zhang, X., Liu, Y., Fang, M., Wu, X., Zhang, J. and Min, X., (2018). Red-shifted emission in $Y_3MgSiAl_3O_{12}$: Ce^{3+}, garnet phosphor for blue light pumped white light emitting diodes. *J. Phys. Chem. C*, 122, 15659–15665. https://doi.org/10.1021/acs.jpcc.8b03940

146. Tong, E., Song, K., Deng, Z., Shen, S., Gao, H., Su, W. and Wang, H., (2018). Ionic occupation sites, luminescent spectra, energy transfer behaviors in $Y_3MgAl_3SiO_{12}$: Ce^{3+}, Mn^{2+} phosphors for warm white LED. *J. Lumin.*, 217, 116787. https://doi.org/10.1016/j.jlumin.2019.116787

147. Shi, Y., Zhu, G., Mikami, M., Shimomura, Y. and Wang, Y., (2015). Novel Ce^{3+} activated $Lu_3MgAl_3SiO_{12}$ garnet phosphor for blue chip light-emitting diodes with excellent performance. *Dalton Trans*, 44, 1775–1781. https://doi.org/10.1039/C4DT03144J

148. Zhou, Y., Zhuang, W., Hu, Y., Liu, R., Xu, H., Chen, M., Liu, Y., Li, Y., Zheng, Y. and Chen, G., (2019). Cyan-green phosphor $(Lu_2M)(Al_4Si)O_{12}$:Ce^{3+} for high-quality LED lamp: Tunable photoluminescence properties and enhanced thermal stability. *Inorg. Chem.*, 58, 1492–1500. https://doi.org/10.1021/acs.inorgchem.8b03017

149. Hu, T., Molokeev, M. S., Xia, Z. and Zhang, Q., (2019). Aliovalent substitution toward reinforced structural rigidity in Ce^{3+}-doped garnet phosphors featuring improved performance. *J. Mater. Chem. C*, 7, 14594–14600. https://doi.org/10.1039/C9TC05354A

150. Zhang, Q., Li, G., Dang, P., Liu, D., Huang, D., Lian, H. and Lin, J., (2021). Enhancing and tuning broadband near-infrared (NIR) photoluminescence properties in Cr^{3+}-doped $Ca_2YHf_2Al_3O_{12}$ garnet phosphors via Ce^{3+}/Yb^{3+}-codoping for LED applications. *J. Mater. Chem. C*, 9, 4815. https://doi.org/10.1039/D0TC05657J

151. Liang, J., Devakumar, B., Sun, L., Wang, S., Sun, Q. and Huang, X., (2020). Full-visible-spectrum lighting enabled by an excellent cyan-emitting garnet phosphor. *J. Mater. Chem. C*, 8, 4934. https://doi.org/10.1039/D0TC00006J

152. Yuan, W., Tan, T., Wu, H., Pang, R., Zhang, S., Jiang, L., Li, D., Wu, Z., Li, C. and Zhang, H., (2021). Intense UV long persistent luminescence benefiting from the coexistence of Pr^{3+}/Pr^{4+} in a praseodymium-doped $BaLu_2Al_2Ga_2SiO_{12}$ phosphor. *J. Mater. Chem. C*, 9, 5206. https://doi.org/10.1039/D1TC00584G

153. Sun, L., Devakumar, B., Liang, J., Wang, S., Sun, Q. and Huang, X., (2020). A broadband cyan-emitting $Ca_2LuZr_2(AlO_4)_3$: Ce^{3+} garnet phosphor for near-ultraviolet-pumped warmwhite light-emitting diodes with an improved color rendering index. *J. Mater. Chem. C*, 8(3), 1095–1103. https://doi.org/10.1039/C9TC04952E

154. Meng, Q., Zhu, Q., Li, X., Sun, X. and Li, J. G., (2021). New Mg^{2+}/Ge^{4+}-stabilized $Gd_3Mg_xGe_xAl_{5-2x}O_{12}$:Ce, garnet phosphor with orange-yellow emission for warm-white LEDs (x=2.0–2.5). *Inorg. Chem.* 60, 9773–9784. https://doi.org/10.1021/acs.inorgchem.1c01072

155. Kang, J. Y., Won, H. I., Hayk, N. and Won, C. W., (2013). Synthesis of BaSrSiO4 phosphors by solid state reaction and its luminescent properties. *Kor. J. Mater. Res.* 23(12), 727–731. Doi: 10.3740/MRSK.2013.23.12.727

156. Thi, M. H., That, P. T. and Anh, N. D., (2020). Eu2+-activated strontium-barium silicate: A positive solution for improving luminous efficacy and color uniformity of white light-emitting diodes. *Materials Science-Poland*, 38(4), 594–600. https://doi.org/10.2478/msp-2020-0069

157. Loan, N. T. and Anh, N. D., (2020). SrBaSiO$_4$:Eu^{2+} phosphor: A novel application for improving the luminous flux and color quality of multi-chip white LED lamps. *Int. J. Electric. Comput. Eng. (IJECE)*, 10(5), 5147–5154. ISSN: 2088-8708, Doi: 10.11591/ijece.v10i5.pp5147-5154

158. Chen, X. and Kim, W. S., (2015). Template-engaged solid-state synthesis of barium-strontium silicate hexagonal tubes. *J. Alloys Compd.*, 647, 1128–1135. https://doi.org/10.1016/j.jallcom.2015.05.185

159. Xue, W., Lei, X., Zhai, K., Wen, W., Jiang, S. and Zhai, S., (2022). Thermal expansion and compressibility of calcium scandate CaSc$_2$O$_4$. *J. Alloys Compd.*, 909, 164756. https://doi.org/10.1016/j.jallcom.2022.164756

160. Carter, J. R. and Feigelson, R. S., (1964). Preparation and crystallographic properties of A^{2+}B$_2$$^{3+}O_4$ type calcium and strontium scandates. *J. Am. Ceram. Soc.*, 47, 141–144. https://doi.org/10.1111/j.1151-2916.1964.tb14373.x

161. Shimomura, Y., Kurushima, T. and Kijima, N., (2007). Photoluminescence and crystal structure of green-emitting phosphor CaSc$_2$O$_4$: Ce^{3+}. *J. Electrochem. Soc.*, 154(8), J234–J238. https://doi.org/10.1149/1.2741172

162. Ma, S., Wang, M., Liu, G., Hu, Y., Hu, C., Wang, S. and Ye, Z., (2020). Site engineering of Ce^{3+}-doped calcium scandate phosphors and understanding of relevant redshifted emitting from green to yellow. *Ceram. Int.*, 46(12), 20004–20011. https://doi.org/10.1016/j.ceramint.2020.05.071

163. Enachi, A., Toma, O. and Georgescu, Ș., (2021). Luminescent Er^{3+} centers in CaSc$_2$O$_4$:Er^{3+}: Yb^{3+} upconversion phosphor. *J. Lumin.*, 231, 117816. https://doi.org/10.1016/j.jlumin.2020.117816

164. Mill, B. V., Belokoneva, E. L., Simonov, M. A. and Belov, N. V., (1977). Refined crystal structures of the scandium garnets Ca$_3$Sc$_2$Si$_3$O$_{12}$, Ca$_3$Sc$_2$Ge$_3$O$_{12}$, and Cd$_3$Sc$_2$Ge$_3$O$_{12}$. *J. Struct. Chem.*, 18(2), 321–323. https://doi.org/10.1007/BF00753987

165. Shimomura, Y. and Kijima, N., (2007). Phosphor, light emitting device using phosphor, and display and lighting system using light emitting device. US patent no 7189340 p B2.

166. Shimomura, Y., Kurushima, T., Shigeiwa, M. and Kijima, N., (2007). Redshift of green photoluminescence of Ca$_3$Sc$_2$Si$_3$O$_{12}$:Ce^{3+} phosphor by charge compensatory additives. *J. Electrochem. Soc.*, 155, J45–J49. https://doi.org/10.1149/1.2814144

167. Kijima, N., Shimomura, Y., Kurushima, T., Watanabe, H., Shimooka, S., Mikami, M. and Uheda, K., (2008). New green and red phosphors for white LEDs. *J. Light Vis. Environ.*, 32(2), 202–207. https://doi.org/10.2150/jlve.32.202

168. Qiao, J., Zhang, J., Zhang, X., Hao, Z., Deng, W., Liu, Y., Zhang, L., Zhang, L., Zhao, H. and Lin, J., (2013). Formation condition of red Ce^{3+} in Ca$_3$Sc$_2$Si$_3$O$_{12}$: Ce^{3+}, N^{3-} as a fullcolor- emitting light-emitting diode phosphor. *Opt. Lett.*, 38(6), 884–886. https://doi.org/10.1364/OL.38.000884

169. Pinelli, S., Bigotta, S., Toncelli, A., Tonelli, M., Cavalli, E. and Bovero, E., (2004). Study of the visible spectra of Ca$_3$Sc$_2$Ge$_3$O$_{12}$ garnet crystals doped with Ce^{3+} or Pr^{3+}. *Opt. Mater.*, 25(1), 91–99. https://doi.org/10.1016/S0925-3467(03)00231-3

170. Ivanovskikh, K. V., Meijerink, A., Piccinelli, F., Speghini, A., Zinin, E. I., Ronda, C. and Bettinelli, M., (2010). Optical spectroscopy of Ca$_3$Sc$_2$Si$_3$O$_{12}$, Ca$_3$Y$_2$Si$_3$O$_{12}$ and Ca$_3$Lu$_2$Si$_3$O$_{12}$ doped with Pr^{3+}. *J. Lumin.*, 130(5), 893–901. https://doi.org/10.1016/j.jlumin.2009.12.031

171. Ivanovskikh, K. V., Meijerink, A., Piccinelli, F., Speghini, A., Ronda, C. and Bettinelli, M. (2012). VUV spectroscopy of Ca$_3$Sc$_2$Si$_3$O$_{12}$: Pr^{3+}: Scintillator optimization by co-doping with Mg^{2+}. *ECS J. Solid State Sci. Technol.*, 1(5), R127–R130. https://doi.org/10.1149/2.009205jss

172. Bettinelli, M., Speghini, A., Piccinelli, F., Neto, A. N. C. and Malta, O. L., (2011). Luminescence spectroscopy of Eu^{3+} in Ca$_3$Sc$_2$Si$_3$O$_{12}$. *J. Lumin.*, 131(5), 1026–1028. https://doi.org/10.1016/j.jlumin.2011.01.016

173. Pasiński, D. and Sokolnicki, J., (2017). Luminescence study of Eu^{3+}-doped garnet phosphors: Relating structure to emission. *J. Alloys Compd.*, 695, 1160–1165. https://doi.org/10.1016/j.jallcom.2016.10.243

174. Chen, Y., Cheah, K. W. and Gong, M., (2011). Low thermal quenching and high-efficiency Ce^{3+}, Tb^{3+}-co-doped Ca$_3$Sc$_2$Si$_3$O$_{12}$ green phosphor for white light-emitting diodes. *J. Lumin.*, 131(8), 1589–1593. https://doi.org/10.1016/j.jlumin.2011.04.002

175. Chen, Y., Feng, J., Pan, Y., Li, J., Zeng, S., Liang, M., Liu, Z., Li, N. and Su, Y., (2014). Sol-gel synthesis and analysis of high efficiency submicron-sized Ca$_3$Sc$_2$Si$_3$O$_{12}$: Ce^{3+}, Tb^{3+} phosphor for white light emitting diodes. *J. Lumin.*, 148, 156–160. https://doi.org/10.1016/j.jlumin.2013.12.015

176. Long, Q., Wang, C., Li, Y., Ding, J., Wang, X. and Wang, Y., (2015). Solid state reaction synthesis and photoluminescence properties of Dy^{3+} doped Ca$_3$Sc$_2$Si$_3$O$_{12}$ phosphor. *Mater. Res. Bull.*, 71, 21–24. https://doi.org/10.1016/j.materresbull.2015.07.001

177. Chepyga, L. M., Osvet, A., Levchuk, I., Ali, A., Zorenko, Y., Gorbenko, V., Zorenko, T., Fedorov, A., Brabec, C. J., and Batentschuk, M., (2018). New silicate based thermographic phosphors $Ca_3Sc_2Si_3O_{12}$: Dy, $Ca_3Sc_2Si_3O_{12}$: Dy, Ce and their photoluminescence properties. *J. Lumin.*, 202, 13–19. https://doi.org/10.1016/j.jlumin.2018.05.039

178. Xiao, W., Wu, D., Zhang, L., Zhang, X., Hao, Z., Pan, G. H., Zhao, H., Zhang, L. and Zhang, J., (2017). Cooperative upconversion luminescence properties of Yb^{3+} and Tb^{3+} heavily codoped silicate garnet obtained by multiple chemical unit cosubstitution. *J. Phys. Chem. C*, 121(5), 2998–3006. https://doi.org/10.1021/acs.jpcc.6b11633

179. Berezovskaya, I. V., Dotsenko, V. P., Voloshinovskii, A. S. and Smola, S. S., (2013). Near infrared emission of Eu^{2+} ions in $Ca_3Sc_2Si_3O_{12}$. *Chem. Phys. Lett.*, 585, 11–14. https://doi.org/10.1016/j.cplett.2013.08.100

180. Fernandez-Gonzalez, R., Velazquez, J. J., Rodriguez, V. D., Rivera-Lopez, F., Lukowiak, A., Chiasera, A., Ferrari, M., Goncalves, R. R., Marrero-Jerez, J., Lahoz, F. and Nunez, P., (2016). Luminescence and structural analysis of Ce^{3+} and Er^{3+} doped and Ce^{3+}-Er^{3+} codoped $Ca_3Sc_2Si_3O_{12}$ garnets: Influence of the doping concentration in the energy transfer processes. *RSC Adv.*, 6(18), 15054–15061. Doi: 10.1039/C5RA22630A

181. Qiao, J., Zhang, J., Zhang, X., Hao, Z., Liu, Y. and Luo, Y., (2014). The energy transfer and effect of doped Mg^{2+} in $Ca_3Sc_2Si_3O_{12}$: Ce^{3+}, Pr^{3+} phosphor for white LEDs. *Dalton Trans.*, 43(10), 4146–4150. https://doi.org/10.1039/C3DT52902A

182. Velazquez, J. J., Fernández-González, R., Marrero-Jerez, J., Rodríguez, V. D., Lukowiak, A., Chiappini, A., Chiasera, A., Ferrari, M. and Núñez, P., (2015). Structural and luminescence study of Ce^{3+} and Tb^{3+} doped $Ca_3Sc_2Si_3O_{12}$ garnets obtained by freeze-drying synthesis method. *Opt. Mater.*, 46, 109–114. https://doi.org/10.1016/j.optmat.2015.03.057

183. Zhou, L., Zhou, W., Pan, F., Shi, R., Huang, L., Liang, H., Tanner, P. A., Du, X., Huang, Y., Tao, Y. and Zheng, L., (2016). Spectral properties and energy transfer of a potential solar energy converter. *Chem. Mater.*, 28(8), 2834–2843. https://doi.org/10.1021/acs.chemmater.6b00763

184. Zhou, L., Tanner, P. A., Zhou, W., Ai, Y., Ning, L., Wu, M. M. and Liang, H., (2017). Unique spectral overlap and resonant energy transfer between europium (II) and ytterbium (III) cations: No quantum cutting. *Angew. Chem.*, 129(35), 10493–10497. https://doi.org/10.1002/ange.201703331

185. Liu, Y., Zhang, X., Hao, Z., Wang, X. and Zhang, J., (2011). Tunable full-color-emitting $Ca_3Sc_2Si_3O_{12}$: Ce^{3+}, Mn^{2+} phosphor via charge compensation and energy transfer. *Chem. Commun.*, 47(38), 10677–10679. https://doi.org/10.1039/C1CC14324G

186. Qiao, J., Zhang, J., Zhang, X., Hao, Z., Liu, Y. and Pan, G., (2014). Red emission of additional Pr^{3+} and adjusting effect of additional Mg^{2+} in $Ca_3Sc_2Si_3O_{12}$: Ce^{3+}, Mn^{2+} phosphor. *Opt. Lett.*, 39(9), 2691–2694. https://doi.org/10.1364/OL.39.002691

187. Liu, Y., Zhuang, W., Hu, Y., Gao, W. and Hao, J., (2010). Synthesis and luminescence of sub-micron sized $Ca_3Sc_2Si_3O_{12}$: Ce green phosphors for white light-emitting diode and field emission display applications. *J. Alloys Compd.*, 504(2), 488–492. https://doi.org/10.1016/j.jallcom.2010.06.007

188. Chen, Y., Gong, M. and Cheah, K. W., (2010). Effects of fluxes on the synthesis of $Ca_3Sc_2Si_3O_{12}$: Ce^{3+} green phosphors for white light-emitting diodes. *Mater. Sci. Eng. B.*, 166(1), 24–27. https://doi.org/10.1016/j.mseb.2009.09.024

189. Chen, Y., Li, J., Zeng, S., Fan, H., Feng, J. and Tan, L., (2014). Use of LiF flux in the preparation of $Ca_3Sc_2Si_3O_{12}$: Ce^{3+} phosphor by sol-combustion method. *Opt. Mater.*, 37, 464–469. https://doi.org/10.1016/j.optmat.2014.07.007

190. Liu, Y., Hao, J., Zhuang, W. and Hu, Y., (2009). Structural and luminescent properties of gel-combustion synthesized green-emitting $Ca_3Sc_2Si_3O_{12}$: Ce^{3+} phosphor for solid-state lighting. *J. Phys. D Appl. Phys.*, 42(24), 245102. https://doi.org/10.1088/0022-3727/42/24/245102

191. Enomoto, N., Sakai, T., Inada, M., Tanaka, Y. and Hojo, J., (2010). Synthesis of $Ca_3Sc_2Si_3O_{12}$: Ce^{3+} phosphor via newly developed emulsion route. *J. Ceram. Soc. Jpn.*, 118(1383), 1067–1070. https://doi.org/10.2109/jcersj2.118.1067

192. Gorbenko, V., Zorenko, T., Witkiewicz, S., Paprocki, K., Iskaliyeva, A., Kaczmarek, A. M., Van Deun, R., Khaidukov, M. N., Batentschuk, M. and Zorenko, Y., (2018). Luminescence of Ce^{3+} multicenters in Ca^{2+}-Mg^{2+}-Si^{4+} based garnet phosphors. *J. Lumin.*, 199, 245–250. https://doi.org/10.1016/j.jlumin.2018.03.058

193. Lohe, P. P., Nandanwar, D. V., Belsare, P. D. and Moharil, S. V., (2019). Cyan emitting $Ca_3Sc_2Si_{1.5}Ge_{1.5}O_{12}$: Ce^{3+} phosphor with 10.4 ns lifetime. *J. Lumin.*, 216, 116744. https://doi.org/10.1016/j.jlumin.2019.116744

194. Petrova, S. A., Mar'evich, V. P., Zakharov, R. G., Selivanov, E. N., Chumarev, V. M. and Udoeva, L. Y., (2003). Crystal structure of zinc calcium oxysulfide. In *Doklady Chemistry* Vol 393, 255–258. Kluwer Academic Publishers-Plenum Publishers. 0012-5008/03/0011

195. Sambrook, T., Smura, C. F., Clarke, S. J., Ok, K. M. and Halasyamani, P. S., (2007). Structure and physical properties of the polar oxysulfide CaZnOS. *Inorg. Chem.*, 46(7), 2571–2574. https://doi.org/10.1021/ic062120z

196. Budde, B., Luo, H., Dorenbos, P. and van der Kolk, E., (2017). Luminescent properties and energy level structure of CaZnOS: Eu²⁺. *Opt. Mater.*, 69, 378–381. Doi: 10.1016/j.optmat.2017.04.045

197. Duan, C. J., Delsing, A. C. A. and Hintzen, H. T., (2009). Photoluminescence properties of novel red-emitting Mn²⁺-activated MZnOS (M= Ca, Ba) phosphors. *Chem. Mater.*, 21(6), 1010–1016. https://doi.org/10.1021/cm801990r

198. Zhang, J. C., Zhao, L. Z., Long, Y. Z., Zhang, H. D., Sun, B., Han, W. P., Yan, X. and Wang, X., (2015). Color manipulation of intense multiluminescence from CaZnOS: Mn²⁺ by Mn²⁺ concentration effect. *Chem. Mater.*, 27(21), 7481–7489.doi: 10.1021/acs.chemmater.5b03570

199. Su, M., Li, P., Zheng, S., Wang, X., Shi, J., Sun, X. and Zhang, H., (2020). Largely enhanced elastico-mechanoluminescence of CaZnOS: Mn²⁺ by co-doping with Nd³⁺ ions. *J. Lumin.*, 217, 116777. https://doi.org/10.1016/j.jlumin.2019.116777

200. Cashmore, A. R., Jarillo, J. A., Wu, Y. J. and Liu, D., (1999). Cryptochromes: Blue light receptors for plants and animals. *Science*, 284(5415), 760–765. Doi: 10.1126/science.284.5415.760

201. Pinho, P., Moisio, O., Tetri, E. and Halonen, L., (2004). Photobiological aspects of crop plants grown under light emitting diodes. In D. Ken Sagawa (ed.), *CIE Expert Symposium on LED Light Sources*, 71–74. CIE Central Bureau, Tokyo, Japan. https://research.aalto.fi/en/publications/photobiological-aspects-of-crop-plants-grown-underlight-emitting

202. Du, H., Fuh, R. C. A., Li, J., Corkan, L. A. and Lindsey, J. S., (1998). Photochem CAD: A computer-aided design and research tool in photochemistry. *Photochem. Photobiol.*, 68(2), 141–142. https://doi.org/10.1111/j.1751-1097.1998.tb02480.x

203. Sager, J. C., Smith, W. O., Edwards, J. L. and Cyr, K. L., (1988). Photosynthetic efficiency and phytochrome photoequilibria determination using spectral data. *Trans. ASAE*, 31(6), 1882–1889. Doi: 10.13031/2013.30952

204. Wheeler, R. M., (2008). A historical background of plant lighting: An introduction to the workshop. *HortScience*, 43(7), 1942–1943 https://doi.org/10.21273/HORTSCI.43.7.1942

205. Okamoto, K., Yanagi, T., Takita, S., Tanaka, M., Higuchi, T., Ushida, Y. and Watanabe, H., (1996). Development of plant growth apparatus using blue and red LED as artificial light source. In *International Symposium on Plant Production in Closed Ecosystems* Vol 440, 111–116. https://doi.org/10.17660/actahortic.1996.440.20

206. Bula, R. J., Morrow, R. C., Tibbitts, T. W., Barta, D. J., Ignatius, R. W. and Martin, T. S., (1991). Light-emitting diodes as a radiation source for plants. *HortScience*, 26(2), 203–205. https://doi.org/10.21273/HORTSCI.26.2.203

207. Fang, S., Lang, T., Cai, M. and Han, T., (2022). Light keys open locks of plant photoresponses: A review of phosphors for plant cultivation LEDs. *J. Alloys Compd.*, 163825. https://doi.org/10.1016/j.jallcom.2022.163825

208. Adachi, S., (2021). Spectroscopy of Cr³⁺ activator: Tanabe– Sugano diagram and Racah parameter analysis. *J. Lumin.*, 232, 117844. https://doi.org/10.1016/j.jlumin.2020.117844

209. Adachi, S., (2022). Temperature dependence of luminescence intensity and decay time in Cr³⁺-activated oxide and fluoride phosphors. *ECS J. Solid State Sci. Technol.*, 11(6), 066001. https://doi.org/10.1149/2162-8777/ac7075

210. Adachi, S., (2022). Negative thermal quenching of Mn4+ luminescence in fluoride phosphors: Effects of the ⁴A₂g→⁴T₂g excitation transitions and normal thermal quenching. *ECS J. Solid State Sci. Technol.*, 11(3), 036001. Doi: 10.1149/2162-8777/ac56c3

211. Crawford, M. H., (2009). LEDs for solid-state lighting: Performance challenges and recent advances. *IEEE J. Select. Top. Quant. Electron.*, 15(4), 1028–1040. Doi: 10.1109/JSTQE.2009.2013476

212. Tsao, J. Y., (2004). Solid-state lighting: Lamps, chips, and materials for tomorrow. *IEEE Circuit. Dev. Magazine*, 20(3), 28–37. Doi: 10.1109/MCD.2004.1304539

213. Hye Oh, J., Ji Yang, S. and Rag Do, Y., (2014). Healthy, natural, efficient and tunable lighting: Four-package white LEDs for optimizing the circadian effect, color quality and vision performance. *Light: Sci. Appl.*, 3(2), 141. https://doi.org/10.1038/lsa.2014.22

214. Yoon, H. C., Oh, J. H., Lee, S., Park, J. B. and Do, Y. R., (2017). Circadian-tunable perovskite quantum dot-based down-converted multi-package white LED with a color fidelity index over 90. *Sci. Rep.*, 7(1), 1–11. Doi: 10.1038/s41598-017-03063-7

215. Boivin, D. B., Duffy, J. F., Kronauer, R. E. and Czeisler, C. A., (1996). Dose-response relationships for resetting of human circadian clock by light. *Nature*, 379(6565), 540–542. https://doi.org/10.1038/379540a0

216. Falchi, F., Cinzano, P., Elvidge, C. D., Keith, D. M. and Haim, A., (2011). Limiting the impact of light pollution on human health, environment and stellar visibility. *J. Environ. Manage.*, 92(10), 2714–2722. https://doi.org/10.1016/j.jenvman.2011.06.029

217. Buijs, R. M., Eden, C. V., Goncharuk, V. D. and Kalsbeek, A., (2003). Circadian and seasonal rhythms-the biological clock tunes the organs of the body: Timing by hormones and the autonomic nervous system. *J. Endocrinol.*, 177(1), 17–26. https://doi.org/10.1677/joe.0.1770017

218. Dijk, D. J. and von Schantz, M., (2005). Timing and consolidation of human sleep, wakefulness, and performance by a symphony of oscillators. *J. Biol. Rhythms*, 20(4), 279–290. https://doi.org/10.1177/0748730405278292

219. Lee, H. S., Billings, H. J. and Lehman, M. N., (2003). The suprachiasmatic nucleus: A clock of multiple components. *J. Biol. Rhythms*, 18(6), 435–449. https://doi.org/10.1177/0748730403259106

220. Klepeis, N. E., Nelson, W. C., Ott, W. R., Robinson, J. P., Tsang, A. M., Switzer, P., Behar, J. V., Hern, S. C. and Engelmann, W. H., (2001). The National Human Activity Pattern Survey (NHAPS): A resource for assessing exposure to environmental pollutants. *J. Expo. Sci. Environ. Epidemiol.*, 11(3), 231–252. https://doi.org/10.1038/sj.jea.7500165

221. Wittmann, M., Dinich, J., Merrow, M. and Roenneberg, T., (2006). Social jetlag:misalignment of biological and social time. *Chronobiol. Int.*, 23(1–2), 497–509. https://doi.org/10.1080/07420520500545979

222. Papatsimpa, C. and Linnartz, J. P., (2020). Personalized office lighting for circadian health and improved sleep. *Sensors*, 20(16), 4569. Doi: 10.3390/s20164569

223. Antypa, N., Vogelzangs, N., Meesters, Y., Schoevers, R. and Penninx, B. W., (2016). Chronotype associations with depression and anxiety disorders in a large cohort study. *Depress. Anxiety*, 33(1), 75–83. https://doi.org/10.1002/da.22422

224. Stevens, R. G., (2009). Light-at-night, circadian disruption and breast cancer: Assessment of existing evidence. *Int. J. Epidemiol.*, 38(4), 963–970. https://doi.org/10.1093/ije/dyp178

225. Reutrakul, S. and Knutson, K. L., (2015). Consequences of circadian disruption on cardiometabolic health. *Sleep Med. Clin.*, 10(4), 455–468. https://doi.org/10.1016/j.jsmc.2015.07.005

226. Baron, K. G. and Reid, K. J., (2014). Circadian misalignment and health. *Int. Rev. Psychiat.*, 26(2), 139–154. https://doi.org/10.3109/09540261.2014.911149

227. Smolders, K. C., De Kort, Y. A. and Cluitmans, P. J. M., (2012). A higher illuminance induces alertness even during office hours: Findings on subjective measures, task performance and heart rate measures. *Physiol. Behav.*, 107(1), 7–16. https://doi.org/10.1016/j.physbeh.2012.04.028

228. Figueiro, M. G., Kalsher, M., Steverson, B. C., Heerwagen, J., Kampschroer, K. and Rea, M. S., (2019). Circadian-effective light and its impact on alertness in office workers. *Light. Res. Technol.*, 51(2), 171–183. https://doi.org/10.1177/1477153517750006

229. Rea, M. S., Bierman, A., Figueiro, M. G. and Bullough, J. D., (2008). A new approach to understanding the impact of circadian disruption on human health. *J. Circadian Rhythm*, 6(1), 1–14. https://doi.org/10.1186/1740-3391-6-7

230. Figueiro, M. G., Nagare, R. and Price, L. L., (2018). Non-visual effects of light: How to use light to promote circadian entrainment and elicit alertness. *Light. Res. Technol.*, 50(1), 38–62. https://doi.org/10.1177/1477153517721598

231. Lucas, R. J., Peirson, S. N., Berson, D. M., Brown, T. M., Cooper, H. M., Czeisler, C. A., Figueiro, M. G., Gamlin, P. D., Lockley, S. W., O'Hagan, J. B. and Price, L. L., (2014). Measuring and using light in the melanopsin age. *Trends Neurosci.*, 37(1), 1–9. https://doi.org/10.1016/j.tins.2013.10.004

232. Berson, D. M., Dunn, F. A. and Takao, M., (2002). Phototransduction by retinal ganglion cells that set the circadian clock. *Science*, 295(5557), 1070–1073. Doi: 10.1126/science.1067262

233. Thapan, K., Arendt, J. and Skene, D. J., (2001). An action spectrum for melatonin suppression: Evidence for a novel non-rod, non-cone photoreceptor system in humans. *J. Physiol.*, 535(1), 261–267. Doi: 10.1111/j.1469-7793.2001.t01-1-00261.x

234. Behar-Cohen, F., Martinsons, C., Viénot, F., Zissis, G., Barlier-Salsi, A., Cesarini, J. P., Enouf, O., Garcia, M., Picaud, S. and Attia, D., (2011). Light-emitting diodes (LED) for domestic lighting: Any risks for the eye. *Prog. Retin. Eye Res.*, 30(4), 239–257. https://doi.org/10.1016/j.preteyeres.2011.04.002

235. Figueiro, M. G., Radetsky, L., Plitnick, B. and Rea, M. S., (2017). Glucose tolerance in mice exposed to light-dark stimulus patterns mirroring dayshift and rotating shift schedules. *Sci. Rep.*, 7(1), 1–7. https://doi.org/10.1038/srep40661

236. Kim H. S. and Lee Y. H., (2019). Correlation analysis of image reproduction and display color temperature change to prevent sleep disorder. *IEEE Access*, 7, 59091–59099. Doi: 10.1109/ACCESS.2019.2914768

237. Richter, K., Acker, J., Adam, S. and Niklewski, G., (2016). Prevention of fatigue and insomnia in shift workers-a review of non-pharmacological measures. *EPMA J.*, 7(1), 1–11. https://doi.org/10.1186/s13167-016-0064-4

238. Saw, Y. J., Kalavally, V. and Tan, C. P., (2020). The spectral optimization of a commercializable multi-channel LED panel with circadian impact. *IEEE Access*, 8, 136498–136511. Doi: 10.1109/ACCESS.2020.3010339

239. Xia, Y., Wan, O. Y. and Cheah, K. W., (2016). OLED for human centric lighting. *Opt. Mater. Express*, 6(6), 1905–1913. https://doi.org/10.1364/OME.6.001905

240. Ding, J., Huang, S., Zheng, H., Huang, L., Zeng, P., Ye, S., Wu, Q. and Zhou, J., (2021). A novel broadband cyan light-emitting oxynitride based phosphor used for realizing the fullvisible- spectrum lighting of WLEDs. *J. Lumin.*, 231, 117786. https://doi.org/10.1016/j.jlumin.2020.117786

241. Wang, X., Zhao, Z., Wu, Q., Li, Y. and Wang, Y., (2016). Synthesis, structure and Photoluminescence properties of $Ca_2LuHf_2(AlO_4)_3$: Ce^{3+}, a novel garnet-based cyan light emitting phosphor. *J. Mater. Chem. C*, 4(48), 11396–11403. https://doi.org/10.1039/C6TC03933B

242. Gong, X., Huang, J., Chen, Y., Lin, Y., Luo, Z. and Huang, Y., (2014). Novel garnet structure $Ca_2GdZr_2(AlO_4)_3$: Ce^{3+} phosphor and its structural tuning of optical properties. *Inorg. Chem.*, 53(13), 6607–6614. https://doi.org/10.1021/ic500153u

243. You, S., Zhuo, Y., Chen, Q., Brgoch, J. and Xie, R. J., (2020). Dual-site occupancy induced broadband cyan emission in $Ba_2CaB_2Si_4O_{14}$: Ce^{3+}. *J. Mater. Chem. C*, 8(44), 15626–15633. https://doi.org/10.1039/D0TC02625E

244. Chew, I., Kalavally, V., Tan, C. P. and Parkkinen, J., (2016). A spectrally tunable smart LED lighting system with closed-loop control. *IEEE Sens. J.*, 16(11), 4452–4459. Doi: 10.1109/JSEN.2016.2542265

245. Tang, S. J. W., Kalavally, V., Ng, K. Y., Tan, C. P. and Parkkinen, J., (2018). Real-time closed-loop color control of a multi-channel luminaire using sensors on board a mobile device. *IEEE Access*, 6, 54751–54759. Doi: 10.1109/ACCESS.2018.2872320

7 Metal Oxides
Spectral Modifiers for Solar Cell Applications

P.K. Tawalare
Jagadamba Mahavidyalaya

P.V. Tumram
Amolakchand Mahavidyalaya

CONTENTS

7.1 INTRODUCTION

Modern civilization heavily depends on energy. Per capita consumption of energy (Figure 7.1) can be taken as an index of development. Even the social progress (SP) index shows reasonable correlation with the energy consumption (Figure 7.2). On the other hand, generation of energy using conventional sources also creates another problem like air pollution and environmental hazards.

Most of the developed and developing countries have turned their attention towards exploring non-conventional, renewable energy sources. Tapping solar energy is at the front of these efforts. This is owing to its abundance, inexhaustible and non-pollutant character. Indeed, the energy problem is solved by science with the discovery of photoelectric effect. As long as the sun shines, we need not look beyond for the source of energy. The problem to be addressed is converting science into technology. The following lines are apt description of the solar energy

अनंत हस्ते कमलावराने देता
किती घेशील दो कराने,

Meaning the God, also known as Sahasrarashmi (having thousand hands), is giving by infinite hands, how much can you accept with your two hands. Solar energy received on the earth is many times

DOI: 10.1201/9781003366232-7

Annual Energy Consumption Per Capita

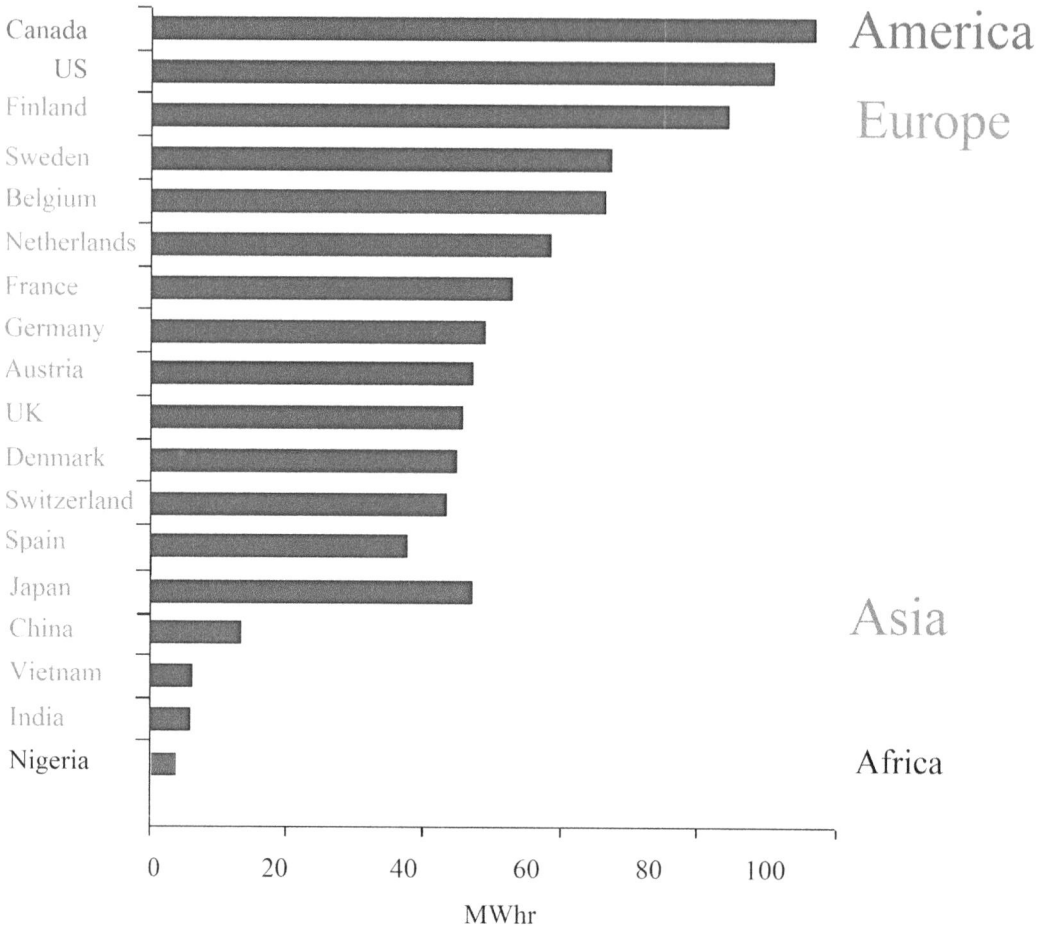

FIGURE 7.1 Per capita consumption of energy.

more than what we need. Limitations are due to inadequacies of the conversion devices. For example, the amount of solar energy that reaches earth is 1.75×10^{17} W; much more than is needed for running the civilizations. Energy needs of entire world can be met by converting the solar radiations falling on merely 0.1% of the earth's surface with an efficiency of 10% [1]. In terms of photovoltaic conversion, it is estimated that by covering only 0.4% of the earth's surface with 15% efficient photovoltaic (PV) panels, our energy demand can be satisfied [2]. If the extra-terrestrial solar radiation is 1,367 watts per square meter (the value when the Earth–Sun distance is 1 astronomical unit), then the direct sunlight at Earth's surface when the Sun is at the zenith is about $1,050 \text{W/m}^2$, but the total amount (direct and indirect from the atmosphere) hitting the ground is around $1,120 \text{W/m}^2$. In photovoltaic, without the aid of any moving devices, there is a direct transformation that changes the sunlight into electricity. It is a massive energy source. The PV elements are easy to design and it is the only system that provides the output from the micro-power to the megapower.

Main negative points of the solar energy are the varying rate, both diurnally and over the year, and high cost of harnessing. Notwithstanding these limitations, a wide range of applications [3,4] including water heating, air heating, solar furnaces [5], air conditioning of buildings, solar refrigeration [6], desalination [7], green houses, power generation [8] and photo-biological conversions have

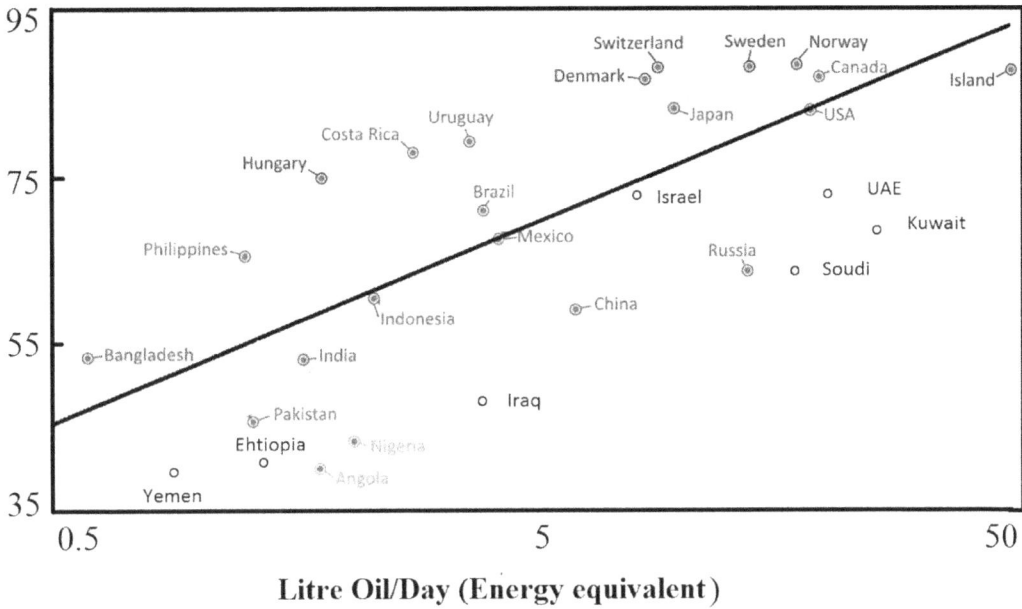

FIGURE 7.2 SP index.

been envisaged for solar energy. There are several ways of harnessing solar energy such as solar thermal, solar thermochemical, solar chemical (photosynthesis), solar photovoltaic, etc.

Of these, solar photovoltaic conversion is most convenient as most of the modern gadgets run on electricity.

Solar photovoltaics (SPV) term is used to describe any process for converting solar radiation (sunlight) into electricity using photovoltaic effect. Photovoltaic conversion of solar energy, though less efficient, is most convenient for the reason that most gadgets in day-to-day use run on electricity. Power generation using photovoltaics can be convenient at any scale. Replacement of conventional power stations working on fossil fuels is of course desirable. For large-scale (>MW) solar photovoltaic plants, the power can be added to the grid for the distribution of conventional electricity. On the small scale (~KW), it is still effective, for example, at remote locations where conventional power is not available. Even at the smallest scale (W), it is useful, e.g., for charging portable devices such as cell phones, tabs, digital cameras, calculators, etc.

7.2 HISTORICAL BACKGROUND

Discovery of photovoltaic effect could be said to be observed as early as 1839 when a French physicist, Edmond Becquerel [9], "found photovoltage between two electrodes in an electrolyte solution". Adams and Day were the first to report Photovoltaic effect in a solid substance in 1877 [10]. They observed variations in electrical properties of selenium when exposed to light. Soon after this, in 1883, Charles Fritts [11] developed a solar cell. This solar cell was in the form of selenium on a thin layer of gold. The efficiency was less than 1%. Many researchers including Lange [12], Grondahl [13] and Schottky [14] did ground-breaking work with cells based on Se and Cu_2O. Till 1950s, selenium and copper/copper oxide cells were the prime photovoltaic converters though the photovoltaic effect in Cadmium Selenide (CdSe) was discovered by Audobert and Stora in 1932. The next important development came from Bell Telephone Laboratories, USA, in 1954. Chapin et al. [15] reported a single-crystal silicon PN junction solar cell with an efficiency of 6%. At about the same time, Reynolds et al. [16] announced the first Cu_2S/CdSe heterostructure solar cell with similar efficiency. Almost immediately after the discovery of III–V semiconductors, Jenny et al. [17] reported the first

GaAs solar cell, with improved efficiency. Russian scientist Zhores Alferov [18] fabricated highly effective heterojunction GaAs solar cells for the first time in 1970. A decade later, The Institute of Energy Conversion at University of Delaware developed the first thin film solar cell exceeding 10% efficiency using Cu_2S/CdS technology. A new type, totally different from the semiconductor-based cells, was demonstrated by Graztel in 1988. In this, light absorption is in a chemical dye, not in semiconductor. These are only a few mile stones which mark the early stages of developments in photovoltaic conversion of solar radiations.

Solar cells based on c-Si remain to date most successfully commercialized device for photovoltaic conversion. The cost of the first solar cell was impracticable for the general use. It was considered adequate in space applications like powering electronic/electrical gadgets on board of a satellite. Realization of the exhaustible and non-renewable nature of the fossil fuels, as well as the fragility of the supply chain witnessed by the Arab-Israel wars, compelled the governments worldwide to consider solar energy as a commercially viable alternative. For achieving this goal, solar cell costs had to be drastically brought down. Efforts in various directions were made towards cost reduction. The important landmarks are reviewed here.

There have been mainly three strategies pursued for making the solar power cost reasonable. The main factor responsible for the high cost is the device grade silicon. This is rather surprising considering its abundance. However, for making solar cell, high-purity, high-perfection single crystals are required. Obvious thing to do for the cost reduction is to look for other low-cost materials and use glass/amorphous materials that will avoid costly crystal growth processes. The degree of perfection of the crystal also affects the photovoltaic conversion efficiency and the cost as well. Lower cost of c-Si solar cells could be achieved by sacrificing the perfection, and in turn, the efficiency. The next strategy for cost reduction is to reduce the quantity of photovoltaic material required for preparing solar cell. Again, this involves replacement of Si. Silicon is an indirect band gap semiconductor. Relatively thick crystal is required for absorbing entire incident radiation. Direct band gap materials like II–VI or III–V compound semiconductors are efficient absorbers, and a few microns thick film is sufficient for photovoltaic conversion. This approach led to research on "thin film solar cells". Third strategy is that, if it is not possible to directly reduce the cost of the solar cell, increase its efficiency, so that the effective cost of power generated decreases. During subsequent stages, efforts are made to combine these strategies as well. The last strategy, particularly, using the phenomenon of luminescence for the cost reduction will be discussed in more details here.

7.3 SOLAR SPECTRUM MODIFICATION CONCEPTS

Experiments on using luminescence for improving performance of solar cells were carried out in two different periods using two different techniques. In late sixties and early seventies, when supply chain and prices were threatened due to Arab-Israel wars, the concept of luminescent solar concentrator (LSC) gained importance [19]. During subsequent years, the situation eased and the research activities in this field also dwindled. With huge increase in the prices at the beginning of the century, the efforts were revived. Particularly, after predictions of Trupke et al. [20,21], the concept of spectrum modification using downconversion was vigorously discussed, and fresh experiments for improving solar cell efficiencies using luminescence gathered momentum.

The concept of "solar spectrum modification" is, in principle, a very attractive way for achieving efficiency increase. Solar radiation reaching earth is distributed over a wide spectrum range, appreciable intensity being located from about 350 to 3,000 nm. Semiconductor-based photovoltaic devices do not utilize the entire spectrum. This is clearly seen from spectral response curves of various solar cells (Figure 7.3). If the radiations are not used or converted, e.g., using luminescence, to suitable spectral region, then the overall efficiency will obviously increase. This is schematically shown in Figure 7.4.

Basically, there are three processes which can be used for spectrum modification (Figure 7.5). Downshifting is the commonly observed Stokes-shifted luminescence, in which photon of high

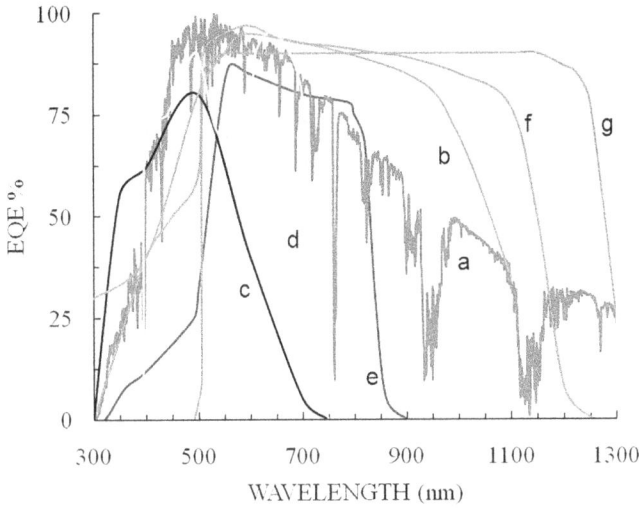

FIGURE 7.3 Solar cell EQE: (a) 1.5 AM solar spectrum, (b) c-Si, (c) a-Si, (d) DSSC, (e) Cds–CdT, (f) CIGS, and (g) CIS.

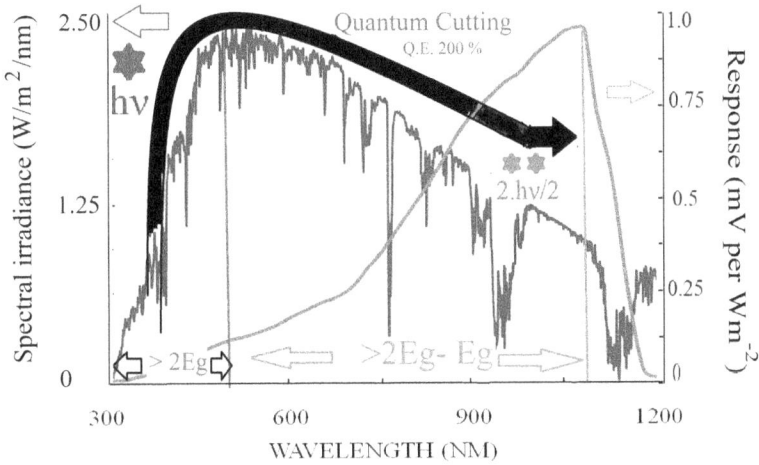

FIGURE 7.4 Spectrum modification concept.

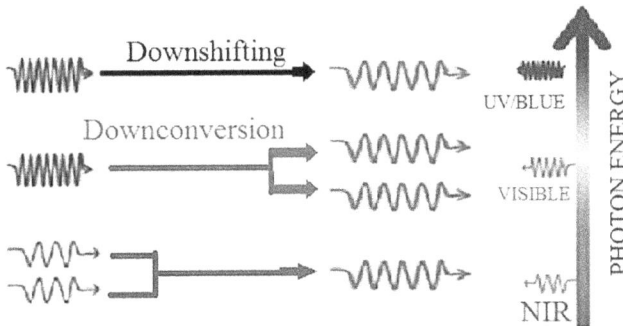

FIGURE 7.5 Downshifting up- and downconversion.

energy (300–500 nm) which is not used up by the photovoltaic material is absorbed by the lumines-cent material and re-emitted as light at longer wavelength (500–1,000 nm) to which the semiconduc-tor is responsive. High-energy photons are not effectively used by the photovoltaic semiconductor due to back-reflection; also at the most, only one e-h pair is generated even if E > 2Eg. The extra energy is lost as phonons.

In downconversion, a high energy photon (E > 2Eg) is converted by the phosphor to two lower energy photons (E < Eg). This is the most attractive mode of spectrum modification. However, it is difficult to achieve over the entire spectral range of interest (300–500 nm). Non-utilization of pho-tons corresponding to E < Eg is rather obvious as such photons cannot be absorbed. On the other hand, there is substantial fraction of solar radiation at energies E < Eg. This can be utilized by the phenomenon of upconversion which is combining two sub-bandgap energy photons to yield a pho-ton with energy E > Eg which can then be used for photovoltaic conversion.

7.3.1 Downshifting Oxide Phosphors

In the photoluminescence phenomenon, most commonly observed is the Stokes-shifted emission, also now known as downshifted luminescence. In this, excitation is by shorter wavelengths and emission at longer. The maximum possible quantum efficiency is 100%. Even with this, the energy conversion efficiency will be short of 100% as the energy of the emitted photon is smaller than that of the absorbed. This was used for improving the performance of solar cells by constructing luminescent solar concentrators (LSC). Several phosphors have been suggested from time to time. Some important oxide phosphors suggested to meet this goal are reviewed in the following.

For using downshifting phosphors towards improvement of photovoltaic conversion efficiencies, the spectral response curves of solar cells are important. Invariably, all the solar cells utilize only a part of the solar spectrum which spreads over a wide range. The shorter wavelengths beginning from about 300 nm up to an upper limit characteristic of the cell material have to be downshifted to a spectral region for which the cell has maximum conversion efficiency. As the spectral response curves of various solar cells differ, we classify the phosphors as per their suitability for a specific cell.

7.3.1.1 Phosphors for Perovskite Solar Cells

Perovskite solar cells (PSC) evolved from dye-sensitized solar cells (DSSC) are developed by Regan and Gratzel [22]. DSSC could not match the efficiency of c-Si solar cells. Moreover, it suffered from other drawbacks like degradation of the dye during the prolonged exposure to UV component in solar radiation. A solid electrolyte is expected to be more stable. Open-circuit voltage is also likely to improve when solid-state hole transfer material (HTM) is used [23,24].

Kojima et al. were the first to report organometal halide perovskites as visible-light sensitizers for photovoltaic cells [25]. Inorganic–organic perovskite compounds ($CH_3NH_3PbX_3$, X = I, Br, and Cl) have since then been considered as light-harvesting materials for hybrid solar cells because of their high extinction coefficients and broader light-absorption [26–28]. Efficiencies were as low as 3.8%. Since then, rapid progress has been made, and the state-of-art perovskite-based solar cells (PSC) have efficiencies as high as 22.7% [29,30]. This is depicted in Figure 7.6. Apart from intense broad-band absorption, the Pb-based perovskites offer several advantages such as high charge car-rier mobility and long charge diffusion length [31].

Perovskite solar cell (PSC) works on the following principle. Perovskite materials act as electro-lytes to absorb sunlight which will excite hole and electrons. Electron in electron transport material (ETM) going towards that acts as an n-type semiconductor and the hole going towards hole trans-port material (HTM) acts as a p-type semiconductor. Fluorine-doped tin oxide is used as anode and gold or silver as cathode. These materials are deposited as a thin film on the top of the perovskite making the structure of the cell [32].

However, PSC is still an emerging technology, and several issues are still to be addressed before it can be commercialized.

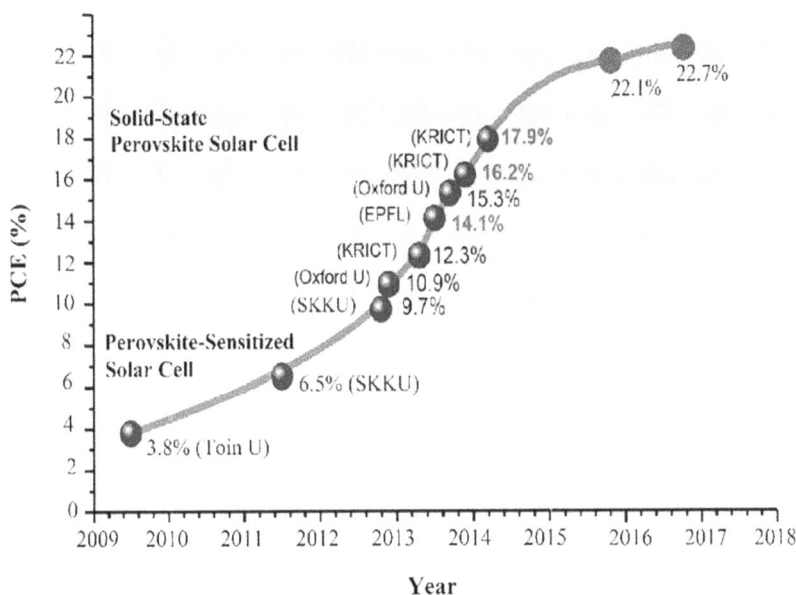

FIGURE 7.6 Improvement in photovoltaic conversion efficiency (PCE) of PSC with time.

Stability in ambient environment, intrinsic toxicity and anomalous hysteresis are some of the drawbacks which need to be overcome. $CH_3NH_3PbX_3$ perovskite compounds absorb light in near-UV as well as visible region. Leijtens et al. [33] showed a loss of charge in PSCs due to UV exposure. They proposed that the UV-degraded cells suffered from a deep trapping of injected electrons within newly available sites in the TiO_2. This instability of PSC under UV exposure can be rectified (i) by TiO_2 surface states pacification, (ii) by completely removing the mesoporous TiO_2 film, and (iii) by prevention of UV light reaching to the mesoporous TiO_2 film. The third option where a suitable down-shifting (DS) material absorbs UV light (<400 nm) falling on the PSC and re-emits visible light which is then utilized by the perovskite absorber can be an effective way of improving the efficiency while reducing the UV light-induced degradation [34].

Typical spectral response of PSC is shown in Figure 7.7. It can be seen that the response dips below 450 nm. Moreover, there is degradation due to long exposures to shorter wavelengths present in the solar spectrum. Hence, if the solar radiations from 350 to 450 nm are absorbed and converted to wavelengths longer than 500 nm, then it will result in increasing the stability and efficiency of PSC. Oxide phosphors meeting these criteria are described in the following.

7.3.1.1.1 BaSrSiO₄:Eu₂₊

Barry was the first to report the luminescence properties of $M_2SiO_4:Eu^{2+}$ (M=Ca, Sr, Ba) [35]. These compounds exhibit efficient luminescence under both near-UV ($\lambda_{max}=405$ nm) and blue ($\lambda_{max}=450$ nm) light excitation, with the emission ranging from green for $Ba_2SiO_4:Eu^{2+}$ [36] to yellow for $Sr_2SiO_4:Eu^{2+}$ [37]. Poort et al. [38] reported finer features of the PL spectra, e.g., presence of two overlapping bands in the emission spectra which are responsible for the observed large width. They tried to resolve emission from Eu^{2+} at two crystallographically non-equivalent sites on the basis of decay constants [39].

Yoo et al. [40] studied solid solutions of $Ba_2SiO_4:Eu^{2+}$ and $Sr_2SiO_4:Eu^{2+}$ with the objective of tuning the excitation and emission spectra to match the requirements of LED lamps based on blue-emitting chips. They found $Sr_{1.06}Eu_{0.04}Ba_{0.9}SiO_4$ to be optimum for such applications. More or less similar results were obtained by Kim et al. [41]. Moreover, while the end members suffer from thermal quenching [42], the solid solutions show much improved thermal stability [43,44]. While most

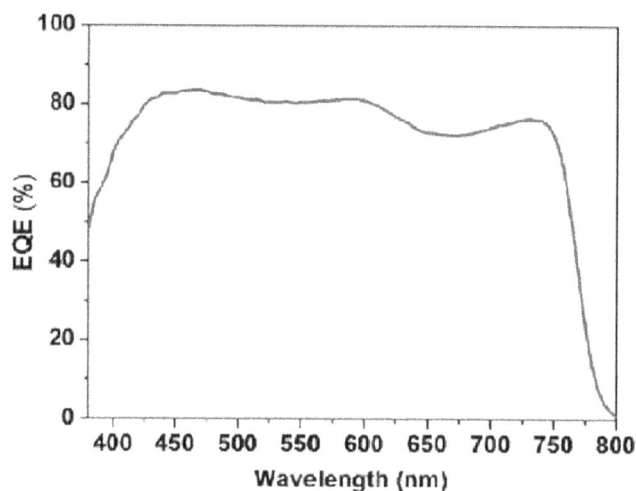

FIGURE 7.7　PSC spectral response.

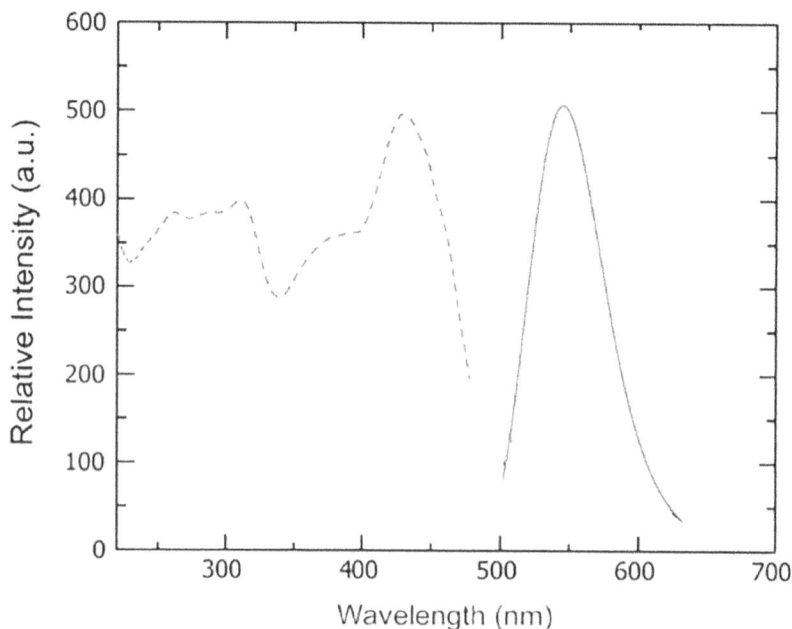

FIGURE 7.8　PL spectra for $BaSrSiO_4$:Eu^{2+}.

workers used solid-state reaction for phosphor synthesis, liquid-phase precursor-driven approach was adopted in recent years [45,46]. Eventually, this phosphor was patented for commercial use [47]. Figure 7.8 shows typical PL spectra for $BaSrSiO_4$:Eu^{2+}. Strong emission is observed for Eu^{2+} in the green region with maximum around 517 nm. Maximum intensity is observed for Eu^{2+} concentration of 2.5 mol.%. Excitation spectrum is very broad, stretching from 250 to 500 nm with broad maxima around 370 nm. The long wavelength tail represents the "staircase" structure characteristic of the Eu^{2+} f–d absorption.

7.3.1.2   For CdS-CdTe Solar Cell

The first significant laboratory CdTe cell was reported in 1972 by Bonnet and Rabnehorst [48] who developed a thin-film-graded gap CdTe-CdS p-n heterojunction solar cell with 6% efficiency. By 1981, Kodak used close-spaced sublimation (CSS) and made the first cells having two-digit efficiency [49]. Monosolar [50] and AMETEK [51] used electrodeposition, a popular early method. Matsushita started with screen printing but shifted in the 1990s to CSS. The next important step was to scale up size and larger area products called modules. These products required higher currents than small cells, and it was found that an additional layer, called a transparent conducting oxide (TCO), could facilitate the movement of current across the top of the cell. CdTe cells achieved above 15% in 1992 by adding a buffer layer to the TCO/CdS/CdTe stack and then thinned the CdS to admit lighter. Developments on CdS-CdTe solar cells have recently been reviewed by Lee and Ebong [52–54]. Cadmium telluride PV is the only thin film technology with lower costs than conventional solar cells made of crystalline silicon in multi-kilowatt systems [55].

The CdS/CdTe solar cell has a high conversion efficiency as well as capability to be produced at low cost. The heterojunction CdS/CdTe solar cell shows a high spectral response in the wavelength region from 500 nm with a sharp rise to 850 nm, while it is non-sensitive for the region below 500 nm due to the absorption in the CdS window layer (Figure 7.9). Accordingly, if sunlight in the wavelength region below 500 nm can be somehow used for the extension of the spectral response, then an increase of short-circuit current which leads to the increase of both output power and conversion efficiency will be achieved [56–58]. An increase in J_{sc} from 19.8 to 22.9 mA/cm^2 was calculated, corresponding to an increase in conversion efficiency from 9.6% to 11.2% [59]. This indicates that a relative increase in the performance of a production CdS/CdTe PV module of nearly 17% can be expected via the application of luminescent downshifting (LDS) layers. This application of luminescent materials for overcoming the poor blue response of solar cells was first described by Hovel et al. in 1979 for Cu_2S/CdS solar cells [60].

Solar spectrum can be shifted using the phosphors which will absorb radiations in the wavelength region 300–500 nm and emit around 650–750 nm. Thus, a layer of luminescent material absorbs the short-wavelength photons, where the PV module has a low IQE, and re-emits

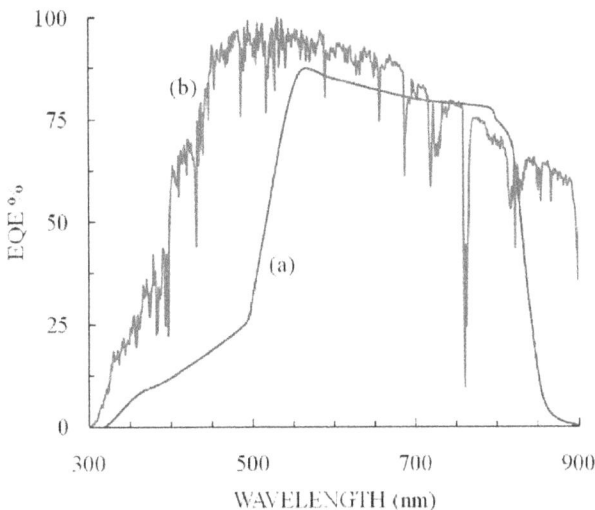

FIGURE 7.9 Mismatch between the spectral response of CdS-CdTe solar cell and the solar spectrum. (a) External quantum efficiency is plotted as a function of wavelength. (b) 1.5 A.M. solar spectrum is given for the comparison. Note the poor response of CdS-CdTe solar cell below 500 nm.

photons of a longer wavelength, where the IQE of the PV device is high. The slight energy losses in the down-shifting process are outweighed by the gains that stand to be made by converting $\lambda < 540$ nm photons into $\lambda > 540$ nm photons. A further advantage of the luminescence downshifting (LDS) layer is that it allows greater freedom of design and optimization of the underlying PV device.

Phosphors with such properties may be found in a variety of ways. Rare-earth activators Eu^{2+} and Sm^{2+} can have strong absorption in 350–500 nm regions in "suitable hosts". In case of Eu^{2+}, for strong nephelauxetic effect, emission is at longer wavelengths, which is usually observed in ligands with weak electronegativity, such as sulphides or nitrides [61]. In case of Sm^{2+}, when levels of $4f^5 5d^1$ configuration are above those of $4f^6$ configuration, $^5D_0 \rightarrow {}^7F_j$, $j = 1,2,3$ transitions yield line emissions around 680, 700 and 720 nm [62]. This situation corresponds to relatively weak crystal fields. Otherwise, band emission due to d–f transition at longer wavelength is obtained. Whether d–f emission is obtained or f–f depends critically on the host lattice as well as temperature of measurement [63]. Cr^{3+} is known to give line emission around 693 nm with excitation near 400 nm or a band emission at longer wavelengths [64–66], depending upon the coordination environment [67]. Line emission corresponding to the $^2E_g \rightarrow {}^4A_{2g}$ transition, which is desired for the application under discussion, is obtained when Cr^{3+} experiences a strong crystal field [68]. Another 3d activator Mn^{4+}, which has same $3d^3$ electronic configuration as Cr^{3+}, also shows excitation covering near-UV-blue region and emission between 630 and 720 nm, depending upon the host [69–74]. In solids, the Mn^{4+} ion always experiences a strong crystal field due to its high effective positive charge with the result that the emission spectrum is always dominated by the sharp emission line corresponding to the spin-forbidden $^2E_g \rightarrow {}^4A_{2g}$ transition. The structure in the emission spectrum is evaluated in terms of the prominent vibronic transitions within this transition. Though the transition $^2E_g \rightarrow {}^4A_{2g}$ is electric dipole forbidden, it gains intensity by the activation of vibronic modes. Emission peak energy of $^2E_g \rightarrow {}^4A_{2g}$ transition is almost independent of the crystal field splitting but is significantly influenced by the covalence of the "Mn^{4+}-ligand" bonding (nephelauxetic effect). The nephelauxetic effect is due to the formation of chemical bonds between impurity 3d ion and ligands. The emission wavelength of Mn^{4+} in highly covalent hosts like oxides is more suitable for the present application than that in ionic hosts like fluorides [75].

7.3.1.2.1 $Al5GeO_{9.5}$:Cr^{3+}

In the system Al_2O_3-GeO_2, several phases which can be represented by a general formula $Al_{4+2x}Ge_{2-2x}O_{10-x}$, termed as germanium mullites, are known [76]. $Al_5GeO_{9.5}$ is an aluminium-rich phase. However, there is no information about luminescence in this host. This is rather surprising considering the wealth of corresponding information on Al_2O_3. Results for emission are shown in Figure 7.10. Intense line attributable to ($^4A_2 \rightarrow {}^2E$) is seen around 699 nm with several components due to splitting of ^2E levels in crystal field. Inset shows variation of 699 nm line intensity with Cr^{3+} concentration. Intensity is maximum for 0.4 mol.%, and quenching is observed for higher concentrations. Intensity is maximum for 0.4 mol.%, and quenching is observed for higher concentrations.

Figure 7.11 shows excitation spectrum for 699 emission of $Al_5GeO_{9.5}$:Cr^{3+}.$^4A_{2g} \rightarrow {}^4T_{1g}(^4F)$, and $^4T_{2g}(^4F)$ bands can again be clearly seen. Splitting of $^4T_{1g}(^4F)$, $^4T_{2g}(^4F)$ levels is barely visible in the spectrum recorded at room temperature. Excitation peaks are clearly seen in at 412 and 541 nm corresponding to transitions $^4A_{2g} \rightarrow {}^4T_{1g}(^4F)$ and $^4T_{2g}(^4F)$, respectively. Transition $^4A_{2g} \rightarrow {}^4E[^4T_{1g}(^4P)]$ expected [77] to be around 256 nm is also partly seen.

$Al_5GeO_{9.5}$:Cr^{3+} show emissions close to 700 nm where the CdS-CdTe solar cell has good response (Figure 7.12, curve b). The suitability of the absorption spectrum for downshifting is shown with the help of Figure 7.15. $Al_5GeO_{9.5}$:Cr^{3+} (Figure 7.12, curve c) and $SrMgAl_{10}O_{17}$:Cr^{3+} (Figure 7.12, curve d) have excitation around 400 nm, where CdS-CdTe has poor response. It is thus suitable for

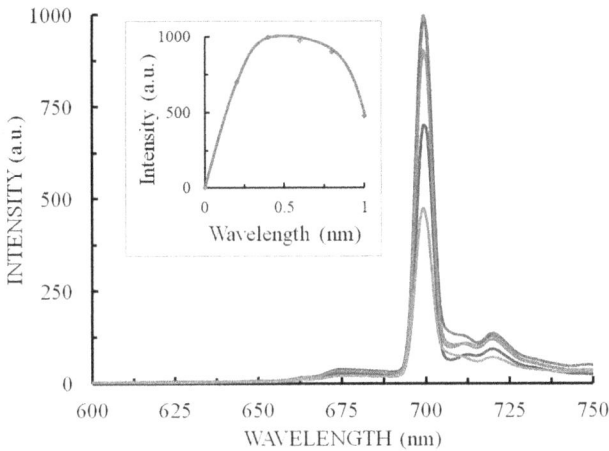

FIGURE 7.10 PL emission spectrum of $Al_5GeO_{9.5}:Cr^{3+}$ for 412 nm excitation.

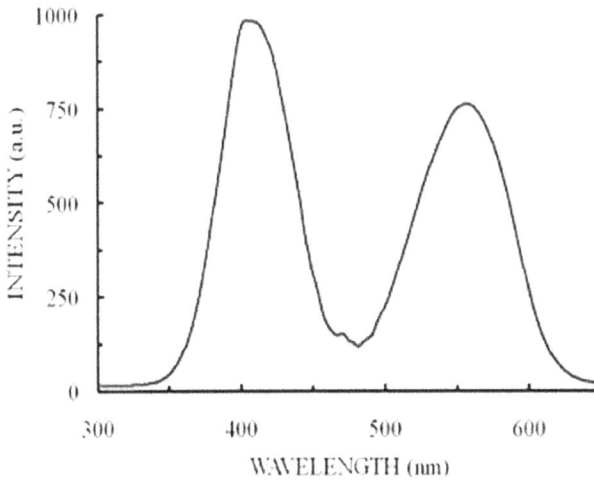

FIGURE 7.11 PL excitation spectrum for 699 nm emission of $Al_5GeO_{9.5}:Cr^{3+}$.

downshifting the solar spectrum (Figure 7.12, curve a) in the region 320–430 nm. An ideal phosphor should absorb in the entire 300–500 nm region (Figure 7.12, shaded portion) and convert to longer wavelengths [78].

7.3.1.2.2 $LiAl_5O_8:Cr$

At least four compounds in Li_2O-Al_2O_3 system, viz., $LiAlO_2$, $LiAl_5O_8$, Li_5AlO_4 and $Li_2Al_4O_7$, are known. Apart from these, Li_3AlO_3 is mentioned by Ginestra et al. [79]. However, it was found to be very unstable and obtainable only in a narrow temperature interval around 400°C. Of these, $LiAlO_2$, $LiAl_5O_8$ and $Li_2Al_4O_7$ have been prepared by the combustion synthesis in earlier work [80].

$LiAlO_2$ exists in three different crystalline phases: alpha, beta and gamma, which are hexagonal, monoclinic and tetragonal, respectively. Gamma-$LiAlO_2$ is the stable phase at high temperature, and alpha or beta $LiAlO_2$ phases transform to gamma-$LiAlO_2$ at an elevated temperature [81]. Since the temperature rises very fast during the combustion synthesis, it is expected that gamma phase will be formed. This was indeed observed. However, we did not observe any Cr^{3+} emission in this host.

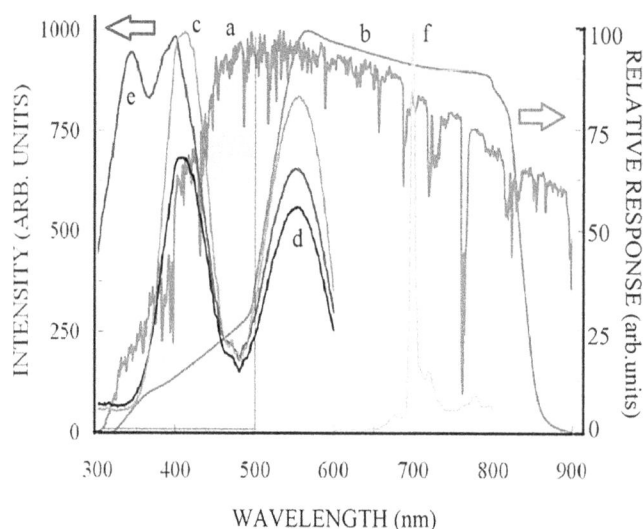

FIGURE 7.12 Modification of solar spectrum using Cr^{3+} doped phosphors. AM1.5 solar spectrum, (b) spectral response of CdS-CdTe solar cell, Cr^{3+} excitation spectrum for (c) $Al_5GeO_{9.5}:Cr^{3+}$, (d) $SrMgAl_{10}O_{17}:Cr^{3+}$, and (e) $SrMgAl_{10}O_{17}: Eu^{2+}, Cr^{3+}$, (f) typical Cr^{3+} emission spectrum.

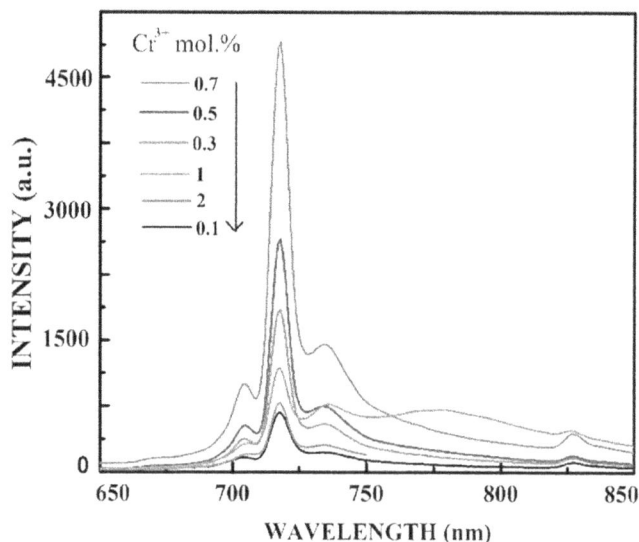

FIGURE 7.13 PL emission for $LiAl_5O_8:Cr^{3+}$ phosphors.

In contrast, very strong emission was observed in $LiAl_5O_8:Cr^{3+}$. There are several previous studies, both theoretical [82–84] and experimental [85–87], on this phosphor.

$LiAl_5O_8$ had been chosen as a host due to its excellent chemical and thermal stability [88]. Temperature dependence of the emission properties has been suggested for application in optical thermometry. Figure 7.13 shows PL emission spectra for $LiAl_5O_8:Cr^{3+}$. Emission is in the form of an intense line around 718 nm with a vibronic component around 704 nm. These are well-known R1 and R2 lines. The emission is due to $^2E_g \rightarrow {}^4A_{2g}$ transition. Another weak line around 734 nm may be due to Cr^{3+}–Cr^{3+} pairs. Intensity is highest for 0.7% Cr^{3+}, and concentration quenching sets in for higher values. In the excitation spectrum (Figure 7.14), two bands are observed around 400 and 563 nm [89]. These can be attributed to $^4A_2 \rightarrow {}^4T_1$ and $^4A_2 \rightarrow {}^4T_2$ transitions, respectively. $LiAl_5O_8:Cr^{3+}$ phosphor appears suitable for the modification of the solar spectrum so as to match the spectral response of CdS-CdTe solar cell.

7.3.1.2.3 Li2Mg₃SnO₆:Mn⁴⁺

$Li_2Mg_3SnO_6$ belongs to cubic crystal system with space- $Fm\overline{3}m$ group (no. 225) and the lattice parameter a=4.246 Å [90] (Figure 7.15).

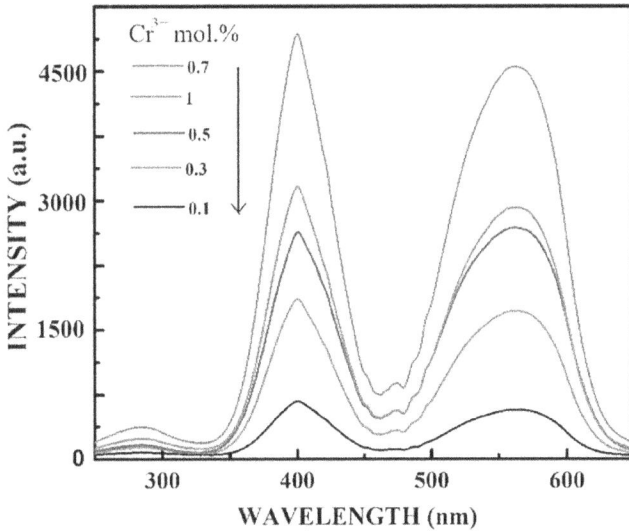

FIGURE 7.14 Excitation spectrum for $LiAl_5O_8:Cr^{3+}$ phosphors.

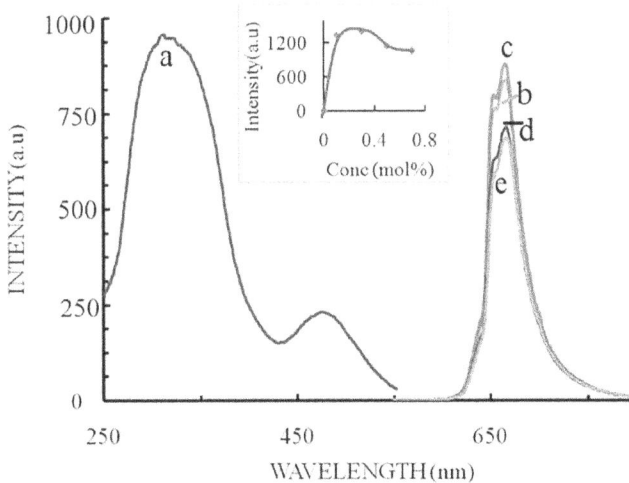

FIGURE 7.15 PL excitation and emission spectra of $Li_2Mg_3SnO_6:Mn^{4+}$. [91] (a) Excitation spectrum for Mn^{4+} for 665 nm emission Mn^{4+} concentrations (b) 0.2, (c) 0.4, (d) 0.6 and (e) 0.8 mol.%.

7.3.1.2.4   Mg₂TiO₄:Mn⁴⁺

Mg_2TiO_4 is an inverse spinel, having the structural formula $Mg[MgTi]O_4$. At least two polymorphs of Mg_2TiO_4 are known. Magnesiotitanate spinel (Mg_2TiO_4) is most closely approached in nature as the mineral qandilite [92]. Barth and Posnjak [93] showed that qandilite has the spinel structure, with @equ_0002.eps@ space group. Several subsequent investigations have confirmed this structure [94,95].

High-temperature cubic form has the same structure as qandilite. Tetragonal polymorph known as low-temperature form was reported by Delamoye and Michel [96], and its detailed structure was described by Wechsler [97].

Figure 7.16 (curve a) shows excitation spectrum for 663 emission of $Mg_2TiO_4:Mn^{4+}$, $^4A_{2g} \rightarrow$ $^4T_{1g}(^4F)$ (350 nm) and $^4T_{2g}(^4F)$ (472 nm) bands can again be clearly seen. Results for emission are also shown in Figure 7.16 (curves b–e). Intense line attributable to $(^4A_2 \rightarrow ^2E)$ is seen around 663 nm with several components due to splitting of ^2E levels in crystal field. Inset shows variation of 663 nm line intensity with Mn^{4+} concentration. Intensity is maximum for 0.1 mol.%, and quenching is observed for higher concentrations [78]. Significance of these results will be discussed in the next section along with that for all Mn^{4+} activated phosphors.

7.3.1.2.5   Ca₁₄Zn₆Al₁₀O₃₅:Mn⁴⁺

Barbanyagre et al. reported two calcium aluminozincates with formulae $Ca_{14}Zn_6Al_{10}O_{35}$ and $Ca_3ZnAl_4O_{10}$ [98]. In recent years, luminescence studies have been done on $Ca_{14}Zn_6Al_{10}O_{35}$. Seki et al. [99] and Lu et al. [100] reported efficient red emission of Mn^{4+} with blue excitation in this host and suggested applications in LEDs for indoor plant cultivation. Energy transfers like $Mn^{4+} \rightarrow Er^{3+}$ [101], $Bi^{3+} \rightarrow Eu^{3+}$ [102], $Bi^{3+} \rightarrow Sm^{3+}$ [103], etc., have also been studied. Figure 7.17 (curve a) shows that excitation spectrum for 720 emission of $Ca_{14}Zn_6Al_{10}O_{35}:Mn^{4+}$. $^4A_{2g} \rightarrow ^4T_{1g}(^4F)$ (340, 410 nm) and $^4T_{2g}(^4F)$ (472 nm) bands can again be clearly seen. Results for emission are also shown in Figure 7.17 (curves b–e). Intense line attributable to $(^4A_2 \rightarrow ^2E)$ is seen around 720 nm with several components due to splitting of ^2E levels in crystal field. Inset shows variation of 720 nm line intensity with Mn^{4+} concentration. Intensity is maximum for 0.1 mol.%, and moderate quenching is observed for higher concentrations. $Ca_{14}Zn_6Al_{10}O_{35}:Mn^{4+}$, $Mg_2TiO_4:Mn^{4+}$ and $Li_2Mg_3SnO_6:Mn^{4+}$ all show emission close to 700 nm where the CdS-CdTe solar cell has good response (Figure 7.18, curve b). The suitability of the absorption spectrum for downshifting is shown with the help of Figure 7.18. $Ca_{14}Zn_6Al_{10}O_{35}:Mn^{4+}$ (Figure 7.18, curve c) and $Mg_2TiO_4:Mn^{4+}$ (Figure 7.18, curve d) have

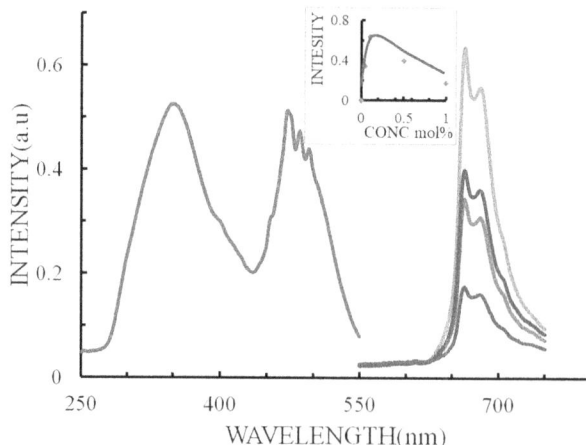

FIGURE 7.16 PL excitation and emission spectra of $Mg_2TiO_4:Mn^{4+}$. (a) Excitation spectrum for Mn^{4+} for 663 nm emission Mn^{4+} concentrations (b) 0.05, (c) 0.1, (d) 0.5 and (e) 1.0 mol.%.

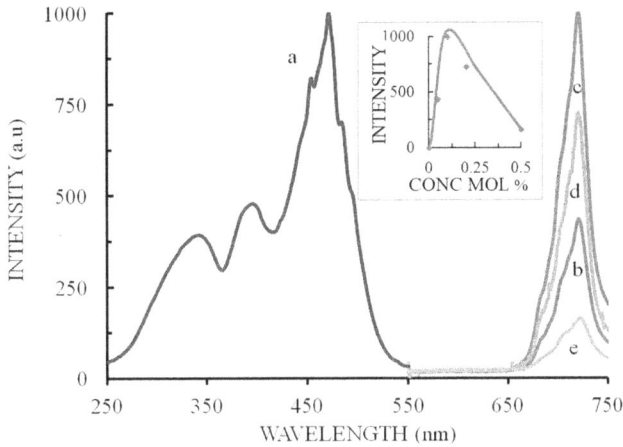

FIGURE 7.17 PL excitation and emission spectra of $Ca_{14}Zn_6Al_{10}O_{35}:Mn^{4+}$. (a) Excitation spectrum for Mn^{4+} for 720 nm emission, (b–e) emission for 472 nm excitation. Mn^{4+} concentrations: (b) 0.05, (c) 0.1, (d) 0.2 and (e) 0.5 mol.%.

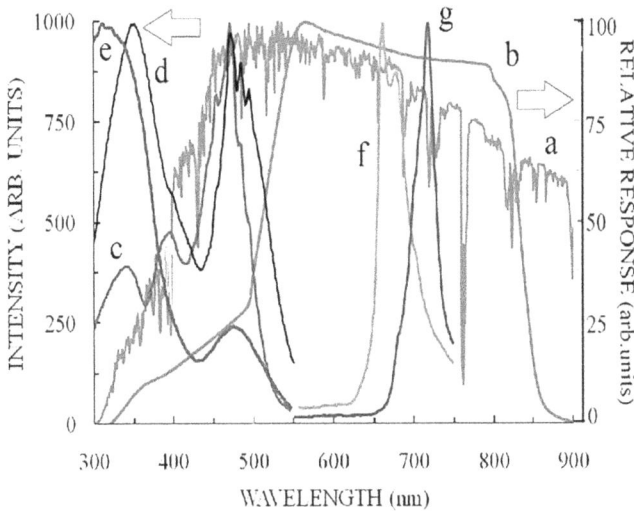

FIGURE 7.18 Modification of solar spectrum using Mn^{4+}-doped phosphors. (a) AM1.5 solar spectrum, (b) spectral response of CdS-CdTe solar cell, Mn^{4+} excitation spectrum for (c) $Ca_{14}Zn_6Al_{10}O_{35}:Mn^{4+}$, (d) $Mg_2TiO_4:Mn^{4+}$ and (e) $Li_2Mg_3SnO_6:Mn^{4+}$. Mn^{4+} emission spectrum (f) $Mg2TiO4:Mn^{4+}$ and (g) $Ca_{14}Zn_6Al_{10}O_{35}:Mn^{4+}$.

adequate excitation around 470 nm, where CdS-CdTe has poor response (Figure 7.18, curve b). $Li_2Mg_3SnO_6:Mn^{4+}$ (Figure 7.18, curve e), on the other hand, has rather weak excitation in this region. $Mg_2TiO_4:Mn^{4+}$ exhibits excitation in a broader range down to 320 nm. It is thus better suitable for downshifting the solar spectrum (Figure 7.18, curve a) in the region 320–430 nm. An ideal phosphor should absorb in the entire 300–500 nm region (Figure 7.18, shaded portion) and convert to longer wavelengths.

All these Mn^{4+}-activated phosphors, unlike Cr^{3+} activation, do not have excitation much beyond 500 nm and hence will not "encroach" the spectral region where CdS-CdTe has good response. On the other hand, Cr^{3+}-activated phosphors cover the near-UV region between 350 and 400 nm more effectively [78].

7.3.2 DOWNCONVERSION

Upconversion is combining energy of two photons to generate one photon of higher energy. In contrast, in downconversion, energy of the donor ion is transferred stepwise or by co-operative energy transfer to two acceptor ions. Figure 7.19 illustrates two ways in which downconversion can take place. Thus, a Pr^{3+} ion in excited state can transfer energy simultaneously to two Yb^{3+} ions via co-operative energy transfer process. Alternatively, it can come to $^1G_4$ state by transferring energy to a Yb^{3+} ion and then from this state to the ground state by transferring energy to the second Yb^{3+} ion.

7.3.2.1 Downconverting Oxide Phosphors

Following seminal paper by Trupke et al. [20], lot of work has been carried out to discover phosphors for modification of solar spectrum [104]. All these efforts are aimed at improving efficiency of solar cells, especially c-Si solar cells. Spectral response curve for commercial c-Si solar cells reveals that there is no efficient photovoltaic conversion for solar radiations in the region 350–500 nm [105,106]. This is also true for many other solar cell materials like CdS-CdTe [107,108], CIGS [109], etc. If this light is absorbed by a phosphor and re-emitted around 1 μm, where c-Si solar cell response is high, then the photovoltaic performance will improve a great deal. This is the concept of solar spectrum modification. Moreover, if "downconversion", which is emission of two or more low energy photons for each high energy photon absorbed, can be achieved then the efficiency increase can be substantial. Most of the initial investigations on solar spectrum modification were focused on obtaining downconversion, and not much attention was paid to the excitation spectrum.

For spectral modification, Yb^{3+} and Nd^{3+} ions are most suitable. Existence of only two levels, $^2F_{5/2}$ and $^2F_{7/2}$, 10,150 cm^{-1} apart makes Yb^{3+} most sought-after activator for NIR emission around 985 nm provided that the absorption of nUV radiations (350–500 nm) followed by downconversion takes place. The emission lies in the spectral region where response of c-Si is maximum. Quantum yield is close to 1. There being no additional levels, no pathways for branching and entire emission have to come through the $^2F_{5/2}{\rightarrow}^2F_{7/2}$ route. In case of Nd^{3+} ions also, energies of most observed $^4F_{3/2}-^4I_{9/2}$ and $^4F_{3/2}-^4I_{11/2}$ transitions lie quite close to the most sensitive spectral region of silicon solar cells. In contrast to Yb^{3+}, there are multiple de-excitation pathways for Nd^{3+}; also the most intense emission around 1,060 nm only marginally overlaps with the response curve of low-cost c-Si solar cells. On the other hand, availability of the number of levels over wider energy range renders sensitization of Nd^{3+} more attainable.

For both these lanthanides, the transitions are f–f and have low oscillator strength. Charge transfer (CT) bands in most cases are at energies which are outside the range of the short wavelength solar emission [110].

Downconversion Mechanisms

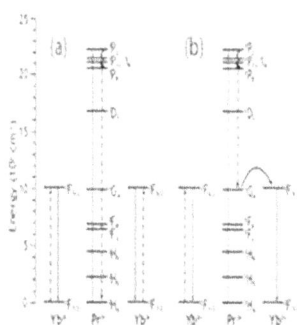

- (a) cooperative energy transfer energy is simultaneously transferred to two Yb^{3+} ions.
- (b) energy is stepwise transferred to Yb $^{3+}$ ions, using the $^1G_4$ level as an intermediate state.

FIGURE 7.19 Downconversion mechanisms.

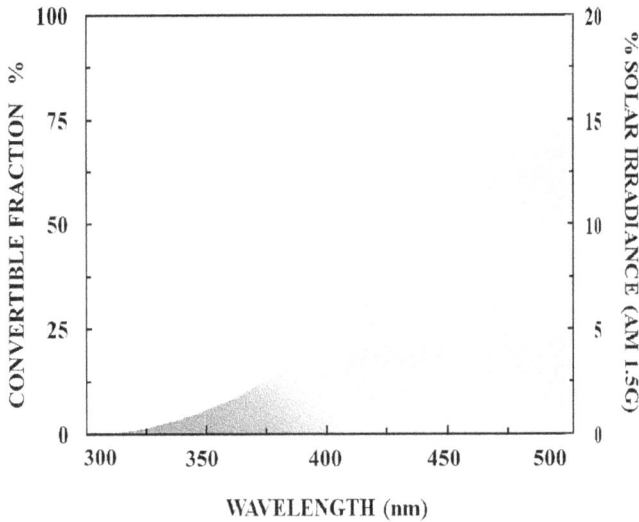

FIGURE 7.20 Fraction of solar radiations between interval 300 nm to wavelength.

A review [111,112] points out that UV to VIS quantum cutting examples in the literature are not useful for improving solar cell efficiency, since the (V)UV excitation wavelengths involved are not (or are minimally) present in the terrestrial solar spectrum. Though these comments have been made a decade back, situation has not significantly changed even today. It may be emphasized that for several solar cells like c-Si, CIS, CIGS, CdS-CdTe, etc., converting spectral region 350–500 nm to NIR is essential for improving the efficiency using the concept of spectrum modification. During the conversion, the preference has to be given to the longer wavelength region of this interval. This is illustrated in Figure 7.20.

The fraction of the solar spectrum that needs to be converted goes on increasing almost exponentially with the wavelength. Thus, it is merely 4% of the total solar emission in the interval 300–400 nm, while 15% for the range 400–500 nm. In terms of the downconvertible solar emission fraction, the corresponding numbers are 20% and 95%, respectively. About 79% of the convertible portion lies in the 400–500 nm region, while only 21% in 300–400 nm. These facts have been overlooked, and majority of the literature describes conversion of 300–400 nm light into NIR around 1 μm. Various solar cells have different characteristics, and thus, same downconverting phosphor will not be suitable as a spectrum modifier for all the solar cells. Hence, we discuss the phosphors for these cells under different sections.

7.3.2.1.1 For c-Si

Developments of c-Si have been thoroughly reviewed frequently [113]. Efficiency improvements are reviewed in Figure 7.21. Solar cell efficiencies are low, even today; the commercially available c-Si solar cells have efficiency no more than 17%. Mismatch between the incident solar spectrum and spectral response of solar cells limits the cell efficiency [114]. Solar cells do not use entire solar spectrum. Photons of energies smaller than the band gap are not absorbed. Thermalization of charge carriers generated by the absorption of high-energy photons profoundly reduces the conversion efficiency of solar cell.

Figure 7.22 shows a typical spectral response curve for c-Si solar cell. It can be seen that EQE drops considerably below 500 nm and above 1,000 nm. This is referred to as the spectral mismatch. Upper limits for conversion efficiencies of c-Si solar cells have been estimated theoretically and referred to as Shockley–Queisser limits [115], which is 33.2% for c-Si.

Solutions to overcome the spectral mismatch have been suggested from time to time. These efforts were vigorously renewed when Trupke et al. predicted that if the light in the region

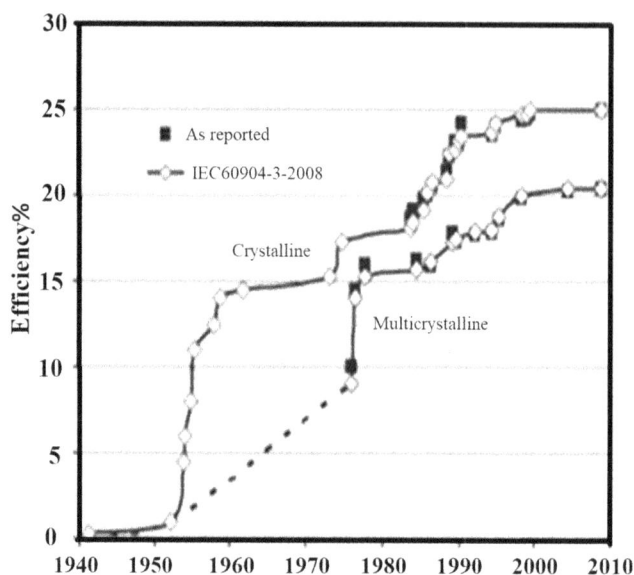

FIGURE 7.21 Evolution of crystalline and multi-crystalline silicon solar cell efficiency.

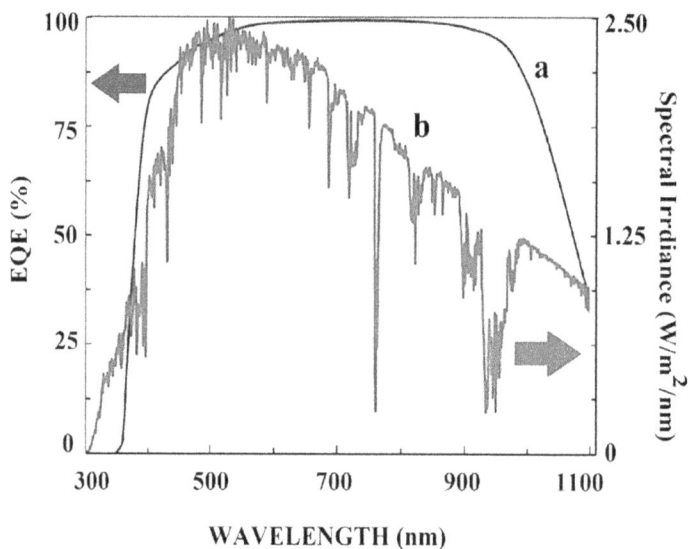

FIGURE 7.22 Typical spectral response curve for (a) c-Si solar cell compared with (b) solar spectrum.

300–500 nm is converted to NIR using quantum cutting, then the Shockley-Queisser limit for c-Si will be pushed from 29.5% to 39.6% [20]. These efforts have been reviewed frequently [111,116–122].

7.3.2.1.1.1 $Bi_2(MoO_4)3:Nd_3+$ and/or Yb^{3+}   Phase diagram of Bi_2O_3-MoO_3 system is very rich; a large variety of phases are known. Among these, $Bi_2(MoO_4)_3$, $Bi_2Mo_2O_9$ and Bi_2MoO_6 are commonly known. These are also designated as α, β and γ bismuth molybdates. Bi_2MoO_6 itself has several polymorphs. Apart from these better-known phases, compounds like Bi_4MoO_9 and Bi_6MoO_{12} have also been prepared [123]. As many as eight formulae have been mentioned by Egashira et al. [124].

α, β and γ bismuth molybdates have a smaller bandgap (2.5–2.9 eV) and can capture visible light (420–500 nm), exhibiting photocatalytic activity for degradation of organic pollutants. These have been studied for more than 50 years for their photocatalytic activities and associated applications like dye degradation, water splitting, and photosynthesis by partial oxidation and ammoxidation [125], etc. For example, $Bi_2(MoO_4)_3$ is used for photocatalytic oxidation of Butene [126,127], degradation of rhodamine B [128], methylene blue [129] and removal of toxic substances [130]. It has also application in the field of acousto-optics [131].

Results on PL emission for $Bi_2(MoO_4)_3$ in the region 850–1,150 nm are presented in Figure 7.23. Excitation wavelength was 400 nm. Nd^{3+} is expected to occupy Bi^{3+} substitutional sites and thus have coordination 8. Nd^{3+} emission is the in form of lines around 885 and 1,069 nm which correspond to transitions $^4F_{3/2} \rightarrow {}^4I_{9/2}$ and $^4F_{3/2} \rightarrow {}^4I_{11/2}$, respectively. The line around 1,069 nm is the strongest. There is considerable splitting in both the lines due to splitting of the levels by crystal field. Inset shows variation of intensity of 1,069 nm line with Nd^{3+} concentration. Intensity of emission increases up to 3 mol.% Nd^{3+} and then decreases due to concentration quenching as a result of $Nd^{3+} \rightarrow Nd^{3+}$ energy transfer (Figure 7.23 inset).

Figure 7.24 shows the excitation spectrum for 1,069 nm emission. The excitation spectrum consists of a broad band around 400 nm which can be attributed to the host absorption. NIR emission of Nd^{3+} is thus host-sensitized. Some sharp kinks are due to overlapping Nd^{3+} lines. Excitation from f–f transitions of Nd^{3+} was observed more distinctly at longer wavelengths. The distinct lines appear at around 883,870 ($^4I_{9/2} \rightarrow {}^4F_{3/2}$), 824 ($^4I_{9/2} \rightarrow {}^4F_{5/2}$), 807 ($^4I_{9/2} \rightarrow {}^4F_{7/2}$), 767, 750, 740 ($^4I_{9/2} \rightarrow {}^4S_{3/2}$), 687 ($^4I_{9/2} \rightarrow {}^4F_{9/2}$), 630 ($^4I_{9/2} \rightarrow {}^2K_{15/2}$), 612, 589, 581 ($^4I_{9/2} \rightarrow {}^4G_{5/2}, {}^2H_{11/2}$), 530, 518 ($^4I_{9/2} \rightarrow {}^2K_{13/2}$) nm.

$Bi_2(MoO_4)_3:Nd^{3+}$ phosphor thus appears suitable for modification of solar spectrum [132]. However, emission of Nd^{3+} is situated at the edge of the response curve of c-Si. It is better suited for thin film solar cells such as CIGS or CIS.

Yb^{3+} emission at relatively shorter wavelengths is more suitable for c-Si solar cell. Figure 7.25 shows emission spectra for $Bi_2(MoO_4)_3$ doped with various concentrations of Yb^{3+} for 400 nm excitation. Strong emission lines are observed around 986 and 997 nm. There are additional, weaker lines also. These are due to splitting of $^2F_{5/2}$ and $^2F_{7/2}$ levels. Figure 7.25 inset shows dependence of emission intensity on Yb^{3+} concentration. Maximum emission is obtained for 1.5 mol.%. Concentration quenching is observed for higher values due to $Yb^{3+} \rightarrow Yb^{3+}$ energy transfer.

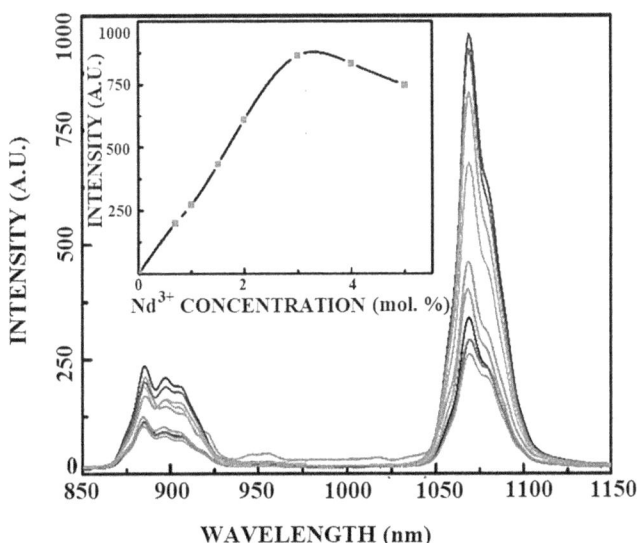

FIGURE 7.23 PL emission spectra for $Bi_2(MoO_4)_3:Nd^{3+}$ phosphors.

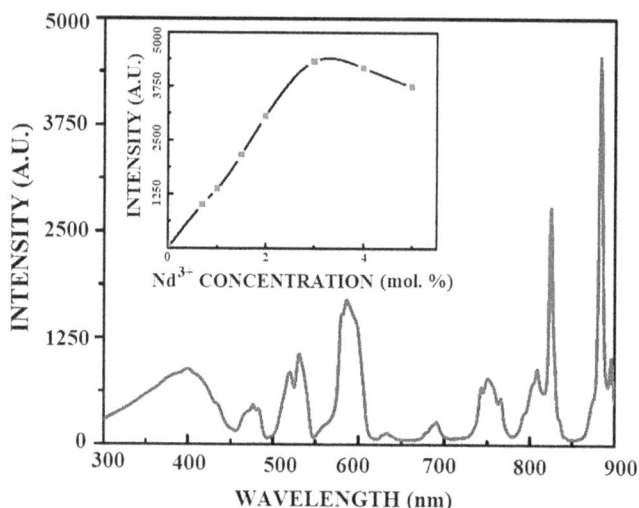

FIGURE 7.24   PL excitation spectra for $Bi_2(MoO_4)_3$:Nd^{3+} phosphors.

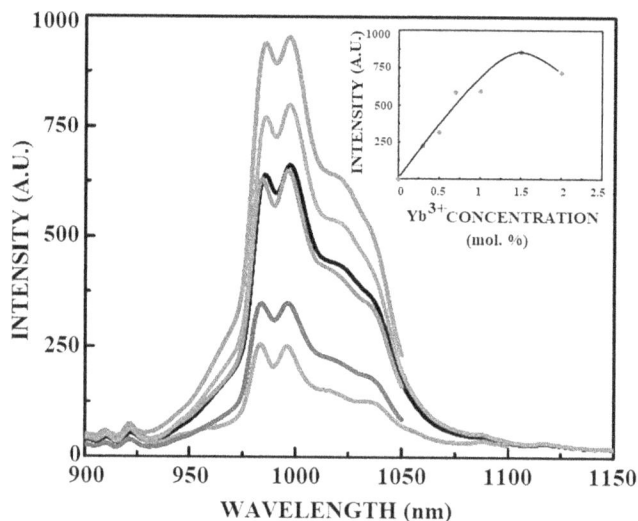

FIGURE 7.25   PL emission spectra for $Bi_2(MoO_4)_3$:Yb^{3+} phosphors.

Both Nd^{3+} and Yb^{3+} show intense host-sensitized, NIR emission. Emission of Yb^{3+} is preferable for applications as it is confined to a narrower region. $Bi_2(MoO_4)_3$:Nd^{3+}, Yb^{3+}phosphors thus appear suitable for modification of solar spectrum.

7.3.2.1.1.2 Bi_2MoO6: Nd^{3+}   Owing to interesting optical properties, Bi_2MoO_6 had been found as a suitable host for activators like Eu^{3+} [133–136], for upconversion studies using Ho-Yb [137] or Er-Yb pairs [138], as a pigment [139], etc.

Nd^{3+} emission in this host is in the form of lines around 887 and 1,069 nm which correspond to transitions $^4F_{3/2} \rightarrow {}^4I_{9/2}$ and $^4F_{3/2} \rightarrow {}^4I_{11/2}$, respectively.

The line around 1,069 nm is the strongest (Figure 7.26). There is considerable splitting in both the lines due to splitting of the levels by crystal field. Intensity of Nd^{3+} emission increases up to 1.5 mol.% Nd^{3+} and then decreases due to concentration quenching as a result of $Nd^{3+} \rightarrow Nd^{3+}$ energy transfer (Figure 7.26 inset). Figure 7.27 shows the excitation spectrum for 1,069 nm emission in

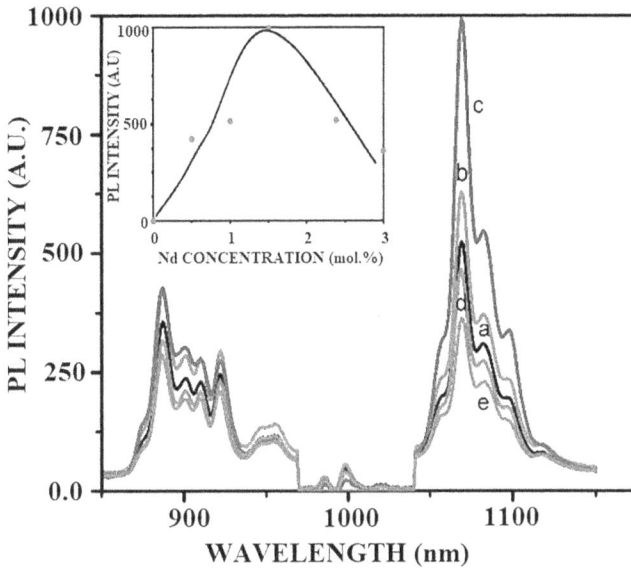

FIGURE 7.26 PL emission for Bi$_2$MoO$_6$:Nd^{3+} phosphors. Nd concentrations (a) 0.5, (b) 1.0, (c) 1.5, (d) 2.0 and (e) 3.0 mol.%.

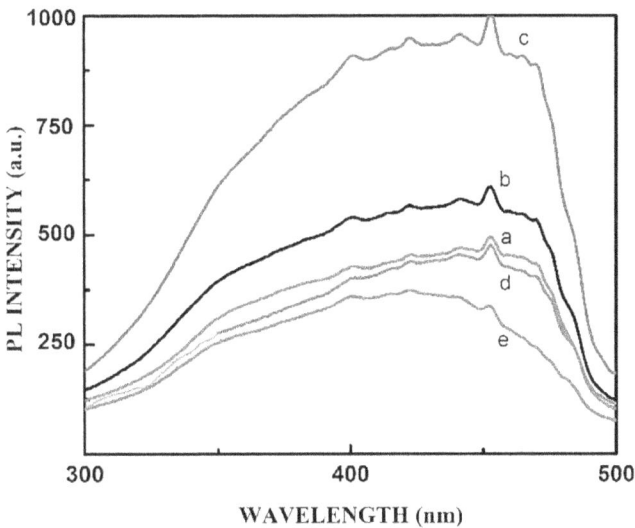

FIGURE 7.27 Host excitation spectra for Nd^{3+} emission in Bi$_2$MoO$_6$. Nd concentrations (a) 0.5, (b) 1.0, (c) 1.5, (d) 2.0 and (e) 3.0 mol.%.

the region 300–500 nm. This excitation falls in the absorption region that can be attributed to the host absorption. NIR emission of Nd^{3+} is thus host-sensitized. Some sharp kinks are due to overlapping Nd^{3+} lines. Excitation from f–f transitions of Nd^{3+} was observed more distinctly at longer wavelengths. Excitation spectra corresponding to these transitions are presented in Figure 7.28. The distinct lines appear at around 883, 870 ($^4I_{9/2}{\rightarrow}^4F_{3/2}$), 824 ($^4I_{9/2}{\rightarrow}^4F_{5/2}$), 807 ($^4I_{9/2}{\rightarrow}^4F_{7/2}$), 767, 750, 740 ($^4I_{9/2}{\rightarrow}^4S_{3/2}$), 687 ($^4I_{9/2}{\rightarrow}^4F_{9/2}$), 630 ($^4I_{9/2}{\rightarrow}^2K_{15/2}$), 612, 589, 581 ($^4I_{9/2}{\rightarrow}^4G_{5/2},^2H_{11/2}$), 530, 518 ($^4I_{9/2}{\rightarrow}^2K_{13/2}$) nm.

Bi$_2$MoO$_6$:Nd^{3+} phosphor thus appears suitable for modification of solar spectrum [104]. By virtue of host$\rightarrow$ Nd^{3+} energy transfer, Bi$_{1.97}$Nd$_{0.03}$MoO$_6$ phosphor appears suitable for modification of solar

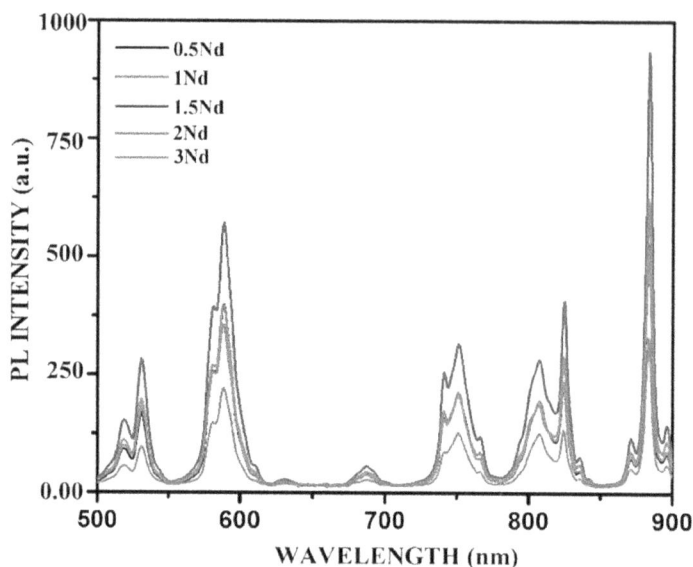

FIGURE 7.28 f–f excitation spectrum for Nd³⁺ emission in Bi₂MoO₆. Nd³⁺ concentrations (a) 0.5, (b) 1.0, (c) 1.5, (d) 2.0 and (e) 3.0 mol.%.

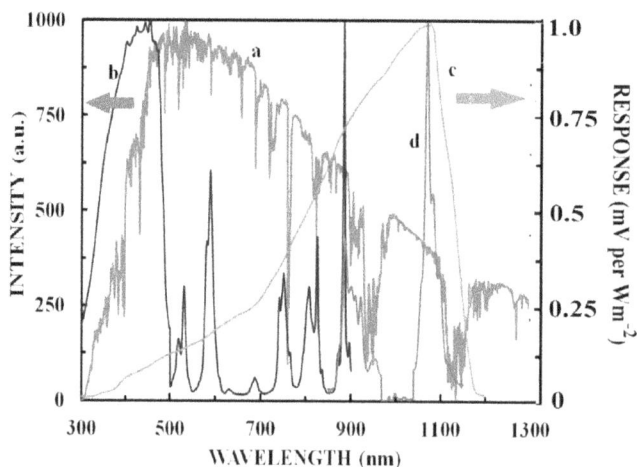

FIGURE 7.29 Modification of solar spectrum using $Bi_{1.97}Nd_{0.03}MoO_6$ phosphor. (a) AM1.5 solar spectrum, (b) Nd³⁺ excitation spectrum, (c) spectral response of c-Si solar cell and (d) Nd³⁺ emission spectrum.

spectrum, which can be relevant for improving performance of solar cells. As an example, this is illustrated for c-Si solar cell in Figure 7.29. As can be seen, c-Si solar cell response decreases rapidly below 500 nm (Figure 7.29, curve c).

If this part of the spectrum is converted by quantum cutting to about 1,000 nm, efficiency of photovoltaic conversion will obviously increase. There is a good overlap between solar spectrum in the 350–500 nm region (Figure 7.29, curve a) and Nd³⁺ excitation in $Bi_{1.97}Nd_{0.03}MoO_6$ phosphor (curve b). near-UV-blue light thus will be converted to Nd³⁺ emission (Figure 7.29, curve d) that is close to the maxima in the response curve of c-Si.

7.3.2.1.2 For CIGS

CIGS has very strong light absorption and a layer of only 1–2 micrometres (µm) are enough to absorb most of the sunlight. By comparison, a much greater thickness of about 160–190 µm is required for crystalline silicon. After a long history of development, the first commercialized CIGS became available in 1998 [140]. Shockley-Queisser limit [115] of the CIGS is 33.4. Refraction effects and the internal quantum efficiency limitations further reduce this number 21.1. The mismatch between the incident solar spectrum and spectral response of solar cells [106] is the main reason to limit the cell efficiency (Figure 7.30).

Kate et al. [109] predicted that converting the short wavelength solar radiations to 1,012 nm will increase the efficiency of CIGS cell to 29.6%. Phosphors which will absorb radiations in the range 350–500 nm and emit in NIR region around 1.01–1.1 μm can be adequate as spectral converters.

Nd^{3+} emission is best suited for this purpose. Nd^{3+} has number of excitation levels covering ultraviolet to near-infrared region and strong near-infrared luminescence at about 1.06 μm [141]. However, Nd^{3+} has only weak optical absorption arising from forbidden f–f transitions. A sensitizer with strong, broad absorption bands becomes necessary.

Two such compositions are discussed next.

7.3.2.1.2.1 Al_2O_3:Cr, Nd   Though near infrared (NIR) luminescence has been observed for Al_2O_3 containing activators like Er^{3+}, Nd^{3+}, Yb^{3+}, the excitation is from f–f transitions which are forbidden in nature and thus weak. Excitation for Cr^{3+}, on the other hand, is in the form of broad and strong bands arising from $^4A_{2g} \rightarrow {}^4T_{1g}(^4F)$ and $^4T_{2g}(^4F)$ transitions [142]. In Al_2O_3, the corresponding absorption bands are around 413 and 556 nm. If energy can be transferred from Cr^{3+}, to the above lanthanide ions, then intense NIR emission can be obtained with excitation in the visible region [143].

Figure 7.31 shows PL emission spectrum of Al_2O_3:Cr^{3+}, Nd^{3+} for 413 nm excitation in the range 850–1,200 nm. Al_2O_3:Nd^{3+} does not show any emission for this excitation wavelength. Strong emission in the NIR region, on the other hand, is observed in the samples co-doped with Cr^{3+}. Similar emission is observed in Al_2O_3:Nd^{3+} when 825 nm excitation is used. Cr^{3+} concentration was fixed at 0.8 mol.%. Emission lines are observed in groups around 915 and 1,103 nm attributable to transitions $^4F_{3/2} \rightarrow {}^4I_{9/2}$ and $^4F_{3/2} \rightarrow {}^4I_{11/2}$, respectively. There is considerable splitting due to strong crystal field. Intensity of emission goes on increasing with Nd^{3+} concentration up to 0.8 mol.%. Concentration quenching is observed for higher values.

Excitation spectra for Al_2O_3:Nd^{3+} contain several weak lines (Figure 7.32, curve a) due to f–f transitions of Nd^{3+}. These lines can be seen at 364 ($^4I_{9/2} \rightarrow {}^4D_{5/2}$), 371 ($^4I_{9/2} \rightarrow {}^4D_{3/2}$), 397,410 ($^4I_{9/2} \rightarrow {}^2P_{1/2}$),

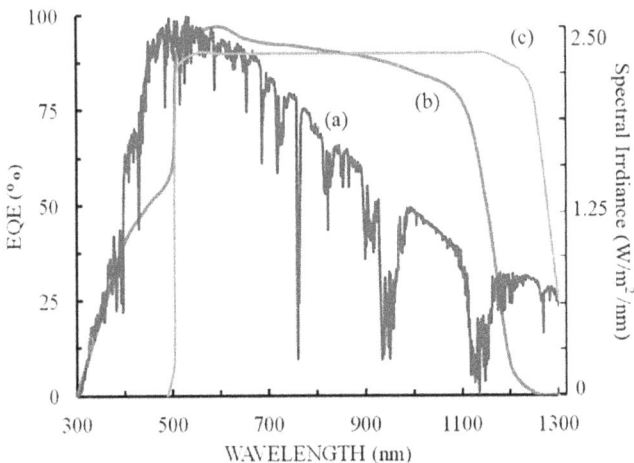

FIGURE 7.30   Spectral mismatch between the AM1.5 solar spectrum (a) and typical spectral responses of CIGS (b) and CIS (c) solar cells.

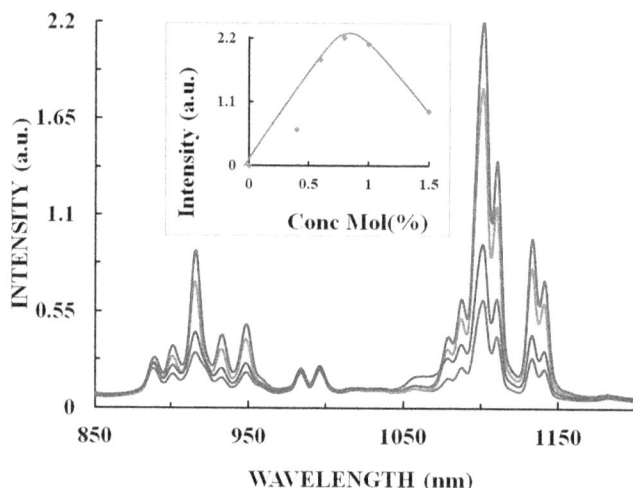

FIGURE 7.31 PL emission spectrum of $Al_2O_3:Cr^{3+}$, Nd^{3+} Nd^{3+} concentrations (a) 0.4, (b) 0.6, (c) 0.8 and (d) 1.5 mol.%.

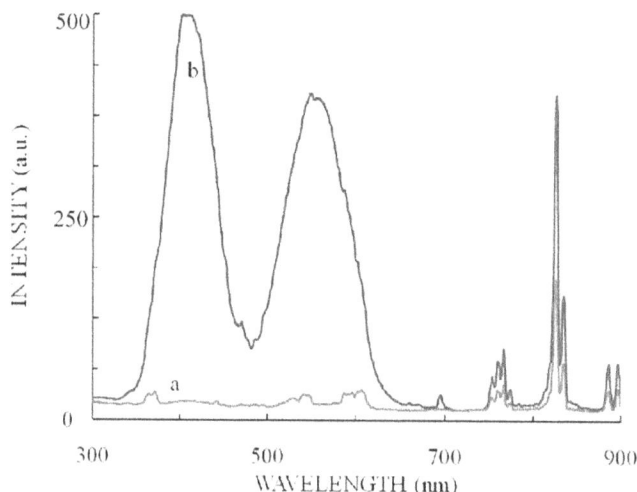

FIGURE 7.32 Excitation spectra for (a) $Al_2O_3:Nd^{3+}$, (b) $Al_2O_3:Cr^{3+}$, Nd^{3+}.

424, 442 ($^4I_{9/2} \rightarrow {}^2D_{5/2}$), 470, 475, 486, 495 ($^4I_{9/2} \rightarrow {}^2K_{11/2}$), 530, 541, 548 ($^4I_{9/2} \rightarrow {}^4G_{7/2}, {}^2K_{13/2}$), 560, 587, 596, 606 ($^4I_{9/2} \rightarrow {}^4G_{5/2}, {}^2H_{11/2}$), 620, 638 ($^4I_{9/2} \rightarrow {}^2K_{15/2}$), 660, 670, 687, 696 ($^4I_{9/2} \rightarrow {}^4F_{9/2}$), 754, 760, 766, 774 ($^4I_{9/2} \rightarrow {}^4F_{7/2}, {}^4S_{3/2}$), 783, 790, 797, 803 ($^4I_{9/2} \rightarrow {}^4F_{5/2}, {}^2H_{9/2}$), 825 ($^4I_{9/2} \rightarrow {}^4F_{5/2}$), 897, 886 ($^4I_{9/2} \rightarrow {}^4F_{3/2}$) nm. Only 825 nm line is of considerable intensity. In the sample co-doped with Cr^{3+}, besides the f–f lines, two broad bands are seen around 413 and 556 nm (Figure 7.32, curve b). These are same $^4A_{2g} \rightarrow {}^4T_{1g}(^4F)$ and $^4T_{2g}(^4F)$ bands as those observed for $Al_2O_3:Cr^{3+}$. Thus, strong Nd^{3+} emission in the NIR region is observed by exciting into Cr^{3+} bands around 413 or 556 nm.

It is thus seen that $Al_2O_3:Cr^{3+}$, Nd^{3+} phosphors can lead to intense NIR emission around 1 μm with broad band excitation in the visible region. Considering the fact that Al_2O_3 is a well-proven material for opto-electronic applications, these phosphors could find interesting applications, e.g., solar pumped lasers in the NIR region. Broad band excitation in the visible region and emission around 1 μm also make these phosphors attractive as a spectrum modifier in solar photovoltaics.

7.3.3 UPCONVERTING PHOSPHORS FOR SOLAR PHOTOVOLTAICS

7.3.3.1 Upconversion

Upconversion is an anti-Stokes process of converting near-infrared (NIR) light to visible light. UC materials are the materials that produce high-energy photons after absorbing low-energy photons. It is a non-linear optical process of absorbing two or more low-energy NIR photons successively, exciting an electron to some intermediate state and ultimately raising it to a high-energy level. The energy of emitted photon may range from the NIR to visible and ultraviolet (UV) regions. The emission of energy in the visible region makes these UC phosphors useful for luminescence applications too. This concept was first proposed by Bloembergen in 1959 [144], and it was Auzel [145], who described the UC process in 1966 for the first time. Since then, researchers have been making numerous efforts to enrich the family of UC luminescent materials continuously. These materials find potential applications in bioimaging [146], solid-state laser [147] and solar cells [148]. In general, four mechanisms are generally involved in the UC process (Figure 7.33). Apart from multi-step excitation due to classical excited-state absorption (ESA), there is a more efficient UC process by sequential energy transfer from a sensitizer to a nearby locate activator called energy transfer upconversion (ETU). The latter phenomenon has to be distinguished from yet another mechanism, namely, cooperative energy transfer (CET) between a pair of sensitizer ions and an emitting activator. Finally, a fourth process, the photon avalanche effect (PAE), is also based on sequential energy transfer, but of downconversion type (cross-relaxation), whereas the UC step itself is due to ESA [149].

Lanthanides have been known to be most efficient upconverters. Yb^{3+}-Er^{3+} pair is most frequently used [150], since Yb^{3+} can absorb radiation from LED emitting around 980 nm. On the other hand, in the context of spectrum modification, an upconverting material which can convert wavelengths in the range of 1,000–2,000 nm is desired. Other pairs are Yb^{3+}-Tm^{3+} [151], Yb^{3+}-Tb^{3+} [152], Yb^{3+}-Pr^{3+} [153], Yb^{3+}-Ho^{3+} [154]. Apart from lanthanide pairs, "organic and organo-metallic chromophores" also exhibit efficient upconversion [155]. These upconverters consist of organic molecules with conjugated p-systems, serving as acceptor and organometallic complexes serving as

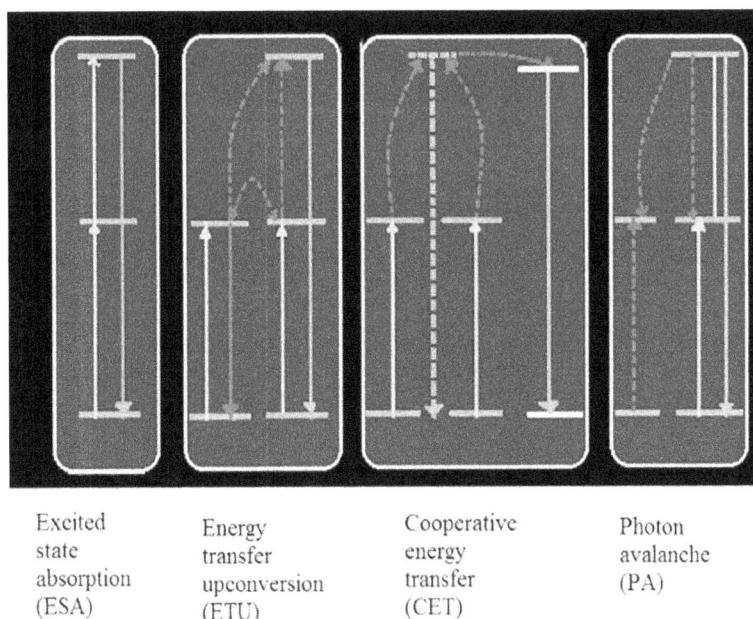

| Excited state absorption (ESA) | Energy transfer upconversion (ETU) | Cooperative energy transfer (CET) | Photon avalanche (PA) |

FIGURE 7.33 Mechanisms involved in the up-conversion (UC) process.

sensitizer. Research on upconversion in these chromophores has gained renewed interest recently [156]. Such materials have been systematically reviewed by Singh-Rachford and Castellano [157], and more recently, by Healy et al. [158].

7.3.3.2 Applications for Spectrum Modification

Using upconversion for improving solar cell performance started in modern times with seminal paper by Trupke et al. [21]. The basic idea is explained in Figure 7.34. Long wavelength limit of various solar cells is determined by the band gap of the semiconductor. Sub-band gap photons are not converted. The losses of photons with energies below the absorption threshold of solar cells are considerable. These sub-band gap losses amount to 19% of the incident power for crystalline silicon solar cells and potentially more than 50% for organic solar cells. The losses for GaAs, perovskite, amorphous silicon and DSSCs are in between these extremes. As a consequence, there is a huge potential for upconversion to increase the efficiency of various solar cell technologies by converting the low-energy photons into photons that can be utilized by the solar cells.

Using upconversion, if these photons are converted to appropriate shorter wavelengths, then much higher fraction of the solar spectrum can be utilized for photovoltaic conversion, thus increasing the overall efficiency. Trupke et al. showed that a system consisting of a solar cell with a single band gap of 2 eV and an ideal upconverter system attached to its rear can achieve a maximum power conversion efficiency of 47.6% under non-concentrated sunlight, approximated by 6,000 K thermal blackbody radiation. Under sunlight concentrated by the maximum possible factor of 46,200, the efficiency limit is 62.18%. An even higher limiting efficiency of 63.17% is found for a system where the emission is restricted to the solid angle of the sun. Badescu and co-authors have carried out in depth analysis of models proposed by Trupke et al. under more refined and realistic conditions [159,160].

A large number of upconverting phosphors have been explored for this purpose, and these have been reviewed from time to time [120,161–164]. Materials have been classified on the basis of type of solar cells, activators, host anions, etc. In the following, we described some oxide hosts employed for this purpose.

7.3.3.3   Upconverting Oxide Phosphors for Solar Photovoltaics

7.3.3.3.1 $Gd_2(MoO_4)_3$:Er^{3+}

Er^{3+} is considered to be suitable for photovoltaic UC purposes since the spectral power from the normalized 1,000 W/m² AM1.5 spectrum yields over 100 W/m² between 1,100 and 1,700 nm, which match well with the energy gap of the $^4I_{15/2} \rightarrow ^4I_{13/2}$ transition (~1,500 nm) of Er^{3+}. Intense UC emissions at 545, 665, 800 and 980 nm with energies greater than the band gap of silicon solar cell have

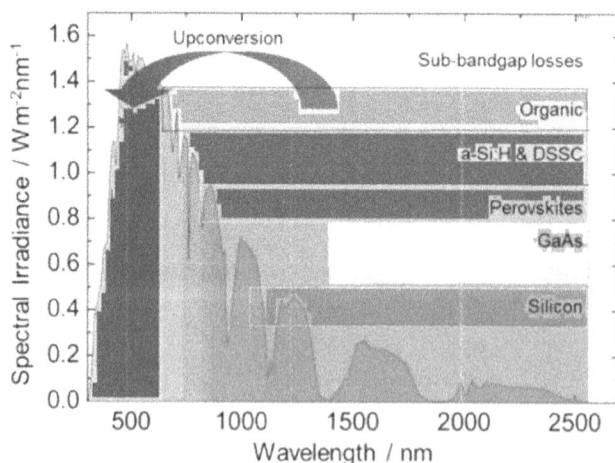

FIGURE 7.34   Upconversion for improving solar cell performance.

been achieved in $Gd_2(MoO_4)_3$:Er^{3+} upon excitation with NIR photons provided by a tunable laser (1,510–1,565 nm); the conversion efficiencies are 0.12%, 0.05%, 0.83% and 1.35%, respectively [165]. Figure 7.35 illustrates the mechanism of upconversion in this phosphor. By absorbing 1,530 nm photon Er^{3+} goes from the ground state $^4I_{15/2}$ to $^4I_{13/2}$ to and by successive absorptions to states $^4I_{9/2}$ and $^2H_{11/2}$. Radiative return to the ground state yields upconverted photon at 525 nm. Alternatively, it may relax to the state $^4S_{3/2}$ before emission. In that case, photon of 545 nm is obtained. If the relaxation is to still lower states of $^4F_{9/2}$ or $^4I_{9/2}$, $^4I_{11/2}$, then photons at 665, 800 and 980 nm are generated. In this way, the solar radiation around 1,530 nm can be converted to the wavelengths which can be used by c-Si solar cells. Co-doping by Yb increases green and red emission fractions by reabsorption of 980 nm by Yb and transfer to another Er^{3+} ion [166].

The sensitivity of c-Si solar cells is better at 545 and 665 nm than at 980 nm. Yb co-doping also can directly cause upconversion of 980 nm light. Further, it had been observed that partial replacement of Gd by Y upto 20% increases the upconversion efficiency [167]. Thin films are most suited for using with solar cells. Upconverting $Gd_2(MoO_4)_3$:Er, Yb thin films have been described [168]. Intense upconversion has been observed in the spin-coated thin films of $Gd_2(MoO_4)_3$:Ho/Yb as well [169]. Such Upconvesion in more complex systems like $Gd_2(MoO_4)_3$:Er, Yb, Eu has also been reported [170]. Though application to photovoltaic is not studied, it appears feasible.

It had also been claimed that the monoclinic variety of $Gd_2(MoO_4)_3$:Er, Yb obtained through sol-gel gives more intense emission than the more frequently studied orthorhombic [171]. Further, incorporation of Au islands in the $Gd_2(MoO_4)_3$:Er, Yb thin films could greatly enhance the upconversion luminescence [172]. $Gd_2(MoO_4)_3$:Er, Yb was also used for increasing efficiency of organic/inorganic hybrid solar cells (HSC) [173]. The broad EQE band covering the spectrum from 350 to 800 nm is

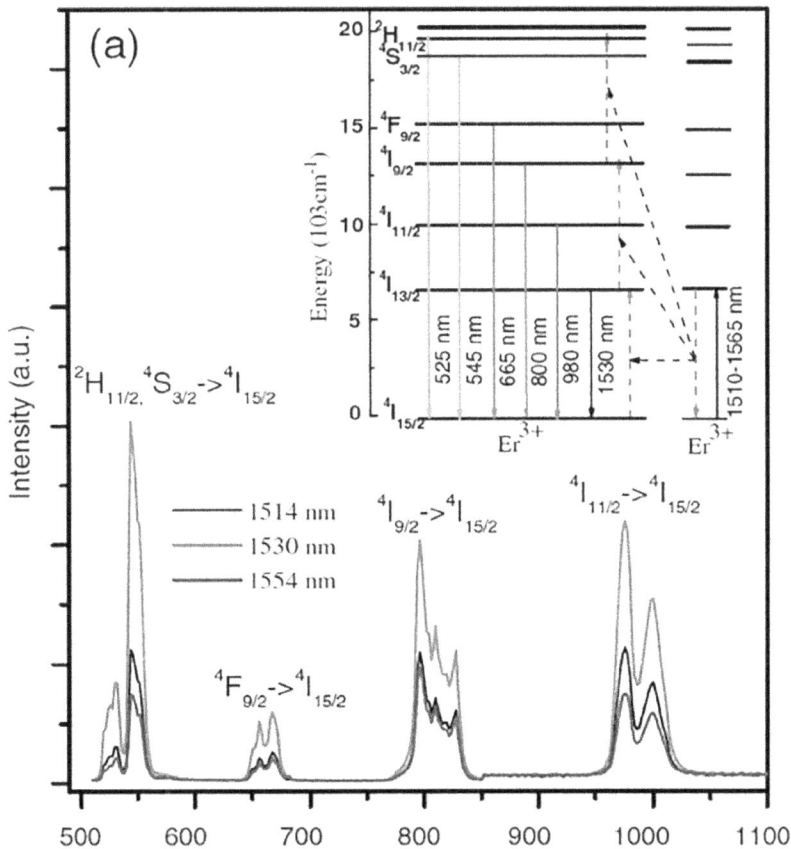

FIGURE 7.35 Mechanism of upconversion in $Gd_2(MoO_4)_3$:Er^{3+}.

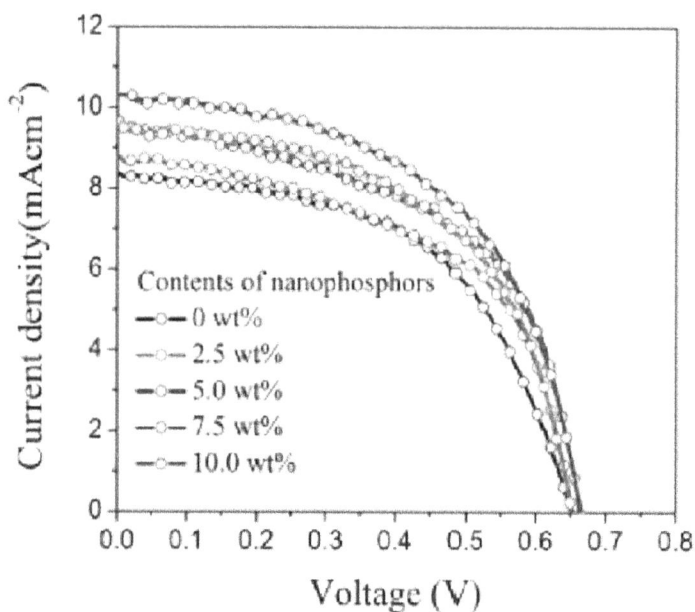

FIGURE 7.36 J-V characteristics of BHJs with different contents of NPs under the illumination of a 1.5 AM solar simulator (100 mW/cm²).

enhanced significantly by $Gd_2(MoO_4)_3$:Er, Yb doping. Under the identical experimental conditions, HSCs with GMO:9Yb^{3+}/1Er^{3+} NPs exhibit better photovoltaic performance than HSCs with pure TiO_2. The JSC increases from 8.38 at 0 wt.% to 10.29 mA/cm² at 7.5 wt.% of $Gd_2(MoO_4)_3$:9Yb^{3+}, 1Er^{3+} NPs (Figure 7.36) whereas both the Voc and filling factor (FF) remain almost constant, thus leading to an improvement by almost 34% than its counterpart without doping, i.e. from 2.94% at 0 wt.% to 3.67% at 7.5 wt.% of $Gd_2(MoO_4)_3$:9Yb^{3+}, 1Er^{3+} NPs. Upconversion by Yb-Tm pair was also exploited for boosting the performance of this cell [174]. $(MoO_4)_3$:Yb/Tm NPs excited at 976 nm exhibit intense blue (460–498 nm) and weak red (627–669 nm) emissions. The lifetime of electron transfer is shortened from 817 to 316 ps after incorporating NPs and correspondingly the electron transfer rate outstrips by three times that of the bare TiO_2. Consequently, a notable power conversion efficiency of 4.15% is achieved as compared to 3.17% of pure TiO_2/PTB7. Apart from upconversion, doping by various lanthanides in $Gd_2(MoO_4)_3$ makes it possible to downshift nUV radiations to green/red region. This can also be effective for enhancing solar cell efficiencies [175].

7.3.3.3.2 Y_2O_3:Yb, Er

Y_2O_3 doped with Er^{3+} itself shows efficient upconversion. This phosphor can help improving efficiency of solar cells in two ways. It exhibits broadband antireflection in the wavelength range from 300 to 1,100 nm, and intense red and infrared emissions under 1.538 nm excitation. The morphologies and transmission ratio of the resulting arrays could be controlled by adjusting micropore diameter sizes of the arrays from 305 to 795 nm. The intensities of red and infrared emissions are enhanced respectively 19 and 3 times by tailoring Er^{3+} ions' local environment with Li$^+$ [176]. The Y_2O_3:5% Er^{3+}, 5% Li$^+$ porous pyramid arrays have potential application in decreasing the reflection and transmission loss for c–Si solar cells, as well as converting 1.538 nm light to red region where c–Si solar cells are effective for photovoltaic conversion.

Wang et al. [177] actually used the Y_2O_3:Er^{3+} nanorods for enhancing performance of dye-sensitized solar cells (DSSC). Y_2O_3:Er^{3+} improves infrared light harvest via upconversion luminescence and increases the photocurrent of the DSSC. The rare earth ions improve the energy level of the TiO_2 electrode through a doping effect and thus increase the photovoltage. The light scattering is ameliorated by the one-dimensional nano-rod structure. The DSSC containing Y_2O_3:Er^{3+} (5 wt.%) in the doping layer achieves a light-to-electric energy conversion efficiency of 7.0%, which is an increase of 19.9% compared to the DSSC lacking of Y_2O_3:Er^{3+} (Figure 7.37).

Using Yb-Er pair further adds to the upconversion by 980 nm light [178]. Jia et al. [179] used (Y_2O_3:Yb–Er)/Bi_2S_3 composite for improving IR response of solar cells through upconversion. The composite films consist of homogeneous crystalline particles of Y_2O_3 with a lamellar morphology covered by Bi_2S_3 nanoparticles. Upon excitation by a 980 nm laser, the Y_2O_3:Yb–Er layer in the composite film converts NIR photons into visible emissions through upconversion, which is absorbed by the covered Bi_2S_3 nanoparticles and leads to the generation of photoelectrons.

Yb effectively absorbs 980 nm light. Three successive energy transfers take Er^{3+} from the ground state $^4I_{15/2}$ to the excited state $^4F_{7/2}$. Following relaxation and return to the ground state from various intermediate levels leads to upconverted green and red emissions (Figure 7.38). In this phosphor also, Li^+ co-doping was found to change the upconversion emission intensity up to 470 times [180].

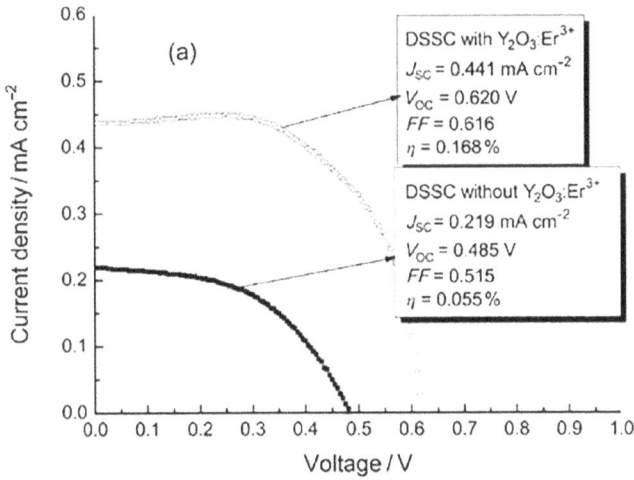

FIGURE 7.37 Er upconversion for dye sensitized solar cells (DSSC).

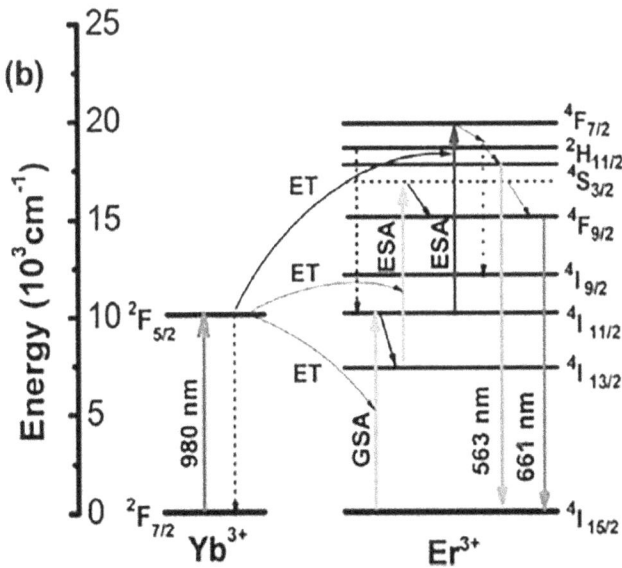

FIGURE 7.38 Yb-Er upconversion.

Du et al. [181] demonstrated the use of upconversion emission in Y_2O_3:Yb–Er for improving the efficiency of dye sensitized solar cells. Under the light excitation at a wavelength of 980 nm, the as-prepared samples exhibited strong upconversion emissions at green and red visible wavelengths. To investigate the influence of Y_2O_3:Er^{3+}/Yb^{3+} nanoparticles on the photovoltaic performance of DSSCs, the phosphor nanoparticles were incorporated into titanium dioxide films to form a composite photoelectrode.

For the resulting DSSCs, the increased power conversion efficiency (PCE) of 6.68% was obtained mainly by the increased photocurrent of JSC = 13.68 mA/cm² due to the light harvesting enhancement via the NIR-to-visible upconversion process (cf., PCE = 5.94%, JSC = 12.74 mA/cm² for the reference DSSCs without phosphor nanoparticles), thus, indicating the PCE increment ratio of ~12.4% (Figure 7.39). The inset of Figure 7.39a shows the schematic diagram of the DSSC with Y_2O_3:Er^{3+}/Yb^{3+} nanoparticles and its photovoltaic performance parameters. Almost at the same time, Kim et al. reported about 34% increase in the conversion efficiency of DSSC by using upconversion layer of Y_2O_3:Yb–Er [182].

Similar efficiency enhancement for a-Si solar cells using Y_2O_3:Yb–Er upconversion layer has been demonstrated by Markose et al. [183].

The potential application of Y_2O_3:Yb–Er phosphor was demonstrated by incorporating it in amorphous silicon thin film solar cells, and short-circuit current J_{sc} of 13.5 mA/cm², as against 13.05 mA/cm² for a reference cell, was achieved when the cell was illuminated with a 980 nm IR laser. A different upconversion channel was proposed by Wu et al. [184] for enhancing conversion efficiency of c-Si solar cell. They used $^4I_{15/2} \to {}^4I_{13/2}$ transition of Er^{3+} for absorbing IR radiation around 1,500 nm. This was converted to Yb^{3+} emission in region 960–1,030 nm (Figure 7.40).

Apart from Yb-Er upconversion that of Yb-Ho pair has also been proposed for spectrum modification. Dutta et al. studied the effect of Y_2O_3:Yb, Ho upconversion layer on electron ejection ability, current density (J_{SC}), open-circuit voltage (V_{OC}) and power conversion efficiency (PCE) of DSSCs. The enhancement of about 31% and 30% for JSC and PCE in DSSCs hybrid film compared with that of DSSCs TiO_2 film has been observed due to improved near-infrared to visible light

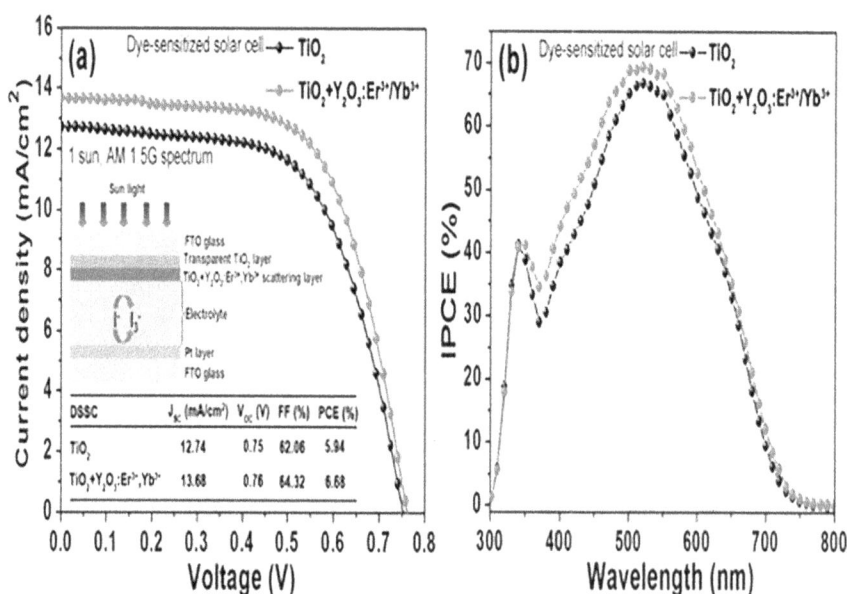

FIGURE 7.39 (a) J-V curves and (b) IPCE spectra of DSSC with (grey line) and without (grey solid line) of Y_2O_3:Er^{3+}/Yb^{3+} nanoparticles.

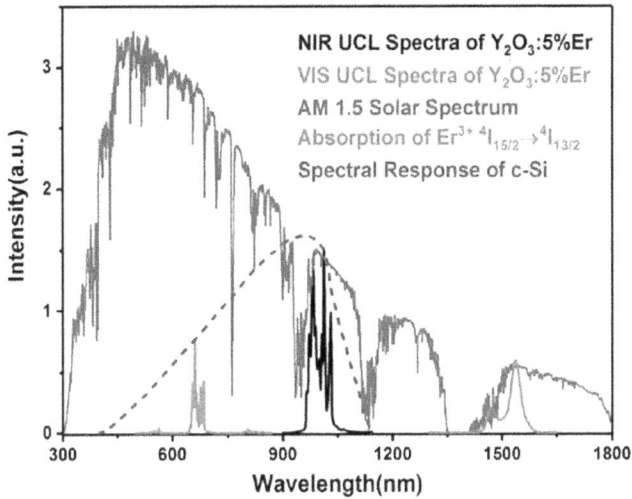

FIGURE 7.40   Er-Yb upconversion for c-Si.

harvesting properties of UC nanophosphors [185]. Tadge et al. [186] performed similar experiments. They observed an increase in efficiency of DSSC from 8.9% to 9.8% using Y_2O_3:Yb, Ho upconversion layer. Correspondingly, J_{SC} increased from 17.49 to 18.97 mA/cm^2.

7.3.3.3.3 *Lu_2O_3:(Tm^{3+}, Yb^{3+})*

Li et al. [187] demonstrated enhancement of DSSC performance using upconverting layer of Lu_2O_3:(Tm^{3+}, Yb^{3+}) phosphor. Yb effectively absorbs 980 nm light. Three successive energy transfers take Tm^{3+} from the ground state $^3H_6$ to the excited state $^1G_4$. Following relaxation and return to the ground state from various intermediate levels leads to upconverted blue and red emissions (Figure 7.41). The $^3H_4$ excited state can be populated by two-step energy transfer (ET) process. $^2F_{7/2} \rightarrow ^2F_{5/2}$ transition of Yb^{3+} occurs by absorption of 980 nm photons. Then, Yb^{3+} in the $^2F_{5/2}$ level transfers its energy to Tm^{3+}, which is excited to the $^3H_5$ level and relaxes non-radiatively to the $^3F_4$ level. The $^3F_4$ level in Tm^{3+} can be excited to the $^3F_2$ levels by ET process after which on radiative relaxation to $^3H_4$ occurs. Therefore, the $^1G_4$ excited state can be populated by energy transfer from Yb^{3+} ions and absorption of 980 nm photons. From the $^1G_4$ level, the Tm^{3+} ions decay radiatively to the $^3H_6$ ground state and the $^3F_4$ metastable state generating blue and red emissions around 476 and 653 nm. By virtue of this upconversion, Lu_2O_3:(Tm^{3+}, Yb^{3+}) improves incident light harvesting and increases photocurrent. Also, as a p-type dopant, the rare-earth ions elevate the energy level of the oxide film and increase the photovoltage. Under a simulated solar light irradiation of 100 mW/cm^2, the light-to-electric energy conversion efficiency of the DSSC with Lu_2O_3:(Tm^{3+}, Yb^{3+}) doping reaches 6.63%, which is an increase of 11.1% compared to the DSSC without Lu_2O_3:(Tm^{3+}, Yb^{3+}) doping (Figure 7.42).

7.3.3.3.4 *Other Lanthanide Oxides*

La_2O_3 was selected as the host for upconversion using Yb^{3+}/Er^{3+} pair [188], owing to its low phonon energy of 450 cm^{-1}, which is among the lowest value achievable in oxide-based materials. UC emission colour of LYE can be finely tuned from green to reddish-orange, while the high UCQY is maintained for the whole tunable colour range. Using this upconversion emission for enhancing solar cell performance was suggested. Upconversion for Yb^{3+}/Ho^{3+} and Yb^{3+}/Tm^{3+} [189] pairs was also studied. However, there are no reports on using these phosphors for enhancing photovoltaic conversion. Upconversion (UC) Er, Yb-CeO_2 hollow spheres were successfully prepared using carbonaceous spheres as removable template via hydrothermal method for improving the

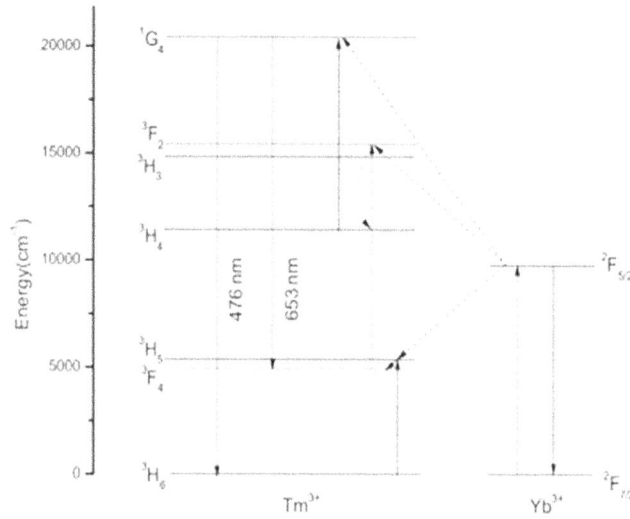

FIGURE 7.41   Yb-Tm upconversion.

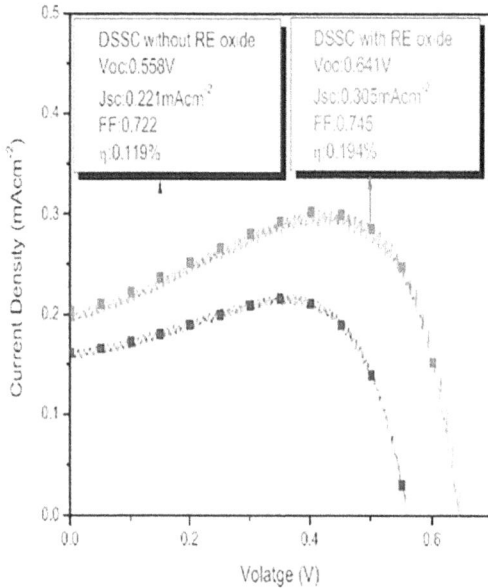

Current-voltage curves of DSSCs under near-infrared irradiation (60 mW cm^{-2}).

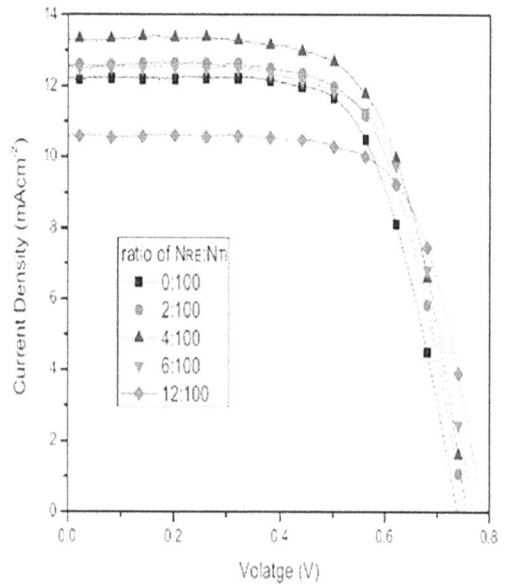

Current-voltage curves of DSSCs with different ratios of Lu$_2$O$_3$:(Tm^{3+}, Yb^{3+}) in TiO$_2$ electrode under a simulated solar light irradiation (100 mW cm^{-2}).

FIGURE 7.42   Current-voltage curves of DSSC for Lu$_2$O$_3$:(Tm^{3+}, Yb^{3+}).

efficiency of dye-sensitized solar cells (DSSCs). Light harvesting efficiency was enhanced 27% due to the upconversion effect, scattering effect and higher dyeloading capacities [190]. Zhao et al. [191] adopted CeO$_2$:Yb, Er nano tubes as an effective assistant layer into DSSCs. The DSSCs fabricated by the complex photoanode of P25 and CeO$_2$:Yb, Er NTs exhibit a higher photoconversion efficiency of 8.67%. Compared to the typical DSSCs constructed by pure P25, PCE and IPCE display an increase of 33.58% and 48.78%, respectively. The enhanced photoconversion performance could be explained as (i) the combination of CeO$_2$:Yb, Er NTs with TiO$_2$ improves the visible light responsive; (ii) the energy bands matching between TiO_2 and CeO_2:Yb, Er NTs enhance the separation and limit the recombination of photo induced electron-hole pairs. Du et al. obtained enhanced photovoltaic performance of DSSC using upconversion layer constituted of Ho^{3+}/Yb^{3+}-co-doped Gd_2O_3 nanoparticles [192]. Yb^{3+} efficiently absorbs 980nm light and by three sequential energy transfer to Ho^{3+}, $^5F_4$ state of the latter is populated. Intense green and red upconversion (UC) emissions corresponding to the $(^5F_4, \,^5S_2) \rightarrow \,^5I_8$ and $^5F_5 \rightarrow \,^5I_8$ transitions of Ho^{3+} ions (Figure 7.43) boost PCE of DSSC by utilizing NIR part of the solar spectrum. With the introduction of Ho^{3+}/Yb^{3+}-co-doped Gd_2O_3 nanoparticles into the TiO_2 porous film of dye-sensitized solar cells (DSSCs), the power conversion efficiency of the cells (7.403%) was ~10.47% higher than that of the DSSCs with pure TiO_2 porous film (6.701%), which is mainly caused by increased short-circuit current density due to their enhanced light-harvesting properties via an efficient UC process.

7.3.3.3.5 TiO_2:Yb, Er/Yb, Tm

Amorphous Er^{3+} and Yb^{3+} co-doped TiO_2 has been used as a scattering layer in DSSC [193], but the upconversion property was not determined or demonstrated. Er-doped TiO_2 nano-rod arrays, on the other hand, were found suitable for improving response of perovskite solar cells (PSC). A higher photocurrent density (Jsc) of 22.97 mA/cm^2 and a power conversion efficiency (PCE) of 10.6% were obtained, showing 20% and 16.5% increment in Jsc and PCE, respectively, when compared with the un-doped device. The performance improvement was ascribed to the upconversion performance, enhanced electron-injection efficiency and reduced charge recombination [194] (Figure 7.44).

Shan and Demopoulos [195] for the first time doped Er^{3+} and Yb^{3+} in LaF_3 and used to make nano-composite with TiO_2 to serve as upconversion layer in DSSC. However, these experiments met with partial success and anticipated results were not obtained. Zhang et al. performed similar experiments with TiO_2 containing Yb-Er upconverting layer. The power conversion efficiency (PCE) of the solar cells without UC-TiO_2 was 14.0%, while the PCE of the solar cells with UC-TiO_2 was increased to 16.5%, which presented an increase of 19% [196]. Wang et al. also obtained similar results. The power conversion efficiency of the solar cells on Er-Yb:TiO_2 increased to 13.4%, which is 20.8% higher than that of the solar cells based on un-doped TiO_2. The experimental results indicated that the larger open circuit voltage could be due to the larger conduction band difference between Er-Yb:TiO_2 and perovskite material. The enlarged short circuit current could be attributed to the faster electron transfer, reduced recombination and the upconversion of Er-Yb:TiO_2 NRs [197]. In a next experiment, Er^{3+}, Yb^{3+} co-doping was noticed to cause changes in the band gap of TiO_2 and lead to 25% increase in the current density. The electrochemical impedance spectroscopy results

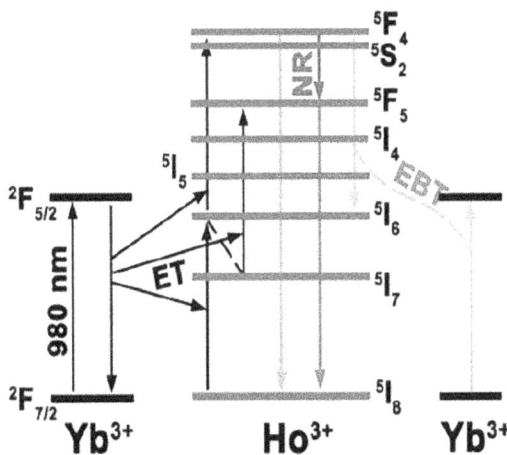

FIGURE 7.43 Upconversion (UC) emissions in Gd_2O_3:Ho^{3+}/Yb^{3+}.

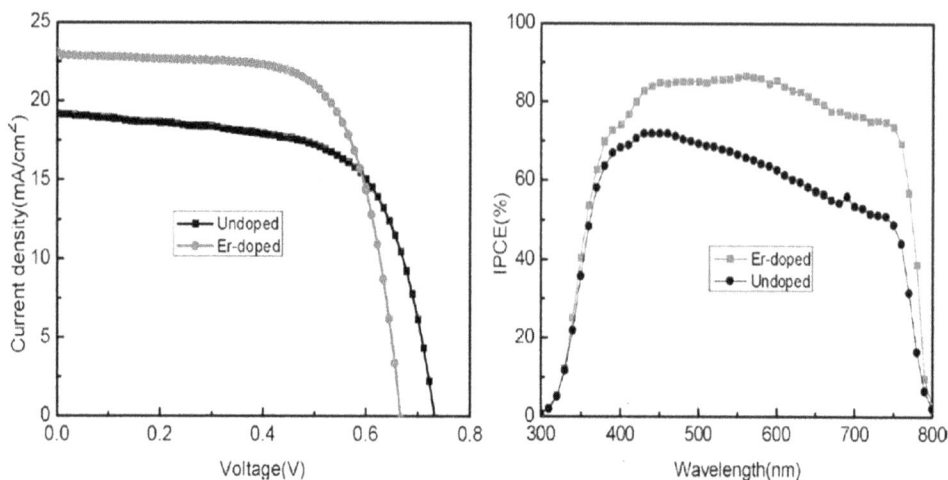

FIGURE 7.44 (a) I-V curves and (b) IPCE curves of perovskite solar cells with and without Er doping.

showed that the device based on the doped TiO_2 NRs has a higher recombination resistance and a lower transfer resistance than those of the undoped device, and thereby, the doped device exhibits a lower electron recombination rate. In addition, the upconversion Er and Yb co-doped device exhibits 25% higher current density and 17% higher photon-to-electron conversion efficiency, as revealed by the J-V test results. Moreover, the optimized efficiency of the TiO_2 NR array-based perovskite solar cell is determined to be 10.02%. Furthermore, the Er and Yb co-doped device exhibits a near-infrared response; an efficiency of 0.1% is achieved under infrared light (800–1,100 nm) irradiation [198]. Xie et al. [199] used Yb, Tm pair for upconversion. Upconversion emission by Yb, Tm pair results in harvesting more incident light and increasing photocurrent for the DSSC. On the other hand, owing to the p-type doping effect by Tm^{3+} and Yb^{3+}, the photovoltage of the DSSC is enhanced. Under a simulated solar light irradiation of 100 mW/cm^2, a DSSC containing Tm^{3+}/Yb^{3+} achieved a conversion efficiency of 7.05%, which is an increase of 10.0% over the DSSC without upconversion layer. With the increase of the amount of RE_2O_3 in the DSSC, short-circuit current J_{SC} increases and then decreases. The increase in J_{SC} with the amount of RE_2O_3 at low concentrations is due to the upconversion luminescence of RE ions, which absorb infrared light and convert to visible light, resulting in an increasing photocurrent in the DSSC. However, at higher concentration, some crystal defects are produced. The defects can capture photo-induced electrons and holes, leading to a decrease in photocurrent. On the other hand, the V_{OC} increases with increase in the percentage of RE_2O_3. When RE^{3+} ions are doped in TiO_2 film and substituted for the Ti^{4+} ion lattice sites, a p-type doping effect occurs, similarly to that trivalent positive ion doped in Si semiconductors, which results in the elevation of the energy level of oxide film and the increase of V_{OC}. Yu et al. achieved as much as 37% increase in PCE of DSSC using upconversion in Ho^{3+}-Yb^{3+}-F^- tri-doped TiO_2 [200]. Under AM1.5G light, open-circuit voltage reached 0.77 ± 0.01 V, short-circuit current density reached 21.00 ± 0.69 mA/cm^2, which resulted in an impressive overall energy conversion efficiency of $9.91\pm0.30\%$, a 37% enhancement compared to DSSCs with pristine TiO_2 photoanode.

7.4 CONCLUSIONS AND OUTLOOK

Solar energy is unquestionably the answer for an ever growing need of mankind for the energy. Photovoltaic conversion of the solar energy is the most convenient was of harnessing as most of the gadgets work on electricity. Various solar cells have been developed to this end. However, all of them suffer from the drawback that they can use only a part of the solar spectrum over which the solar radiations are distributed. Various schemes implementing the concept of spectrum modification have been suggested from time to time to overcome this problem. The phosphors used for this purpose may be divided into three categories as downshifting, downconverting and upconverting. First two are useful for utilizing short wavelength portion of the solar spectrum which is not adequately used for photovoltaic conversion by the existing devices. Upconverting phosphors, on the other hand, can help in using infrared spectrum. Various phosphors under these three categories, suggested from time to time, have been reviewed here. Downshifting phosphors are most easily achievable; however, it gives only one longer wavelength photon for the absorbed shorter wavelength photon and hence intrinsically of lower energy conversion efficiency. Downconversion is much attractive proposal as in principle quantum efficiencies approaching 200% are feasible. More often than not, the downconversion has been achieved for a very narrow spectral region, thus limiting its utility. It is important to downconvert radiations in the region 400–500 nm. Phosphors capable of doing this are limited in number. Moreover, these are to be coated on the solar cell, without attracting the losses from scattering, back reflection, etc. There is still a long way to go before getting practical systems employing downconverting phosphors.

Efficiency improvement using upconversion is easy to demonstrate, as there is zero response of solar cell for sub-band gap radiations, and small conversion efficiency becomes perceptible. On the other hand, upconversion efficiencies are low. Theoretically, the quantum efficiency of the process cannot exceed 50%. In practice, efficiencies exceeding 10% have not been observed. Moreover, all the upconversion schemes are based on absorbers which operate over very narrow spectral range. Ideally, an upconverting phosphor which can use all the radiations over broad IR region, typically 1,000–3,000 nm, is required. A different mechanism of upconversion which can ensure such utilization of solar radiations in IR region needs to be developed.

REFERENCES

1. Gratzel, M. (2001). *Nature*, 414, 338.
2. Adrian, G. (2015). *Key World Energy Statistics 2013* (Paris: International Energy Agency).
3. Kalogirou, S.A., Karellas, S., Badescu, V., & Braimakis, K. (2016). Exergy analysis on solar thermal systems: A better understanding of their Sustainability, *Renew. Energ.*, 85, 1328–1333. https://doi.org/10.1016/j.renene.2015.05.037.
4. Saidur, R., Boroumand Jazi, G., Mekhlif, S., & Jameel, M. (2012). Exergy analysis of solar energy applications, *Renew. Sust. Energ. Rev.*, 16(1), 350–356. https://doi.org/10.1016/j.rser.2011.07.162.
5. Nwosu, N.P. (2010). Employing exergy-optimized pin fins in the design of an absorber in a solar air heater, *Energy*, 35(2), 571–575. https://doi.org/10.1016/j.energy.200.10.027.
6. Koroneos, C., Nanaki, E., & Xydis, G. (2010). Solar air conditioning systems and their applicability-An exergy approach, *Resour. Conserv. Recycl.*, 55(1), 74–82. https://doi.org/10.1016/j.resconrec.2010.07.005.
7. Gomri, R. (2009). Energy and exergy analyses of seawater desalination system integrated in a solar heat transformer, *Desalination*, 249(1), 188–196. https://doi.org/10.1016/j.desal.2009.01.021.
8. Suresh, M.V.J.J., Reddy, K.S., & Kolar, A.K. (2010). 4-E (energy, exergy, environment, and economic) analysis of solar thermal aided coal-fired power plants, *Energy. Sustain. Dev.*, 14(4), 267–279. https://doi.org/10.1016/j.esd.2010.09.002.
9. Becquerel, E. (1839). On electric effects under the influence of solar radiation, *Compt. Rend.*, 9, 561. https://cir.nii.ac.jp/crid/1570009750806226944.
10. Adams, W.G., & Day, R.E. (1877). The action of light on selenium, *Proc. R. Soc. Lond.*, 25(171–178), 113–117. https://doi.org/10.1098/rspl.1876.0024.
11. https://en.wikipedia.org/wiki/Timeline_of_solar_cells.
12. Lange, B. (1930). New photoelectric cell, *Zeitschrift fur Physik*, 31, 139. https://scholar.google.com/scholar_lookup?title=&journal=Physik.%20Z.&volume=31&publication_year=1930&author=Lange%2CB.
13. Grondahl, L.O. (1933). The copper-cuprous-oxide rectifier and photoelectric cell, *Rev. Mod. Phys.*, 5(2), 141–168. https://doi.org/10.1103/RevModPhys.5.141.
14. Schottky, W. (1930). On the origins of photoelectrons in Cu2 O-Cu photocells, *Phys. Z*, 31, 913–925. https://scholar.google.com/scholar?hl=en&as_sdt=0%2C5&q=W.+Schottky%2C+Phys.+Z.+31%2C+913+%281930%29&btnG=

15. Destriau, G. (1954). Brightness waves and transitory phenomena in the quenching of luminescence by alternating electric fields, *J. Appl. Phys.*, 25(1), 67–71. https://doi.org/10.1063/1.1721522.

16. Reynolds, D.C., Leies, G., Antes, L.L., & Marburger, R.E. (1954). Photovoltaic effect in cadmium sulfide, *Phys. Rev.*, 96(2), 533–534. https://doi.org/10.1103/PhysRev.96.533.

17. Jenny, D.A., Loferski, J.J., & Rappaport, P. (1956). Photovoltaic effect in GaAs p-n junctions and solar energy conversion, *Phys. Rev.*, 101(3), 1208–1209. https://doi.org/10.1103/PhysRev.101.1208.

18. Alferov, Z.I., Andreev, V.M., Kagan, M.B., Protasov, I.I., & Trofim, V.G. (1971). Solar-energy converters based on p-n AlxGal-x As-GaAs heterojunctions, *Sov. Phys.-Semicond.* (Engl. Transl.); (United States), 4(12). https://www.osti.gov/biblio/5083845.

19. Weber, W.H., & Lambe, J. (1976). Luminescent greenhouse collector for solar radiation, *Appl. Opt.*, 15(10), 2299–2300. https://doi.org/10.1364/AO.15.002299.

20. Trupke, T., Green, M.A., & Wurfel, P. (2002). Improving solar cell efficiencies by down-conversion of high-energy photons, *J. Appl. Phys.*, 92(3), 1668–1674. https://doi.org/10.1063/1.1492021.

21. Trupke, T., Green, M.A., & Wurfel, P. (2002). Improving solar cell efficiencies by up-conversion of sub-band-gap light, *J. Appl. Phys.*, 92(7), 4117–4122. https://doi.org/10.1063/1.1505677.

22. Regan, B.O., & Gratzel, M. (1991). A low-cost, high-efficiency solar cell based on dye sensitized colloidal TiO_2 films, *Nature*, 353(6346), 737–740. https://doi.org/10.1038/353737a0.

23. Kim, H.S., Lee, C.R., Jang, I.H., Kang, W., and Park, N.G. (2012). Effect of overlayer thickness of hole transport material on photovoltaic performance in solid-sate dye sensitized solar cell, *Bull. Korean Chem. Soc.*, 33(2), 670–674. https://doi.org/10.5012/bkcs.2012.33.2.670.

24. Kalaiselvi, C.R., Muthukumarasamy, N., Velauthapillai, D., Kang, M., & Senthil, T.S. (2018). Importance of halide perovskites for next generation solar cells – A review, *Mater. Lett.*, 219, 198–200. https://doi.org/10.1016/j.matlet.2018.02.089.

25. Kojima, A., Teshima, K., Shirai, Y., & Miyasaka, T. (2009). Organometal halide perovskites as visible-light sensitizers for photovoltaic cells, *J. Am. Chem. Soc.*, 131(17), 6050–6051. https://doi.org/10.1021/ja809598r.

26. Kim, H.S., Lee, C.R., Im, J.H., Lee, K.B., Moehl, T., Marchioro, A., Moon, S.J., Baker, R.H., Yum, J.H., Moser, J.E., Gratzel, M., & Park, N.G. (2012). Lead iodide perovskite sensitized all-solid-state submicron thin film mesoscopic solar cell with efficiency exceeding 9%, *Sci. Rep.*, 2(1), 591, 1–7. https://doi.org/10.1038/srep00591.

27. Lee, M.M., Teuscher, J., Miyasaka, T., Murakami, T.N., & Snaith, H.J. (2012). Efficient hybrid solar cells based on meso-superstructured organometal halide perovskites, *Science*, 338(6107), 643–647. https://doi.org/10.1126/science.1228604.

28. Etgar, L., Gao, P., Xue, Z., Peng, Q., Chandiran, A.K., Liu, B., Nazeeruddin, M.K., & Gratzel, M. (2012). Mesoscopic $CH_3NH_3PbI_3/TiO_2$ heterojunction solar cells, *J. Am. Chem. Soc.*, 134(42), 17396–17399. https://doi.org/10.1021/ja307789s.

29. Yang, W.S., Park, B.W., Jung, E.H., Jeon, N.J., Kim, Y.C., Lee, D.U., Shin, S.S., Seo, J., Kim, E.K., & Noh, J.H. (2017). Iodide management in formamidinium-lead-halide-based perovskite layers for efficient solar cells, *Science*, 356(6345), 1376–1379. https://doi.org/10.1126/science.aan2301.

30. Xu, Q., Yang, D., Lv, J., Sun, Y.Y., & Zhang, L. (2018). Perovskite solar absorbers: Materials by design, *Small Methods*, 2(5), 1700316. https://doi.org/10.1002/smtd.201700316.

31. Liu, C., Li, W., Fan, J., & Mai, Y. (2018). A brief review on the lead element substitution in perovskite solar cells, *J. Energy Chem.*, 27(4), 1054–1066. https://doi.org/10.1016/j.jechem.2017.10.028.

32. Jung, H.S., & Park, N.G. (2015). Perovskite solar cells: From materials to devices, *Small*, 11(1), 10–25. https://doi.org/10.1002/smll.201402767.

33. Leijtens, T., Eperon, G.E., Pathak, S., Abate, A., Lee, M.M., & Snaith, H.J. (2013). Overcoming ultraviolet light instability of sensitized TiO_2 with meso-superstructured organometal tri-halide perovskite solar cells, *Nat. Commun.*, 4(1), 2885, 1–8. https://doi.org/10.1038/ncomms3885.

34. Chander, N., Khan, A.F., Chandrasekhar, P.S., Thouti, E., Swami, S.K., Dutta, V., & Komarala, V.K. (2014). Reduced ultraviolet light induced degradation and enhanced light harvesting using YVO_4:Eu^{3+} down-shifting nano-phosphor layer in organometal halide perovskite solar cells, *Appl. Phys. Lett.*, 105(3), 033904. https://doi.org/10.1063/1.4891181.

35. Barry, T.L. (1968). Fluorescence of Eu^{2+} activated phases in binary alkaline earth orthosilicate systems, *J. Electrochem. Soc.*, 115(11), 1181–1184. https://doi.org/10.1149/1.2410935.

36. Blasse, G., & Wanmaker, W.L. (1968). Fluorescence of Eu^{2+} activated silicates, *Philips Res. Rep.*, 23, 189–189. https://www.semanticscholar.org/paper/FLUORESCENCE-OF-EU%2B2-ACTIVATED-SILICATES-Blasse-Wanamaker/e61b011ef31545b282cd96fec4132bd79b6d971c.

37. Park, J.K., Lim, M.A., Kim, C.H., Park, H.D., Park, J.T., & Choi, S.Y. (2003). White light-emitting diodes of GaN-based Sr_2SiO_4:Eu and the luminescent properties, *Appl. Phys. Lett.*, 82(5), 683–685. https://doi.org/10.1063/1.1544055.

38. Poort, S.H.M., Janssen, W., & Blasse, G. (1997). Optical properties of Eu^{2+} activated orthosilicates and orthophosphates, *J. Alloys Compd.*, 260(1–2), 93–97. https://doi.org/10.1016/S0925-8388(97)00140-0.

39. Poort, S.H.M., Meyerink, A., & Blasse, G. (1997). Lifetime measurements in Eu^{2+} doped host lattices, *J. Phys. Chem. Solids*, 58(9), 1451–1456. https://doi.org/10.1016/S0022-3697(97)00010 3.

40. Yoo, J.S., Kim, S.H., Yoo, W.T., Hong, G.Y., Kim, K.P., Rowland, J., & Holloway, P.H. (2005). Control of spectral properties of strontium-alkaline earth-silicate-europium phosphors for LED applications, *J. Electrochem. Soc.*, 152(5), G382–G385. https://doi.org/10.1149/1.1888365.

41. Kim, J.S., Jeon, P.E., Choi, J.C., & Park, H.L. (2005). Emission color variation of M_2SiO_4:Eu^{2+} (M=Ba, Sr, Ca) phosphors for light-emitting diode, *Solid State Commun.*, 133(3), 187–190. https://doi.org/10.1016/j.ssc.2004.10.017.

42. Kim, J.S., Park, Y.H., Kim, S.M., Choi, J.C., & Park, H.L. (2005). Temperature-dependent emission spectra of M_2SiO_4:Eu^{2+} (M=Ca, Sr, Ba) phosphors for green and greenish white LEDs, *Solid State Commun.*, 133(7), 445–448. https://doi.org/10.1016/j.ssc.2004.12.002.

43. Denault, K.A., Brgoch, J., Gaultois, M.W., Mikhailovsky, A., Petry, R., Winkler, H., DenBaars, S.P., & Seshadri, R. (2014). Consequences of optimal bond valence on structural rigidity and improved luminescence properties in $Sr_xBa_{2-x}SiO_4$:Eu^{2+} orthosilicate phosphors, *Chem. Mater.*, 26(7), 2275–2282. https://doi.org/10.1021/cm500116u.

44. He, L., Song, Z., Xiang, Q., Xia, Z., & Liu, Q. (2016). Relationship between thermal quenching of Eu^{2+} luminescence and cation ordering in $(Ba_{1-x}Sr_x)_2SiO_4$:Eu phosphors, *J. Lumin.*, 180, 163–168. https://doi.org/10.1016/j.jlumin.2016.08.031.

45. Tyagi, N., Jo, D.S.U., Abe, T., Toda, K., Masaki, T., & Yoon, D.H. (2014). Liquid phase precursor driven approach for the synthesis of Ca-Sr-Ba orthosilicate phosphors for white LED applications, *J. Ceram. Soc. Jpn.*, 122(1429), 806–809. https://doi.org/10.2109/jcersj2.122.806.

46. Humayoun, U.B., Song, Y.H., Lee, M., Masato, K., Abe, H., Toda, K., Sato, Y., Masaki, T. & Yoon, D.H., (2016). Synthesis of high intensity green emitting (Ba, Sr)SiO_4:Eu^{2+} phosphors through cellulose assisted liquid phase precursor process, *Opt. Mater.*, 51, 110–114. https://doi.org/10.1016/j.optmat.2015.11.036.

47. Fiedler, T., Hempel, W., & Jermann, F. (2009). Patent, US 7,489,073.

48. Bonnet, D., & Rabenhorst, H. (1972). 9th Photovoltaic Specialists Conference Silver Spring Md 129.

49. Tyan, Y.S., & Perez-Albuerne, E.A. (1982). U.S. Patent 4,315,096.

50. Basol, B., Tseng, E., & Rod, R.L. (1981). U.S. Patent 4,388,483.

51. Meyers, P. (1987). U.S.Patent 4,710,589

52. Lee, T.D., & Ebong, A.U. (2017). A review of thin film solar cell technologies and challenges, *Renew. Sust. Energ. Rev.*, 70, 1286–1297. https://doi.org/10.1016/j.rser.2016.12.028.

53. Ramanujam, J., Verma, A., Gonzalez-Diaz, B., Guerrero-Lemus, R., del Canizo, C., Garcia-Tabares, E., Rey-Stolle, I., Granek, F., Korte, L., Tucci, M., & Rath, J. (2016). Inorg. photovoltaics-planar and nanostructured devices, *Prog. Mater. Sci.*, 82, 294–404. https://doi.org/10.1016/j.pmatsci.2016.03.005.

54. Miles, R.W., Hynes, K.M., & Forbes, I. (2005). Photovoltaic solar cells: An overview of state-of-the-art cell development and environmental issues, *Prog. Cryst. Growth Charact. Mater.*, 51(1–3), 1–42. https://doi.org/10.1016/j.pcrysgrow.2005.10.002.

55. Zweibel, K., Mason, J., & Fthenakis, V. (2008). A solar grand plan, *Sci. Am.*, 298(1), 64–73. https://www.jstor.org/stable/26000377.

56. Hong, B.C., & Kawano, K. (2003). PL and PLE studies of $KMgF_3$:Sm crystal and the effect of its wavelength conversion on CdS/CdTe solar cell, *Sol. Energy Mater. Sol. Cells*, 80(4), 417–432. https://doi.org/10.1016/j.solmat.2003.06.013.

57. Kawano, K., Hong, B.C., Sakamoto, K., Tsuboi, T., & Seo, H.J. (2009). Improvement of the conversion efficiency of solar cell by rare earth ion, *Opt. Mater.*, 31(9), 1353–1356. https://doi.org/10.1016/j.optmat.2008.10.012.

58. Streudel, F., Dyrba, M., & Schweizer, S. (2012). Fluorescent borate glass superstrates for high efficiency CdTe solar cells, photonics for solar energy systems IV, *Proc. SPIE*, 8438, 843803. https://doi.org/10.1117/12.921590.

59. Richards, B.S., & McIntosh, K.R. (2007). Overcoming the poor short wavelength spectral response of CdS/CdTe photovoltaic modules via luminescence down-shifting: Ray-tracing simulations, *Prog. Photovolt.*, 15(1), 27–34. https://doi.org/10.1002/pip.723.

60. Hovel, H.J., Hodgson, R.T., & Woodall, J.M. (1979). The effect of fluorescent wavelength shifting on solar cell spectral response, *Sol. Energy Mater.*, 2(1), 19–29. https://doi.org/10.1016/0165-1633(79)90027-3.

61. Dorenbos, P. (2003). Energy of the first $4f^7 \rightarrow 4f^6 5d$ transition of Eu^{2+} in inorganic compounds, *J. Lumin.*, 104(4), 239–260. https://doi.org/10.1016/S0022-2313(03)00078-4.

62. Dorenbos, P. (2003). f → d transition energies of divalent lanthanides in inorganic compounds, *J. Phys. Condens. Matter*, 15(3), 575–594. https://doi.org/10.1088/0953-8984/15/3/322.

63. Blasse, G. (1987). Progress in rare-earth and actinide spectroscopy. In: Di Bartolo, B. (ed) *Spectroscopy of Solid-State Laser-Type Materials* (Boston, MA: Springer). https://doi.org/10.1007/978-1-4613-0899-7_7.

64. Blasse, G. (1988). Luminescence of inorganic solids: From isolated centres to concentrated systems, *Prog. Solid State Chem.*, 18(2), 79–171. https://doi.org/10.1016/0079-6786(88)90004-0.

65. Hush, N.S., & Hobbs, R.J.M. (1968). *Progress in Inorganic Chemistry*, vol. 10, Albert Cotton, F. (ed) (John Wiley). ISBN-13: 978–0470176696.

66. Grinberg, M., Lesniewski, T., Mahlik, S., & Liu, R.S. (2017). $3d^3$ system-comparison of Mn^{4+} and Cr^{3+} in different lattices, *Opt. Mater.*, 74, 93–100. https://doi.org/10.1016/j.optmat.2017.03.057.

67. Bersuker, I.B. (1996). *Electronic Structure and Properties of Transition Metal Compounds, Introduction to the Theory* (New York, Chichester, Brisbane, Toronto, Singapore: John Wiley & Sons). ISBN: 978-0-470-18023-5.

68. Aramburu, J.A., Garcia- Fernandez, P., Garcia Lastra, J.M., Barriuso, M.T., & Moreno, M. (2013). Colour due to Cr^{3+} ions in oxides: A study of the model system $MgO:Cr^{3+}$, *J. Phys. Condens. Matter*, 25(17), 175501. https://doi.org/10.1088/0953-8984/25/17/175501.

69. Long, J., Ma, C., Wang, Y., Yuan, X., Du, M., Ma, R., Wen, Z., Zhang, J., & Cao, Y. (2017). Luminescent performances of Mn^{4+} ions during the phase evolution from $MgTiO_3$ to Mg_2TiO_4, *Mater. Res. Bull.*, 85, 234–239. https://doi.org/10.1016/j.materresbull.2016.09.015.

70. Cao, R., Wang, W., Zhang, J., Jiang, S., Chen, Z., Li, W., & Yu, X. (2017). Synthesis and luminescence properties of $Li_2SnO_3:Mn^{4+}$ red-emitting phosphor for solid-state lighting, *J. Alloys Compd.*, 704, 124–130. https://doi.org/10.1016/j.jallcom.2017.02.079.

71. Ogasawara, K., Alluqmani, F., & Nagoshi, H., (2016). Multiplet energy level diagrams for Cr^{3+} and Mn^{4+} in oxides with O_h site symmetry based on first-principles calculations, *ECS J. Solid State Sci. Tech.*, 5(1), R3191–R3196. https://doi.org/10.1149/2.0231601jss.

72. Brik, M.G., & Srivastava, A.M. (2013). On the optical properties of the Mn^{4+} ion in solids, *J. Lumin.*, 133, 69–72. https://doi.org/10.1016/j.jlumin.2011.08.047.

73. Zhou, Z., Zhou, N., Xia, M., Yokoyama, M., & Hintzen, H.B. (2016). Research progress and application prospects of transition metal Mn^{4+} activated luminescent materials, *J. Mater. Chem. C*, 4(39), 9143–9161. https://doi.org/10.1039/C6TC02496C.

74. Lin, Y., Hu, Y., Wu, H., Duan, H., Chen, L., Yu, Y., u, G., Mu, Z., & He, M. (2016). A deep red phosphor $Li_2MgTiO_4:Mn^{4+}$ exhibiting abnormal emission: Potential application as color converter for warm w-LEDs, *Chem. Eng. J.*, 288, 596–607. https://doi.org/10.1016/j.cej.2015.12.027.

75. Brik, M., Camardello, S., & Srivastava, A. (2015). Influence of covalency on the $Mn^{4+}\,^2E_g \rightarrow\,^4A_{2g}$ emission energy in crystals, *ECS J. Solid State Sci. Technol.*, 4(3), R39–R43. https://doi.org/10.1149/2.0031503jss.

76. Kahn-Harari, A., Abolhassani, S., Michel, D., Mazerolles, L., Portier, R., & Perez-Ramirezs, J.G. (1991). Observation of ordering in silicon and germanium mullites, *J. Solid State Chem.*, 90(2), 234–248. https://doi.org/10.1016/0022-4596(91)90139-9.

77. Macfarlane, R.M. (1965). On the ground-state splitting in ruby, *J. Chem. Phys.*, 42(1), 442–442. https://doi.org/10.1063/1.1695719.

78. Tawalare, P.K., Belsare, P.D., & Moharil, S.V. (2020). Spectral converters for CdS-CdTe solar cell, *J. Alloys Compd.*, 825, 154007. https://doi.org/10.1016/j.jallcom.2020.154007.

79. La Ginestra, A., Lo Jacono, M., & Porta, P. (1972). The preparation, characterization, and thermal behaviour of some lithium aluminum oxides: Li_3AlO_3 and Li_5AlO_4, *J. Therm. Anal.*, 4(1), 5–17. https://doi.org/10.1007/BF02100945.

80. Kale, M.A., Joshi, C.P., & Moharil, S.V. (2012). Combustion synthesis of some compounds in the $Li_2O-Al_2O_3$ system, *Int. J. Self Propag. High Temp. Synth.*, 21(1), 19–24. https://doi.org/10.3103/s1061386212010062.

81. Ribeiro, R.A., Silva, G.G., & Mohallem, N.D.S. (2001). The influences of heat treatment on the structural properties of lithium aluminates, *J. Phys. Chem. Solids*, 62(5), 857–864. https://doi.org/10.1016/S0022-3697(00)00239-0.

82. Sviridov, D.T., & Sviridova, R.K. (1980). Energy levels and wave functions of $Cr^{3+}(3d^3)$ ions in the inert spinels $LiAl_5O_8$ and $LiGa_5O_8$, *J. Appl. Spectrosc.*, 33, 1007–1011. https://doi.org/10.1007/BF00619502.

83. Avram, C.N., Brik, M.G., & Gruia, A.S. (2010). Theoretical calculations of energy levels scheme of Cr^{3+} – Doped $LiAl_5O_8$ spinel, Optoelectron. *Adv. Mater., Rapid Commun.*, 4, 1127–1130. https://www.researchgate.net/profile/Adrian-Sorin-Gruia/publication/233755870_Theoretical_calculations_of_energy_levels_scheme_of_Cr3_-_doped_LiAl5O8_spinel/links/09e4150b39b242d8d9000000/Theoretical-calculations-of-energy-levels-scheme-of-Cr3-doped-LiAl5O8-spinel.pdf

84. Liao, B.T., Mei, Y., & Zheng, W.C. (2018). A study on the thermal red-shift of R1-line for $LiAl_5O_8$: Cr^{3+} crystal from the static and vibrational effects, *Optik*, 155, 213–215. https://doi.org/10.1016/j.ijleo.2017.10.139.

85. Samartsev, V.V., Usmanov, R.G., Khadiev, I.K., Kustov, E.F., & Baranov, M.N. (1976). Photon echo in $CaWO_4$:Nd^{3+} and $LiAl_5O_8$: Cr^{3+}, *Physica Status Solidi (b)*, 76(1), 55–66. https://doi.org/10.1002/pssb.2220760105.

86. Abritta, T., Melamed, N.T., Neto, J.M., & Barros, F.D.S. (1979). The optical properties of Cr^{3+} in $LiAl_5O_8$ and $LiGa_5O_8$, *J. Lumin.*, 18–19, 179–182. https://doi.org/10.1016/0022-2313(79)90098-X.

87. Singh, V., Chakradhar, R.P.S., Rao, J.L., & Kwak, H.Y. (2009). Characterization, EPR and photoluminescence studies of $LiAl_5O_8$:Cr phosphors, *Solid State Sci.*, 11(4), 870–874. https://doi.org/10.1016/j.solidstatesciences.2009.01.009.

88. Li, X., Jiang, G., Zhou, S., Wei, X., Chen, Y., Duan, C.K., & Yin, M. (2014). Luminescent properties of chromium (III)-doped lithium aluminate for temperature sensing, *Sens. Actuators B Chem.*, 202, 1065–1069. https://doi.org/10.1016/j.snb.2014.06.053.

89. Tumram, P.V., Kautkar, P.R., Wankhede, S.P., Belsare, P.D., & Moharil, S.V. (2021). Combustion synthesis of some Cr^{3+} activated aluminate phosphors, *Phys. Solid State*, 63(7), 1104–1112. https://doi.org/10.1134/S1063783421070234.

90. Chaban, N., Porotnikov, N., Margolin, L., & Petrov, K. (1985) *Russ. J. Inorg. Chem. (Engl.Transl.)*, 30, 1663.

91. Cao, R., Zhang, J., Wang, W., Hu, Z., Chen, T., Ye, Y., & Yu, X. (2017). Preparation and photoluminescence characteristics of $Li_2Mg_3SnO_6$:Mn^{4+} deep red phosphor, Mater. *Res. Bull.*, 87, 109–113. https://doi.org/10.1016/j.materresbull.2016.11.031.

92. Millard, R.L., Peterson, R.C., & Hunter, B.K., (1995). Study of the cubic to tetragonal transition in Mg_2TiO_4 and Zn_2TiO_4 spinels by 17O MAS NMR and Rietveld refinement of X-ray diffraction data, *Am. Miner.*, 80(9–10), 885–896. https://doi.org/10.2138/am-1995-9-1003.

93. Barth, Tom, F.W., & Posnjak, E. (1932). Spinel structures: With and without variate atom equipoints, *Zeitschrift für Kristallographie – Crystalline Materials*, 82(1–6), 325–341. https://doi.org/10.1524/zkri.1932.82.1.325.

94. De Grave, E., De Sitter, J., & Vandenberghe, R. (1975). On the cation distribution in the spinel system-$yMg_2TiO_{4-(1-y)MgFe2}O_4$, *Appl. Phys.*, 7(1), 77–80. https://doi.org/10.1007/BF00900525

95. Agranovskaya, A.I., & Saksonov, Yu, G. (1966). *Sovt. Phys. Crystallogr.*, 11, 196.

96. Delamoye, P., & Michel, A. (1969). *C. R. Acad. Sci. Ser. C*, 269, 837.

97. Wechsler, B.A., & Von Dreele, R.B. (1989). Structure refinements of Mg_2TiO_4, $MgTiO_3$ and $MgTi_2O_5$ by time-of-flight neutron powder diffraction, *Acta Cryst.*, B45(6), 542–549. https://doi.org/10.1107/S010876818900786X.

98. Barbanyagre, V., Timoshenko, T., Ilyinets, A., & Shamshurov, V. (1997). Calcium aluminozincates of CaxAlyZnkOn composition, *Powder Diffract.*, 12(1), 22–26. https://doi.org/10.1017/S0885715600009398.

99. Seki, K., Uematsu, K., Toda, K., & Sato, M. (2014). Novel deep red emitting phosphors $Ca_{14}Zn_6M_{10}O_{35}$:Mn^{4+} (M=Al^{3+} and Ga^{3+}), *Chem. Lett.*, 43(8), 1213–1215. https://doi.org/10.1246/cl.140227.

100. Lu, W., Lv, W., Zhao, Q., Jiao, M., Shao, B., & You, H. (2014). A novel efficient Mn^{4+} activated $Ca_{14}Al_{10}Zn_6O_{35}$ phosphor: Application in red-emitting and white LEDs, *Inorg. Chem.*, 53(22), 11985–11990. https://doi.org/10.1021/ic501641q.

101. Gao, X., Li, W., Yang, X., Jin, X., & Xiao, S., (2015). Near-infrared emission of Er^{3+} sensitized by Mn^{4+} in $Ca_{14}Zn_6Al_{10}O_{35}$ matrix, *J. Phys. Chem. C*, 119(50), 28090–28098. https://doi.org/10.1021/acs.jpcc.5b05825.

102. Li, L., Pan, Y., Huang, Y., Huang, S., & Wu, M. (2017). Dual-emissions with energy transfer from the phosphor $Ca_{14}Al_{10}Zn_6O_{35}$:Bi^{3+}, Eu^{3+} for application in agricultural lighting, *J. Alloys Compd.*, 724, 735–743. https://doi.org/10.1016/j.jallcom.2017.07.047.

103. Zhang, Y., Zhang, X., Zheng, L., Zeng, Y., Lin, Y., Liu, Y., Lei, B., & Zhang, H. (2018). Energy transfer and tunable emission of $Ca_{14}Al_{10}Zn_6O_{35}$:Bi^{3+}, Sm^{3+} phosphor, *Mater. Res. Bull.*, 100, 56–61. https://doi.org/10.1016/j.materresbull.2017.12.003.

104. Tumram, P.V., Kautkar, P.R., Wankhede, S.P., & Moharil, S.V. (2019). NIR emitting Bi_2MoO_6: Nd^{3+}/Yb^{3+} phosphor as a spectral converter for solar cells, *J. Lumin.*, 206, 39–45. https://doi.org/10.1016/j.jlumin.2018.10.022.

105. Richards, B.S. (2006). Luminescent layers for enhanced silicon solar cell performance: Down-conversion, *Sol. Energy Mater. Sol. Cells*, 90(9), 1189–1207. https://doi.org/10.1016/j.solmat.2005.07.001.

106. Van Sark, W.G.J.H.M., Meijerink, A., Schropp, R.E.I., Van Roosmalen, J.A.M., & Lysen, E.H. (2005). Enhancing solar cell efficiency by using spectral converters, *Sol. Energy Mater. Sol. Cells*, 87(1–4), 395–409. https://doi.org/10.1016/j.solmat.2004.07.055.

107. Nakata, R.N.R., Hashimoto, N.H.N., & Kawano, K.K.K. (1996). High-conversion-efficiency solar cell using fluorescence of rare-earth ions, *Jpn. J. Appl. Phys.*, 35(1B), L90. https://doi.org/10.1143/JJAP.35.L90.

108. Steudel, F., Loos, S., Ahrens, B., & Schweizer, S. (2015). Luminescent borate glass for efficiency enhancement of CdTe solar cells, *J. Lumin.*, 164, 76–80. https://doi.org/10.1016/j.jlumin.2015.03.022

109. Ten Kate, O.M., De Jong, M., Hintzen, H.T., & Van Der Kolk, E. (2013). Efficiency enhancement calculations of state-of-the-art solar cells by luminescent layers with spectral shifting, quantum cutting, and quantum tripling function, *J. Appl. Phys.*, 114(8), 084502. https://doi.org/10.1063/1.4819237.

110. Van Pieterson, L., Heeroma, M., De Heer, E., & Meijerink, A. (2000). Charge transfer luminescence of Yb^{3+}, *J. Lumin.*, 91(3–4), 177–193. https://doi.org/10.1016/S0022-2313(00)00214-3.

111. Van Der Ende, B.M., Aarts, L., & Meijerink, A. (2009). Lanthanide ions as spectral converters for solar cells, *Phys. Chem. Chem. Phys.*, 11(47), 11081–11095. https://doi.org/10.1039/B913877C.

112. Strumpel, C., McCann, M., Beaucarne, G., Arkhipov, V., Slaoui, A., Svrcek, V., Del Canizo, C., & Tobias, I. (2007). Modifying the solar spectrum to enhance silicon solar cell efficiency-An overview of available materials, *Sol. Energy Mater. Sol. Cells*, 91(4), 238–249. https://doi.org/10.1016/j.solmat.2006.09.003.

113. Solanki, C.S., & Singh, H.K. (2017). c-Si solar cells: Physics and technology, *Green Energy Technol.*, 17–42. https://www.ese.iitb.ac.in/publications/author/37.

114. Richards, B.S. (2006). Enhancing the performance of silicon solar cells via the application of passive luminescence conversion layers, *Sol. Energy Mater. Sol. Cells*, 90(15), 2329–2337. https://doi.org/10.1016/j.solmat.2006.03.035.

115. Shockley, W., & Queisser, H.J. (1961). Detailed balance limit of efficiency of p-n junction solar cells, *J. Appl. Phys.*, 32(3), 510–519. https://doi.org/10.1063/1.1736034.

116. Zhang, Q.Y., & Huang, X.Y. (2010). Recent progress in quantum cutting phosphors. *Prog. Mater. Sci.*, 55(5), 353–427. https://doi.org/10.1016/j.pmatsci.2009.10.001.

117. Bunzli, J.C.G., & Eliseeva, S.V. (2010). Lanthanide NIR luminescence for telecommunications, bioanalyses and solar energy conversion, *J. Rare Earths*, 28(6), 824–842. https://doi.org/10.1016/S1002-0721(09)60208-8.

118. Chen, D., Wang, Y., & Hong, M. (2012). Lanthanide nanomaterials with photon management characteristics for photovoltaic application, *Nano Energy*, 1(1), 73–90. https://doi.org/10.1016/j.nanoen.2011.10.004.

119. Lian, H., Hou, Z., Shang, M., Geng, D., Zhang, Y., & Lin, J. (2013). Rare earth ions doped phosphors for improving efficiencies of solar cells, *Energy*, 57, 270–283. https://doi.org/10.1016/j.energy.2013.05.019.

120. Huang, X., Han, S., Huang, W., & Liu, X. (2013). Enhancing solar cell efficiency: The search for luminescent materials as spectral converters, *Chem. Soc. Rev.*, 42(1), 173–201. https://doi.org/10.1039/C2CS35288E.

121. De la Mora, M.B., Amelines-Sarria, O., Monroy, B.M., Hernandez-Perez, C.D., & Lugo, J.E. (2017). Materials for downconversion in solar cells: Perspectives and challenges, *Sol. Energy Mater. Sol. Cells*, 165, 59–71. https://doi.org/10.1016/j.solmat.2017.02.016.

122. Kumar, P., Singh, S., Lahon, R., Gundimeda, A., Gupta, G., & Gupta, B.K. (2018). A strategy to design lanthanide doped dual-mode phosphor mediated spectral convertor for solar cell applications, *J. Lumin.*, 196, 207–213. https://doi.org/10.1016/j.jlumin.2017.12.035.

123. Rastogi, R.P., Singh, A.K., & Shukla, C.S. (1982). Kinetics and mechanism of solid-state reaction between bismuth (III) oxide and molybdenum (VI) oxide, *J. Solid State Chem.*, 42(2), 136–148. https://doi.org/10.1016/0022-4596(82)90260-2.

124. Egashira, M., Matsuo, K., Kagawa, S., & Seiyama, T. (1979). Phase diagram of the system Bi_2O_3段MoO_3, *J. Catal.*, 58(3), 409–418. https://doi.org/10.1016/0021-9517(79)90279-3.

125. Licht, R.B., Getsoian, A.B., & Bell, A.T. (2016). Identifying the unique properties of α-$Bi_2Mo_3O_{12}$ for the activation of propene, *J. Phys. Chem. C*, 120(51), 29233–29247. https://doi.org/10.1021/acs.jpcc.6b09949.

126. Bleijenberg, A.C.A.M., Lippens, B.C., & Schuit, G.C.A. (1965). Catalytic oxidation of 1-butene over bismuth molybdate catalysts: I. The system Bi2O3 MoO3, *J. Catal.*, 4(5), 581–585. https://doi.org/10.1016/0021-9517(65)90163-6.

127. Jung, J.C., Kim, H.S., Choi, A.S., Chung, Y.M., Kim, T.J., Lee, S.J., Oh, S.H., & Song, I.K. (2007). Preparation and characterization of bismuth molybdate catalyst for oxidative dehydrogenation of n-butene into 1, 3-butadiene, *Solid State Phenomena*, 119, 251–254. Trans Tech Publications Ltd. https://doi.org/10.4028/www.scientific.net/SSP.119.251.

128. Martinez-de La Cruz, A., Villarreal, S.M., Torres-Martinez, L.M., Cuellar, E.L., & Mendez, U.O. (2008). Photoassisted degradation of rhodamine B by nanoparticles of α-$Bi_2Mo_3O_{12}$ prepared by an amorphous complex precursor, *Mater. Chem. Phys.*, 112(2), 679–685. https://doi.org/10.1016/j.matchemphys.2008.06.035.

129. Chen, T., Wang, M.H., & Ma, X.Y. (2016). Preparation in acidic and alkaline conditions and characterization of α-$Bi_2Mo_3O_{12}$ and γ-Bi_2MoO_6 powders, *J. Elec. Mater.*, 45(8), 4375–4379. https://doi.org/10.1007/s11664-016-4648-5.

130. Hipolito, E.L., Martinez-de la Cruz, A., Yu, Q., & Brouwers, H.J.H. (2013). Photocatalytic removal of nitric oxide by $Bi_2Mo_3O_{12}$ prepared by co-precipitation method, *Appl. Catal. A: General*, 468, 322–326. https://doi.org/10.1016/j.apcata.2013.09.013.

131. Biegelsen, D.K., Chen, T., & Zesch, J.C. (1975). Acousto-optic parameters of $Bi_2(MoO_4)_3$, *J. Appl. Phys.*, 46(2), 941–942. https://doi.org/10.1063/1.321620.

132. Tumram, P.V., Kautkar, P.R., Wankhede, S.P., Belsare, P.D., & Moharil, S.V. (2019). NIR emission and energy transfer phenomena in $Bi_2(MoO_4)_3$ doped with Nd^{3+} and/or Yb^{3+}, *AIP Adv.*, 9(10), 105113. https://doi.org/10.1063/1.5112135.

133. Zhang, Z.J., & Chen, X.Y. (2016). Sb_2MoO_6, Bi_2MoO_6, Sb_2WO_6, and Bi_2WO_6 flake-like crystals: Generalized hydrothermal synthesis and the applications of Bi_2WO_6 and Bi_2MoO_6 as red phosphors doped with Eu^{3+} ions, *J. Mater. Sci. Eng. B.*, 209, 10–16. https://doi.org/10.1016/j.mseb.2015.12.003.

134. Feng, L., Li, M., Pan, K., Li, R., Fan, N., & Wang, G. (2016). Photoluminescence and photocatalytic activity of Bi_2MoO_6: Ln^{3+} nanocrystals, *J. Nanoscience Nanotech.*, 16(4), 3781–3785. https://doi.org/10.1166/jnn.2016.11850.

135. Zhang, J., Liu, Y., Li, L., Zhang, N., Zou, L., & Gan, S. (2015). Hydrothermal synthesis, characterization, and color-tunable luminescence properties of Bi_2MoO_6: Eu^{3+} phosphors, *RSC Adv.*, 5(37), 29346–29352.

136. Han, B., Zhang, J., Li, P., Li, J., Bian, Y., & Shi, H. (2015). Synthesis and luminescence properties of Eu^{3+} doped high temperature form of Bi_2MoO_6, *J. Electron. Mater.*, 44(3), 1028–1033. https://doi.org/10.1007/s11664-014-3621-4.

137. Jin, S., Hao, H., Gan, Y., Guo, W., Li, H., Hu, X., Hou, H., Zhang, G., Yan, S., Gao, W., & Liu, G. (2017). Preparation and improved photocatalytic activities of Ho^{3+}/Yb^{3+} co-doped Bi_2MoO_6, *Mater. Chem. Phys.*, 199, 107–112. https://doi.org/10.1016/j.matchemphys.2017.06.053.

138. Adhikari, R., Gyawali, G., Cho, S.H., Narro-Garcia, R., Sekino, T., & Lee, S.W. (2014). Er^{3+}/Yb^{3+} co-doped bismuth molybdate nanosheets upconversion photocatalyst with enhanced photocatalytic activity, *J. Solid State Chem.*, 209, 74–81. https://doi.org/10.1016/j.jssc.2013.10.028.

139. Kumari, L.S., Gayathri, T.H., Sameera, S.F., & Rao, P.P. (2011). Y-doped Bi_2MoO_6 yellow pigments for the coloration of plastics, *J. Am. Cer. Soc.*, 94(2), 320–323. https://doi.org/10.1111/j.1551-2916.2010.04268.x.

140. Pearton, S.J., Abernathy, C.R., Overberg, M.E., Thaler, G.T., Norton, D.P., Theodoropoulou, N., Hebard, A.F., Park, Y.D., Ren, F., Kim, J., & Boatner, L.A. (2003). Wide band gap ferromagnetic semiconductors and oxides, *J. Appl. Phys.*, 93(1), 1–13. https://doi.org/10.1063/1.1517164.

141. Gruber, J.B., Sardar, D.K., Yow, R.M., Allik, T.H., & Zandi, B. (2004). Energy-level structure and spectral analysis of Nd^{3+} ($4f^3$) in polycrystalline ceramic garnet $Y_3Al_5O_{12}$, *J. Appl. Phys.*, 96(6), 3050–3056. https://doi.org/10.1063/1.1776320.

142. Garcia-Lastra, J.M., Barriuso, M.T., Aramburu, J.A., & Moreno, M. (2008). Microscopic origin of the different colors displayed by $MgAl_2O_4$: Cr^{3+} and emerald, *Phys. Rev. B*, 78(8), 085117. https://doi.org/10.1103/PhysRevB.78.085117

143. Tawalare, P.K., Bhatkar, V.B., Omanwar, S.K., & Moharil, S.V. (2019). Cr^{3+} sensitized near infrared emission in Al_2O_3:Cr, Nd/Yb phosphors, *J. Alloys Compd.*, 790, 1192–1200. https://doi.org/10.1016/j.jallcom.2019.03.201.

144. Bloembergen, N. (1959). Solid state infrared quantum counters, *Phys. Rev. Lett.*, 2, 84–85. https://doi.org/10.1103/PhysRevLett.2.84.

145. Auzel, F. (1966). Compteur quantique par transfert d'energie entre deux ions de terres rares dans un tungstate mixte et dans un verre, *CR Acad. Sci. Paris*, 262, 1016–1019. https://cir.nii.ac.jp/crid/1573950399692310784.

146. Haase, M., & Schafer, H. (2011). Upconverting nanoparticles. *Angew. Chem. Int. Ed.*, 50(26), 5808–5829. https://doi.org/10.1002/anie.201005159.

147. Bowman, S.R., Shaw, L.B., Feldman, B.J., & Ganem, J. (1996). A 7-/spl mu/m praseodymium-based solid-state laser, *IEEE J. Quant. Elect.*, 32(4), 646–649. https://doi.org/10.1109/3.488838.

148. Wang, H.Q., Batentschuk, M., Osvet, A., Pinna, L., & Brabec, C.J. (2011). Rare-earth ion doped up-conversion materials for photovoltaic applications, *Adv. Mater.*, 23(22–23), 2675–2680. https://doi.org/10.1002/adma.201100511.

149. Khare, A. (2020). A critical review on the efficiency improvement of upconversion assisted solar cells, *J. Alloys Compd.*, 821, 153214. https://doi.org/10.1016/j.jallcom.2019.153214.

150. Wang, S., Zhang, B., Li, Y., Zhang, J., & Chen, B. (2021). High quality upconversion luminescence and excellent thermal stability of Yb^{3+}, Er^{3+} codoped $Sr_3NaY(PO_4)_3F$ phosphors, *Mater. Res. Bull.*, 140, 111306. https://doi.org/10.1016/j.materresbull.2021.111306.

151. Dogan, A., & Erdem, M. (2021). Investigation of the optical temperature sensing properties of up-converting TeO_2-ZnO-BaO activated with Yb^{3+}/Tm^{3+} glasses, *Sens. Actuator A Phys.*, 322, 112645. https://doi.org/10.1016/j.sna.2021.112645.

152. Deshmukh, P., Deo, R.K., Ahlawat, A., Khan, A.A., Singh, R., Karnal, A.K., & Satapathy, S. (2021). Spectroscopic investigation of upconversion and downshifting properties LaF_3: Tb^{3+}, Yb^{3+}: A dual mode green emitter nanophosphor, *J. Alloys. Compd.*, 859, 157857. https://doi.org/10.1016/j.jallcom.2020.157857.

153. Wang, W., Tian, J., Dong, J., Xue, Y., Hu, D., Hou, W., Tang, H., Wang, Q., Xu, X., & Xu, J. (2021). Growth, spectroscopic properties and up-conversion of Yb, Pr co-doped CaF_2 crystals, *J. Lumin.*, 233, 117931. https://doi.org/10.1016/j.jlumin.2021.117931.

154. Sung Lim, C., Aleksandrovsky, A., Atuchin, V., Molokeev, M., & Oreshonkov, A. (2020). Microwave-employed sol-gel synthesis of scheelite-type microcrystalline $AgGd(MoO_4)_2$: Yb^{3+}/Ho^{3+} upconversion yellow phosphors and their spectroscopic properties, *Crystals*, 10(11), 1000. https://doi.org/10.3390/cryst10111000.

155. Ansari, A.A., Nazeeruddin, M.K., & Tavakoli, M.M. (2021). Organic-inorganic upconversion nanoparticles hybrid in dye-sensitized solar cells, *Coord. Chem. Rev.*, 436, 213805. https://doi.org/10.1016/j.ccr.2021.213805.

156. de Wild, J., Rath, J.K., Meijerink, A., Van Sark, W.G.J.H.M., & Schropp, R.E.I. (2010). Enhanced near-infrared response of a-Si: H solar cells with β-$NaYF_4$: Yb^{3+} (18%), Er^{3+} (2%) upconversion phosphors, *Sol. Energy Mater. Sol. Cells*, 94(12), 2395–2398. https://doi.org/10.1016/j.solmat.2010.08.024.

157. Singh-Rachford, T.N., & Castellano, F.N. (2010). Photon upconversion based on sensitized triplet-triplet annihilation, *Coord. Chem. Rev.*, 254(21–22), 2560–2573. https://doi.org/10.1016/j.ccr.2010.01.003.

158. Healy, C., Hermanspahn, L., & Kruger, P.E. (2021). Photon upconversion in self-assembled materials, *Coord. Chem. Rev.*, 432, 213756. https://doi.org/10.1016/j.ccr.2020.213756.

159. Badescu, V., & Badescu, A.M. (2009). Improved model for solar cells with up-conversion of low-energy photons, *Renew. Energy*, 34(6), 1538–1544. https://doi.org/10.1016/j.renene.2008.11.006.

160. De Vos, A., Szymanska, A., & Badescu, V. (2009). Modelling of solar cells with down-conversion of high energy photons, anti-reflection coatings and light trapping, *Energy Convers. Manage.*, 50(2), 328–336. https://doi.org/10.1016/j.enconman.2008.09.012.

161. Goldschmidt, J.C., & Fischer, S. (2015). Upconversion for photovoltaics – A review of materials, devices and concepts for performance enhancement, *Adv. Opt. Mater.*, 3(4), 510–535. https://doi.org/10.1002/adom.201500024.

162. Ghazy, A., Safdar, M., Lastusaari, M., Savin, H., & Karppinen, M. (2021). Advances in upconversion enhanced solar cell performance, *Sol. Energy Mater. Sol. Cells*, 230, 111234. https://doi.org/10.1016/j.solmat.2021.111234.

163. Richards, B.S., Hudry, D., Busko, D., Turshatov, A., & Howard, I.A. (2021). Photon Upconversion for photovoltaics and photocatalysis: A critical review, *Chem. Rev.*, 121(15), 9165–9195. https://doi.org/10.1021/acs.chemrev.1c00034.

164. Chen, C., Zheng, S., & Song, H. (2021). Photon management to reduce energy loss in perovskite solar cells, *Chem. Soc. Rev.*, 50(12), 7250–7329. https://doi.org/10.1039/d0cs01488e.

165. Liang, X.F., Huang, X.Y., & Zhang, Q.Y. (2009). $Gd_2(MoO_4)_3$: Er^{3+} nanophosphors for an enhancement of silicon solar-cell near-infrared response, *J. Fluoresc.*, 19(2), 285–289. https://doi.org/10.1007/s10895-008-0414-2.

166. Fan, X., Ying, W., Xu, S., Gu, J., & Liu, S. (2022). Multiple logic gates system based on dual-wavelength triggered enhancing upconversion luminescence of $Gd_2(MoO_4)_3$: Yb^{3+}/Er^{3+}, *J. Am. Ceram. Soc.*, 105(1), 402–411. https://doi.org/10.1111/jace.18076.

167. Chen, S., Pang, T., & Mao, J. (2020). Upconversion temperature sensing assists the understanding of light-emitting properties in $Gd_2(MoO_4)_3$: Y^{3+}/Yb^{3+}/Er^{3+}-based LED, *Appl. Phys. A*, 126(6), 1–7. https://doi.org/10.1007/s00339-020-03610-6.

168. Hao, H., Chen, Z., Yang, J., Ao, G., Song, Y., Wang, Y., & Zhang, X. (2018). Multi-photon up-conversion enhancement from $Gd_2(MoO_4)_3$: Er/Yb thin film via the use of sandwich structure, *J. Lumin.*, 202, 77–82. https://doi.org/10.1016/j.jlumin.2018.05.018.

169. Li, L., Xing, F., Zhang, X., Hao, H., & Wang, Y. (2021). Emission enhancement and color modulation of Tm (Ho)/Yb codoped $Gd_2(MoO_4)_3$ thin films via the use of multilayered structure, *J. Rare Earths*, 39(7), 765–771. https://doi.org/10.1016/j.jre.2020.09.014.

170. Kumari, A., Mukhopadhyay, L., & Rai, V.K. (2019). Energy transfer and dipole-dipole interaction in $Er^{3+}/Eu^{3+}/Yb^{3+}$: $Gd_2(MoO_4)_3$ upconverting nanophosphors, *New J. Chem.*, 43(16), 6249–6256. https://doi.org/10.1039/C9NJ00463G

171. Xu, W., Li, D., Hao, H., Song, Y., Wang, Y., & Zhang, X. (2018). Optical thermometry through infrared excited green upconversion in monoclinic phase $Gd_2(MoO_4)_3$: Yb^{3+}/Er^{3+} phosphor, *Opt. Mater.*, 78, 8–14. https://doi.org/10.1016/j.optmat.2018.02.001.

172. Hao, H., Lu, H., Meng, R., Nie, Z., Ao, G., Song, Y., Wang, Y., & Zhang, X. (2017). Thermometry via Au island-enhanced luminescence of Er^{3+}/Yb^{3+} co-doped $Gd_2(MoO_4)_3$ thin films, *J. Alloys Compd.*, 695, 2065–2071. https://doi.org/10.1016/j.jallcom.2016.11.045

173. Jin, X., Li, H., Li, D., Zhang, Q., Li, F., Sun, W., Chen, Z., & Li, Q. (2016). Role of ytterbium-erbium co-doped gadolinium molybdate ($Gd_2(MoO_4)_3$: Yb/Er) nanophosphors in solar cells, *Opt. Exp.*, 24(18), A1276–A1287. https://doi.org/10.1364/OE.24.0A1276.

174. Sun, W., Chen, Z., Zhang, Q., Zhou, J., Li, F., Jin, X., Li, D., & Li, Q. (2016). Efficient charge transfer and utilization of near-infrared solar spectrum by ytterbium and thulium co-doped gadolinium molybdate ($Gd_2(MoO_4)_3$:Yb/Tm) nanophosphor in hybrid solar cells, *Phys. Chem. Chem. Phys.*, 18(44), 30837–30844. https://doi.org/10.1039/c6cp04963j.

175. Kumar, P., & Gupta, B.K. (2015). New insight into rare-earth doped gadolinium molybdate nanophosphor assisted broad spectral converters from UV to NIR for silicon solar cells, *RSC Adv.*, 5(31), 24729–24736. https://doi.org/10.1039/C4RA15383A.

176. Wang, X., Yan, X., & Kan, C. (2011). Controlled synthesis and optical characterization of multifunctional ordered Y_2O_3: Er^{3+} porous pyramid arrays, *J. Mater. Chem.*, 21(12), 4251–4256. https://doi.org/10.1039/c0jm03761c.

177. Wang, J., Wu, J., Lin, J., Huang, M., Huang, Y., Lan, Z., Xiao, Y., Yue, G., Yin, S., & Sato, T., (2012). Application of Y_2O_3: Er^{3+} nanorods in dye-sensitized solar cells, *ChemSusChem.*, 5(7), 1307–1312. https://doi.org/10.1002/cssc.201100596.

178. Vetrone, F., Boyer, J.C., Capobianco, J.A., Speghini, A., & Bettinelli, M. (2004). Significance of Yb^{3+} concentration on the upconversion mechanisms in codoped Y_2O_3: Er^{3+}, Yb^{3+} nanocrystals, *J. Appl. Phys.*, 96(1), 661–667. https://doi.org/10.1063/1.1739523.

179. Jia, H., Ping, C., Xu, C., Zhou, J., Sang, X., Wang, J., Liu, C., Liu, X., & Qiu, J. (2015). Fabrication of the (Y_2O_3: Yb-Er)/Bi_2S_3 composite film for near-infrared photoresponse, *J. Mater. Chem. A.*, 3(11), 5917–5922. https://doi.org/10.1039/c5ta00692a.

180. Huerta, E.F., Carmona-Tellez, S., Gallardo-Hernandez, S., Cabanas-Moreno, J.G., & Falcony, C. (2016). Up and down conversion photoluminescence from Er, Yb and Li doped Y_2O_3 phosphors and composites films with PMMA, *ECS J. Solid State Sci. Technol.*, 5(7), R129. https://doi.org/10.1149/2.0261607jss.

181. Du, P., Lim, J.H., Leem, J.W., Cha, S.M., & Yu, J.S. (2015). Enhanced photovoltaic performance of dye-sensitized solar cells by efficient near-infrared sunlight harvesting using upconverting Y_2O_3: Er^{3+}/Yb^{3+} phosphor nanoparticles, *Nanoscale Res. Lett.*, 10(1), 321, 1–6. https://doi.org/10.1186/s11671-015-1030-0.

182. Kim, Y.M., Kim, K.H., Bark, C.W., & Choi, H.W. (2015). Improving photovoltaic performance of dye-sensitized solar cell by effect of Y_2O_3: Yb^{3+}, Er^{3+}. *J. Nanoelectron. Optoelectron.*, 10(1), 126–130. https://doi.org/10.1166/jno.2015.1705.

183. Markose, K.K., Anjana, R., Antony, A., & Jayaraj, M.K. (2018). Synthesis of Yb^{3+}/Er^{3+} codoped Y_2O_3, YOF and YF_3 UC phosphors and their application in solar cell for subbandgap photon harvesting, *J. Lumin.*, 204, 448–456, https://doi.org/10.1016/j.jlumin.2018.08.005.

184. Wu, H., Hao, Z., Zhang, L., Zhang, X., Pan, G.-H., Luo, Y., Wu, H., Zhao, H., Zhang, H., & Zhang, J. (2020). Enhancing IR to NIR upconversion emission in Er^{3+}-sensitized phosphors by adding Yb^{3+} as a highly efficient NIR-emitting center for photovoltaic applications, *CrystEngComm.*, 22(2), 229–236. https://doi.org/10.1039/c9ce01386e.

185. Dutta, J., Rai, V.K., Durai, M.M., & Thangavel, R. (2019). Development of Y_2O_3:Ho^{3+}/Yb^{3+} upconverting nanophosphors for enhancing solar cell efficiency of dye-sensitized solar cells, *IEEE J. Photovolt.*, 9(4), 1040–1045. https://doi.org/10.1109/JPHOTOV.2019.2912719.

186. Tadge, P., Yadav, R.S., Vishwakarma, P.K., Rai, S.B., Chen, T.M., Sapra, S., & Ray, S. (2020). Enhanced photovoltaic performance of Y_2O_3:Ho^{3+}/Yb^{3+} upconversion nanophosphor based DSSC and

investigation of color tunability in $Ho^{3+}/Tm^{3+}/Yb^{3+}$ tridoped Y_2O_3, *J. Alloys Compd.*, 821, 153230. https://doi.org/10.1016/j.jallcom.2019.153230.

187. Li, Q., Lin, J., Wu, J., Lan, Z., Wang, Y., Peng, F., & Huang, M. (2011). Enhancing photovoltaic performance of dye-sensitized solar cell by rare-earth doped oxide of Lu_2O_3:(Tm^{3+}, Yb^{3+}), *Electrochimica Acta*, 56(14), 4980–4984. https://doi.org/10.1016/j.electacta.2011.03.125.

188. Gao, G., Busko, D., Kauffmann-Weiss, S., Turshatov, A., Howard, I.A., & Richards, B.S. (2017). Finely-tuned NIR-to-visible up-conversion in La_2O_3:Yb^{3+}, Er^{3+} microcrystals with high quantum yield, *J. Mater. Chem. C*, 5(42), 11010–11017. https://doi.org/10.1039/c6tc05322j.

189. Xu, Z., Zhao, Q., Sun, Y., Ren, B., You, L., Wang, S., & Ding, F. (2013). Synthesis of hollow La_2O_3:Yb^{3+}/Er^{3+}/Tm^{3+} microspheres with tunable up-conversion luminescence properties, *RSC Adv.*, 3(22), 8407–8416, https://doi.org/10.1039/c3ra40414e.

190. Han, G., Wang, M., Li, D., Bai, J., & Diao, G. (2017). Novel upconversion Er, Yb-CeO_2 hollow spheres as scattering layer materials for efficient dye-sensitized solar cells, *Sol. Energy Mater. Sol. Cells*, 160, 54–59. https://doi.org/10.1016/j.solmat.2016.10.021.

191. Zhao, R., Huan, L., Gu, P., Guo, R., Chen, M., & Diao, G. (2016). Yb, Er-doped CeO_2 nanotubes as an assistant layer for photoconversion-enhanced dye-sensitized solar cells, *J. Power Sources*, 331, 527–534. https://doi.org/10.1016/j.jpowsour.2016.09.039.

192. Du, P., Lim, J.H., Kim, S.H., & Yu, J.S. (2016). Facile synthesis of Gd_2O_3:Ho^{3+}/Yb^{3+} nanoparticles: An efficient upconverting material for enhanced photovoltaic performance of dyesensitized solar cells, *Opt. Mater. Exp.*, 6(6), 1896–1904. https://doi.org/10.1364/OME.6.001896.

193. Han, C.H., Lee, H.S., Lee, K.W., Han, S.D., & Singh, I. (2009). Synthesis of amorphous Er^{3+}-Yb^{3+} co-doped TiO_2 and its application as a scattering layer for dye-sensitized solar cells, *Bull. Korean Chem. Soc.*, 30(1), 219–223. https://doi.org/10.5012/bkcs.2009.30.1.219.

194. Zhang, H., Zhang, Q., Lv, Y., Yang, C., Chen, H., & Zhou, X. (2018). Upconversion Er-doped TiO_2 nanorod arrays for perovskite solar cells and the performance improvement, *Mater. Res. Bull.*, 106, 346–352. https://doi.org/10.1016/j.materresbull.2018.06.014.

195. Shan, G.B., & Demopoulos, G.P. (2010). Near-infrared sunlight harvesting in dye-sensitized solar cells via the insertion of an upconverter-TiO_2 nanocomposite layer, *Adv. Mater.*, 22(39), 4373–4377. https://doi.org/10.1002/adma.201001816.

196. Zhang, Z., Qin, J., Shi, W., Liu, Y., Zhang, Y., Liu, Y., Gao, H., & Mao, Y., (2018). Enhanced power conversion efficiency of perovskite solar cells with an up-conversion material of Er^{3+}-Yb^{3+}-Li^+ tri-doped TiO_2, *Nanoscale Res. Lett.*, 13(1), 147, 1–8. https://doi.org/10.1186/s11671-018-2545-y.

197. Wang, X., Zhang, Z., Qin, J., Shi, W., Liu, Y., Gao, H., & Mao, Y. (2017). Enhanced photovoltaic performance of perovskite solar cells based on Er-Yb Co-doped TiO_2 nanorod arrays, *Electrochim. Acta*, 245, 839–845. https://doi.org/10.1016/j.electacta.2017.06.032.

198. Wang, L., Chen, K., Tong, H., Wang, K., Tao, L., Zhang, Y., & Zhou, X. (2020). Inverted pyramid Er^{3+} and Yb^{3+} Co-doped TiO_2 nanorod arrays based perovskite solar cell: Infrared response and improved current density, *Ceram. Int.*, 46(8), 12073–12079. https://doi.org/10.1016/j.ceramint.2020.01.249.

199. Xie, G., Wei, Y., Fan, L., & Wu, J. (2012). Application of doped rare-earth oxide TiO_2:(Tm^{3+}, Yb^{3+}) in dye-sensitized solar cells, *J. Phys.: Conf. Ser.*, 339(1), 012010. https://doi.org/10.1088/1742-6596/339/1/012010.

200. Yu, J., Yang, Y., Fan, R., Liu, D., Wei, L., Chen, S., Li, L., Yang, B., & Cao, W. (2014). Enhanced near-infrared to visible upconversion nanoparticles of Ho^{3+}-Yb^{3+}-F^- tri-doped TiO_2 and its application in dye-sensitized solar cells with 37% improvement in power conversion efficiency, *Inorg. Chem.*, 53 (15), 8045–8053. https://doi.org/10.1021/ic501041h.

8 Metal Oxides
Long Persistent Luminescence

P.V. Tumram
Amolakchand Mahavidyalaya

G.B. Nair
University of the Free State

CONTENTS

DOI: 10.1201/9781003366232-8

8.1   INTRODUCTION

Persistent luminescence (PersL, or PerL), also commonly known as long afterglow (LAP or LAG) long persistent (LPP), or long-lasting phosphorescence (LLP), is an intriguing optical phenomenon resulting in the luminescence of a phosphor that lasts for several seconds to a few days after switching off the excitation sources. LLP phosphors have been reported with emissions ranging from ultraviolet (UV) to near infrared (NIR).

Such materials attract interest of researchers from several fields; chemists, physicists, and even biologists. This is mainly due to the widespread applications. The applications of LPPs have spread from the initial common place uses, that is, decoration, safety displays, dials, etc., to a wide range of sophisticated techniques in life sciences, biomedicine, clinical medicine, and energy and environmental engineering. Moreover, scientists have proposed many theoretical models to understand the basic mechanism. These developments are briefly reviewed in this chapter.

8.1.1   EARLY HISTORY

"Intriguingly, the effect of energy storage resulting in generation of persistent emission was not only observed but also deliberately exploited for the first time not later than in 140–88 B.C. This we know from a Chinese publication dated in the Song dynasty (960–1279). In that text, a painting with a cow grazing outside is described. Yet, in dark the same painting presented the cow in a cowshed [1–3]. This effect could obviously be obtained using a paint or ink containing a persistent phosphor, the light from which produced the night-specific picture after dusk. Whatever the persistent luminescence material was, it was the first artificially prepared phosphor to which a written document refers. A great number of equally intriguing stories the reader may find in a fascinating book by Harvey [4]. Here, let us yet refer shortly to the history of Bologna stone. It was discovered in 1603 by Vincenzo Cascariolo, a cobbler, in Volcanic rock of Mount Paderno near Bologna, Italy [5]. After heat treatment and exposure to sunlight, it could then glow in dark for hours. At that time, it was a real mystery and the discoverer hoped that he found the Philosopher's Stone. It took some 400 years to understand that Cascariolo found a barite, $BaSO_4$. This, heat treated at reducing atmosphere got converted to BaS and the impurities it contained, like Cu, granted the described glowing in dark. Nowadays, Cascariolo is considered the discoverer of a first persistent phosphor in Europe." Goethe in his work "The Sorrows of Young Werther," described "It is said that the Bonona stone, when placed in the sun, attracts the rays, and for a time appears luminous in the dark" [6].

Almost equally accidental were many later discoveries of persistent and storage phosphors until the end of 20th century. Among them, the most efficient was ZnS activated with Cu^+ discovered in 1920 and next further improved in efficiency and duration of the persistent emission by codoping with Co^{2+} [7–9]. A special role in the development of understanding of the energy storage processes is connected with research on the alkaline earth metals (M) aluminates, MAl_2O_4, activated with lanthanide ions.

In 1966 Lange [10] and Abbruscato in 1971 [11] reported on a bright green luminescence followed by a significant afterglow in $SrAl_2O_4:Eu^{2+}$. Similar results were reported for analogous compositions with Ca and Ba by Blasse and Bril in 1968 [12]. At that time, all those observations were very disappointing, as the afterglow precluded practical application of these phosphors in lighting.

It has taken almost 100 years for the persistence time to extend from minutes to tens of hours. One reason is that it is hard to produce suitable traps in these materials for storing incident energy; another reason is that the nature of the traps and their mechanisms for capturing and releasing energy are complicated and have not been fully understood yet.

In the early time, red LAP was mainly made from rare earth ion-activated sulfides but they are unstable. The afterglow time of red $CaTiO_3:Pr^{3+}$ is too short [13]. Murazaki et al. [14] reported $Y_2O_2S:Eu, Ti, Mg$ which shows better stability and longer red afterglow time above 5 hours, and thereafter the oxysulfides system was studied extensively. The afterglow time of $Y_2O_2S:Eu, Ti$ is above 5 hours [15]. Generally, such oxysulfides are obtained under controlled reducing atmosphere which requires better equipment and complicated technique. Moreover, harmful gases are released in the process of synthesis. In subsequent years, red LAP $Y_2O_3:Eu^{3+}, Mg^{2+}, Ti^{4+}$ [16], $Zn_3(PO_4)_2:Mn^{2+}$ [17], and $MgSiO_3:Mn^{2+}, Eu^{2+}, Dy^{3+}$ [18] were reported.

8.1.2 BASIC MECHANISMS

LLP process can be divided into four parts: (i) Excitation of charged carriers. Upon effective optical charging (UV, vis, or NIR light), electronic implantation (electron beam) or high energy-ray irradiation (X-ray, beta ray, or gamma radiation, etc.), charge carriers (electrons or holes) are generated, with the delocalization and migration of charge carriers. (ii) Storage of charged carriers. The excited carriers are firmly captured on the trapping states, while the storage capacity of trapping states strongly depends on the types and numbers of carriers and defects. (iii) Release of charged carriers. Captured carriers escape from the traps. In addition to the depths of traps in hosts, the release rate of carriers captured in traps also can be influenced by the disturbance in the external fields (thermal, optical, or mechanical disturbance), giving a control of persistent duration. (iv) Recombination process of charged carriers. Though different mechanisms have been suggested, ranging from the basic theoretical models to complex experimental models, there is still controversy in explaining the trapping and detrapping process of carriers. In this part, we review the three most widely accepted models, that is, conduction band (CB)–valence band (VB) model, quantum tunneling model, and oxygen vacancy model. The mechanism for LLP is essentially same as those for thermoluminescence. The basic difference is that in the former traps are filled by radiations like X, β, γ, etc. rays, and in LLP UV/visible light does this job.

Characterization of persistent phosphors is mainly performed by optical spectroscopy methods. Parameters of practical interest, such as the spectral composition of emission as well as its intensity and duration, can be determined from conventional photoluminescence measurements. Fundamental information about energy storage in the material is obtained from thermally stimulated luminescence (TSL) analysis. The concentration and stability of charge carriers accumulated in the trap centers not only determine the intensity and duration of luminescence, but also provide valuable insights into the mechanisms of persistent luminescence.

8.2 OXIDE PHOSPHORS EXHIBITING LONG PERSISTENCE

8.2.1 SIMPLE OXIDE

8.2.1.1 Lu_2O_3

$Lu_2O_3:Tb^{3+}$ was found to exhibit LLP when exposed to light shorter than 330 nm [19]. LLP was observable by the naked eye for a time that depended on Tb concentration. IR-stimulated emission was also observed in this phosphor even after storage period of a year [20]. $Lu_2O_3:R^{3+}$, M (Pr, Hf;

Eu; or Tb, Ca^{2+}) materials show moderate to strong LLP [21–23]. In addition, the lutetia host lattice possesses an exceptionally high absorption coefficient for ionizing radiation. This high stopping power is due to the high density of the material (9.42 g/cm^3). These phosphors are thus suitable as scintillator as well. Thus, there are additional applications such as medical diagnosis imaging. As a drawback, on the top of being expensive, Lu_2O_3-based LLP materials are difficult to prepare. The synthesis involves high temperatures (~1,700°C) in high vacuum or in H_2-N_2.

Synthesis route affects LLP properties a great deal. This was noticed by Zych et al. [24–26]. The samples prepared in air did not produce any thermoluminescence, while those made in vacuum, and especially in a more reducing atmosphere of a forming gas produced a persistent green emission lasting for hours.

Persistent luminescence emission (267 nm irradiation for 3 minutes) of Lu_2O_3:Pr^{3+}, HfIV (0.05 and 0.1 mol.%) is in the red and near-IR region. The spectra are composed of emission bands assigned to $^1D_2 \rightarrow ^3F_2$, $^3H_{4-6}$ and $^3P_0 \rightarrow ^1G_4$, $^3F_{3,4}$, $^3H_5$ transitions of Pr^{3+}. In general, the emission spectra are similar to the UV-excited luminescence ones although the $^1D_2 \rightarrow ^3H_4$ transition has small differences in crystal field fine structure and the intensities of $^1D_2 \rightarrow ^3F_2$, $^3H_{5,6}$ and $^3P_0 \rightarrow ^1G_4$, $^3F_{3,4}$ transitions are higher for persistent luminescence than under UV excitation.

The Lu_2O_3:Eu^{3+} (0.2 mol.%) persistent luminescence spectra show emission assigned to the $^5D_0 \rightarrow ^7F_{0-4}$ transitions of Eu^{3+}.

Lu_2O_3:Tb^{3+}, Ca^{2+} (0.5 and 1.5 mol.%) persistent luminescence spectra also exhibit narrow emission lines assigned to $^5D_4 \rightarrow ^7F_{6-0}$ transitions of the Tb^{3+} ion. Lu_2O_3:Tb^{3+}, Ca^{2+} (0.5 and 1.5 mol.%) prepared by MASS with H_3BO_3 flux yielded the highest intensity and longest duration of persistent luminescence. The persistent luminescence of this material can achieve 8 hours, although qualitatively the persistent luminescence was detected until 30 hours by a charge-coupled device (CCD) camera [27].

It was postulated that in Lu_2O_3:Tb, M (M=Ca, Sr, Ba) ceramics oxygen vacancies (VO″) were involved in the energy trapping. Charge compensating codopant decreased vacancy concentration and LLP intensity. Properties of Lu_2O_3-based LLP phosphors and the mechanism of LLP in the same are nicely reviewed by Kulesza et al. [23].

8.2.1.2 SnO_2

Tin oxide (SnO_2) is an outstanding optical host material for its wide band gap (Eg = 3.6 eV at 300 K), good conductivity, and transparency in the visible spectrum. Its electrical conductivity is dependent on the concentration of oxygen ions adsorbed on its surface [28]. It is already known for various applications such as gas sensing [29], hydrogen generation [30], photocatalyst [31], supercapacitor electrode [32], in flexible photonics [33], latent fingerprint detection [34], etc. Luminescence of Sm^{3+} in this host had been studied frequently [35,36]. The role of SnO_2 as Sm^{3+} luminescence sensitizers was studied and reported in several works on SnO_2: Sm^{3+} nanoparticles [37–39]. Feng et al. [40] developed this phosphor as a LLP material by using Zr^{4+} codopant. The afterglow lasted for about 542 seconds. Zhang et al. [41] also had observed similar effect of Zr^{4+} in inducing LLP in SnO_2:Sm^{3+}. Zr^{4+} induces some shallow traps which are negatively charged. These were tentatively attributed to tin vacancies. The depth of the traps was calculated as 0.61 eV, which is very suitable for LLP.

8.2.2   Aluminate Family

8.2.2.1 $SrAl_2O_4$:Eu^{2+}, Dy^{3+}

Schulze and Muller-Buschbaum in 1981 reported [42] the monoclinic structure of $SrAl_2O_4$ for the first time. "The $SrAl_2O_4$ host crystallizes in the stuffed tridymite type of structure. $SrAl_2O_4$ belongs to monoclinic structure with space group $P2_1$. The three-dimensional network consists of corner-sharing AlO_4 tetrahedron containing large voids, in which the Sr^{2+} ions locate on two types of nine-fold coordinated sites which differ only due to slight distortion of their square planes. Average distances of these nine oxygen ions are 2.8776 and 2.8359 Å for Sr1 and Sr2 respectively [43]." Aluminum forms AlO_4 tetrahedra. $SrAl_2O_4$:Eu^{2+} is an established green-emitting material with emission maximum at 520 nm. This phosphor was rigorously investigated during 1960s and early 1970s. The interest stemmed from uses in lamp and cathode-ray tube (CRT) [11]. Apart from green emission, there is another emission of lesser strength in the blue region at about 435 nm which has low temperature for thermal quenching.

Various hypotheses have been proposed by different researchers for explaining the two emission bands in $SrAl_2O_4$:Eu^{2+}. These are nicely summed by Botterman et al. [44]. Most obvious explanation is that the two emission bands arise from two Eu^{2+} centers that are at two non-equivalent strontium sites in the host. Differences in the coordination environment will change the crystal field strength and centroid shift and hence the emission band positions. Howsoever logical it may seem, but several authors discarded this view arguing that the coordination environment of two sites do not differ as mush so as to explain the large energy difference observed in the two emission bands [45]. Another possible explanation was offered by Poort et al. [46]. They postulated that "the two emission bands result from a possible preferential orientation of the d orbitals of Eu^{2+} on Sr sites that appear to line up." Clabau et al. [47] suggested that "the blue emission band arises from the charge transfer from the fundamental level of the $4f^7$ configuration of Eu^{2+} to the VB and is associated to a hole detrapping mechanism." Holsa et al. [43] explained "the blue emission band as anomalous low-temperature luminescence and proposed that it originates from a higher Eu^{2+} 5d state that may be observed due to the absence of high energy lattice vibrations at low temperatures."

Palilla et al. [48] had noticed "that $SrAl_2O_4$:Eu^{2+} phosphor exhibits initial rapid decay followed by long persistence at very low light level." In 1996, Matsuzawa et al. [49] attempted increasing decay time and phosphorescence intensity by introducing traps (Figure 8.1). They succeeded in this endeavor by introducing Dy^{3+} as a codopant. "The phosphorescence of this new phosphor was so bright and long-lasting that it can be perceived almost throughout the entire night." The intensity and persistence of the phosphorescence of this phosphor were an order of magnitude more as compared to the ZnS:Cu, Co which was prevailing commercial phosphor at that time. Later, afterglow time could be increased by optimizing the composition, for example, by codoping with boron [50,51]. Optimum performance was obtained by using 1 mol.% Eu and 2 mol.% Dy. The emission

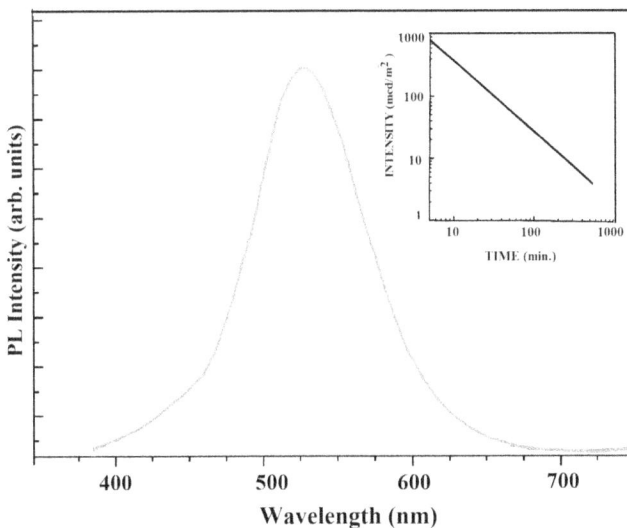

FIGURE 8.1 Long-lasting emission in SrAl2O4:Eu, Dy. Inset shows double log plot of phosphorescence intensity and time.

color is due to Eu^{2+}. The green color is best suited for the long-lasting phosphorescence (LLP) applications, as the eye sensitivity is maximum for this color. While considering the mechanism of LLP, it was suggested "that Dy^{3+} codopant introduces hole-trapping levels at the optimal depth of 0.65 eV to show the long phosphorescence at room temperature. The mechanism involves photoconductivity due to holes and by the trapping and thermal release of holes. It is shown that Dy^{3+} ion introduced as an auxiliary activator acts as the hole trap in the system and creates the highly dense trapping level located at a suitable depth in relation to the thermal release rate at room temperature."

This discovery "marked the beginning of the second generation of persistent or afterglow phosphors, also called glow-in-the-dark materials [52]. Although many new persistent phosphors have been reported since then [8], rare earth-doped strontium aluminates are still by far the most studied and commercialized compounds. $SrAl_2O_4$:Eu, Dy is widely used in various glow-in-the-dark applications, ranging from write-in-the-dark papers [53], road markings [54–56], emergency signage, and watch dials to toys."

Although the LLP material $SrAl_2O_4$:Eu, Dy has more than 25 years of history, the mechanism of LLP is still far from fully understood. Especially, detailed knowledge of the trapping and detrapping processes is still lacking.

First of all, it is still unclear whether the traps are related to host defects, the trivalent codopant, or a combination of both. Different opinions also exist on the trapping and detrapping routes, regarding the role of the CB and whether tunneling processes also play a role [57]. Matsuzawa et al. [49] had suggested that irradiation causes redox reactions

$$Eu^{2+} + Dy^{3+} \rightarrow Eu^+ + Dy^{4+}$$

During phosphorescence, reverse reaction takes place,

$$Eu^+ + Dy^{4+} \rightarrow Eu^{2+} + Dy^{3+}$$

Phosphorescence is thus characteristic of Eu^{2+} while Dy^{3+} acts as a hole trap. However, the species Eu^+ and Dy^{4+} appear too artificial and never been observed. Aitasalo et al. [58,59] suggested that the traps are various lattice defects. However, the role of Dy^{3+} remains unexplained. Hence, a different kind of redox reaction was suggested next [60].

$$Eu^{2+} + Dy^{3+}\, (\text{irradiation}) \rightarrow Eu^{3+} + Dy^{2+}$$

During phosphorescence, reverse reaction takes place,

$$Eu^{3+} + Dy^{2+} \rightarrow Eu^{2+} + Dy^{3+}$$

Subsequently, experiments were performed to detect the valence changes. $Eu^{2+} \rightarrow Eu^{3+}$ change was detected; however, no direct evidence could be obtained for $Dy^{3+} \rightarrow Dy^{2+}$. Hence, Clabau [47] suggested involvement of lattice defects along with $Eu^{2+} \leftrightarrow Eu^{3+}$ redox reactions. The debate is far from over. Various models have been reviewed in details recently by Vitola et al. [61].

Besides commercial applications [62], LLP $SrAl_2O_4$:Eu, Dy have also been envisaged for in vivo imaging applications [63–65], retinal tissue repair [66], latent fingerprint detection [67]. Green emission that can be excited by nUV/blue light can also be useful in developing LED lamps. Phosphorescence comes handy for reducing the flickering [68].

$SrAl_2O_4$:Eu^{2+}, Dy^{3+} films with reproducible luminescent signals can be used in phosphorescent devices for thermometry requiring low power. Moreover, such devices are environment friendly. They play a major role in developing infrastructure required for community safety [69].

8.2.2.2 $Sr_4Al_{14}O_{25}$:Eu^{2+}, Dy^{3+}

Smets et al. [70] were among the early workers who studied $Sr_4Al_{14}O_{25}$:Eu^{2+} phosphor. However, they did not report LAG in this phosphor. Akiyama et al. [71] were the first to prepare $Sr_4Al_{14}O_{25}$:Eu^{2+}, Dy^{3+} phosphor during their quest for photostimulable phosphors for X-ray imaging. They also made a reference to long-lasting phosphorescence (LLP) in this material. Lin et al. [72] developed the phosphorescent material based on this composition and published a systematic report. Dy^{3+} was used as a codopant. UV exposure resulted in LLP. They suggested the following mechanism for LLP in which Dy^{3+} acts as a hole trap. UV exposure creates holes which are trapped at Dy^{3+}, creating Eu^+-Dy^{4+} pairs. When excitation is removed, holes are gradually released into VB. They recombine with Eu^+, converting it to Eu^{2+}, and Eu^{2+} emission is observed. Subsequently, they studied effect of Eu^{2+} concentration on LLP intensity [73]. The composition $Sr_{3.968}Eu_{0.012}Dy_{0.02}Al_{14}O_{25}$ gave highest LLP. The phosphorescence intensity was above detection threshold (0.32 mcd/m^2) for more than 20 hours. Yuan et al. [74] investigated the influence of Sr/Al ratio on LLP properties. Compositions rich in Al have better long phosphorescence in terms of emission intensity and decay time as compared to the Sr-rich phosphors. The improvement was attributed to the presence of different trap levels and trap concentrations. Al:Sr ratio 3.7 yielded highest intensity and longest decay time for LLP. On the other hand, Suriyamurthy et al. [75] found that "photoluminescence intensity of both the strontium-deficit and strontium-rich phosphors was enhanced, but no definite correlation was observed between the afterglow intensity and non-stoichiometry. Substitution of strontium by calcium resulted in white afterglow emission at higher calcium concentration. The emission centers in case of photoluminescence and afterglow emission also appeared to be different. Addition of silver significantly enhanced the afterglow intensity due to increased trap density." Nakazawa et al. [76] studied all the lanthanide as codopants in $Sr_4Al_{14}O_{25}$:Eu^{2+} phosphor and their effect on LLP properties. Codoping with Pr, Nd, Dy, Ho, and Er was found to lead to the LLP. Almost no LLP was observed for Ln = Y, La, Ce, Sm, Eu, Gd, Tb, Tm, Yb, and Lu. They observed afterglow when these samples were exposed to 365 nm UV light. These studies enabled them to suggest a better mechanism of LLP. The theoretically estimated activation energies on the basis of hole trapping by Ln^{3+} model did not agree with the experimental results; hence, this model is not correct. Next, they considered hole-trapping model of divalent lanthanide ions. These predictions are consistent with the results. However, better agreement was reached with "the electron trapping by trivalent rare earth ions model." According to this model, when a 4f electron is excited by absorbing 365 nm light to the states of the $4f^65d^1$ configuration of the Eu^{2+} ion, the excited electron is donated to the neighboring ion of another lanthanide. This takes place via the composite state of the 5d states of the Eu^{2+} and Ln ions. Some of the Eu ions which have transferred the electron to the trapping lanthanide ions, that is, may create holes into the VB and induce photoconductivity. During the afterglow, the path is retraced; the trapped electrons are thermally released into the transition state. This has associated activation energy of about 0.8 eV. From the transition state, electron can recombine radiatively with the hole in the 4f state of Eu ions or get retrapped. The retrapping prolongs the phosphorescence decay.

Zhao et al. [77] attempted combustion synthesis of this phosphor. However, the results indicate that there is $SrAl_2O_4$ also in the synthesized compound. Later, Sharma et al. [78] successfully prepared the phosphor using this method. They suggested using this phosphor for "latent fingerprint detection for application in forensic science." Various other applications have also been suggested in recent years. Excitation spectrum extending into violet region makes it possible to use this phosphor for enhancing photovoltaic conversion efficiency of the CdS quantum dot-sensitized solar cell [79]. Application as a photocatalyst had also been claimed [80]. Degradation of toxic materials [81] such as organic dyes [82] to some harmless form can be achieved using photocatalysis with LLP materials. Carlos et al. [83] found enhancement in photocatalytic activity due to Bi^{3+} codoping. However, this is detrimental for LLP.

8.2.3 Stannate Family

Metal stannates ($M_xSn_yO_z$, where M = Mg, Ca, Sr, Ba or Zn, x = 1 or 2; y = 1; z = 3 or 4) represent a promising group of oxides with tunable persistent luminescence characteristics. Luminescence wavelength can be tailored via the introduction of appropriate emission centers. Stannate hosts doped with optically active ions such as Sm^{3+}, Eu^{3+}, Pr^{3+}, Tb^{3+}, Nd^{3+}, Dy^{3+}, etc., have been reported to exhibit efficient persistent luminescence in various spectral regions. In addition, the optimal dopant ion content is typically miniscule (around 0.1%) in these hosts. Some dopants, for example, Gd^{3+}, Tb^{3+}, Yb^{3+}, and Lu^{3+}, stimulate $Sn^{4+} \rightarrow Sn^{2+}$ reduction of the host ions, enabling long-lasting cyan luminescence of the material. The concentration of trap centers and their activation energies determine the intensity and duration of persistent luminescence. Both of these key performance parameters can be optimized during synthesis via the introduction of codopants which either serve as trap centers or stimulate the formation of intrinsic defects in the lattice. In this regard, stannates are especially interesting due to the heterovalent nature of tin ions; that is, they can be stabilized in several oxidation states, and serve both as emission as well as trap centers. Moreover, these complex oxide materials exhibit high thermal and chemical stability, are relatively inexpensive and simple to prepare, and, thus, are attractive candidates for applications.

8.2.3.1 Ca_2SnO_4:RE^{3+}

Ca_2SnO_4 itself exhibits intrinsic luminescence. Doping by La converts it into a cyan-emitting LLP phosphor [84]. LLP capability of stannates was pointed out as back as 1991 by Ropp [85]. Lei et al. [86] reported phosphorescence lasting for more than 1 hour in Ca_2SnO_4:Sm^{3+}. Around same time, Ju et al. [87] reported detailed study of LLP in Sm^{3+}-activated Ca_2SnO_4. The afterglow time is more than 7 hours which was the longest red or reddish orange LAG phosphor at that time. The defects V_{Ca} and structure defects Sn^{3+} and Sn^{2+} may act as hole-trapping centers, while the defects Sm_{Ca} may act as electron-trapping centers, and trapping centers play an essential role for photoenergy storage in persistent phosphors. Codoping by Gd^{3+} increases the number of shallow traps of charges which enhance the intensity of persistent luminescence [88]. The concept of using Gd^{3+} codoping for enhancing LLP was used for Ca_2SnO_4:Eu^{3+}. Decay time could be extended from about 20 minutes to close to 2 hours [89]. Apart from Eu and Sm, Pr^{3+} also gives red emission and LLP in Ca_2SnO_4:Pr^{3+} was soon discovered [90]. However, the persistence time was only 20 minutes. Pr^{3+} at Ca^{2+} site was suggested as an electron trap. Green-emitting Ca_2SnO_4:Tb^{3+} LLP was reported around the same time [91]. The decay time is about 3 hours. Two kinds of traps were identified from thermoluminescence experiment. The one at lower temperature was assigned to defects Tb^{3+} at Sn^{4+} site and Ca^{2+} vacancies, and the one at higher temperature was attributed to Sn^{3+}, Sn^{2+} which were regarded as hole-trapping centers. Following this, a yellow-emitting Ca_2SnO_4:Dy^{3+} LLP was described by Shi et al. [92]. The negative defects such as cation vacancies were assigned the role of hole-trapping centers, while the positive defects like Dy^{3+} at Ca^{2+} site the electron-trapping centers. The decay time of only a few minutes limits the utility of this phosphor. Similar short persistence was observed in green-emitting Ca_2SnO_4:Er^{3+} [93]. Tm^{3+} which yields persistent blue emission, on the other hand, possesses a decent decay time of 5 hours [94].

8.2.3.2 Mg_2SnO_4:Ti^{4+} and Related

Mg_2SnO_4 is a chemically stable host with a cubic inverse spinel crystal structure that belongs to the space group Fd3m. The unit cell has 96 cation sites with 24 being occupied by the cations; 8 of 64 tetrahedral sites by Mg^{2+}, and 16 of 32 octahedral sites by other Mg^{2+} and Sn^{4+}. Mg_2SnO_4 shows good host-related photoluminescence, afterglow, and photostimulated luminescence [95]. In addition to the large number of unoccupied sites, the degree of cation disorder can be considerable (as large as 30%), so that a large number of defects can be created. Some of these defects serve as electron or hole traps. For example, an Mg^{2+} at Sn^{4+} site can trap a hole; while a Sn^{4+} at an Mg^{2+} site will form an electron trap. Oxygen vacancies may also serve as F-center-like electron traps. Therefore, this oxide matrix is expected to be a good candidate for LLP phosphor.

Mg_2SnO_4:Mn^{2+} was the first stannate phosphor to be developed as LLP [96]. The phosphorescence can be observed by the naked eyes (0.32 mcd/m²) in the dark clearly for over 5 hours after the 5 minutes of UV irradiation. Oxygen vacancies act as electron traps and Sn vacancies or Mg^{2+} at Sn^{4+} site as hole traps.

Mg_2SnO_4 exhibits green photoluminescence and persistent luminescence, which originate from oxygen vacancies. When Ti^{4+} ions are doped, an interesting Mg_2SnO_4:Ti^{4+} phosphor with bluish white photoluminescence under ultraviolet irradiation and with green persistent luminescence was obtained. Two emission centers exist in Mg_2SnO_4:Ti^{4+}. The centers responsible for the green emission are considered to be the F centers (oxygen vacancies), and the blue centers are the TiO_6 complex. Trap clusters in the band gap with different depths, such as $[Sn^{\bullet}_{Mg}-O''_i]$, $[Sn^{\bullet\bullet}_{Mg}-V'_O]$, $[Sn^{\bullet\bullet}_{Mg}-V^{\times}_O]$, and Mg''_{Sn}, correspond to the components at 85°C, 146°C, and 213°C of the thermoluminescence curve [97].

Zhang et al. [98] reported a green LLP in Mg_2SnO_4 which could be considerably enhanced by Eu^{3+} doping. Eu^{3+} doping introduces vacancies which are part of shallow traps responsible for LLP.

8.2.3.3 Zn_2SnO_4:Cr^{3+} and Zn_2SnO_4:Eu^{3+}

Zn_2SnO_4 also has inverse spinel structure. Coupled with the stannate cation, persistent luminescence behavior was anticipated and explored. In 2015, Yang Li et al. developed an NIR phosphorescent Zn_2SnO_4:Cr^{3+}. They obtained a broad emission band from 650 to 1,200 nm that peaks at 800 nm and a persistence time of > 35 hours [99]. NIR emission is considered suitable for bioimaging applications. Codoping with Al enables manipulating emission wavelength and persistence time. A detailed analysis of the results has been provided by Taktak et al. [100]. Fan et al. [101] studied the influence of different codopants M^{3+} (M^{3+} = Tm^{3+}, Ga^{3+}, Pr^{3+}, Yb^{3+}, Nd^{3+}) on LLP properties of Zn_2SnO_4:Cr^{3+} and concluded that Tm^{3+} is the best choice.

Zn_2SnO_4:Eu^{3+} phosphors displayed an orange afterglow [102]. The decay time is just 4 seconds. Oxygen vacancies are part of the traps responsible for the phosphorescence. The photoluminescence and persistent luminescence decay properties of Cr^{3+}, Eu^{3+}-codoped samples were reported by Zhang et al. [103]. The photoluminescence and photoluminescence excitation spectra revealed that the NIR emission (centered at 800 nm) of Cr^{3+} was significantly enhanced after the incorporation of Eu^{3+} ions, which was attributed to the energy transfer from Eu^{3+} ions to Cr^{3+}. Moreover, the incorporation of Eu^{3+} ions could effectively delay the decay rate and enhance the NIR persistent luminescence. The analysis of persistent decay curves and thermoluminescence spectra indicated that the Eu^{3+} ions acting as effective deep-trap centers contributed to improved Cr^{3+} NIR persistent luminescence. Emission is characteristic of Cr^{3+}. Eu^{3+} codoping doubled the intensity of persistent emission and the decay time as long as 17 hours was observed. Redox reaction of $Eu^{3+} \longleftrightarrow Eu^{2+}$ was suggested as the trapping and LLP mechanism.

8.2.3.4 Sr_2SnO_4:RE^{3+}

Xu et al. [104] and Lei et al. [105], almost around the same time, reported LLP in Sr_2SnO_4:Sm^{3+}. The afterglow decay curve of the Sm^{3+}-doped Sr_2SnO_4 phosphor contains a fast decay component and another slow decay one. Due to the presence of the slow decay component, the afterglow can be seen with the naked eye in the dark clearly for more than 1 hour after removal of the excitation source. Effect of Eu codoping and the trap structure using thermoluminescence glow curves were explored by Wang et al. [106]. Marginal improvement in LLP properties was also observed after codoping by alkali atoms [107]. The codoping of alkaline ions (K^+, Na^+) increased not only the initial trap concentration, but also the depth of electron traps, which can be regarded as the main reasons for the enhanced afterglow.

Enhancement in intensity and decay time was also obtained by codoping with Dy^{3+} [108]. In particular, the afterglow time could be extended beyond 1 hour. The enhancement arises from the thermal activation of holes which are trapped in the hole trap levels created by the Dy^{3+}. The increase of the appropriate traps could be related to the creation of Sr vacancy as efficient trapping center.

LLP was observed in Tm^{3+}-doped Sr_2SnO_4. LLP emission in NIR region is also exhibited by this phosphor [109]. The decay lasts only for several minutes. A tentative mechanism of LLP suggested by the authors goes as follows. "Under 254 nm UV light excitation, some electrons from the valence band (VB) of the host are excited into the conduction band (CB), which can be captured by different traps, and holes are formed, which transfer through VB and are trapped in hole traps. Afterward, when we remove the UV light, both the trapped electrons and holes may severally escape slowly to the CB and VB with the help of thermal disturbances at the appropriate temperature and transfer to Tm^{3+} ions in the vicinity by the relaxation of the defect level. Finally, electrons recombine with holes near Tm^{3+} ions, and the energy released from the recombination will excite Tm^{3+} ions to generate intense PPL emission in the Sr_2SnO_4:Tm^{3+} phosphor."

Kamimura et al. [110] discovered a new NIR persistent luminescent material Sr_2SnO_4:Nd^{3+}. The NIR persistent luminescence observed in Sr_2SnO_4:Nd^{3+} ranged from 850 to 1,400 nm with a major peak at around 1,079 nm, which was ascribed to the electron transition of Nd^{3+} ions. The intensity of NIR persistent luminescence, which penetrates through a human finger and can be used to visualize a finger vein pattern by using a charge coupled device camera, reached approximately 1.1 mW/m^2 ($\lambda = 900$ nm) at a decay time of 10 seconds after ceasing the excitation.

8.2.4 GERMANATE FAMILY

Germanate hosts can lead to LLP on account of the low phonon energy and abundant lattice defects [111]. For LPP, gallate and gallogermanate-based host are proved to be suitable hosts. Due to the creation of deep traps, afterglow persists for hours after the stoppage of excitation at room temperature. These are produced due to the structural disorder, antisite defect, oxygen vacancy, and codopant. Wang et al. firstly demonstrated a Bi^{3+}-doped UV PerL germinate $NaLuGeO_4$. After Eu^{3+} codoping, the decay time was prolonged to 63 hours, but half of the emission band extends into the visible region [112]. After that, Shi et al. reported an intense UVA persistent phosphor $LiYGeO_4$:Bi^{3+} with a super-long decay time up to 300 hours. It has the longest persistence time among the reported UV persistent phosphors [113]. Recently, Bi^{3+}-doped $LiScGeO_4$ was developed by Zhang et al. [114] and Zhou et al. [115]. Oxygen vacancies were demonstrated to be effective in $LiScGeO_4$ during the PerL procedure. However, Lyu and Dorenbos opined that Bi^{3+} itself forms hole traps (Bi^{4+}) or electron traps (Bi^{2+}) in $ARE(Si, Ge)O_4$ (A = Li, Na; RE = Y, Lu) by controlling the composition [116,117].

8.2.4.1 $Zn_3Ga_2Ge_2O_{10}$:Cr^{3+}

Pan et al. first reported the Cr^{3+}-doped $Zn_3Ga_2Ge_2O_{10}$ NIR LPP phosphor, a broadband emission range 650–1,100 nm and afterglow time of more than 360 hours [111]. They suggested that the NIR LPP materials can be used as a taggant in night vision surveillance and in vivo imaging applications. Later, Shen et al. investigated the effect of Cr^{3+} doping concentration on its luminescence properties [118]. Wu et al. investigated the luminescence properties by sintering in different surrounding atmospheres and found that the longer afterglow of sample sintered in air compared to reducing atmospheres [119]. Xu et al. improved the emission profile by Ca^{2+} doping and they found a 15-fold enhancement of emission intensity [120]. Zhu et al. synthesized the same materials by hydrothermal method assisted with citric anions and as a result, monodisperse, uniform spherical nanoparticles obtained [121]. The obtained nanoparticles showed NIR afterglow for more than 120 minutes and afterglow signal was detected for 60 minutes when used as a luminescent probe for in vivo imaging. Eu^{3+} codoping increased the persistent luminescence intensity [122]. Similarly, Pr^{3+} codoping [123] resulted in improved performance by creating suitable Zn deficiency. The Sr@Cr-ZGGO also enhances the LLP properties as a result of increasing light traps and generation of new paths between the energy levels of trapping centers [124]. Das et al. applied the materials in night vision surveillance for defense and security applications [125]. King et al. used for detection of latent finger prints [126]. Li et al. used nanoparticles of this phosphor for targeted optical probes for labeling human breast cancer cells [127].

The reason for long red persistent luminescence in $Zn_3Ga_2Ge_2O_{10}$ (ZGGO) as hosting material could be attributed to the presence of tetrahedral and octahedral sites available from the formed solid solution of two crystal structures, spinel $ZnGa_2O_4$ and willemite Zn_2GeO_4.

8.2.4.2 $La_3Ga_5GeO_{14}$:Cr^{3+}

$La_3Ga_5GeO_{14}$ crystal (LGG) belongs to langasite family [128]. The structure of LGG was determined in 1983 [129]. The growth and laser properties of these crystals were also described there; optical and spectroscopic properties were measured by Burkov et al. [130], and optical characteristics were summarized by Veremeichik and Simonov [131]. The recent results on luminescence in LGG: Cr^{3+}, Yb^{3+}, Er^{3+} offer promise for application of the crystals in medicine [132].

Jia et al. discovered NIR LLP in Cr^{3+}-activated LGG. The phosphor with Dy^{3+} codopant is observed to have a persistent IR emission for more than 8 hours [133]. The persistent emission was obtained by irradiation at 254 nm. It can also be observed after charging at a longer wavelength up to 400 nm. Almost at the same time, NIR (660–1,300 nm) long-persistent phosphorescence from Cr^{3+} ions with persistence time of more than 1 hour was reported in $La_3Ga_5GeO_{14}$:Cr^{3+} phosphor by Yan et al. [134]. The NIR phosphorescence can be effectively achieved under UV illumination (~240–360 nm). A probable mechanism of the phosphorescence in LGG:Cr, Zn phosphor under different irradiation energies was given. Under UV irradiation, the ground-state electrons of Cr^{3+} ions are promoted to the $^4T_1(te^2)$ or some energy levels with higher energies. The excited electrons are trapped by the neighboring oxygen vacancies (V_O), which are close to the bottom of the CB and locate in the vicinity of the photogenerated Cr^{4+} ions. These trapped electrons are then thermally released to the photogenerated Cr^{4+} ions, leading to effective Cr^{3+} phosphorescence. For the case of blue light irradiation, the energy is perhaps too low to induce the above phosphorescence process. However, the blue light energy may activate the F centers (i.e., $Cr^{3+}e$-) in the phosphor, which were found to exist in the lanthanum-based crystals. Electrons are thermally released from the F centers to the neighbor traps at the Cr^{4+} levels, leading to persistent emission of Cr^{3+} ions.

In subsequent work, Zhang et al. [135] used for bioimaging-guided in vivo drug delivery to gut.

8.2.4.3 $CaZnGe_2O_6$:Lanthanide or 3d Transition

Redhammer and Roth [136] solved the crystal structure of $CaZnGe_2O_6$ for the first time. It crystallizes in space group C2/c at room temperature and adopts the general structural topology of the clinopyroxenes. This consists of infinite chains of corner sharing GeO_4 tetrahedra running parallel to the c-axis, zigzag chains of edge-sharing M1O6 octahedra (M1 = Zn^{2+}), and eight-coordinate M2 sites hosting Ca^{2+} ions.

Che et al. reported white LLP in $CaZnGe_2O_6$ activated with Dy. After irradiation under 240-nm UV light, $CaZnGe_2O_6$:2% Dy^{3+} emits intense white emission with color coordinate values of x = 0.36, y = 0.39, and its afterglow can be seen with the naked eye in the dark clearly for more than 3 hours after removal of the excitation source [137]. In the same year, Che et al. also discovered LLP in Mn^{2+}-doped $CaZnGe_2O_6$. This phosphor emitting in the red spectral region had decay time almost same as that for Dy^{3+}-doped phosphor [138]. Ye et al. observed enhancement in the LLP properties by codoping Bi^{3+} [139]. On the other hand, Bi^{3+} is also an effective dopant for imparting LLP property to $CaZnGe_2O_6$ [140]. This phosphor shows a broad emission band from 300 to 700 nm, a bright persistent luminescence over 12 hours after ceasing the excitation and the photostimulated luminescence under the excitation of visible to near-infrared light. Kang et al. investigated LLP by doping with Cr^{3+} for bioimaging application [141]. After being exposed under a 254 UV lamp for 30 minutes, a LPP luminescence with broad emission band at 800 nm is detected from $CaZnGe_2O_6$:0.5% Cr^{3+} powder. The persistent emission could be detected for > 3 hour. By codoping with Mn^{2+}, tunable LLP could be observed. Liu et al. used Tb^{3+} activator to induce LLP [142]. Intense green LLP could be observed for >3 hour after UV irradiation for 5 minutes.

8.2.4.4 LiLuGeO$_4$:Bi^{3+}, Yb^{3+}

LiLuGeO$_4$ compound crystallizes in an orthorhombic system with the Pnma (62) space group and cell parameters of a = 11.0226, b = 6.2315, c = 5.0426 Å. The host lattice forms an olivine-type structure. [LiO6] octahedra and [GeO4] tetrahedra together form parallel layers with [LuO6] octahedra inserted between the layers. The average bond lengths for [LiO6], [LuO6], and [GeO4] are 2.235, 2.241, and 1.753 Å, respectively. Compared to the ionic radius of Li$^+$ (r = 0.76 Å, CN = 6) and Lu^{3+} (r = 0.861 Å, CN = 6), it is more favorable for Bi^{3+} (r = 1.03 Å, CN = 6) to occupy the Lu^{3+} site. Furthermore, Lu^{3+} and Bi^{3+} share the same trivalent valance state, which also supports the replacement of Lu^{3+} by Bi^{3+}. Bi^{3+} is an efficient activator for producing UV light. UV LLP of LiLuGeO$_4$:Bi^{3+} (LLG) could be used for antibacterial activity and wound healing by coupling with TiO$_2$. The phosphor yields UV LLP for more than 100 hours. UV light can be effectively absorbed by TiO$_2$. Through photocatalysis, TiO$_2$ generates reactive oxygen species (ROS) needed for the destruction of bacteria. Because of the persistent antibacterial ability of LLG/T, hydrogel loaded (H@LLG/T) effectively promoted wound repair [143]. Cai et al. [144] suggested LiLuGeO$_4$:Bi^{3+}, Yb^{3+} as an in situ UVA photodynamic therapy (PDT) light source in targeted cancer treatments. They further generalized the conclusion that LiLuGeO$_4$:Bi^{3+}, Yb^{3+} is a prominent persistent UVA-emitting candidate to trigger photosensitizers or fluorescent probes when used for biophotonic applications.

8.2.5 Silicate Family

8.2.5.1 Sr$_2$MgSi$_2$O$_7$:Eu^{2+}, Dy^{3+}

Since the work by Matsuzawa et al. on aluminates, many other compounds based on the simultaneous doping of Eu^{2+} and Dy^{3+} were found to exhibit intense persistent luminescence, and these include M$_2$MgSi$_2$O$_7$ (M = Sr, Ba) [145]. Sr$_2$MgSi$_2$O$_7$:Eu was first studied by Poort et al. in 1996 [146]. They have not reported any LLP. LLP in Dy-codoped material was first reported by Lin et al. However, the slowest component had decay time of about only 1 minute. Emission was cyan, characteristic of Eu^{2+}, Dy^{3+} acting as a trap. In a subsequent work, the decay time improved to about 5 minutes and the phosphorescence reached the human detection limit of 0.32 mcd/m^2 after about 5 hours [147]. Dorenbos discussed various possible mechanisms for LLP in this material [60,148]. This was further elaborated by Setlur et al. [149]. They also studied other codopants like Gd and Sm. Xiao et al. [150,151] synthesized Sr$_2$MgSi$_2$O$_7$:Eu, Dy phosphor, and its afterglow lasts till 20 hours. This also became available commercially (Yada Luminescent Co.Ltd., YD-#7506). Glass ceramics were also successfully prepared using this phosphor retaining the LLP properties [152]. Partial replacement of Sr by Ca retained the long-lasting properties, shifting the emission to longer wavelength at the same time [153]. Jia et al. studied various phases in SrO-MgO-SiO$_2$ system and found that the phosphorescence time depends strongly on the phase. They used Nd as a codopant to induce LLP [154]. Efforts were made to improve LLP properties by different codopings such as Er^{3+}, Ho^{3+}. Er^{3+} codoping was claimed to have LLP properties even better than Dy^{3+} [155]. Part of Ho^{3+} added was reduced to divalent form and it resulted in three-fold increase in LLP intensity [156]; however, decay became faster at the same time which is not desirable.

The Sr$_2$MgSi$_2$O$_7$:Eu^{2+}, Dy^{3+} persistent luminescence mechanism can be described as following [157]: (i) the excitation energy can promote an electron from the 4f^7(^{8}S$_{7/2}$) ground state to 4f^{6}5d^1 excited states of the Eu^{2+} ion; (ii) the electron is transferred to the conduction band (CB) followed by its storage in electron traps (Dy^{3+} ions or oxygen vacancies); (iii) absorbing thermal energy, the electron can return to the CB, thus recombining with Eu^{2+} excited 4f^{6}5d^1 levels; (iv) lastly, the electron radiative decays to the ground-state 4f^7(^{8}S$_{7/2}$) of the activator ion leads to the emission of light.

8.2.5.2 Sr$_3$MgSi$_2$O$_8$:Eu, Dy

Sr$_3$MgSi$_2$O$_8$, was first studied in 1957 for its structural properties by Klasens et al. [158]. Sr$_3$MgSi$_2$O$_8$ has monoclinic crystal structure [159]. Later, phosphors based on the Me$_3$MgSi$_2$O$_8$ (Me = Ca, Sr, Ba) formulae were studied. A shift in emission from blue to green region depending on the size of Me^{2+}

ion was observed. Subsequently, these phosphors have been explored for various applications such as UV excitable white light-emitting diodes (LEDs), plasma display panels, and dosimetry application [160]. After success of $Sr_2MgSi_2O_7$ Eu, Dy, as a LLP material, Lin et al. explored $Ca_3MgSi_2O_8$ host and discovered LLP lasting for more than 5 hours. Better phosphor in form of $Sr_3MgSi_2O_8$:Eu, Dy was discovered in 2003 [161]. Alvani et al. improved the phosphorescence time to 10 hours [162]. $Sr_3MgSi_2O_8$:Eu^{2+}, Dy^{3+} afterglow NPs were used in order to conjugate with variety of biomolecules, to qualitatively investigate cancer cells targeting. Cellular uptake and targeting of luminescent NPs could be enhanced, by conjugation to variety of biomolecules which are specific for breast and prostate cancer [163].

8.2.5.3   MgSiO$_3$:Eu^{2+}, Mn^{2+}

LLP in $MgSiO_3$:Mn^{2+}, Eu^{2+}, Dy^{3+} was reported by Wang et al. [164]. Persistence time over 4 hours was observed. The function of Dy^{3+} is to create hole-trapped centers in the $MgSiO_3$:Mn^{2+}, Eu^{2+}, Dy^{3+} system, where the Mn^{2+} received energy from the excitation of the electron–hole pairs. The excitation spectrum of $MgSiO_3$:Mn^{2+} consists of several bands in the range of 300–500 nm, attributed to the transitions $^6A_1(^6S) \rightarrow ^4T_1(^4P)$ (317 nm), $^6A_1(^6S) \rightarrow ^4E(^4D)$ (347 nm), $^6A_1(^6S) \rightarrow ^4T_2(^4D)$ (364 nm), $^6A_1(^6S) \rightarrow ^4A_1$ (^{4} G) (413 nm), and $^6A_1(^6S) \rightarrow ^4E(^4G)$ (426 nm), according to the Tanabe-Sugano diagram. The emission spectra of Mn^{2+}-doped $MgSiO_3$ samples consist of a broad band peaked at about 650 nm, which is attributed to the $^4T_1(^4G) \rightarrow ^6A_1(^6S)$ transitions of Mn^{2+} ions [165]. A detailed model of LLP process was presented by Lin et al. [166]. They proposed that Eu^{2+} and Dy^{3+} ions act as sensitizers and trap centers, respectively. Emission comes from Mn^{2+}. They further postulated that Dy^{3+} at Mg substitutional site is an electron trap. Mn^{2+} ions are mainly excited through the effective energy transfer via Eu^{2+} under 254 nm excitation. Under this excitation, a few electrons at the Mn^{2+} ground state can be directly ionized to the CB. In contrast, the electrons excited to the Mn^{2+} excited states due to the energy transfer of $Eu^{2+} \rightarrow Mn^{2+}$ can also be stimulated into the CB. Afterward, these electrons in the CB can be captured by the electron-trapped centers (Dy^{3+} at Mg substitutional site). Thus, the codoped Eu^{2+} ions cannot only raise emission intensity of Mn^{2+}, but also increase the sum of trap centers capturing electrons, that is, postpone the afterglow of Mn^{2+}.

In a later work, Zhu and Ge [167] studied LLP properties of $MgSiO_3$:Mn^{2+}, Nd^{3+}. However, afterglow time of $MgSiO_3$:Mn^{2+}, Nd^{3+} was only 15 minutes as against 4 hours of $MgSiO_3$:Mn^{2+}, Eu^{2+}, Dy^{3+}. Recently, $MgSiO_3$ has also been studied as a photocatalyst [168].

8.2.5.4   Sr$_3$Al$_{10}$SiO$_{20}$:Eu^{2+}, Ho^{3+}

Compound $Sr_3Al_{10}SiO_{20}$ was first studied as a luminescent material by Kubota et al. [169]. Its crystal structure was solved by Rief and Kubel [170]. Kuang et al. studied LLP in $Sr_3Al_{10}SiO_{20}$:Eu^{2+}, Ho^{3+} [171]. The phosphor emits blue light and shows long-lasting phosphorescence after it is excited with 254/365 nm ultraviolet light. The LLP lasts for nearly 6 hours. Ho^{3+} doping introduces trap levels responsible for long-lasting emission. Eu^{2+} is the luminescence center. Ho^{3+} at Sr substitutional site can produce Sr vacancies. These can act as traps. Kuang et al. further studied effect of various lanthanides as codopant on LLP properties [172]. However, none surpassed Ho^{3+} in performance.

8.2.5.5   CdSiO$_3$:Tb^{3+}

$CdSiO_3$ is a novel self-activated phosphor [173]. When rare earth ions such as Y^{3+}, La^{3+}, Gd^{3+}, Lu^{3+}, Ce^{3+}, Nd^{3+}, Ho^{3+}, Er^{3+}, Tm^{3+}, and Yb^{3+} are introduced into the $CdSiO_3$ host, one broadband centered at about 420 nm resulted from traps can be observed. In the case of other earth ions which show emissions at the visible spectrum region, such as Pr^{3+}, Sm^{3+}, Eu^{3+}, Tb^{3+}, and Dy^{3+}, the mixture of their characteristic line emissions with the ~420 nm strong broadband luminescence results in various emitting colors. As a consequence, different emitting colors can be attained via introducing certain appropriate active ions into the $CdSiO_3$ matrix. In addition, this kind of phosphors shows good long-lasting properties when excited by UV light. The white afterglow from the $CdSiO_3$:Dy^{3+} samples lasted for more than 5 hours after the removal of the 254-nm UV light [174].

8.2.6 Phosphate Family

8.2.6.1 $Sr_2P_2O_7$:Eu^{2+}, Y^{3+}

Eu^{2+}-activated strontium pyrophosphate had been known as an efficient blue phosphor used in lamps for phototherapy. It was firstly reported by Wanmaker and ter Vrugt [175], but no long-lasting phenomenon of $Sr_2P_2O_7$:Eu^{2+} was reported. When Y^{3+} ions were added into $Sr_2P_2O_7$:Eu^{2+} to serve as charge traps, a blue-emitting LLP phosphor resulted [176]. LLP could be observed by naked eyes even 8 hours after the excitation source was switched off. Emission is characteristic of Eu^{2+}, Y^{3+} which provides traps. A tentative mechanism was suggested as "defects were generated by non-equivalent substitution, i.e. two Y^{3+} ions substituted three Sr^{2+} ions and created two $Y^{\cdot}_{Sr}$ positive and a V''_{Sr} negative defects, which act as electron and hole traps, respectively. In the process of excitation, electrons were pumped to the 5d orbit of Eu^{2+}, part of which were captured by positive charge defects Y_{Sr}. Under the thermal activation, these trapped electrons escaped from the traps and transferred to Eu^{2+} followed by the characteristic emission of Eu^{2+}."

8.2.6.2 $Ca_9Bi(PO_4)_7$:Eu^{2+}, Dy^{3+}

The whitlockite (or β-$Ca_3(PO_4)_2$) compounds are characterized by multi-cation sites. $Ca_9A(PO_4)_7$ (A = Al, Sc, Y, Ln, etc. trivalent metal) compounds are materials among the whitlockite family with remarkable properties. Recently, the $Ca_9Bi(PO_4)_7$ activated with rare earth ions was reported as LLP phosphor. The host was reported by Golubev and Lazoryak and shown to be isostructural with β- $Ca_3(PO_4)_2$ [177]. Jia et al. developed this as a LLP material with doping Eu^{2+}, Dy^{3+} [178]. The bluish green phosphorescence of CBP:0.06Eu^{2+}, 0.02Dy^{3+} could be observed for 5 hours by naked eyes after the excitation source was switched off. The LLP emission arises from the 5d–4f transition of Eu^{2+}. Dy^{3+} introduces traps required for LLP.

8.2.6.3 $SrZn_2(PO_4)_2$

$SrZn_2(PO_4)_2$ was first prepared wayback in 1961 as a phosphor host incorporating Sn^{2+} [179]. This compound has good thermal and chemical stability. In 1990, Hemon and Courbion [180] determined the crystal structure of $SrZn_2(PO_4)_2$ of a Hurlbutite type in which each ZnO_4 tetrahedron and four PO_4 tetrahedra share with four oxygen atoms to form a network structure. Luminescence of various activators such as Eu^{3+} [181], Bi^{3+} [182], Sm^{3+} [183], Tb^{3+}, etc. has been studied in subsequent years. LLP in this host was only recently reported with Dy^{3+} doping [184]. It exhibits yellowish green LPP emission with decay time of about 6 hours. Authors suggested the following tentative mechanism of LLP. "Electrons in the V''_{Zn} defects with excessive negative charges are excited by UV light, generating holes which will be caught by negatively charged V'' Zn defects to form V'_{Zn} or V_{Zn}. Some of the excited electrons are captured and stored by positively charged V_O defects to form V_O or V_O. The pairs of electron–hole keep migrating in the material lattice until they meet and recombine with the release of photons. After the Dy_{Zn} defect catches an electron to become Dy_{Zn} in company of the release of a photon, the Dy_{Zn} is reduced to be $Dy^{\cdot}_{Zn}$ by capturing a hole. Therefore, the decay time of phosphors is dependent on the lifespan of the electron–hole pairs in the crystal."

8.2.7 Gallate Family

At present, gallium salt has become a suitable matrix for LPL phosphors because of its low synthesis temperature, good stability, and excellent electrical conductivity. Several gallate-based LLP phosphors have been reported, such as $SrLaGaO_4$:Tb^{3+} [185], $Ba_2Ga_2GeO_7$:Tb^{3+}, Bi^{3+} [186], $SrGa_{12}O_{19}$ [187], $MgGa_2O_4$ [188,189], $Y_3Ga_5O_{12}$:Bi^{3+} [190], $Y_3Al_2Ga_3O_{12}$:Cr^{3+} [191], $CaGa_2O_4$:Bi^{3+} [192], $ZnGa_2O_4$:Cr^{3+} [193], $LiGa_5O_8$:Mn^{2+} [194], $Ca_5Ga_6O_{14}$:Pr^{3+} [195], etc. LLP has also been discovered for $A_3Ga_4O_9$ (A = Ca, Sr) series such as $Ca_3Ga_4O_9$:Tb^{3+}/Zn^{2+} [196], $Sr_3Ga_4O_9$:Bi^{3+} [197], $Sr_3Ga_4O_9$:Tb^{3+} [198], $Ca_3Ga_4O_9$:Bi^{3+} [199]. The LPL time reported for $Sr_3Ga_4O_9$:Bi^{3+} is about 6.5 hours.

8.2.7.1 ZnGa$_2$O$_4$

ZnGa$_2$O$_4$ belongs to the cubic crystal system with a spinel structure in the Fd$\bar{3}$m (O$_h$ 7) space group and band gap of about 4.4 eV. Zn^{2+} ions occupy the tetrahedral A sites and Ga^{3+} the octahedral B-sites. It has excellent thermal and chemical stability, and it can be easily synthesized. Some kinds of defect centers depending on synthesis condition exist in ZnGa$_2$O$_4$ crystals, which are responsible for self-activated ultraviolet (UV) and blue luminescence in non-doped ZnGa$_2$O$_4$ ceramics [200]. It has proved to be a good host for activators such as Cr^{3+}, Mn^{2+}, Ni^{2+}, Eu^{3+}, etc. [201]. Cr^{3+}-doped zinc gallium (ZnGa$_2$O$_4$:Cr^{3+}) is phosphor that can convert ultraviolet/visible light to persistent NIR afterglow at ~700 nm [202]. When doped with Cr^{3+}, it gives rise to a strong far red long-lasting luminescence at 696 nm [203,204]. Owing to its remarkable properties, ZGO:Cr has become the best candidate to date to prepare LLP nanoprobes for a novel method of in vivo optical imaging. Bessiere et al. [205] have shown that a small Zn deficiency could lead to a more intense red long-lasting luminescence compared to the stoichiometric composition. ZnGa$_2$O$_4$ contains approximately 3% antisite defects, which is considered to be the main reason for the LAG of ZnGa$_2$O$_4$:Cr^{3+}. They proposed a phosphorescence mechanism whereby the positive Ga^{3+} antisite defect, responsible for the distortion of Cr^{3+}, acts as a deep trap. A more detailed mechanism was proposed by Gourier et al. [206]. Pan et al. [111] published a breakthrough with the synthesis of a new Cr^{3+}-doped zinc gallogermanate (Zn$_3$Ga$_2$Ge$_2$O$_{10}$:Cr^{3+} nominal formula) exhibiting an impressive superlong red afterglow of more than 360 hours, although measured with the use of a night vision monocular. Improvement in LLP properties was also noticed by Bi [207], Li [208], Sn [209,210], Al [211], or Si codoping [212]. In 2014, new generation of NIR persistent nano-phosphors (Cr^{3+}: ZnGa$_2$O$_4$) was successfully prepared and applied for in vivo bioimaging [213]. Compared with other persistent nano-phosphors, the novel Cr^{3+}:ZnGa$_2$O$_4$ one exhibits much stronger NIR PersL originating from highly efficient Cr^{3+}:^{2}E→^{4}A$_2$ transition (R-line) and from some other Cr^{3+} ions, having an antisite defect in the first cationic coordination sphere. Thus, the Cr^{3+}: ZnGa$_2$O$_4$ PersL materials attract more and more attentions and show great potential in ultra-sensitive biological detection and imaging, due to their excellent red and NIR PersL properties [214]. Codoping by Yb, Er improved the LLP performance [215].

Cr^{3+} gives LLP at the edge of the visible window. Pr^{3+} also gives red LLP, but at shorter wavelength [216]. LLP emission in Mn-doped ZnGa$_2$O$_4$ is green (505 nm). This could be induced by 285 nm light that corresponds to Mn^{2+} CT band [217]. On the other hand, Bi^{3+} can give white emission from sites with various coordinations. Three emission bands at 410, 480, and 540 nm are observed giving overall white appearance to the afterglow [218]. Ni^{2+} persistent emission is in the second biological window (1,000–1,350 nm) and thus can be useful for bioimaging [201].

8.2.7.2 CaO-Ga$_2$O$_3$

Four compounds CaGa$_2$O$_4$, Ca$_3$Ga$_4$O$_9$, Ca$_3$Ga$_2$O$_6$, and CaGa$_4$O$_7$ have been found in the subsolidus region of the phase diagram of the CaO-Ga$_2$O$_3$ system [219]. LLP was discovered in the first two. Later, another compound Ca$_5$Ga$_6$O$_{14}$ was discovered through Raman spectroscopy [220] and afterglow properties were studied [195].

Three polymorphs of CaGa$_2$O$_4$ are known. CaGa$_2$O$_4$-I is orthorhombic, with a=7.73, b=9.14, c=10.36 Å; CaGa$_2$O$_4$-II is orthorhombic with a=5.32, b=14.38, c=16.82 Å; m-CaGa$_2$O$_4$ is monoclinic with a=7.86, b=8.90, c=10.55 Å, β=93° [221]. Crystallographic studies continued till 1980, and first luminescence investigations were carried out in 1997 [222]. LLP was reported as late as 2017 in CaGa$_2$O$_4$:Bi^{3+} [192]. Yellow emission around 584 nm with persistence of 36 hours could be observed. Yellow LLP was observed with Pr^{3+} activator [223]. LLP in NIR region is obtainable in this host with Cr, Yb, Ge dopants [224].

Ca$_3$Ga$_4$O$_9$ exhibits blue self-activated LPL due to the creation of intrinsic traps. When Tb^{3+} is doped, the photoluminescence (PL) and LPL colors change from blue to green with their intensities significantly enhanced. Codoping with Zn^{2+} improves the PL and LPL performances of the Ca$_3$Ga$_4$O$_9$ and Ca$_3$Ga$_4$O$_9$:Tb^{3+} [196].

$Ca_5Ga_6O_{14}$ crystallizes in orthorhombic space group $Cmc2_1$ (36) with lattice parameters a = 11.511 Å, b = 11.282 Å, c = 10.489 Å, and V = 1,362.18 Å^3. Doped sample exhibits white LLP emission. The optimum initial brightness of the sample $Ca_5Ga_6O_{14}:0.07Pr^{3+}$ is about 41.49 mcd/m^2. The duration of $Ca_5Ga_6O_{14}:0.07Pr^{3+}$ at the particular intensity level (0.32 mcd/m^2) was up to 10.3 hours.

8.3 APPLICATIONS

Most commonplace applications of LLP are related with the light emission without any power source in the dark surroundings for example, signage, road signs, luminescent clothings for workers in hazardous traffic which will make them conspicuous. The idea to use LLP for road markings has been proposed a few years ago as a way to avoid the use of street lighting without any loss of their visibility [55]. More decorative applications are related to create "night sky" in dark room, children toys, etc. (Figure 8.2).

Long-lasting phosphors can be mixed with polymers to yield phosphorescent fibers and films. Luminous fiber is a sort of long phosphorescent functional fiber. It is prepared by making a uniformly dispersed mixture of persistent phosphors and a suitable polymer. Such fibers are routinely used for making phosphorescent clothing, toys, and signage, as well as, the anticounterfeiting applications [225] (Figure 8.3). Polymers such as polyester, nylon, and polyacrylonitrile which have adequate mechanical strength and which can be spun to yield fibers are most commonly used [226].

The phosphorescent flexible films are outstanding elements for thermometry [227]. The temperature sensing become possible due to dependence of PL intensity on the temperature in the range 300–420 K.

FIGURE 8.2   'Glow in the dark' applications of long-lasting phosphorescence.

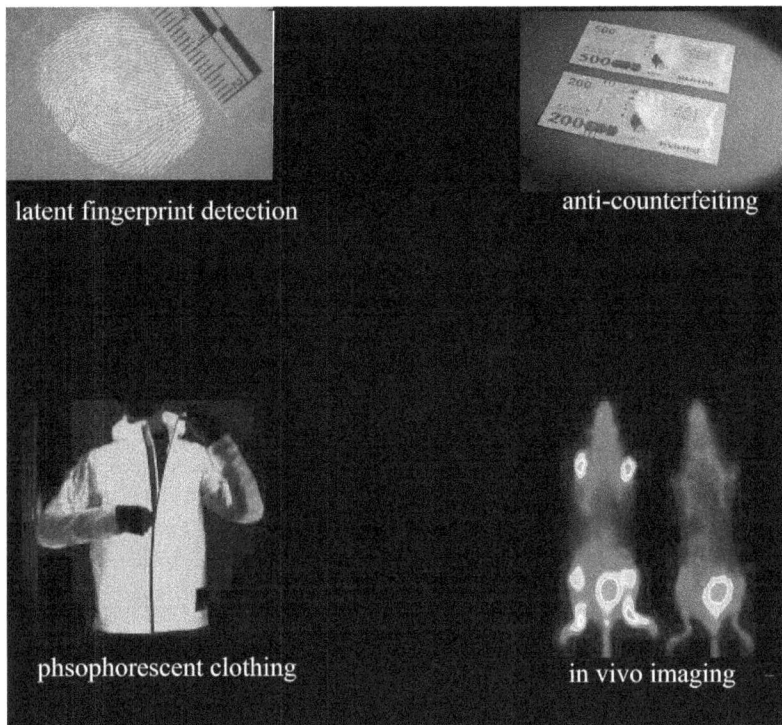

FIGURE 8.3 Various applications of the long-lasting phosphorescence.

8.3.1 FLICKER REDUCTION IN AC-LEDS

With introduction of solid-state lighting sources, there is priority for AC-driven LEDs. This eliminates need for rectifying electronic circuits and the associated power losses. However, LEDs have short response time. LED current and light output would vary with AC frequency. This will cause undesired flickering which strains the eye. Using LLP materials as conversion phosphors will reduce the flicker [228]. Moreover, the luminescence being isotropic, use of scatterer for uniformly distributing the light over a given volume can be eliminated.

8.3.2 OPTICAL STORAGE, SENSOR, AND DETECTOR

The ability of persistent phosphors to store and emit back optical energy after external stimulation makes them interesting to develop materials for optical data storage. Due to exceptional re-writability and high storage capacity, which come with low ecological impact, persistent phosphors are being investigated as optical information storage materials. Recording is performed using a high energy source that leads to charge separation and trapping, whereas reading is through light emission triggered by thermal, optical, or mechanical stimulation [229,230].

8.3.3 ANTICOUNTERFEITING

The growing worldwide exchanges of money and valuables have undoubtedly led to an increased circulation of counterfeits, making the development of anticounterfeiting agents a priority in the bank security field. UV fluorescent inks are well known for this security purpose, as, for example, present in the form of glowing stars and threads in currencies all over the world, to limit their illegal reproduction. The security they offer is though limited due to their widespread usage and the single

UV fluorescence check mode. Using persistent phosphor-based anticounterfeiting agents enables a supplementary time-gated security check mode, which is the continuous light emission once the UV excitation is stopped. Combining persistent phosphor-based anticounterfeiting agents with regular UV fluorescent ones can significantly increase the security [231.]

8.3.4 BIOMEDICAL APPLICATIONS

Various biomedical applications of LLP have recently been reviewed [232]. Viana et al. introduced the persistent luminescence in Cr^{3+}-doped spinels and their applications in vivo bioimaging [233]. They further provided a comprehensive summary about the bioimaging applications of chemically engineered persistent luminescence nanoprobes [234].

8.3.4.1 In Vivo Imaging

Phenomenon of fluorescence had already been put to use for bioimaging. However, fluorescence imaging suffers from a suboptimal signal-to-noise ratio (SNR) and a shallow detection depth (<1 cm). These drawbacks are mainly associated with tissue autofluorescence, which is caused by the fluorescence from tissues under continuous external excitation. Thus, ideally, an optical probe that can illuminate without external excitation would be free of autofluorescence and, hence, enable significantly improved imaging sensitivity and depth. Thus, the use of LLP. The number of papers devoted to bioimaging applications using LLP has drastically increased since the pioneer work published in 2007. Nanosized LLP materials are most suitable for this application. The desirable properties are as follows:

1. Size ranging from 50 to 200 nm. Recent studies have shown that nanomaterials (NMs) with small sizes exhibited enhanced performance in vivo, such as greater tissue penetration, particularly those with size around or smaller than 50 nm.
2. Water resistant to prevent rapid dissolution in biological fluids and allow imaging for hours or days
3. Emission in the tissue transparency window, between 650 and 1,350 nm, to be able to pass through the animal body and be detected by a photon-counting system.

8.3.4.2 Photodynamic Therapy (PDT)

"LLP materials have been studied as light source for photodynamic activation on cancer treatment and animal imaging. When the nanoparticle-photosensitizer conjugates are targeted to the tumor, the light from afterglow nanoparticles will activate the photosensitizers, so that no external light is required for PDT. More importantly, it can be used to treat deep tumor such as breast cancer because the light source is attached to the photosensitizers, so the efficiency of cancer diagnosis and treatment is enhanced. Upon exposure to ionizing radiation such as X-rays, nanoparticles emit luminescence which activates the photosensitizers; thus, singlet oxygen is produced to kill cancer cells along with ionizing radiation. Therefore, the combination of conventional radiation therapy and PDT enables the use of lower doses of radiation. PDT could result in selective and irreversible destruction of diseased tissues such as cancerous tissue, without damaging healthy cells. Another important advantage of PDT as compared to conventional methods is outpatient therapeutic property of PDT. Moreover, PDT can induce immunity, even against less immunogenic tumors, resulting in long-term tumor control [163]."

8.3.4.3 Biodetection

LLP is not only chosen as an optical label to trace cancer cells, but also as a bioprobe to monitor and identify the variation of biotic habitat. Using nanoprobes consisting of LLP materials, detection is possible without using external excitation, thus increasing the sensitivity. For example, Niu et al. [235] developed a novel nanoprobe for the determination and screening of ascorbic acid (AA) in living cells and in vivo by using CoOOH-modified persistent luminescence nanoparticles. Based on the specific reaction of CoOOH and AA, which could detect and image AA in living cells and in vivo, NPPs were employed as the optical unit, and CoOOH nanoflakes were used as the quencher. Numerous such LLP-based bioprobes have been demonstrated for detection of various reagents of strategic importance.

8.3.5 PHOTOCATALYSIS

"Over the last decade, there has been an increasing interest for using LPP phosphors in photoca-talysis applications. Here, typically, a LPP phosphor is combined with a photocatalyst, where, if illumination has stopped, the former activates the latter, resulting in prolonged photocatalysis. The most used photocatalyst nowadays is the anatase phase of TiO_2, despite a lot of research investi-gating alternatives. Using LLP materials like $Sr_{2.90}Eu_{0.03}Dy_{0.07}Al_4SiO_{11}$ with N-doped TiO_2 for the photocatalytic degradation of crystal violet dye in solution [236], and $CdSiO_3$:Gd^{3+}, Bi^{3+} coated with TiO_2 nanoparticles for degradation of methylene blue [237] was demonstrated."

During last 10 years, there have been growing recognition for employing LPP phosphors for photocatalysis-related uses. For such applications, a LPP phosphor is mixed with a photocatalyst. Photocatalyst functions by itself as long as there is incident light. When it ceases, the phosphores-cent material provides the illumination required for functioning of the catalyst, thus prolonging the photocatalysis duration [238]. For example, "Degradation tests of Rhodamine B (RhB) and photocatalytic hydrogen evolution from water showed that the γC_3N_4@Au@$SrAl_2O_4$:Eu^{2+}, Dy^{3+} composite maintained some photocatalytic activity in dark conditions after irradiation for 10 minutes [239]."

8.3.6 SOLAR CELL: NIGHT-TIME SOLAR ENERGY

Use of phosphors as spectrum modifiers for boosting the solar cell performance had already been attempted frequently [240]. It may be speculated that using LLP phosphors can make solar cells active even after the sun has set. LLP phosphors can provide light for 12 hours to generate photo-current. However, the amount of light stored and released by LLP phosphors is order of magnitude smaller than the sunlight. Even then small devices not needing much power for their working can function in this way. Some reported performances of solar cells using a persistent phosphor during the night were a nighttime solar cell efficiency of 0.0076% [241], or a generated power density of 56 µW/cm² [242]. Under short-circuit conditions, the device can output 300 mV for 30 hours in the dark.

LLP can also be used to convert nUV light to which solar cells are not sensitive to green light which can be used for photovoltaic conversion, thus increasing sensitivity [241,243]. Using phospho-rescence can enable obtaining small solar power during night time as well.

Various applications of LLP have been elaborated in some recent reviews [62].

8.4 CONCLUSIONS AND OUTLOOK

LLP had been known and used for various routine applications for a long time. After introduction of $SrAl_2O_4$:Eu^{2+}, Dy^{3+} phosphors in 1996 by Matsuzawa et al. research on LLP materials witnessed steep rise. Various applications such as solid-state lighting, optical storage, optical thermometry, anticounterfeiting, bioimaging, and detection, etc., were envisaged. While phosphors for the appli-cations involving mass usage such as road signs, toys, decoration, etc. are commercially available, those required for more sophisticated applications like anticounterfeiting, and bioimaging are still at research stage. Particularly, for bio-related applications, LLP phosphors in nanoforms are required. Most of the methods attempted for obtaining nanoparticles of LLP materials are bottom up. On the other hand, for mass-scale production, top-down methods are better suited. In future, efforts may be directed toward designing top-down syntheses for LLP nanoparticles.

For studying the mechanism of LLP and identifying the shallow traps, thermoluminescence glow curves are recorded. The process may be turned inside out for obtaining LLP materials. Large number of TL materials have been discarded for TL applications due to shallow nature of the traps which lead to fast fading. On the other hand, these could be potential LLP materials provided that these shallow traps can be filled by UV exposure. Thus, exploring the discarded TL phosphors can be a fruitful method for obtaining new LLP phosphors.

REFERENCES

1. Yen, W.M., Shionoya, S., & Yamamoto, H. (2007). *Phosphor Handbook.* 2nd ed. CRC Press, Boca Raton, FL, USA.
2. Yen, W.M., & Weber, M.J. (2004). *Inorganic Phosphors: Compositions, Preparation and Optical Properties.* CRC Press, Boca Raton, FL, USA.
3. Smet, P.F., Moreels, I., Hens, Z., & Poelman, D. (2010). Luminescence in sulfides: A rich history and a bright future. *Materials* 3(4), 2834–2883. https://doi.org/10.3390/ma3042834
4. Harvey, E.N. (1957). *A History of Luminescence from the Earliest Times until 1900.* American Philosophical Society, Philadelphia.
5. Lastusaari, M., Laamanen, T., Malkamaki, M., Eskola, K.O., Kotlov, A., Carlson, S., Welter, E., Brito, H.F., Bettinelli, M., Jungner, H., & Holsa, J. (2012). The Bologna Stone: History's first persistent luminescent material. *Eur. J. Mineral.* 24(5), 885–890. https://doi.org/10.1127/0935-1221/2012/0024-2224
6. Von Goethe, J.W., & Dole, N.H. (Eds.), Translated by Carlyle, T., & Boylan, R.D. (2002). *The Sorrows of Young Werther.* Dover Publications, Inc., Mineola, New York.
7. Holsa, J. (2009). Persistent luminescence beats the afterglow: 400 years of persistent luminescence. *The Electrochem. Soc. Interface*, 42–45. https://www.electrochem.org/dl/interface/wtr/wtr09/wtr09_p042-045.pdf
8. Van Den Eeckhout, K., Poelman, D., & Smet, P.F. (2013). Persistent luminescence in non-Eu²⁺-doped compounds: A review. *Materials* 6(7), 2789–2818. https://doi.org/10.3390/ma6072789
9. Bredol, M., Merikhi, J., & Ronda, C. (1992). Defect chemistry and luminescence of ZnS: Cu, Au, Al. *Ber. Bunsenges. Phys. Chem.* 96(11), 1770–1774. https://doi.org/10.1002/bbpc.19920961149
10. Lange, H. (1966). Luminescent europium activated strontium aluminate. US patent 3,294,699.
11. Abbruscato, V. (1971). Optical and electrical properties of $SrAl_2O_4$:Eu^{2+}. *J. Electrochem. Soc.* 118(6), 930–932. https://doi.org/10.1149/1.2408226
12. Blasse, G., & Bril, A. (1968). Fluorescence of Eu²⁺-activated alkaline-earth aluminates. *Philips Res. Reps.* 23, 201–206. https://scholar.google.com/scholar?hl=en&as_sdt=0%2C5&q=Fluorescence+of+Eu2%2B+activated+alkaline-earth+aluminates.+&btnG=
13. Diallo, P. T., Boutinaud, P., Mahiou R., & Cousseins, J.C. (1997). Red luminescence in Pr³⁺-doped calcium titanates. *Phys. Status Solidi A* 160(1), 255–263. https://doi.org/10.1002/1521-396X(199703)160:1<255::AID-PSSA255>3.0.CO;2-Y
14. Murazaki, Y., Arai, K., & Ichinomiya, K. (1999). A new long persistence red phosphor. *Rare Earths Jpn* 35(1), 41–45. https://scholar.google.com/scholar?hl=en&as_sdt=0%2C5&q=Murazaki%2C+Y.%2C+Arai%2C+K.%2C+%26+Ichinomiya%2C+K.+%281999%29.+A+new+long+persistence+red+phosphor.+Rare+Earths+Jpn%2C+35%2C+41-45&btnG=
15. Hong, Z., Zhang, P., Fan, X., & Wang, M. (2007). Eu³⁺ red long afterglow in Y_2O_2S:Ti, Eu phosphor through afterglow energy transfer. *J. Lumin.* 124(1), 127–132. https://doi.org/10.1016/j.jlumin.2006.02.008
16. Lin, Y. H., Nan, C. W., Cai, N., Zhou, X. S., Wang, H. F., & Chen, D.P. (2003). Anomalous afterglow from Y_2O_3-based phosphor. *J. Alloys Compd.* 361(1–2), 92–95. https://doi.org/10.1016/S0925-8388(03)00432-8
17. Wang, J., Wang, S. B., & Su, Q. (2004). The role of excess Zn²⁺ ions in improvement of red long lasting phosphorescence (LLP) performance of β-$Zn_3(PO_4)_2$:Mn phosphor. *J. Solid State Chem.* 177(3), 895–900. https://doi.org/10.1016/j.jssc.2003.09.027
18. Chen, Y. H., Cheng, X. R., Liu, M., Qi, Z. M., and Shi, C.S. (2009). Comparison study of the luminescent properties of the white-light long afterglow phosphors: $Ca_xMgSi_2O_{5+x}$:Dy³⁺(x=1, 2, 3). *J. Lumin.* 129(5), 531–535. https://doi.org/10.1016/j.jlumin.2008.12.008
19. Zych, E., Trojan-Piegza, J., Hreniak, D., & Strek, W. (2003). Properties of Tb-doped vacuum-sintered Lu_2O_3 storage phosphor. *J. Appl. Phys.* 94(3), 1318–1324. https://doi.org/10.1063/1.1587891
20. Trojan-Piegza, J., Niittykoski, J., Holsa, J., & Zych, E. (2008). Thermoluminescence and kinetics of persistent luminescence of vacuum-sintered Tb³⁺-doped and Tb³⁺, Ca²⁺-codoped Lu_2O_3 materials. *Chem. Mater.* 20(6), 2252–2261. https://doi.org/10.1021/cm703060c

21. Wiatrowska, A., & Zych, E. (2013). Traps formation and characterization in long-term energy storing Lu_2O_3:Pr, Hf luminescent ceramics. *J. Phys. Chem. C* 117(22), 11449–11458. https://doi.org/10.1021/jp312123e

22. Trojan-Piegza, J., & Zych, E. (2010). Afterglow luminescence of Lu_2O_3:Eu ceramics synthesized at different atmospheres. *J. Phys. Chem. C* 114(9), 4215–4220, https://doi.org/10.1021/jp910126r

23. Kulesza, D., Bolek, P., Bos, A.J.J., & Zych, E. (2016). Lu_2O_3-based storage phosphors. An (in)harmonious family. *Coord. Chem. Rev.* 325, 29–40. https://doi.org/10.1016/j.ccr.2016.05.006

24. Trojan-Piegza, J., Zych, E., Holsa, J., & Niittykoski, J. (2009). Spectroscopic properties of persistent luminescence phosphors: Lu_2O_3:Tb^{3+}, M^{2+} (M=Ca, Sr, Ba). *J. Phys. Chem. C* 113(47), 20493–20498. https://doi.org/10.1021/jp906127k

25. Kulesza, D., Wiatrowska, A., Trojan-Piegza, J., Felbeck, T., Geduhn, R., Motzek, P., Zych, E., & Kynast, U. (2013). The bright side of defects: Chemistry and physics of persistent and storage phosphors. *J. Lumin.* 133, 51–56. https://doi.org/10.1016/j.jlumin.2011.12.019

26. Chen, S., Yang, Y., Zhou, G., Wu, Y., Liu, P., Zhang, F., Wang, S., Trojan-Piegza, J., & Zych, E. (2012). Characterization of afterglow-related spectroscopic effects in vacuum sintered Tb^{3+}, Sr^{2+} co-doped Lu_2O_3 ceramics. *Opt. Mater.* 35(2), 240–243. https://doi.org/10.1016/j.optmat.2012.08.001

27. Pedroso, C.C., Carvalho, J.M., Rodrigues, L.C., Holsa J., & Brito, H.F. (2016). Rapid and energy-saving microwave-assisted solid-state synthesis of Pr^{3+}-, Eu^{3+}-, or Tb^{3+}-doped Lu_2O_3 persistent luminescence materials. *ACS Appl. Mater. Interfaces* 8(30), 19593–19604. https://doi.org/10.1021/acsami.6b04683

28. Cheng, J.P., Wang, J., Li, Q.Q., Liu, H.G., & Li, Y. (2016). A review of recent developments in tin dioxide composites for gas sensing application. *J. Ind. Eng. Chem.* 44, 1–22. https://doi.org/10.1016/j.jiec.2016.08.008

29. Nascimento, E. P., Firmino, H.C.T., Neves, G. A., & Menezes, R.R. (2022). A review of recent developments in tin dioxide nanostructured materials for gas sensors. *Ceram. Int.* 48(6), 7405–7440. https://doi.org/10.1016/j.ceramint.2021.12.123

30. Bhawna, Kumar, S., Sharma, R., Gupta, A., Tyagi, A., Singh, P., Kumar, A., & Kumar, V. (2022). Recent insights into SnO_2-based engineered nanoparticles for sustainable H_2 generation and remediation of pesticides. *New J. Chem.* 46, 4014–4048. https://doi.org/10.1039/D1NJ05808H

31. Pham, V.V., Tran, H.H., Truong, T.K., Cao, T.M., & Beilstein. (2022). Tin dioxide nanomaterial-based photocatalysts for nitrogen oxide oxidation: A review. *Beilstein J. Nanotechnol.* 13(96), 96–113. https://doi.org/10.3762/bjnano.13.7

32. Babu, B., Kim, J., & Yoo, K. (2022). Nanocomposite of SnO_2 quantum dots and Au nanoparticles as a battery-like supercapacitor electrode material. *Mater. Lett.* 309, 131339. https://doi.org/10.1016/j.matlet.2021.131339

33. Tran, T.N.L., Szczurek, A., Varas, S., Armellini, C., Scotognella, F., Chiasera, A., Ferrari, M., Righini, G.C., & Lukowiak, A. (2022). Rare-earth activated SnO_2 photoluminescent thin films on flexible glass: Synthesis, deposition and characterization. *Opt. Mater.* 124, 111978. https://doi.org/10.1016/j.optmat.2022.111978

34. Kumar, K.N, Dagupati, R., Lim, J., & Choi, J. (2022). Bright red luminescence from Eu^{3+}-activated non-cytotoxic SnO_2 quantum dots for latent fingerprint detection. *Ceram. Int.* 48(12), 17738–17748. https://doi.org/10.1016/j.ceramint.2022.03.044

35. Lee, C.T., & Lu, C.H. (2011). Preparation and characterization of microwave solvothermally derived SnO_2:Sm^{3+} phosphors. *Int. J. Appl. Ceram. Technol.* 8(4), 718–724. https://doi.org/10.1111/j.1744-7402.2010.02513.x

36. Tran, T.N.L., Szczurek, A., Lukowiak, A., Chiasera, A. (2022). A review on rare-earth activated SnO_2-based photonic structures: Synthesis, fabrication and photoluminescence properties. *Opt. Mater.: X* 13, 100140. https://doi.org/10.1016/j.omx.2022.100140

37. Singh, L.P., Luwang, M.N., & Srivastava, S.K. (2014). Luminescence and photocatalytic studies of Sm^{3+} ion doped SnO_2 nanoparticles. *New J. Chem.* 38(1), 115–121. https://doi.org/10.1039/C3NJ00759F

38. Chen, C, Li, Y., & Shen, M. (2020). Structure-related luminescent properties induced by doping in Sm-doped SnO_2 hollow spheres. *Ceram. Int.* 46(10), Part B, 17025–17033. https://doi.org/10.1016/j.ceramint.2020.03.289

39. Faustino, B.M., Foot, P.J.S., & Kresinski, R.A. (2016). Synthesis and photoluminescent properties of Sm^{3+}-doped SnO_2 nanoparticles. *Ceram. Int.* 42(16), 18474–18478. https://doi.org/10.1016/j.ceramint.2016.08.183

40. Feng, P., Zhang, J., Qin, Q., Hu, R., & Wang, Y. (2014). Hydrothermal synthesis and afterglow luminescence properties of hollow SnO_2:Sm^{3+}, Zr^{4+} spheres for potential application in drug delivery. *Mater. Res. Bull.* 50, 365–368. http://dx.doi.org/10.1016/j.materresbull.2013.11.011

41. Zhang, J., Ma, X., Qin, Q., Shi, L., Sun, J., Zhou, M., Liu, B., & Wang, Y. (2012). The synthesis and afterglow luminescence properties of a novel red afterglow phosphor: SnO_2:Sm^{3+}, Zr^{4+}. *Mater. Chem. Phys.* 136(2–3), 320–324. http://dx.doi.org/10.1016/j.matchemphys.2012.08.033

42. Schulze A. R., & Buschbaum, H.M. (1981). Zur Verbindungsbildung von MeO: M_2O_3. IV. Zur Struktur von monoklinem $SrAl_2O_4$. *Z. Anorg. Allg. Chem.* 475(4), 205–210. https://doi.org/10.1002/zaac.19814750423

43. Holsa, J., Laamanen, T., Lastusaari, M., Niittykoski, J., & Novak, P. (2009). Electronic structure of the $SrAl_2O_4$:Eu^{2+} persistent luminescence material. *J. Rare Earths* 27(4), 550–554. https://doi.org/10.1016/S1002-0721(08)60286-0

44. Botterman, J., Joos, J.J., & Smet, P.F. (2014). Trapping and detrapping in $SrAl_2O_4$:Eu, Dy persistent phosphors: Influence of excitation wavelength and temperature. *Phys. Rev. B* 90(8), 085147. https://doi.org/10.1103/PhysRevB.90.085147

45. Dorenbos, P., Bos, A.J.J., Poolton, N.R.J., & You, F. (2013). Photon controlled electron juggling between lanthanides in compounds. *J. Lumin.* 133, 45–50. https://doi.org/10.1016/j.jlumin.2011.12.020

46. Poort, S.H.M., Blokpoel, W. P., & Blasse, G. (1995). Luminescence of Eu^{2+} in barium and strontium aluminate and gallate. *Chem. Mater.* 7(8), 1547–1551. https://doi.org/10.1021/cm00056a022

47. Clabau, F., Rocquefelte, X., Jobic, S., Deniard, P., Whangbo, M. H., Garcia, A., & Le Mercier, T. (2005). Mechanism of phosphorescence appropriate for the long-lasting phosphors Eu^{2+}-doped $SrAl_2O_4$ with codopants Dy^{3+} and B^{3+}. *Chem. Mater.* 17(15), 3904–3912. https://doi.org/10.1021/cm050763r

48. Palilla, F.C., Levine, A.K., & Tomkus, M.R. (1968). Fluorescent properties of alkaline earth aluminates of the type MAl_2O_4 activated by divalent europium. *J. Electrochem. Soc.* 115(6), 642–644. https://doi.org/10.1149/1.2411379

49. Matsuzawa, T., Aoki, Y., Takeuchi, N., & Murayama, Y. (1996). A new long phosphorescent phosphor with high brightness, of $SrAl_2O_4$:Eu^{2+}, Dy^{3+}. *J. Electrochem. Soc.* 143(8), 2670–2673. https://doi.org/10.1149/1.1837067

50. Nag A., & Kutty, T.R.N. (2003). Role of B_2O_3 on the phase stability and long phosphorescence of $SrAl_2O_4$:Eu, Dy. *J. Alloys Compd.* 354(1–2), 221–231. https://doi.org/10.1016/S0925-8388(03)00009-4

51. Vitola, V., Lahti, V., Bite, I., Spustaka, A., Millers, D., Lastusaari, M., Petit, L., & Smits, K. (2021). Low temperature afterglow from $SrAl_2O_4$:Eu, Dy, B containing glass. *Scr. Mater.* 190, 86–90. https://doi.org/10.1016/j.scriptamat.2020.08.023

52. Van den Eeckhout, K., Smet, P. F., & Poelman, D. (2010). Persistent luminescence in Eu^{2+}-doped compounds: A review. *Materials* 3(4), 2536–2566. https://doi.org/10.3390/ma3042536

53. Khattab, T.A., Abou-Yousef, H., & Kamel, S. (2018). Photoluminescent spray-coated paper sheet: Write-in-the-dark. *Carbohydr. Polym.* 200, 154–161. https://doi.org/10.1016/j.carbpol.2018.07.094

54. Botterman, J., & Smet, P.F. (2015). Persistent phosphor $SrAl_2O_4$:Eu, Dy in outdoor conditions: Saved by the trap distribution. *Opt. Express* 23(15), A868–A881. https://doi.org/10.1364/OE.23.00A868

55. Nance, J., & Sparks, T.D. (2020). Comparison of coatings for of $SrAl_2O_4$:Eu^{2+}, Dy^{3+} powder in waterborne road striping paint under wet conditions. *Prog. Org. Coat.* 144, 105637. https://doi.org/10.1016/j.porgcoat.2020.105637

56. Nance, J., & Sparks, T.D. (2020). From streetlights to phosphors: A review on the visibility of roadway markings. *Prog. Org. Coat.* 148, 105749. https://doi.org/10.1016/j.porgcoat.2020.105749

57. Vedda, A., Martini, M., Meinardi, F., Chval, J., Dusek, M., Mares, J., Mihokova, E., & Nikl, M. (2000). Tunneling process in thermally stimulated luminescence of mixed $Lu_xY_{1-x}AlO_3$:Ce crystals. *Phys. Rev. B* 61(12), 8081. https://doi.org/10.1103/PhysRevB.61.8081

58. Aitasalo, T., Holsa, J., Jungner, H., Krupa, J.C., Lastusaari, M., Legendziewicz, J., & Niittykoski, J. (2004). Effect of temperature on the luminescence processes of $SrAl_2O_4$:Eu^{2+}. *Radiat. Meas.* 38(4–6), 727–730. https://doi.org/10.1016/j.radmeas.2004.01.031

59. Aitasalo, T., Deren, P., Holsa, J., Jungner, H., Krupa, J.C., Lastusaari, M., Legendziewicz, J., Niittykoski, J., & Strek, W. (2003). Persistent luminescence phenomena in materials doped with rare earth ions. *J. Solid State Chem.* 171(1–2), 114–122, https://doi.org/10.1016/S0022-4596(02)00194-9

60. Dorenbos P. (2005). Mechanism of persistent luminescence in Eu^{2+} and Dy^{3+} codoped aluminate and silicate compounds. *J. Electrochem. Soc.* 152(7), H107–H110. https://doi.org/10.1149/1.1926652

61. Vitola, V., Millers, D., Bite, I., Smits, K., Spustaka, A. (2019). Recent progress in understanding the persistent luminescence in $SrAl_2O_4$:Eu, Dy. *Mater. Sci. Technol. (United Kingdom)* 35(14), 1661–1677. https://doi.org/10.1080/02670836.2019.1649802

62. Poelman, D., Van Der Heggen, D., Du, J., Cosaert, E., & Smet, P.F. (2020). Persistent phosphors for the future: Fit for the right application. *J. Appl. Phys.* 128(24), 240903. https://doi.org/10.1063/5.0032972

63. Can-Uc, B., Montes-Frausto, J.B., Juarez-Moreno, K., Licea-Rodriguez, J., Rocha-Mendoza, I., & Hirata, G.A. (2018). Light sheet microscopy and SrAl$_2$O$_4$ nanoparticles codoped with Eu^{2+}/Dy^{3+} ions for cancer cell tagging. *J. Biophotonics* 11(6), e201700301. https://doi.org/10.1002/jbio.201700301

64. Maldiney, T., Sraiki, G., Viana, B., Gourier, D., Richard, C., Scherman, D., Bessodes, M., Van den Eeckhout, K., Poelman, D., & Smet, P.F. (2012). In vivo optical imaging with rare earth doped Ca$_2$Si$_5$N$_8$ persistent luminescence nanoparticles. *Opt. Mater. Express* 2(3), 261–268. https://doi.org/10.1364/OME.2.000261

65. Sun, M., Li, Z.J., Liu, C.L., Fu, H.X., Shen, J.S., & Zhang, H.W. (2014). Persistent luminescent nanoparticles for super-long time in vivo and in situ imaging with repeatable excitation. *J. Lumin.* 145, 838–842. https://doi.org/10.1016/j.jlumin.2013.08.070

66. Sepahvandi, A., Eskandari, M., & Moztarzadeh, F. (2016). Fabrication and characterization of SrAl$_2$O$_4$: Eu^{2+} Dy^{3+}/CS-PCL electrospun nanocomposite scaffold for retinal tissue regeneration. *Mater. Sci. Eng. C* 66, 306–314. https://doi.org/10.1016/j.msec.2016.03.028

67. Sharma, V., Das, A., & Kumar, V. (2016). Eu^{2+}, Dy^{3+} codoped SrAl$_2$O$_4$ nanocrystalline phosphor for latent fingerprint detection in forensic applications. *Mater. Res. Express* 3(1), 015004. https://doi.org/10.1088/2053-1591/3/1/015004

68. Li, B., Zhang, J., Zhang, M., Long, Y., & He, X. (2015). Effects of SrCl2 as a flux on the structural and luminescent properties of SrAl$_2$O$_4$:Eu^{2+}, Dy^{3+} phosphors for AC-LEDs. *J. Alloys Compd.* 651, 497–502. https://doi.org/10.1016/j.jallcom.2015.08.161

69. Wang, L., Shang, Z., Shi, M., Cao, P., Yang, B., & Zou, J. (2020). Preparing and testing the reliability of long-afterglow of SrAl$_2$O$_4$:Eu^{2+}, Dy^{3+} phosphor flexible films for temperature sensing. *RSC Adv.* 10(19), 11418–11425. https://doi.org/10.1039/D0RA00628A

70. Smets, B., Rutten, J., Hocks, G., & Verlijsdonk, J. (1989). 2SrO.3Al$_2$O$_3$:Eu^{2+} and 1.29(Ba, Ca)O, 6Al$_2$O$_3$:Eu^{2+}, two new blue-emitting phosphors. *J. Electrochem. Soc.* 136(7), 2119–2123. https://doi.org/10.1149/1.2097210

71. Akiyama, M., Xu, C., Matsui, H., Nonaka, K., & Watanabe, T. (2000). Photostimulated luminescence phenomenon of Sr$_4$Al$_{14}$O$_{25}$:Eu, using only visible lights. *J. Mater. Sci. Lett.* 19(13), 1163–1165. https://doi.org/10.1023/A:1006763310775

72. Lin, Y., Tang, Z., & Zhang, Z. (2001). Preparation of long-afterglow Sr$_4$Al$_{14}$O$_{25}$ based luminescent material and its optical properties. *Mater. Lett.* 51(1), 14–18. https://doi.org/10.1016/S0167-577X(01)00257-9

73. Lin, Y., Tang, Z., Zhang, Z., & Nan, C.W. (2002). Anomalous luminescence in Sr$_4$Al$_{14}$O$_{25}$:Eu, Dy phosphors. *Appl. Phys. Lett.* 81(6), 996–998. https://doi.org/10.1063/1.1490631

74. Yuan, Z.X., Chang, C.K., Maoa, D.L., & Ying, W. (2004). Effect of composition on the luminescent properties of Sr$_4$Al$_{14}$O$_{25}$:Eu^{2+}, Dy^{3+} phosphors. *J. Alloys Compd.* 377(1–2), 268–271. https://doi.org/10.1016/j.jallcom.2004.01.063

75. Suriyamurthy, N., & Panigrahi, B.S. (2008). Effects of non-stoichiometry and substitution on photoluminescence and afterglow luminescence of Sr$_4$Al$_{14}$O$_{25}$:Eu^{2+}, Dy^{3+} phosphor. *J. Lumin.* 128(11), 1809–1814. https://doi.org/10.1016/j.jlumin.2008.05.001

76. Nakazawa, E., Murazaki, Y., & Saito, S. (2006). Mechanism of the persistent phosphorescence in Sr$_4$Al$_{14}$O$_{25}$:Eu and SrAl$_2$O$_4$: Eu codoped with rare earth ions. *J. Appl. Phys.* 100(11), 113113. https://doi.org/10.1063/1.2397284

77. Zhao, C., Chen, D., Yuan, Y., & Wu, M. (2006). Synthesis of Sr$_4$Al$_{14}$O$_{25}$:Eu^{2+}, Dy^{3+} phosphor nanometer powders by combustion processes and its optical properties. *Mater. Sci. Eng. B* 133(1–3), 200–204. https://doi.org/10.1016/j.mseb.2006.04.042

78. Sharma, V., Das, A., Kumar, V., Ntwaeaborwa, O. M., & Swart, H.C. (2014). Potential of Sr$_4$Al$_{14}$O$_{25}$:Eu^{2+}, Dy^{3+} inorganic oxide-based nanophosphor in Latent fingermark detection. *J. Mater. Sci.* 49(5), 2225–2234. https://doi.org/10.1007/s10853-013-7916-2

79. Sun, H., Pan, L., Zhu, G., Piao, X., Zhang, L., & Sun, Z. (2014). Long afterglow Sr$_4$Al$_{14}$O$_{25}$:Eu^{2+}, Dy^{3+} phosphors as both scattering and down converting layer for CdS quantum dot-sensitized solar cells. *Dalton Trans.* 43(40), 14936–14941. https://doi.org/10.1039/C4DT01276C

80. Sung, H.J., Jung-Sik Kim, S.C., & Ki, B.M. (2016). Photocatalytic characteristics for the nanocrystalline TiO$_2$ supported on Sr$_4$Al$_{14}$O$_{25}$:Eu^{2+}, Dy^{3+} phosphor beads. *Adv. Mater. Lett.* 7(1), 36–41. https://doi.org/10.5185/amlett.2016.6106

81. Eun, S.R., Mavengere, S., & Kim, J.S. (2021). Preparation of Ag-TiO$_2$/ Sr$_4$Al$_{14}$O$_{25}$:Eu^{2+}, Dy^{3+} photocatalyst on phosphor beads and its photoreaction characteristics. *Catalysts* 11(2), 261. https://doi.org/10.3390/catal11020261

82. Menon, S.G., Bedyal, A.K., Pathak, T., Kumar, V., & Swart, H.C. (2021). $Sr_4Al_{14}O_{25}$:Eu^{2+}, Dy^{3+}@ZnO nanocomposites as highly efficient visible light photocatalysts for the degradation of aqueous methyl orange. *J. Alloys Compd.* 860, 158370. https://doi.org/10.1016/j.jallcom.2020.158370

83. Garcia, C.R., Oliva, J., Romero, M.T., & Diaz-Torres, L.A. (2016). Enhancing the photocatalytic activity of $Sr_4Al_{14}O_{25}$:Eu^{2+}, Dy^{3+} persistent phosphors by codoping with Bi^{3+} ions. *Photochem. Photobiol.* 92(2), 231–237. https://doi.org/10.1111/php.12570

84. Antuzevics, A., Krieke, G., Doke, G., & Berzina, B. (2022). The origin of bright cyan persistent luminescence in Ca_2SnO_4:La^{3+}. *Materialia* 21, 101374. https://doi.org/10.1016/j.mtla.2022.101374

85. Ropp, R.C. (1991). *Luminescence and the Solid State.* Elsevier, Amsterdam.

86. Lei, B., Zhang, H., Mai, W., Yue, S., Liu, Y., & Man, S.Q. (2011). Luminescent properties of orange emitting long-lasting phosphorescence phosphor Ca_2SnO_4:Sm^{3+}. *Solid State Sci.* 13(3), 525–528. https://doi.org/10.1016/j.solidstatesciences.2010.12.019

87. Ju, Z., Wei, R., Zheng, J., Gao, X., Zhang, S., & Liu, W. (2011). Synthesis and phosphorescence mechanism of a reddish orange emissive long afterglow phosphor Sm^{3+}-doped Ca_2SnO_4. *Appl. Phys. Lett.* 98(12), 121906. https://doi.org/10.1063/1.3567511

88. Krieke, G., Antuzevics, A., Smits, K., & Millers, D. (2021). Enhancement of persistent luminescence in Ca_2SnO_4:Sm^{3+}. *Opt. Mater.* 113, 110842. https://doi.org/10.1016/j.optmat.2021.110842

89. Gao, X., Zhang, Z., Wang, C., Xu, J., Ju, Z., An, Y., & Liu, W. (2011). The persistent energy transfer and effect of oxygen vacancies on red long-persistent phosphorescence phosphors Ca_2SnO_4:Gd^{3+}, Eu^{3+}. *J. Electrochem. Soc.* 158(12), J405–J408. https://doi.org/10.1149/2.062112jes

90. Jin, Y., Hu, Y., Chen, L., Wang, X., & Ju, G. (2013). Luminescent properties of a red afterglow phosphor Ca_2SnO_4:Pr^{3+}. *Opt. Mater.* 35(7), 1378–1384. https://doi.org/10.1016/j.optmat.2013.02.008

91. Jin, Y., Hu, Y., Chen, L., Wang, X., Ju, G., & Mu, Z. (2013). Luminescent properties of Tb^{3+}-doped Ca_2SnO_4 phosphor. *J. Lumin.* 138, 83–88. https://doi.org/10.1016/j.jlumin.2013.01.023

92. Shi, M., Zhang, D., & Chang, C. (2015). Dy^{3+}: Ca_2SnO_4, a new yellow phosphor with afterglow behavior. *J. Alloys Compd.* 639, 168–172. https://doi.org/10.1016/j.jallcom.2015.02.068

93. Zhang, D., Shi, M., Sun, Y., Guo, Y., & Chang, C. (2016). Long afterglow property of Er^{3+} doped Ca_2SnO_4 phosphor. *J. Alloys Compd.* 667, 235–239. https://doi.org/10.1016/j.jallcom.2016.01.081

94. Li, H.F., Sun, W.Z., Jia, Y.L., Ma, T.F., Fu, J.P., Li, D., Zhang, S., Jiang, L.H., Pang, R., & Li, C.Y. (2015). Investigation on luminescence properties of a long afterglow phosphor Ca_2SnO_4:Tm^{3+}. *Chem. Asian J.* 10(11), 2361–2367. https://doi.org/10.1002/asia.201500580

95. Zhang, J., Yu, M., Qin, Q., Zhou, H., Zhou, M., Xu, X., & Wang, Y. (2010). The persistent luminescence and up conversion photostimulated luminescence properties of nondoped Mg_2SnO_4 material. *J. Appl. Phys.* 108(12), 123518. https://doi.org/10.1063/1.3524280

96. Lei, B., Li, B., Wang, X., & Li, W. (2006). Green emitting long lasting phosphorescence (LLP) properties of Mg_2SnO_4:Mn^{2+} phosphor. *J. Lumin.* 118(2), 173–178. https://doi.org/10.1016/j.jlumin.2005.08.010

97. Zhang, J.C., Qin, Q.S., Yu, M.H., Zhou, H.L., & Zhou, M.J. (2011). Photoluminescence and persistent luminescence properties of non-doped and Ti^{4+}-doped Mg_2SnO_4 phosphors. *Chinese Phys. B* 20(9), 094211. https://doi.org/10.1088/1674-1056/20/9/094211

98. Zhang, J., Qin, Q., Yu, M., Zhou, M., & Wang, Y. (2012). The photoluminescence, afterglow and up conversion photostimulated luminescence of Eu^{3+} doped Mg_2SnO_4 phosphors. *J. Lumin.* 132(1), 23–26. https://doi.org/10.1016/j.jlumin.2011.07.022

99. Li, Y., Li, Y., Chen, R., Sharafudeen, K., Zhou, S., Gecevicius, M., Wang, H., Dong, G., Wu, Y., Qin, X., & Qiu, J. (2015). Tailoring of the trap distribution and crystal field in Cr^{3+}-doped non-gallate phosphors with near-infrared long-persistence phosphorescence. *NPG Asia Mater.* 7, e180. https://doi.org/10.1038/am.2015.38

100. Taktak, O., Souissi, H., & Kammoun, S. (2020). Optical properties of the phosphors Zn_2SnO_4:Cr^{3+} with near-infrared long-persistence phosphorescence for bio-imaging applications. *J. Lumin.* 228, 117563. https://doi.org/10.1016/j.jlumin.2020.117563

101. Fan, L., Li, Y., Hu, Y., Xue, F., Zhang, S., Ju, G., & Lin, X. (2017). A co-doping influence towards enhanced persistent duration of long persistent phosphors. *J. Mater. Sci.: Mater. Electron.* 28(22), 16842–16846. https://doi.org/10.1007/s10854-017-7600-4

102. Fu, J., Sun, L., Hu, X., Du, H., Li, G., Kuang, Y., Li, M., & Guo, D. (2020). Preparation and fluorescence properties of Zn_2SnO_4:Eu^{3+} orange emitting afterglow phosphor. *Ceram. Int.* 46(12), 20277–20283. https://doi.org/10.1016/j.ceramint.2020.05.110

103. Zhang, Y., Huang, R., Lin, Z., Song, J., Wang, X., Guo, Y., Song, C., & Yu, Y. (2016). Co-dopant influence on near-infrared luminescence properties of Zn_2SnO_4:Cr^{3+}, Eu^{3+} ceramic discs. *J. Alloys Compd.* 686, 407–412. http://dx.doi.org/10.1016/j.jallcom.2016.06.041

104. Xu, X., Wang, Y., Gong, Y., Zeng, W., & Li, Y. (2010). Effect of oxygen vacancies on the red phosphorescence of Sr_2SnO_4:Sm^{3+} phosphor. *Opt. Express* 18(16), 16989–16994. https://doi.org/10.1364/OE.18.016989

105. Bing-Fu, L., Song, Y., Yong-Zhe, Z., & Ying-Liang, L. (2010). Luminescence properties of Sr_2SnO_4:Sm^{3+} afterglow phosphor. *Chinese Phys. Lett.* 27(3), 037201. https://doi.org/10.1088/0256-307X/27/3/037201

106. Wang, C., Zheng, Z., Zhang, Y., Liu, Q., Deng, M., Xu, X., Zhou, Z., & He, H. (2020). Modulating trap properties by Nd^{3+}- Eu^{3+} co-doping in Sr_2SnO_4 host for optical information storage. *Opt. Exp.* 28(3), 4249–4257. https://doi.org/10.1364/OE.386164

107. Zhang, B., Shi, M., Li, X., Guo, Y., & Chang, C. (2017). Enhanced red afterglow in Sr_2SnO_4:Sm^{3+} by co-doping Na^+/K^+. *J. Mater. Sci.: Mater. Electron.* 28(23), 17647–17654. https://doi.org/10.1007/s10854-017-7703-y

108. Yu, X., Xu, X., & Qiu, J. (2011). Enhanced long persistence of Sr_2SnO_4:Sm^{3+} red phosphor by co-doping with Dy^{3+}. *Mater. Res. Bull.* 46(4), 627–629. https://doi.org/10.1016/j.materresbull.2010.12.028

109. Xiong, P., & Peng, M. (2019). Visible to near-infrared persistent luminescence from Tm^{3+}-doped two-dimensional layered perovskite Sr_2SnO_4. *J. Mater. Chem., C* 7(27), 8303–8309. https://doi.org/10.1039/C9TC02378J

110. Kamimura, S., Xu, C.N., Yamada, H., Terasaki, N., & Fujihala, M. (2014). Long-persistent luminescence in the near-infrared from Nd^{3+}-doped Sr_2SnO_4 for in vivo optical imaging. *Jpn. J. Appl. Phys.* 53(9), 092403. http://dx.doi.org/10.7567/JJAP.53.092403

111. Pan, Z., Lu, Y.Y., & Liu, F. (2012). Sunlight-activated longpersistent luminescence in the near-infrared from Cr^{3+}-doped zinc gallogermanates. *Nat. Mater.* 11(1), 58–63. https://doi.org/10.1038/nmat3173

112. Wang, W., Sun, Z., He, X., Wei, Y., Zou, Z., Zhang, J., Wang, Z., Zhang, Z., & Wang, Y. (2017). How to design ultraviolet emitting persistent materials for potential multifunctional applications: A living example of a $NaLuGeO_4$:Bi^{3+}, Eu^{3+} phosphor. *J. Mater. Chem. C* 5(17), 4310–4318. https://doi.org/10.1039/C6TC05598B

113. Shi, J., Sun, X., Zheng, S., Fu, X., Yang, Y., Wang, J., & Zhang, H. (2019). Super-long persistent luminescence in the ultraviolet A region from a Bi^{3+}-doped $LiYGeO_4$ phosphor. *Adv. Opt. Mater.* 7, 1900526. https://doi.org/10.1002/adom.201900526

114. Zhang, Y., Chen, D., Wang, W., Yan, S., Liu, J., & Liang, Y. (2020). Long-lasting ultraviolet-A persistent luminescence and photostimulated persistent luminescence in Bi^{3+}-doped $LiScGeO_4$ phosphor. *Inorg. Chem. Front.* 7(17), 3063–3071. https://doi.org/10.1039/D0QI00578A

115. Zhou, Z., Xiong, P., Liu, H., & Peng, M. (2020). Ultraviolet-A persistent luminescence of a Bi^{3+}-activated $LiScGeO_4$ material. *Inorg. Chem.* 59(17), 12920–12927. https://doi.org/10.1021/acs.inorgchem.0c02007

116. Lyu, T., & Dorenbos, P. (2020). Vacuum-referred binding energies of bismuth and lanthanide levels in $ARE(Si, Ge)O_4$ (A = Li, Na; RE = Y, Lu): Toward designing charge-carrier-trapping processes for energy storage. *Chem. Mater.* 32(3), 1192–1209. https://doi.org/10.1021/acs.chemmater.9b04341

117. Lyu, T., & Dorenbos, P. (2020). Towards information storage by designing both electron and hole detrapping processes in bismuth and lanthanide-doped $LiRE(Si, Ge)O_4$ (RE = Y, Lu) with high charge carrier storage capacity. *Chem. Eng. J.* 400, 124776. https://doi.org/10.1016/j.cej.2020.124776

118. Shen, F., Deng, C., Wang, X., & Zhang, C. (2016). Effect of Cr on long persistent luminescence of near-infrared phosphor $Zn_3Ga_2Ge_2O_{10}$:Cr^{3+}. *Mater. Lett.* 178, 185–189. https://doi.org/10.1016/j.matlet.2016.03.108

119. Wu, Y., Li, Y., Qin, X., Chen, R., Wu, D., Liu, S., & Qiu, J. (2015). Near-infrared long-persistent phosphor of $Zn_3Ga_2Ge_2O_{10}$:Cr^{3+} sintered in different atmosphere. *Spectrochim. Acta-A: Mol. Biomol.* 151, 385–389. https://doi.org/10.1016/j.saa.2015.06.117

120. Xu, H., & Chen, G. (2017). Enhancing near infrared persistent luminescence from Cr3+ activated zinc gallogermanate powders through Ca^{2+} doping. *Opt. Mater.* 7(8), 2783–2792. https://doi.org/10.1364/OME.7.002783

121. Zhu, Q., Xiahou. J., Guo, Y., Li, H., Ding, C., Wang, J., Li, X., Sun, X., & Li, J.G. (2019). $Zn_3Ga_2Ge_2O_{10}$:Cr^{3+} uniform microspheres: Template free synthesis, tunable bandgap/trap depth, and in vivo rechargeable near infrared-persistent luminescence. *ACS Appl. Bio Mater.* 2(1), 577–587. https://doi.org/10.1021/acsabm.8b00734

122. Wang, Q., Zhang, S., Li, Z., & Zhu, Q. (2018). Near infrared-emitting Cr^{3+}/Eu^{3+} co-doped zinc gallogermanate persistence luminescent nanoparticles for cell imaging. *Nanoscale Res. Lett.* 13(1), 1–9. https://doi.org/10.1186/s11671-018-2477-6

123. Abdukayum, A., Chen, J.T., Zhao, Q., & Yan, X.P. (2013). Functional near infrared-emitting Cr^{3+}/Pr^{3+} co-doped zinc gallogermanate persistent luminescent nanoparticles with superlong afterglow for in vivo targeted bioimaging. *J. Am. Chem. Soc.* 135(38), 14125–14133. https://doi.org/10.1021/ja404243v

124. Sanad, M. (2021). Extending the luminescence properties of zinc gallogermanate via co-doping with cost-effective metals ions (Cr^{3+}/Mg^{2+}, Cr^{3+}/Ca^{2+}, and Cr^{3+}/Sr^{2+}). *J. Mater. Sci.: Mater. Electron.* 32(8), 9929–9937. https://doi.org/10.1007/s10854-021-05650-x

125. Das, S., Sharma, S. K., & Manam, J. (2022). Near infrared emitting Cr^{3+} doped $Zn_3Ga_2Ge_2O_{10}$ long persistent phosphor for night vision surveillance and anti-counterfeit applications. *Ceram. Int.* 48(1), 824–831. https://doi.org/10.1016/j.ceramint.2021.09.163

126. King, R. S., & Skros, D.A. (2017). Sunlight-activated near-infrared phosphorescence as a viable means of latent fingermark visualisation. *Forensic Sci. Int.* 276, e35–e39. https://doi.org/10.1016/j.forsciint.2017.04.012

127. Li, J., Shi, J., Shen, J., Man, H., Wang, M., & Zhang, H. (2015). Specific recognition of breast cancer cells in vitro using near infrared-emitting long-persistence luminescent $Zn_3Ga_2Ge_2O_{10}$:Cr^{3+} nanoprobes. *Nano-Micro Lett.* 7(2), 138–145. https://doi.org/10.1007/s40820-014-0026-0

128. Dudka, A.P. (2017). Multicell model of $La_3Ga_5GeO_{14}$ crystal: A new approach to the description of the short-range order of atoms. *Crystallogr. Rep.* 62(3), 374–381. https://doi.org/10.1134/S106377451703004X

129. Kaminskii, A.A., Mill, B.V., Belokoneva, E.L., & Khodzhabagyan, G.G. (1983). Growth and crystal structure of new inorganic laser material $La_3Ga_5GeO_{14}$-Nd^{3+}. *Izv Akad Nauk SSSR, Neorg Mater* 19(10), 1762–1764. https://inis.iaea.org/search/search.aspx?orig_q=RN:15035335

130. Burkov, V.I., Konstantinova, A.F., Mill, B.V., Veremeichik, T.F., Pyrkov, Y.N., Orekhova, V.P., & Fedotov, E.V. (2009). The absorption and circular dichroism spectra of langasite family crystals doped with chromium ions. *Crystallogr. Rep.* 54(4), 613–618. https://doi.org/10.1134/S1063774509040129

131. Veremeichik, T.F., & Simonov, V.I. (2010). Comparison of the spectral characteristics of chromium-containing crystals of the calcium gallogermanate type with their structural features. *Crystallogr. Rep.* 55(6), 976–982. https://doi.org/10.1134/S106377451006012X

132. Yi, X., Chen, Z., Ye, S., Li, Y., Song, E., & Zhang, Q. (2015). Multifunctionalities of near-infrared upconversion luminescence, optical temperature sensing and long persistent luminescence in $La_3Ga_5GeO_{14}$:Cr^{3+}, Yb^{3+}, Er^{3+} and their potential coupling. *RSC Adv.* 5(61), 49680–49687. https://doi.org/10.1039/C5RA09095D

133. Jia, D., Lewis, L.A., & Wang, X.J. (2010). Cr^{3+}-Doped lanthanum gallogermanate phosphors with long persistent IR emission. *Electrochem. Solid-State Lett.* 13(4), J32–J34. https://doi.org/10.1149/1.3294520

134. Yan, W., Liu, F., Lu, Y.Y., Wang, X.J., Yin, M., & Pan, Z. (2010). Near infrared long-persistent phosphorescence in $La_3Ga_5GeO_{14}$:Cr^{3+} phosphor. *Opt. Express* 18(19), 20215–20221. https://doi.org/10.1364/OE.18.020215

135. Zhang, D.D., Liu, J.M., Song, N., Liu, Y.Y., Dang, M., Fang, G.Z., & Wang, S. (2018). Fabrication of mesoporous $La_3Ga_5GeO_{14}$:Cr^{3+}, Zn^{2+} persistent luminescence nanocarriers with super-long afterglow for bioimaging-guided in vivo drug delivery to gut. *J. Mater. Chem. B* 6(10), 1479–1488. https://doi.org/10.1039/C7TB02759A

136. Redhammer, G.J., & Roth, G. (2005). A comparison of the clinopyroxene compounds $CaZnSi_2O_6$ and $CaZnGe_2O_6$. *Acta Crystallogr. C Struct. Chem.* 61(2), i20–i22. https://doi.org/10.1107/S0108270104033153

137. Che, G., Liu, C., Wang, Q., & Xu, Z. (2008). White-light-emission afterglow phosphor $CaZnGe_2O_6$: Dy^{3+}. *Chem. Lett.* 37(2), 136–137. https://doi.org/10.1246/cl.2008.136

138. Che, G., Liu, C., Li, X., Xu, Z., Liu, Y., & Wang, H. (2008). Luminescence properties of a new Mn^{2+}-activated red long-afterglow phosphor. *J. Phys. Chem. Solids* 69(8), 2091–2095. https://doi.org/10.1016/j.jpcs.2008.03.006

139. Ye, K., Yang, X., & Xiao, S. (2021). Improving red afterglow properties of $CaZnGe_2O_6$: Mn^{2+} by co-doping Bi3+. *Optik* 246, 167799. https://doi.org/10.1016/j.ijleo.2021.167799

140. Dou, X., Xiang, H., Wei, P., Zhang, S., Ju, G., Meng, Z., Chen, L., Hu, Y., & Li, Y. (2018). A novel phosphor $CaZnGe_2O_6$:Bi^{3+} with persistent luminescence and photo-stimulated luminescence. *Mater. Res. Bull.* 105, 226–230. https://doi.org/10.1016/j.materresbull.2018.04.047

141. Kang, R., Nie, J., Dou, X., Zhang, S., Ju, G., Chen, L., Hu, Y., & Li, Y. (2018). Tunable NIR long persistent luminescence and discovery of trap-distribution-dependent excitation enhancement in transition metal doped weak-crystal-field $CaZnGe_2O_6$. *J. Alloys Compd.* 735, 692–699. https://doi.org/10.1016/j.jallcom.2017.11.182

142. Liu, C., Che, G., Xu, Z., & Wang, Q. (2009). Luminescence properties of a Tb^{3+} activated long-afterglow phosphor. *J. Alloys Compd.* 474(1–2), 250–253. https://doi.org/10.1016/j.jallcom.2008.06.055

143. Liu, L., Shi, J., Sun, X., Zhang, Y., Qin, J., Peng, S., Xu, J., Song, L., & Zhang, Y. (2022). Thermo-responsive hydrogel-supported antibacterial material with persistent photocatalytic activity for continuous sterilization and wound healing. *Compos. B. Eng.* 229, 109459. https://doi.org/10.1016/j.compositesb.2021.109459

144. Cai, H., Song, Z., & Liu, Q. (2021). Infrared-photostimulable and long-persistent ultraviolet-emitting phosphor LiLuGeO$_4$:Bi^{3+}, Yb^{3+} for biophotonic applications. *Mater. Chem. Front.* 5(3), 1468–1476. https://doi.org/10.1039/D0QM00932F

145. Lin, Y., Tang, Z., Zhang, Z., Wang, X., & Zhang, J. (2001). Preparation of a new long afterglow blue-emitting Sr$_2$MgSi$_2$O$_7$-based photoluminescent phosphor. *J. Mater. Sci. Lett.* 20(16), 1505–1506. https://doi.org/10.1023/A:1017930630889

146. Poort, S.H.M., Reijnhoudt, H. M., Van der Kuip, H.O.T., & Blasse, G. (1996). Luminescence of Eu^{2+} in silicate host lattices with alkaline earth ions in a row. *J. Alloys Compd.* 241(1–2), 75–81. https://doi.org/10.1016/0925-8388(96)02324-9

147. Lin Y., Nan C.W., Zhou X., Wu J., Wang H., Chen D., & Xu S. (2003). Preparation and characterization of long afterglow M$_2$MgSi$_2$O$_7$-based (M: Ca, Sr, Ba) photoluminescent phosphors. *Mater. Chem. Phys.* 82(3), 860–863. https://doi.org/10.1016/j.matchemphys.2003.07.015

148. Dorenbos, P. (2005). Mechanism of persistent luminescence in Sr$_2$MgSi$_2$O$_7$:Eu^{2+}; Dy^{3+}. *Phys. Status Solidi.* 242(1), R7–R9. https://doi.org/10.1002/pssb.200409080

149. Setlur, A.A., Srivastava, A.M., Pham, H.L., Hannah, M.E., & Happek, U. (2008). Charge creation, trapping, and long phosphorescence in Sr$_2$MgSi$_2$O$_7$:Eu^{2+}, RE^{3+}. *J. Appl. Phys.* 103(5), 053513, https://doi.org/10.1063/1.2844473

150. Xiao, Z., & Xiao, Z. (1998). Long afterglow silicate luminescent material and its manufacturing method. US patent 6,093,346.

151. Xiao, Z., & Xiao, Z. EP 972815A1.

152. Fernandez-Rodriguez, L., Balda, R., Fernandez, J., Duran, A., & Pascual, M.J. (2022). Role of Eu^{2+} and Dy^{3+} concentration in the persistent luminescence of Sr$_2$MgSi$_2$O$_7$ glass-ceramics. *Materials* 15(9), 3068 (15 pages). https://doi.org/10.3390/ma15093068

153. Liu, B., Shi, C., Yin, M., Dong, L., & Xiao, Z. (2005). The trap states in the Sr$_2$MgSi$_2$O$_7$ and (Sr, Ca) MgSi$_2$O$_7$ long afterglow phosphor activated by Eu^{2+} and Dy^{3+}. *J. Alloys Compd.* 387(1–2), 65–69. https://doi.org/10.1016/j.jallcom.2004.06.061

154. Jia, D., Jia, W., & Jia, Y. (2007). Long persistent alkali-earth silicate phosphors doped with Eu^{2+}, Nd^{3+}. *J. Appl. Phys.* 101(2), 023520. https://doi.org/10.1063/1.2409767

155. Wu, H., Hu, Y., Chen, L., & Wang, X. (2011). Enhancement on the afterglow properties of Sr$_2$MgSi$_2$O$_7$: Eu^{2+} by Er^{3+} cooping. *Mater. Lett.* 65(17–18), 2676–2679. https://doi.org/10.1016/j.matlet.2011.05.079

156. Wu, H., Hu, Y., Wang, Y., & Fu, C. (2010). The luminescent properties of the substitution of Ho^{3+} for Dy^{3+} in the M$_2$MgSi$_2$O$_7$: Eu^{2+}, Dy^{3+} (M: Sr, Ca) long afterglow phosphors. *Mater. Sci. Eng. B* 172(3), 276–282. https://doi.org/10.1016/j.mseb.2010.05.030

157. Merizio, L.G., Bonturim, E., Ichikawa, R.U., Silva, I.G.N., Teixeira, V.C., Rodrigues, L.C. V., & Brito, H.F. (2021). Toward an energy-efficient synthesis method to improve persistent luminescence of Sr$_2$MgSi$_2$O$_7$:Eu^{2+}, Dy^{3+} materials. *Materialia* 20, 101226. https://doi.org/10.1016/j.mtla.2021.101226

158. Klasens, H.A., Hoekstra, A.H., & Cox, A.P.M. (1957). Ultraviolet fluorescence of some ternary silicates activated with lead. *J. Electrochem. Soc.* 104(2), 93–100. https://doi.org/10.1149/1.2428519

159. Yonesaki, Y., Takei, T., Kumada, N., & Kinomura, N. (2009). Crystal structure of Eu^{2+}-doped M$_3$MgSi$_2$O$_8$ (M: Ba, Sr, Ca) compounds and their emission properties. *J. Solid State Chem.* 182(3), 547–554. https://doi.org/10.1016/j.jssc.2008.11.032

160. Bidwai, D., Parauha, Y.R., Sahu, M.K., Dhoble, S.J., Jayasimhadri, M., & Swati, G. (2022). Synthesis and luminescence characterization of aqueous stable Sr$_3$MgSi$_2$O$_8$: Eu^{2+}, Dy^{3+} long afterglow nano-phosphor for low light illumination. *J. Solid State Chem.* 310, 123089. https://doi.org/10.1016/j.jssc.2022.123089

161. Lin, Y., Tang, Z., Zhang, Z., & Nan, C.W. (2003). Luminescence of Eu^{2+} and Dy^{3+} activated R$_3$MgSi$_2$O$_8$-based (R=Ca, Sr, Ba) phosphors. *J. Alloys Compd.* 348(1–2), 76–79. https://doi.org/10.1016/S0925-8388(02)00796-X

162. Alvani, A.S., Moztarzadeh, F., & Sarabi, A.A. (2005). Effects of dopant concentrations on phosphorescence properties of Eu/Dy-doped Sr$_3$MgSi$_2$O$_8$. *J. Lumin.* 114(2), 131–136. https://doi.org/10.1016/j.jlumin.2004.12.012

163. Rashidi, L.H., Homayoni, H., Zou, X., Liu, L., & Chen, W. (2016). Investigation of the strategies for targeting the afterglow nanoparticles to tumor cells. *Photodiagnosis Photodyn. Ther.* 13, 244–254. https://doi.org/10.1016/j.pdpdt.2015.08.001

164. Wang, X.J., Jia, D., & Yen, W.M. (2003). Mn^{2+} activated green, yellow, and red long persistent phosphors. *J. Lumin.* 102–103, 34–37. https://doi.org/10.1016/S0022-2313(02)00541-0

165. Lin, L., Min, Y., Chaoshu, S., Weiping, Z., & Baogui, Y. (2006). Synthesis and luminescence properties of red phosphors: Mn^{2+} doped MgSiO$_3$ and Mg$_2$SiO$_4$ prepared by sol-gel method. *J. Rare Earths* 24(1), SUPPL. 1, 104–107. https://doi.org/10.1016/S1002-0721(07)60334-2

166. Lin, L., Shi, C., Wang, Z., Zhang, W., & Yin, M. (2008). Kinetics model of red long-lasting phosphorescence in $MgSiO_3$:Eu^{2+}, Dy^{3+}, Mn^{2+}. *J. Alloys Compd.* 466(1–2), 546–550. https://doi.org/10.1016/j.jallcom.2007.11.093

167. Zhu, Y., & Ge, M. (2013). Factors affecting afterglow properties of red-emitting phosphor $MgSiO_3$:Mn^{2+}, Nd^{3+} for luminescent fiber. *J. Rare Earths* 31(7), 660–664. https://doi.org/10.1016/S1002-0721(12)60338-X

168. Mejia-Bernal, J.R., Mumanga, T.J., Diaz-Torres, L.A., Vallejo-Hernandez, M.A., & Gomez-Solis, C. (2021). Synthesis and evaluation of $MSiO_3$ (M = Ba, Sr, Mg) for photocatalytic hydrogen generation under UV irradiation. *Mater. Lett.* 295, 129851. https://doi.org/10.1016/j.matlet.2021.129851

169. Kubota, S., Yamane, H., & Shimada, M. (2002). A new luminescent material, $Sr_3Al_{10}SiO_{20}$:Tb^{3+}. *Chem. Mater.* 14(10), 4015–4016. https://doi.org/10.1021/cm025607o

170. Rief, A., & Kubel, F. (2007). $Sr_3Al_{10}SiO_{20}$ from single-crystal data. *Acta Crystallogr. Sect. E Struct. Rep.*, Online 63(1), i19–i21. https://doi.org/10.1107/S1600536806054936

171. Kuang, J., Liu, Y., Zhang, J., Huang L., Rong, J., & Yuan, D. (2005). Blue-emitting long-lasting phosphor, $Sr_3Al_{10}SiO_{20}$:Eu^{2+}, Ho^{3+}. *Solid State Comm.* 136(1), 6–10. https://doi.org/10.1016/j.ssc.2005.06.030

172. Kuang, J.Y., Liu, Y.L., & Zhang, J.X. (2006). Effects of RE3+ as a co-dopant in blue-emitting long-lasting phosphors, $Sr_3Al_{10}SiO_{20}$:Eu^{2+}. *J. Mater. Sci.* 41(17), 5500–5503. https://doi.org/10.1007/s10853-006-0244-z

173. Bingfu, L., Yingliang, L., Jie, L., Zeren, Y., & Chunshan, S. (2004). Luminescence properties of rare earth ions in cadmium metasilicate. *J. Rare Earth Sci.* 22(4), 443–446. http://www.cqvip.com/qk/84120x/200404/11038147.html

174. Liu, Y., Lei, B., & Shi, C. (2005). Luminescent properties of a white afterglow phosphor $CdSiO_3$:Dy^{3+}. *Chem. Mater.* 17(8), 2108–2113. https://doi.org/10.1021/cm0496422

175. Wanmaker, W.L., & ter Vrugt, J.W. (1967). Lanthanide activated tube lighting and cathode ray tube phosphors. *Philips Res. Rept.* 22, 355–359.

176. Pang, R., Li, C., Shi, L., & Su, Q. (2009). A novel blue-emitting long-lasting pyrophosphate phosphor $Sr_2P_2O_7$:Eu^{2+}, Y^{3+}. *J. Phys. Chem. Solids* 70(2), 303–306. https://doi.org/10.1016/j.jpcs.2008.10.016

177. Golubev, V.N., & Lazoryak, B.I. (1991). Double phosphates $Ca_9R(PO_4)_7$. *Russ J. Inorg. Mater.* 27, 480–483.

178. Jia, Y., Li, H., Zhao, R., Sun, W., Su, Q., Pang, R., & Li, C. (2014). Luminescence properties of a new bluish green long-lasting phosphorescence phosphor $Ca_9Bi(PO_4)_7$:Eu^{2+}, Dy^{3+}. *Opt. Mater.* 36(11), 1781–1786. http://dx.doi.org/10.1016/j.optmat.2014.04.006

179. Sarver, J.F., Hoffman, M.V., & Hummel, F.A. (1961). Phase equilibria and Tin-activated luminescence in strontium orthophosphate systems. *J. Electrochem. Soc.* 108(12), 1103–1110. https://doi.org/10.1149/1.2427964

180. Hemon, A., & Courbion, G. (1990). The crystal structure of α-$SrZn_2(PO_4)_2$: A hurlbutite type. *J. Solid State Chem.* 85(1), 164–168. https://doi.org/10.1016/S0022-4596(05)80072-6

181. Yang, W.J., & Chen, T.M. (2006). White-light generation and energy transfer in $SrZn_2(PO_4)_2$: Eu, Mn phosphor for ultraviolet light-emitting diodes. *Appl. Phys. Lett.* 88(10), 101903. https://doi.org/10.1063/1.2182026

182. Xu, Y., Zhang, K., & Chang, C. (2020). Effects of Bi3+ co-doping on structure and luminescence of $SrZn_2(PO_4)_2$-based phosphor. *J. Mater. Sci. Mater. Electron.* 31(13), 10072–10077. https://doi.org/10.1007/s10854-020-03552-y

183. Lu, W., Liu, Y.F., Wang, Y., Wang, Z.J., & Pang, L.B. (2015). Improved luminescent properties of $SrZn_2(PO_4)_2$:Sm^{3+} doped with Li^+, Na^+ and K^+ ions. *Optoelectron. Lett.* 11(15), 366–369. https://doi.org/10.1007/s11801-015-5116-9

184. Sun, Y., Chen, W., Liu, S., Yan, S., Zhang, S., Huang, L., & Zheng, Z. (2021). Effect of Zn deficiency on the enhancement of yellowish green persistent phosphor Dy^{3+}-activated $SrZn_2(PO_4)_2$. *Solid State Sci.* 121, 106754. https://doi.org/10.1016/j.solidstatesciences.2021.106754

185. Fu, X., Zheng, S., Shi, J., Li, Y., & Zhang, H. (2017). Long persistent luminescence property of a novel green emitting $SrLaGaO_4$: Tb^{3+} phosphor. *J. Lumin.* 184, 199–204. https://doi.org/10.1016/j.jlumin.2016.12.047

186. Tian, S., Zhao, L., Chen, W., Liu, Z., Fan, X., Min, Q., Yu, H., Yu, X., Qiu, J., & Xu, X. (2018). Abnormal photo-stimulated luminescence in $Ba_2Ga_2GeO_7$: Tb^{3+}, Bi^{3+}. *J. Lumin.* 202, 414–419. https://doi.org/10.1016/j.jlumin.2018.06.007

187. Xu, J., Chen, D., Yu, Y., Zhu, W.J., Zhou, J., & Wang, Y. (2014). $Cr^{3+}:SrGa_{12}O_{19}$: A broadband near-infrared long-persistent phosphor. *Chem. Asian J.* 9(4), 1020–1025. https://doi.org/10.1002/asia.201400009

188. Basavaraju, N., Priolkar, K.R., Gourier, D., Sharma, S.K., Bessiere, A., & Viana, B. (2015). The importance of inversion disorder in the visible light induced persistent luminescence in Cr^{3+} doped AB_2O_4 (A = Zn or Mg and B = Ga or Al). *Phys. Chem. Chem. Phys.* 17(3), 1790–1799. https://doi.org/10.1039/C4CP03866E

189. Basavaraju, N., Priolkar, K.R., Gourier, D., Bessiere, A., & Viana, B. (2015). Order and disorder around Cr^{3+} in chromium doped persistent luminescent AB_2O_4 spinels. *Phys. Chem. Chem. Phys.* 17(16), 10993–10999. https://doi.org/10.1039/C5CP01097G

190. Nikl, M., Novoselov, A., Mihokova, E., Polak, K., Dusek, M., McClune, B., Yoshikawa, A., & Fukuda, T. (2005). Photoluminescence of Bi^{3+} in $Y_3Ga_5O_{12}$ single-crystal host. *J. Phys.: Condens. Matter* 17(21), 3367–3376. https://doi.org/10.1088/0953-8984/17/21/029

191. Dai, Z., Boiko, V., Grzeszkiewicz, K., Markowska, M., Ursi, F., Holsa, J., Saladino, M.L., & Hreniak, D. (2021). Effect of annealing temperature on persistent luminescence of $Y_3Al_2Ga_3O_{12}:Cr^{3+}$ co-doped with Ce^{3+} and Pr^{3+}. *Opt. Mater.* 111,110522. https://doi.org/10.1016/j.optmat.2020.110522

192. Wang, S., Chen, W., Zhou, D., Qiu, J., Xu, X., & Yu, X. (2017). Long persistent properties of $CaGa_2O_4:Bi^{3+}$ at different ambient temperature. *J. Am. Ceram. Soc.* 100(8), 3514–3521. https://doi.org/10.1111/jace.14875

193. Xiahou, J., Zhu, Q., Zhu, L., Li, S., & Li, J.G. (2021). Local structure regulation in near-infrared persistent phosphor of $ZnGa_2O_4:Cr^{3+}$ to fabricate natural-light rechargeable optical thermometer. *ACS Appl. Electron. Mater.* 3(9), 3789–3803. https://doi.org/10.1021/acsaelm.1c00305

194. Lin, S., Lin, H., Ma, C., Cheng, Y., Ye, S., Lin, F., Li, R., Xu, J., & Wang, Y. (2020). High-security level multi-dimensional optical storage medium: Nanostructured glass embedded with $LiGa_5O_8:Mn^{2+}$ with photostimulated luminescence. *Light Sci. Appl.* 9(1), 1–10. https://doi.org/10.1038/s41377-020-0258-3

195. Li, N., Zhang, P., Wang, Z., Wei, Z., Jiang, Z., Shang, Y., Zhang, M., Qiang, Q., Zhao, L., & Chen, W. (2021). Novel UV and X-ray irradiated white-emitting persistent luminescence and traps distribution of $Ca_5Ga_6O_{14}:Pr^{3+}$ phosphors. *J. Alloys Compd.* 858, 157719. https://doi.org/10.1016/j.jallcom.2020.157719

196. Long, Z., Zhou, J., Qiu, J., Wang, Q., Zhou, D., Xu, X., Yu, X., Wu, H., & Li, Z. (2018). Thermally stable photoluminescence and long persistent luminescence of $Ca_3Ga_4O_9$: Tb^{3+}/Zn^{2+}. *J. Rare Earths* 36(7), 675–679. https://doi.org/10.1016/j.jre.2017.11.016

197. Wang, X., Boutinaud, P., Li, L., Cao, J., Xiong, P., Li, X., Luo, H., & Peng, M. (2018). Novel persistent and tribo-luminescence from bismuth ion pairs doped strontium gallate. *J. Mater. Chem.* 6(38), 10367–10375. https://doi.org/10.1039/c8tc04012e

198. Wang, J., Chen, W., Peng, L., Han, T., Liu, C., Zhou, Z., Qiang, Q., Shen, F., Wang, J., & Liu, B. (2022). Long persistent luminescence property of green emitting $Sr_3Ga_4O_9:Tb^{3+}$ phosphor for anti-counterfeiting application. *J. Lumin.* 250, 119066. https://doi.org/10.1016/j.jlumin.2022.119066

199. Liu, D., Yun, X., Li, G., Dang, P., Molokeev, M.S., Lian, H., Shang, M., & Lin, J. (2020). Enhanced cyan emission and optical tuning of $Ca_3Ga_4O_9:Bi^{3+}$ for high-quality full-spectrum white light-emitting diodes. *Adv. Opt. Mater.* 8(22), 2001037. https://doi.org/10.1002/adom.202001037

200. Kim, J.S., Kang, H.I., Kim, W.N., Kim, J.I., Choi, J.C., Park, H.L., Kim, G.C., Kim, T.W., Hwang, Y.H., Mho, S.I., Jung, M.C., & Han, M. (2003). Color variation of $ZnGa_2O_4$ phosphor by reduction-oxidation processes. *Appl. Phys. Lett.* 82(13), 2029–2031. https://doi.org/10.1063/1.1564632

201. Jin, M. Li, F., Xiahou, J., Zhu, L., Zhu, Q., & Li, J.G. (2023). A new persistent luminescence phosphor of $ZnGa_2O_4:Ni^{2+}$ for the second near-infrared transparency window. *J. Alloys Compd.* 931, 167491. https://doi.org/10.1016/j.jallcom.2022.167491

202. Allix, M., Chenu, S., Veron, E., Poumeyrol, T., Kouadri-Boudjelthia, E.A., Alahrache, S., Porcher, F., Massiot, D., & Fayon, F. (2013). Considerable improvement of long-persistent luminescence in germanium and tin substituted $ZnGa_2O_4$. *Chem. Mater.* 25(9), 1600–1606. https://doi.org/10.1021/cm304101n

203. Wang, X., Qiao, H., Wang, X., Xu, Y., Liu, T., Song, F., An, Z., Zhang, L., & Shi, F., (2022). NIR photoluminescence of $ZnGa_2O_4:Cr$ nanoparticles synthesized by hydrothermal process. *J. Mater. Sci.: Mater. Electron.* 33(24), 19129–19137. https://doi.org/10.1007/s10854-022-08750-4

204. Rudolf, M.M., Bortel, G., Markus, B.G., Jegenyes, N., Verkhovlyuk, V., Kamaras, K., Simon, F., Gali, A., & Beke, D. (2022). Optimization of chromium-doped zinc gallate nanocrystals for strong near-infrared emission by annealing. *ACS Appl. Nano Mater.* 5(7), 8950–8961. https://doi.org/10.1021/acsanm.2c01156

205. Bessiere, A., Jacquart, S., Priolkar, K., Lecointre, A., Viana, B., & Gourier, D. (2011). $ZnGa_2O_4:Cr^{3+}$: A new red long-lasting phosphor with high brightness. *Opt. Express* 19, 10131–10137. https://doi.org/10.1364/OE.19.010131

206. Gourier, D., Bessiere, A., Sharma, S.K., Binet, L., Viana, B., Basavaraju, N., & Priolkar, K.R. (2014). Origin of the visible light induced persistent luminescence of Cr^{3+}-doped zinc gallate. *J. Phys. Chem. Solids* 75(7), 826–837. http://dx.doi.org/10.1016/j.jpcs.2014.03.005

207. Zhuang, Y., Ueda, J., & Tanabe, S. (2013). Enhancement of red persistent luminescence in Cr^{3+}-doped $ZnGa_2O_4$ phosphors by Bi2O3 codoping. *Appl. Phys. Express* 6(5), 052602. https://doi.org/10.7567/APEX.6.052602

208. Li, D., Wang, Y., Xu, K., Li, L., Hu, Z., & Zhao, H. (2015). Enhancement of photoluminescence, persistent luminescence and photocatalytic activity in $ZnGa_2O_4$ phosphors by lithium ion doping. *Opt. Mater.* 42, 313–318. https://doi.org/10.1016/j.optmat.2015.01.020

209. Pan, Z., Castaing, V., Yan, L., Zhang, L., Zhang, C., Shao, K., Zheng, Y., Duan, C., Liu, J., Richard, C., & Viana, B. (2020). Facilitating low-energy activation in the near-infrared persistent luminescent phosphor $Zn_{1+x}Ga_{2-2x}Sn_xO_4:Cr^{3+}$ via crystal field strength modulations. *J. Phys. Chem. C* 124(15), 8347–8358. https://doi.org/10.1021/acs.jpcc.0c01951

210. Shi, J., Sun, X., Zheng, S., Song, L., Zhang, F., Madl, T., Zhang, Y., Zhang, H., & Hong, M. (2020). Tin-doped near-infrared persistent luminescence nanoparticles with considerable improvement of biological window activation for deep tumor photodynamic therapy. *ACS Appl. Bio Mater.* 3(9), 5995–6004. https://doi.org/10.1021/acsabm.0c00644

211. Zhang, W., Zhang, J., Chen, Z., Wang, T., & Zheng, S. (2010). Spectrum designation and effect of Al substitution on the luminescence of Cr^{3+} doped $ZnGa_2O_4$ nanosized phosphors. *J. Lumin.* 130(10), 1738–1743. https://doi.org/10.1016/j.jlumin.2010.04.002

212. Hu, Z., Ye, D., Lan, X., Zhang, W., Luo, L., & Wang, Y. (2016). Influence of co-doping Si ions on persistent luminescence of $ZnGa_2O_4$: Cr^{3+} red phosphors. *Opt. Mater. Express* 6(4), 1329–1338. https://doi.org/10.1364/OME.6.001329

213. Maldiney, T., Bessiere, A., Seguin, J., Teston, E., Sharma, S.K., Viana, B., Bos, A.J., Dorenbos, P., Bessodes, M., Gourier, D., & Scherman, D. (2014). The in vivo activation of persistent nanophosphors for optical imaging of vascularization, tumours and grafted cells. *Nat. Mater.* 13(4), 418–426, https://doi.org/10.1038/nmat3908

214. Dai, Z., Mao, X., Liu, Q., Zhu, D., Chen, H., Xie, T., Xu, J., Hreniak, D., Nikl, M., & Li, J. (2022). Effect of dopant concentration on the optical characteristics of $Cr^{3+}:ZnGa_2O_4$ transparent ceramics exhibiting persistent luminescence. *Opt. Mater.* 125, 112127. https://doi.org/10.1016/j.optmat.2022.112127

215. Zhang, X., Zhang, J., & Zhu, Q. (2022). Doping upconversion ion pair of Yb^{3+}/Er^{3+} in $ZnGa_2O_4:Cr^{3+}$ for multimode luminescence and advanced anti-counterfeiting. *Opt. Mater.* 125, 112100. https://doi.org/10.1016/j.optmat.2022.112100

216. Noto, L.L., Shaat, S.K.K., Poelman, D., Dhlamini, M.S., Mothudi, B.M., & Swart, H.C. (2016). Structure, photoluminescence and thermoluminescence study of a composite $ZnTa_2O_6/ZnGa_2O_4$ compound doped with Pr^{3+}. *Opt. Mater.* 55, 68–72. http://dx.doi.org/10.1016/j.optmat.2016.03.029

217. Luchechko, A., Zhydachevskyy, Y., Ubizskii, S., Kravets, O., Popov, A.I., Rogulis, U., Elsts, E., Bulur, E., & Suchocki, A. (2019). Afterglow, TL and OSL properties of Mn^{2+}-doped $ZnGa_2O_4$ phosphor. *Sci. Rep.* 9(1), 9544 (8 pages). https://doi.org/10.1038/s41598-019-45869-7

218. Zhuang, Y., Ueda, J., & Tanabe, S. (2012). Photochromism and white long-lasting persistent luminescence in Bi^{3+}-doped $ZnGa_2O_4$ ceramics. *Opt. Mater. Express* 2(10), 1378–1383. https://doi.org/10.1364/OME.2.001378

219. Scolis, Y.Y., Levitskii, V.A., Lykova, L.N., & Kalinina, T.A. (1981). Thermodynamics of double oxides. III. Study of the $CaO-Ga_2O_3$ system by the emf method and X-ray analysis. *J. Solid State Chem.* 38(1), 10–18. https://doi.org/10.1016/0022-4596(81)90466-7

220. Nosenko, A.E., Bily, A.I., & Grechukh, T.Z. (1992). Oscillatory spectra of Ca5Ga6O14 and $Ca_3Ga_4O_9$ crystals. *Ukrainskii Khimicheskii Zhurnal* 58(3), 127–131. https://www.elibrary.ru/item.asp?id=31104656

221. Jeevaratnam, J., Glasser, F.P., & Glasser, S.D. (1963). Crystallography of the $CaGa_2O_4$ Polymorphs. *Zeitschrift für Kristallographie-Crystalline Materials* 118(1–6), 257–262. https://doi.org/10.1524/zkri.1963.118.3-4.257

222. Minami, T., Yamada, H., Kubota, Y., & Miyata, T. (1997). Mn-activated $CaO\text{-}Ga_2O_3$ phosphors for thin-film electroluminescent devices. *Jpn. J. Appl. Phys., Part 2: Lett.* 36(9A), L1191–L1194. https://doi.org/10.1143/JJAP.36.L1191

223. Wang, Y., Feng, P., Ding, S., Tian, S., & Wang, Y. (2021). A promising route for developing yellow long persistent luminescence and mechanoluminescence in $CaGa_2O_4{:}Pr^{3+}$, Li^+. *Inorg. Chem. Front.* 8(15), 3748–3759. https://doi.org/10.1039/D1QI00326G

224. Rai, M., Mishra, K., Rai, S.B., & Morthekai, P. (2018). Tailoring UV-blue sensitization effect in enhancing near infrared emission in X, Yb^{3+}: $CaGa_2O_4$ (X = 0, Eu^{3+}, Bi^{3+}, Cr^{3+}) phosphor for solar energy conversion. *Mater. Res. Bull.* 105, 192–201. https://doi.org/10.1016/j.materresbull.2018.04.051

225. Chen, Z., Cheng, Q., Ke, H., Li, Y., Wei, Q., & Ge, M. (2020). Preparation and luminescent properties of multicolored of $SrAl_2O_4{:}Eu^{2+}$, Dy^{3+}/SiO_2-coated red-emitting coumarin color converter/polyamide 6 luminous fiber with warm-toned luminescence. *Text. Res. J.* 90(15–16), 1783–1791. https://doi.org/10.1177/0040517519900934

226. Xue, H., Ge, M., & Zhu, Y. (2021). Preparation and properties research of a warm tone luminous polyacrylonitrile coaxial fiber based on SrAl2O4 phosphor. *J. Lumin.* 231, 117777. https://doi.org/10.1016/j.jlumin.2020.117777

227. Zhao, L., Mao, J., Jiang, B., Wei, X., Chen, Y., & Yin, M. (2018). Temperature-dependent persistent luminescence of $SrAl_2O_4{:}Eu^{2+}$, Dy^{3+}, Tb^{3+}: A strategy of optical thermometry avoiding real-time excitation. *Opt. Lett.* 43(16), 3882–3884. https://doi.org/10.1364/OL.43.003882

228. Ueda, J., Kuroishi, K., & Tanabe, S. (2014). Yellow persistent luminescence in $Ce^{3+}\text{–}Cr^{3+}$-codoped gadolinium aluminum gallium garnet transparent ceramics after blue-light excitation. *Appl. Phys. Express* 7(6), 062201. https://doi.org/10.7567/APEX.7.062201

229. Castaing, V., Arroyo, E., Becerro, A.I., Ocana, M., Lozano, G., & Miguez, H. (2021). Persistent luminescent nanoparticles: Challenges and opportunities for a shimmering future. *J. Appl. Phys.* 130(8), 080902. https://doi.org/10.1063/5.0053283

230. Xu, J., & Tanabe, S. (2019). Persistent luminescence instead of phosphorescence: History, mechanism, and perspective. *J. Lumin.* 205, 581–620. https://doi.org/10.1016/j.jlumin.2018.09.047

231. Ren, W., Lin, G., Clarke, C., Zhou, J., & Jin, D. (2020). Optical nanomaterials and enabling technologies for high-security-level anticounterfeiting. *Adv. Mater.* 32(18), 1901430. https://doi.org/10.1002/adma.201901430

232. Wang, J., Ma, Q., Wang, Y., Shen, H., & Yuan, Q. (2017). Recent progress in biomedical applications of persistent luminescence nanoparticles. *Nanoscale* 9(19), 6204–6218. https://doi.org/10.1039/C7NR01488K

233. Viana, B., Sharma, S.K., Gourier, D., Maldiney, T., Teston, E., Scherman, D., & Richard, C. (2016). Long term in vivo imaging with Cr^{3+} doped spinel nanoparticles exhibiting persistent luminescence. *J. Lumin.* 170, Part 3, 879–887. https://doi.org/10.1016/j.jlumin.2015.09.014

234. Lecuyer, T., Teston, E., Ramirez-Garcia, G., Maldiney, T., Viana, B., Seguin, J., Mignet, N., Scherman, D., & Richard, C. (2016). Chemically engineered persistent luminescence nanoprobes for bioimaging. *Theranostics* 6(13), 2488–2524. https://doi.org/10.7150/thno.16589

235. Niu, J.Y., Wang, X., Lv, J.Z., Li, Y., & Tang, B. (2014). Luminescent nanoprobes for in-vivo bioimaging. *Trends Analyt. Chem.* 58, 112–119. https://doi.org/10.1016/j.trac.2014.02.013

236. Vaiano, V., Sacco, O., & Sannino, D. (2019). Electric energy saving in photocatalytic removal of crystal violet dye through the simultaneous use of long-persistent blue phosphors, nitrogen-doped TiO_2 and UV-light emitting diodes. *J. Clean. Prod.* 210, 1015–1021. https://doi.org/10.1016/j.jclepro.2018.11.017

237. Feng, P., Wei, Y., Wang, Y., Zhang, J., Li, H., & Ci, Z. (2016). Long persistent phosphor $CdSiO_3{:}Gd^{3+}$, Bi^{3+} and potential photocatalytic application of $CdSiO_3{:}Gd^{3+}$, $Bi^{3+}@TiO_2$ in dark. *J. Am. Ceram. Soc.* 99(7), 2368–2375. https://doi.org/10.1111/jace.14142

238. Aliabadi, H. M., Zargoosh, K., Afshari, M., Dinari, M., & Maleki, M.H. (2021). Synthesis of a luminescent $g\text{-}C_3N_4\text{-}WO_3\text{-}Bi_2WO_6/SrAl_2O_4{:}Eu^{2+}$, Dy^{3+} nanocomposite as a double z-scheme sunlight activable photocatalyst. *New J. Chem.* 45(10), 4843–4853. https://doi.org/10.1039/D0NJ05529H

239. Liu, X., Chen, X., Li, Y., Wu, B., Luo, X., Ouyang, S., Luo, S., Al Kheraif, A. A., & Lin, J. (2019). A $g\text{-}C_3N_4@Au@SrAl_2O_4{:}Eu^{2+}$, Dy^{3+} composite as an efficient plasmonic photocatalyst for round-the-clock environmental purification and hydrogen evolution. *J. Mater. Chem. A* 7(32), 19173–19186. https://doi.org/10.1039/C9TA06423K

240. Tawalare, P.K. (2021). Optimizing photovoltaic conversion of solar energy. *AIP Adv.* 11(10), 100701. https://doi.org/10.1063/5.0064202

241. Zhang, J., Lin, J., Wu, J., Zhang, S., Zhou, P., Chen, X., & Xu, R. (2016). Preparation of long persistent phosphor $SrAl_2O_4:Eu^{2+}$, Dy^{3+} and its application in dye-sensitized solar cells. *J. Mater. Sci.: Mater. Electron.* 27(2), 1350–1356. https://doi.org/10.1007/s10854-015-3896-0

242. Puntambekar, A., & Chakrapani, V. (2016). Excitation energy transfer from long-persistent phosphors for enhancing power conversion of dye-sensitized solar cells. *Phys. Rev., B* 93(24), 245301. https://doi.org/10.1103/PhysRevB.93.245301

243. Znajdek, K., Szczecinska, N., Sibinski, M., Czarnecki, P., Wiosna-Salyga, G., Apostoluk, A., Mandrolo, F., Rogowski, S., & Lisik, Z. (2018). Energy converting layers for thin-film flexible photovoltaic structures. *Open Phys.* 16(1), 820–825. https://doi.org/10.1515/phys-2018-0102

9 Metal Oxides
Solid-State Lasers

S.G. Revankar
Priyadarshani Bhagwati College of Engineering

CONTENTS

9.1 INTRODUCTION

9.1.1 EARLY HISTORY

A solid-state laser is a laser that uses a gain medium that is a solid, rather than a liquid as in dye lasers or a gas as in gas lasers. Semiconductor-based lasers are also in the solid state, but are generally considered as a separate class from solid-state lasers. Most solid-state lasers (excluding diode and semiconducting lasers) need optical pumping, and hence, the phenomenon may be treated as photoluminescence. A large number of solid-state lasers have been found. The solid-state laser materials should have high absorption coefficient, population inversion, high quantum efficiency (QE), high cross section for stimulated laser emission, and quick non-radiative relaxation of the lower laser levels.

DOI: 10.1201/9781003366232-9

Lasing in solid state is most conveniently achieved by introducing lanthanide or 3d transition metal ions in a suitable insulating, transparent host. These ions under some stringent conditions provide light amplification upon optical pumping which induces population inversion. In semiconductor lasers, population inversion is achieved by charged carrier injection under high current density. A large number of host–activator pairs have been proposed for as solid-state laser systems. These span very broad spectral range. Solid-state lasers can in principle be constructed for wavelengths from UV to far infrared.

Notwithstanding these proposals, practical solid-state lasers are few in numbers. The main reason for this is that the practical laser system must satisfy several stringent conditions.

These include but are not limited to the following [1]:

1. chemical and photostability of the ion–host combination,
2. resistance of the host to fracture during high-power operation,
3. resistance of the host to moisture,
4. ability to dope the host matrix at the desired laser-active ion concentration,
5. high luminescence QE at room temperature,
6. high QE at the desired dopant concentration,
7. absence of absorption and scattering losses in the host at the lasing wavelengths,
8. absence of power degrading mechanisms such as excited-state absorption (ESA) or Auger-type energy-transfer upconversion (ETU) at the lasing wavelength [10–12],
9. absence of ESA at the pump wavelength [10,13], and
10. availability of practical pump lasers overlapping with the absorption band of the gain medium, and repeatable synthesis protocol of the active medium.

Al_2O_3:Cr was the first system in which lasing in solid state was demonstrated. Since then, this topic received attention of researchers owing to applications in several fields such as homeland defense, documents security, forgery detection, forensic science, medicine and surgery, fabrication of metal parts, printing, holography, remote control and sensing, optical memories, data storage and retrieval, practically in every walk of life. SSL enjoys several advantages over its predecessor gas laser such as compact size, portability, and robust. Initially, SSL media needed large, transparent single crystals which are rather difficult to grow. Subsequently, transparent ceramics and glasses have also been successfully used as laser media.

9.2　OPTICALLY PUMPED OXIDE LASERS

9.2.1　Al_2O_3-BASED SOLID-STATE LASERS

Aluminum oxide (Al_2O_3) is a dielectric material found in both crystalline and amorphous phases. It has attractive features making it an excellent candidate for integrated photonic devices. Al_2O_3 exhibits a wide transparency window extending from the ultraviolet to the mid-infrared (150–5,500 nm) [2]. Moreover, it has RI contrast of 0.2 with respect to silica. This permits construction of miniature-integrated optoelectronics. Al_2O_3 can also be deposited on silicon wafers, which makes it compatible with associated photonics technologies. Though Al is small in size compared to trivalent lanthanides that put solubility limits, these are still better than those for silica-based systems [3]. Amplification and lasing have been demonstrated at a variety of wavelengths, including ~0.88, ~1.06 and ~1.33 μm for Nd^{3+}:Al_2O_3 [4], ~1.03 μm for Yb^{3+}:Al_2O_3 [5], ~1.55 μm for Er^{3+}:Al_2O_3 [6,7], ~1.8–1.9 μm for Tm^{3+}:Al_2O_3 [8], and at ~2 μm for Ho^{3+}:Al_2O_3 [9].

9.2.1.1　Al_2O_3:Cr^{3+}, Ruby

In ruby, some of the Al^{3+} ions (usually much less than 1%) of the Al_2O_3 crystal are replaced by Cr^{3+} ions with three unpaired electrons in their outer $3d^3$ shell. Each Cr^{3+} ion is surrounded by six

O_2^- ions in the form of a distorted octahedron. The electrostatic crystal field, which the chromium ions are subject to, lifts the 3d orbital degeneracy.

Aluminum oxide (Al_2O_3) is an excellent electrical insulator with high thermal conductivity and high optical transparency; also, it has considerable mechanical strength and chemical stability even at several hundreds of degree Celsius. Al_2O_3 is also an important laser material for transition metal ions (ruby and Ti:sapphire lasers) [10]. Therefore, this material has been the subject of considerable research effort even now due to the technological importance for use of it in a variety of solid-state device applications.

The luminescence spectrum of α-Al_2O_3 consists of the sharp Eg $\rightarrow$ $^4A_{2g}$ transition lines without any strong broad emission band either in the Stokes or in the anti-Stokes region.

The sharp emission lines are due to the split of the $2E_g$ state caused by the spin–orbit interactions (R1 and R2 lines). The splitting energy of the R1 and R2 doublet is ~4 meV in the case of trigonal distortion [11]. The R doublet energies observed are 1.791 eV (R1) and 1.788 eV (R2).

The PLE spectrum shows the large, broad excitation bands peaking at ~2.2 and 3.1 eV, which are mainly caused by the $^4A_{2g} \rightarrow {}^4T_{2g}$ and $^4A_{2g} \rightarrow$ 4T1g transitions, respectively. The experimental PLE spectrum further reveals fine peak structures at ~2.6 eV that may be due to the spin-forbidden $^4A_{2g} \rightarrow {}^2T_{2g}$ transitions [12].

Alpha-alumina α-Al_2O_3 is a remarkable host material for rare-earth ions because it presents excellent mechanical hardness combined with chemical stability and solubility besides a high transparency window from the ultraviolet to the infrared [13]. It plays a major role in many technologies due to its remarkable physical properties, such as a high melting point, hydrophobicity, high elastic modulus, high optical transparency, high refractive index (RI), thermal and chemical stability, low surface acidity, and fine optical and dielectric characteristics [14–16]. The good adhesion to Si surface makes Al_2O_3 attractive in the microelectronics and optoelectronics [17,18]. Owing to some of these properties, it is also an excellent laser host. The ruby (Al_2O_3:Cr^{3+}) is the first crystal found to exhibit a laser emission as demonstrated by Maiman at the Hughes Lab in 1960 [19,20], which marked the beginning of the solid-state laser technology. Al_2O_3:Cr^{3+} has also been recently suggested for optical thermometry application [21]. Apart from celebrated α-Al_2O_3, various polymorphs are known and Adachi has described PL of Cr^{3+} in these polymorphs [12].

Following the success of ruby and sapphire which are based on 3d activators Cr^{3+} [22] and Ti^{3+} [23], studies on other 3d dopants like Mn^{2+} [24], V [25], and lanthanide-doped Al_2O_3 were initiated.

9.2.1.2 Al_2O_3:Ti, Sapphire

Ti:sapphire lasers (also known as Ti:Al_2O_3 lasers, titanium–sapphire lasers, or Ti:sapphire) are tunable lasers which emit red and near-infrared light in the range from 650 to 1,100 nm. These lasers are mainly used in scientific research because of their tunability and their ability to generate ultrashort pulses. Lasers based on Ti:sapphire were first constructed and invented in June 1982 by Peter Moulton at the MIT Lincoln Laboratory [26].

Titanium–sapphire refers to the lasing medium, a crystal of sapphire (Al_2O_3) that is doped with Ti^{3+} ions. A Ti:sapphire laser is usually pumped with another laser with a wavelength of 514–532 nm, for which argon-ion lasers (514.5 nm) and frequency-doubled Nd:YAG, Nd:YLF, and Nd:YVO lasers (527–532 nm) are used. They are capable of laser operation from 670 to 1,100 nm wavelength [27]. Ti:sapphire lasers operate most efficiently at wavelengths near 800 nm [28].

The Ti:sapphire laser was invented by Peter Moulton in June 1982 at MIT Lincoln Laboratory in its continuous-wave (CW) version. Subsequently, these lasers were shown to generate ultrashort pulses through Kerr-lens mode-locking [29]. Strickland and Mourou, in addition to others, working at the University of Rochester, showed chirped pulse amplification of this laser within a few years [30], for which these two shared in the 2018 Nobel Prize in physics [31] (along with Arthur Ashkin for optical tweezers). The cumulative product sales of the Ti:sapphire laser have amounted to more than $600 million, making it a big commercial success that has sustained the solid-state laser industry for more than three decades.

Ti-doped Al_2O_3 is an exceptionally important material of quantum electronics extensively used as an active medium in tunable lasers which are capable of delivering terawatt light pulses of femtosecond duration [32]. As a consequence of this, Al_2O_3 has been at the focus of research for decades. However, the overwhelming majority of all optical studies of Ti-doped Al_2O_3 have been carried out using mainly visible and ultraviolet light. Two dominant emission bands are usually identified in the luminescence spectra of Al_2O_3–Ti: the blue 420 nm band and the near-infrared emission centered at about 750 nm. Apparently, the nature of the near-IR emission band is well established as being the $2E$-$2T2$ radiative transition of Ti^{3+} ions, whereas there is no clear agreement on the origin of the former band. The dominant opinion is that the 420 nm band of Al_2O_3–Ti is the characteristic emission of Ti^{4+} which can be excited through $^2O^{2-3}dTi^{4+}$ charge-transfer transitions [33]. The wide gain and high QE of sapphire (Al_2O_3:Ti) makes this material the optimum ultrafast laser crystal [23].

9.2.1.3 Al_2O_3 Thin Film Lasers

To integrate solid-state lasers with electronic circuitry, laser media need to be integrated into the silicon-on-insulator (SOI) and the silicon nitride (Si_3N_4) passive platforms. Lanthanide-doped Al_2O_3 is particularly suitable for this purpose. The integration does not involve complex assembly steps.

First such integration was illustrated with SOI waveguides top-coated with Al_2O_3:Er layers [34]. Similar integration was achieved on Si_3N_4 platform during recent years. These are of use in fabrication of devices like "distributed feedback and distributed Bragg reflector lasers". To this end, a two-layer system is designed. In one-layer Si_3N_4, waveguides are accommodated. The other layer encloses waveguides comprising of lanthanide-activated Al_2O_3. The layers are separated by a SiO_2 spacer layer. Al_2O_3-based microring lasers [35] were designed on this principle. Recently, erbium-doped Al_2O_3 lasers integrated with Si_3N_4 waveguides were introduced onto a CMOS-compatible silicon photonic platform to produce a fully integrated system-on-a-chip [36].

Low-propagation losses coupled with high optical gain makes Al_2O_3-based media highly suitable for designing on-chip lasers. It took nearly 10 years for moving from these prototype designs to the first integrated laser in Er^{3+}:Al_2O_3 [5]. De Goede et al. achieved a better miniaturized arrangement comprising of Yb^{3+}:Al_2O_3 microdisk/microring coupled to a bus waveguide. This device exhibited an output power of ~25 µW with a ~2% on-chip slope efficiency. Single-mode laser emission was obtained at a wavelength of 1,024 nm with a line width of 250 kHz [37]. Similar methodology was adopted, in design of Al_2O_3:Tm microring laser [38]. "This laser exhibited an on-chip output power of ~220 µW and double-side slope efficiency of 24% with 1.6 µm resonant pumping and lasing in the wavelength range of 1.8–1.9 µm". For optimizing absorption of the pumping light in the microcavity, it is crucial to minimize coupling losses. Otherwise, the laser threshold will increase. This design was first demonstrated in Yb^{3+}-doped Al_2O_3 for a biosensing application with a H_2O top cladding. A 2 mW multimode laser power was measured at the output fiber, corresponding to ~6 mW on-chip laser output power.

Thin-film Al_2O_3-based lasers integrated with various platforms have been recently reviewed by Hendriks et al. [39].

9.2.2   Rare-Earth Oxide

Sesquioxide ceramics (Y_2O_3, Lu_2O_3, and Sc_2O_3) along with their solid solutions show great potential as hosts for solid-state laser gain materials due to their great suitability for various rare-earth ions, comparably physical properties to single crystals and far lower preparation temperature than the melting point (above 2,400°C). "Recently, sesquioxides as excellent laser hosts and scintillator materials including Sc_2O_3, Y_2O_3, and Lu_2O_3 have gained increasing popularity due to their good physical and chemical properties such as high thermal conductivity, large Stark splitting, low phonon energies, and simple cubic structure [40]". These oxides are completely miscible in each other. (Lu, Sc)$_2O_3$ [41], (Lu, Y)$_2O_3$ [42], and (Y, Sc)$_2O_3$ [43] are well known. The solid solutions introduce disorder in otherwise ordered structure. This enables a broader gain bandwidth which is conducive for ultrashort pulses.

The sesquioxides Ln_2O_3 (Ln = Y, Lu, Sc) are cubic crystals with bixbyite structure in space group Ia 3. A unit cell possesses lattice constants around 10 Å and contains 32 cation sites, of which 24 are C2 symmetry and the remaining 8 sites exhibit C3i symmetry (24d and 8b Wyckoff positions, respectively). Two out of five ions in the chemical formula are trivalent rare-earth (RE^{3+}) ions, which leads to cation densities around $3 \times 10^{22} cm^{-3}$. The Ln^s ions in the C2 and C3i sites can be substituted by RE^{3+} ions, and a random distribution on both sites is a good approximation in most cases. For the C3i site, the electric dipole transitions are forbidden due to inversion symmetry. Usually, the cross sections of the RE^{3+} ions implemented on the C3i are significantly lower than those of ions on C2 symmetry sites.

Therefore, the optical properties of RE-doped sesquioxides are largely determined by RE^{3+} ions on C2 sites. Table 9.1 summarizes crystallographic properties of the sesquioxides.

All sesquioxides possess a broad optical transparency wavelength range between ~0.2 and 8.5 μm. Their refractive indices which were calculated by Sellmeier equation lie around 1.9 in the wavelength range of 1~3 μm, and their dependence on wavelength and temperature is a vital part of virtually any calculation in lasers, frequency conversion devices, and optical fibers. These sesquioxides exhibit a wide bandgap of around 6.0 eV, making them suitable laser host materials for many active ions. Meanwhile, three sesquioxides have relatively low maximum phonon energies in the order of 600 cm^{-1} compared to YAG (857 cm^{-1}), which can increase the possibility of various radiative transitions.

It is worth to notice that Y_2O_3 has been identified as the best host doped with Er^{3+} for mid-IR laser operation due to Y_2O_3 offering the lowest maximum phonon energy of 597 cm^{-1}. For pure sesquioxides, Sc_2O_3 has the highest thermal conductivity, however, which would decrease greatly with the increasing doping concentration, but though Lu_2O_3 has the lowest thermal conductivity, the value almost remains unchanged for heavy doping with various rare-earth active ions [44].

Apart from the crystallographic properties, other properties relevant to ceramic laser fabrication are listed in Table 9.2.

9.2.2.1 Y_2O_3

Y_2O_3 crystals and ceramics have been investigated for a long time as active host materials for lanthanide ions. Yttrium oxide has many properties which are desirable in a laser host material. Among these are its refractory nature (mp 2,450°C), stability, ruggedness, and optical clarity over a broad spectral region. It is particularly attractive as a host for trivalent rare-earth ions because isomorphic substitution in the lattice can be affected without the complication of charge compensation.

The properties of neodymium in yttrium oxide (Y_2O_3) have been studied extensively. Hoskins and Soffer [45] reported pulsed stimulated emission, at 1.073 and 1.078 μm wavelength, in Nd:Y_2O_3 held at a temperature of 77 0K and pumped with a flash lamp. Nd:Y_2O_3 ceramic-based laser was demonstrated as early as 1973 [46,47]. Lu et al. demonstrated the CW lasing of Nd:Y_2O_3 ceramics

TABLE 9.1
Crystallographic Properties of the Sesquioxides

Lattice	Cubic	Cubic	Cubic
Space group	Ia3	Ia3	Ia3
Site symmetry	C_2/C_{3i}	C_2/C_{3i}	C_2/C_{3i}
Coordination number	6	6	6
Cation radius (Å)	0.90	0.86	0.75
Lattice constant (Å)	10.602	10.391	9.857
Cation density ($10^{22} cm^{-3}$)	2.687	2.852	3.355
Melting point (°C)	2,450	2,490	2,430
Density (g/cm³)	5.01	9.42	3.85

TABLE 9.2

Thermal, Mechanical, and Optical Properties of Sesquioxide Single Crystal or Ce ramics

	Y_2O_3	Lu_2O_3	Sc_2O_3
Thermal expansion coefficient	8.5	8.6	9.6
Thermal conductivity (W/mK)	9.8 (323 K)	9.5	18
Bandgap (eV)	5.6±0.1	5.8±0.1	6.0
Max. phonon energy (cm^{-1})	591	612	672
Thermal capacity	0.44	0.24	0.70
Young's modulus (GPa)	~180	180	200
Mohs hardness	6½	6½	6½
Thermal-optic coefficient dn/dT	8.5	9.1	8.9
Transparency range (μm)	0.220~8	0.225~8	0.210~8
RI at 1 μm	1.889	1.911	1.966
RI at 2 μm	1.874	1.896	1.946
RI at 3 μm	1.860	1.883	1.926

supplied by Konoshima Chemical Co. with 160 mW output power and 37% slope efficiency at 1,074 and 1,078 nm simultaneously [48] and Yb:Y_2O_3 ceramics with 750 mW output power and 9.6% slope efficiency at 1,078 nm [49,50]. Recently, Yin et al. fabricated 0.6 at.% Nd:Y_2O_3 transparent ceramics by vacuum sintering plus HIP and achieved CW laser operation with a maximum output power of 3.6 W with 45.2% slope efficiency at 1.08 μm [51]. Kong et al. used the Yb:Y_2O_3 ceramic disks of the same company to observe random wavelength emission and central lasing wavelength shift [52] and realize diode-end-pumped CW and passively mode-locked laser. They achieved the slope efficiencies of 82.4% and 57.1% at 1,078 and 1,040 nm laser emission, respectively, by high-efficiency diode-end-pumping. In this year, Xie et al. realized the first self-mode-locked ceramic laser without any additional active or passive element in the cavity [53]. In 2005, Takaichi et al. performed the tunable laser of Yb:Y_2O_3 ceramics in two regions including 1,034–1,050 and 1,071–1,078 nm wavelength, with pumping power of 3.9 W [54]. Laser properties of Yb:Y_2O_3 ceramics were studied at room and cryogenic temperatures by Merkle et al. in 2008, and the laser performance was very substantially improved by cooling to liquid nitrogen temperature due to the great reduction in ground-state absorption [55]. Last year, David et al. reported CW and pulsed laser characteristics of Yb:Y_2O_3 transparent ceramic at cryogenic temperatures [44,56].

Suitability of Y_2O_3:Tm for lasing was pointed out as early as 2007 [57]. A waveguide laser was reported in 2013 conference [58]. Near 2-μm lasing at the $^3F_4 \rightarrow {}^3H_6$ transition in Y_2O_3:Tm^{3+}ceramics was achieved by Ryabochkina et al. in 2016 [59]. These were at 1.95 and 2.05 μm with the maximum output power of 2.4 and 0.3 W, respectively. This was followed by the CW and passively Q-switched Tm:Y_2O_3 ceramic laser emitting at 2.1 μm with a VBG [60]. Huang et al. then improved the output power of CW operation and the pulse width for Q-switched operation. Most recently, Yue et al. reported 3 at.% Tm:Y_2O_3 transparent ceramic at cryogenic temperatures and with the VBG pump diode. The maximum output power of 6.4 W was achieved at 80 K corresponding to a slope efficiency of 52.0% [61].

The first CW ceramic laser based on Ho^{3+} emission in Y_2O_3 host was reported by Newburgh et al. This was resonantly diode-pumped at 1.93 μm and gave output at 2.085 μm with 29% slope efficiency [62]. It was further improved to 35% next year at 77 K. Yang et al. prepared similar device. They used highly transparent Yb, Ho-doped (Y, La)$_2O_3$ ceramics which lased 2.1 μm [63]. A device operating at room temperature could be produced only recently. "Wang et al. first realized the room temperature laser oscillation from Ho:Y_2O_3 ceramics which were in-band pumped at 1,941 nm with a maximum output power of 1.3 W at 2,116.8 nm [64]".

In 2008 conference, the first laser operation of $Er:Y_2O_3$ ceramics was announced. Laser performance of cryogenically cooled $Er:Y_2O_3$ ceramics at 1,535.7 nm was first reported in 2008 [65]. Later, lasing at ~2.7 μm was successful with direct diode pumping at room temperature. The spectral output consisted of four emission lines. The line at 2,707 nm was most intense [66]. Subsequently, ten-fold power increase to 14 W CW output power at ~2.7 μm was attained. In 2007, Ren et al. achieved a repetitively short-pulse-width [67] and high-peak-power acousto-optically Q-switched laser at 2.7-μm.

Cryogenically cooled laser experiment aims to utilize the longer upper laser level lifetime as well as a much higher emission cross section of $Er:Y_2O_3$ at reduced sample temperatures. Hence, much higher net gain even with the same sample of marginal quality is the next logical step toward high power, quantum defect limited, and laser operation. More works focus on the cryogenic laser operation of $Er:Y_2O_3$ ceramics at ~2.7 μm. Sanamyan et al. demonstrated the first cryogenically cooled performance of Er^{3+} in Y_2O_3 ceramics laser at ~2.7 μm with a CW output power of over 1.6 W and 27.5% slope efficiency [44,68].

9.2.2.2 Lu_2O_3

In 2002, the efficient CW laser oscillation in $Nd:Lu_2O_3$ ceramics was obtained. The output corresponded to two wavelengths of $^4F_{3/2} \rightarrow ^4I_{11/2}$ channel [69]. With a laser diode as pump source, a CW laser output power of 2.81 W was achieved, and its wavelength is also found to be dual wavelength. Because the emission cross sections at 1,076 and 1,080 nm are almost identical, laser oscillation for such two wavelengths can be obtained simultaneously. All the properties showed that $Nd:Lu_2O_3$ is an excellent crystal for laser applications [70]. Toci et al. later used SPS method to prepare $Nd:Lu_2O_3$ transparent ceramics. Laser oscillations were observed at 1,076.3 and 1,080.5 nm [71]. Next year, using same technique Xu et al. attained a slope efficiency of 38% with a maximum output of 1.25 W. Besides two wavelengths around 1,080 nm, emission at 1,359.7 nm was also observed with a maximum output of 200 mW [72]. Liu et al. used two-step approach, co-precipitation followed by sintering to obtain $Nd:Lu_2O_3$ ceramics. QCW laser was fabricated using the same [73].

Passively mode-locked $Yb:Lu_2O_3$ laser was first reported by Griebner et al. by use of a semiconductor saturable absorber mirror (SESAM). The laser emitted up to 470 mW in the picosecond regime, corresponding to a pump efficiency as high as 32%. With dispersion compensation, pulses as short as 220 fs at an average power of 266 mW were obtained at 1,033.5 nm [74]. For the $Yb:Lu_2O_3$ ceramics, Takaichi et al. firstly utilized the samples from Konoshima Chemical Co. to realize laser operation with a 0.7 and 0.9 W CW output power at 1,035 and 1,079 nm, respectively, [75]. The diode-pumped passively mode-locked laser oscillation was firstly demonstrated by Tokurakawa et al. in 2006 [76]. They employed a Z-shaped cavity configuration. In quick succession, they achieved 65 fs pulses at 1,032 nm, and the spectral bandwidth and the time bandwidth product were 18.9 and 0.345 nm, respectively [77]. They utilized a Kerr-lens to perform laser operation. Highest performing laser using hot-pressed $Yb:Lu_2O_3$ ceramics was achieved in 2010. This laser had an optical-to-optical slope efficiency of up to 74% [78]. Similar results were achieved at room temperature elsewhere as well. David et al. studied the CW laser performances of $Yb:Lu_2O_3$ ceramics at low temperatures [79].

The first diode-pumped laser operation of thulium-doped Lu_2O_3 was reported by Koopmann et al. With a very compact setup, an output power of 75 W and slope efficiencies of around 40% with respect to the incident pump power were achieved at room temperature. Free running laser operation was observed at wavelengths of 2,065 and 1,965 nm. With a birefringent filter, the wavelength could continuously be tuned from 1,922 to 2,134 nm [80]. Antipov et al. obtained the CW lasing of a 2.0 at.% $Tm:Lu_2O_3$ ceramic laser at 2,066 nm under diode laser pumping at 796 and 811 nm, respectively [81]. Later, they achieved continuous Q-switched operation [82] which led to improved performance.

A broadly tunable femtosecond mode-locking laser was devised by Lagatsky et al. They made use of SESAM in 2012 [83]. Lasing with reasonable power became possible by combining with thin-disk cavity configurations. Soon after, they fabricated ~410 fs mode-locking laser by single-layer graphene. A 1.9-μm waveguide laser based on Tm:Lu$_2$O$_3$ ceramics was assembled in 2017. Ultrafast laser inscription technique was used for the first time in its construction. The device provided a maximum output power of 81 mW at 1,942 nm [84]. Baylam et al. recently made a detailed experimental investigation of the energy efficiency and rich temporal dynamics of Tm:Lu$_2$O$_3$ ceramic laser pumped near 800 nm using a plane-wave rate equation method, which could provide predictions in qualitative agreement with experimental data [85]. First laser based on Ho:Lu$_2$O$_3$ was announced in 2011 [86]. Later, multiwatt lasing operation was achieved at 2.12 μm [87]. Kim et al. in 2013 used for the first time Ho:Lu$_2$O$_3$ ceramics obtained by hot pressing in construction of a laser around 2 μm [88].

Diode-pumped laser action of Er:Lu$_2$O$_3$ at 2.7 μm was demonstrated as early as 1998. A fiber-coupled laser diode with an emission wavelength of 973 nm and a maximum output power of 3 W was used as pumped source. Up to an absorbed pumped power of 1.3 W, linear increase of the output power with a sloped efficiency of 3.2% is observed. These results suggested that Er:Lu$_2$O$_3$ is a possible laser material for 3 μm laser applications [89]. Pumping with an optically pumped semiconductor laser at 971 nm, 1.4 W of CW output power with a slope efficiency of ~36% at 2.85 μm was obtained at room temperature. Under diode pumping, 5.9 W of output power with 27% of slope efficiency was achieved with an M2 of 1.2–1.4 [90]. CW operation of diode-end-pumped Er:Y$_2$O$_3$ and Er:Lu$_2$O$_3$ ceramic lasers operating at 2.7 μm at room temperature was demonstrated subsequently. The maximum output power of 611 mW was obtained from the Er:Lu$_2$O$_3$ ceramic lasers, with slope efficiency of 7.6% [91]. Lasing around 1.6 μm in bulk Er^{3+}-doped Lu$_2$O$_3$ has seldom been reported even for low Er^{3+} concentrations. In 2015, Qiao et al. demonstrated the CW laser operation of diode pumped 3 at.% Er:Lu$_2$O$_3$ ceramic laser at 2.7 μm with a maximum output power of 189 mW corresponding to a slope efficiency of 2.16% [92]. The same researchers subsequently attained sustainable self-pulsed operation at 2.74 μm. The output power was 1.3 W with a slope efficiency of 11.9%. Pumping was done by a wavelength-locked narrow bandwidth, 976-nm laser diode. In 2018, Wang et al. demonstrated high power and short pulse width operation of passively Q-switched laser by Bragg-reflector-based SESAM and achieved stable Q-switched laser pulses of 70 ns in duration and 0.14 kW in peak power, which were the shortest pulse duration and the highest peak power ever produced from a Q-switched Er:Lu$_2$O$_3$ laser [93]. All above cited examples go on to show that the Er:Lu$_2$O$_3$ ceramics are a potential laser gain medium that can lead to efficient and high-power 2.7-μm laser.

9.2.2.3 Sc$_2$O$_3$

Lasing in Yb:Sc$_2$O$_3$ was reported in 2000. Yb ($\approx$6.5$\times$1,020 cm^{-3}):Sc$_2$O$_3$-doped crystal showed efficient laser operation when pumped with a Ti:sapphire laser at 941 nm. A maximum output power of 970 mW at 1,041 nm was obtained with approximately 1.85 W of incident power [94]. CW lasing in Yb:Sc$_2$O$_3$ ceramic plate was achieved in 2003 [95]. The output was around 1,041 and 1,094 nm. The latter was about 1.5 times broader than that obtained in YAG media. First, room temperature lasing in this system was reported in 2007. Using laser diode as a pump CW operation at 1.0405~1.0426 and 1.0905~1.0965 μm was demonstrated [96]. Almost simultaneously, Tokurakawa et al. reported for the first time the Kerr-lens mode-locked laser generating pulses as short as 92 fs with an average power of 850 mW [97]. Later, they constructed mode-locked laser. This was accomplished using commercial Yb:Sc$_2$O$_3$ and Yb:Y$_2$O$_3$ multi-gain-media. Output power of 1 W could be derived for 53 fs pulses. For 66 fs pulses, it was even higher, 1.5 W. In 2012, Pirri et al. reported the first oscillation at 1,040.5 nm based on 1 at.% Yb-doped Sc$_2$O$_3$ under QCW pumping at 968 nm. The maximum output power was 2.2 W, and the slope efficiency was 59% with a continuous tunability range of above 41 nm [98]. This was further improved to 6.62 W output power with a slope efficiency of 50% [99]. More or less similar results were obtained by Dai et al.; they first obtained powders by precipitation and went on to fabricate Yb:Sc$_2$O$_3$ transparent ceramics by vacuum sintering [100]. Self-propagating high-temperature method was used by Permin et al. to obtain Yb:Sc$_2$O$_3$ powders.

These were converted to ceramics by hot-pressing. However, the slope efficiency was only moderate [101]. The branching ratios for the $^4F_{3/2} \rightarrow ^4I_{9/2}$ and $^4F_{3/2} \rightarrow ^4I_{11/2}$ transitions are almost equal (~0.45) in $Nd:Sc_2O_3$. The crystal field interaction and nephelauxetic effect is large owing to the large mismatch between the Nd^{3+} and Sc^{3+} ionic radii, which determines the possible laser emission wavelengths at 966 nm ($^4F_{3/2} \rightarrow ^4I_{9/2}$), 1,072, and 1,078 nm ($^4F_{3/2} \rightarrow ^4I_{11/2}$) [102]. Up to now, no lasing in $Nd:Sc_2O_3$ ceramics could be achieved till date.

Fornasiero et al. demonstrated lasing in $Er:Sc_2O_3$ with low power [103]. Lasing could be obtained only in cryogenically cooled system. QCW IR output was obtained at 1,605.5 nm with the slope efficiency of 77% at the level of 2.35 W. However, there is hardly any mention of the laser operation near 3 μm in this system. For achieving these long wavelengths, Er has to be incorporated at high concentrations. However, this deteriorates the optical quality of the ceramics. With $Ho:Sc_2O_3$ as the active material, the accessible wavelength range could be expanded to 2,158 nm in a diode-pumped setup [104]. However, future efforts need to be focused on $Tm:Sc_2O_3$ and $Ho:Sc_2O_3$ ceramics.

9.2.2.4 Conclusion and Prospects

It is thus seen that the oxides Sc_2O_3, Y_2O_3, and Lu_2O_3, even in ceramic form are good laser media for lanthanide activators. Their solid solutions offer better opportunities for the power and efficiency scaling, ultrashort pulses, provided that better optical quality ceramics can be produced. Among these, Lu_2O_3 seems to be the best. There is still substantial scope to improve the optical quality and size of these ceramics. Purity and morphology of the ingredient raw powders is critical to achieve this. It is not easy to obtain the required nanopowders with high sinterability, dispersivity, and chemical uniformity in large scale.

9.2.3 $BeAl_2O_4:Cr^{3+}$ Alexandrite

The perfect crystal $BeAl_2O_4$ is the mineral chrysoberyl. Chrysoberyl can be regarded as a close-packed analog to the spinel structure and is isomorphic with olivine. The gemstone alexandrite is a Cr-doped variation of chrysoberyl. It has an orthorhombic cell with a space group of Pnma. The crystal structure was refined by Farrel et al. [105] based on the earlier measurement of Bragg and Brown [106]. There are two nonequivalent Al sites denoted by Al_1 and Al_2. Both Al ions are octahedrally coordinated with Al-O, bond lengths ranging from 1.861 to 2.017 Å. The Be atom is tetrahedrally coordinated with relatively short Be-O bond lengths of 1.579, 1.631, and 1.687 Å. All1 and Al2 have local point group symmetry of Cs and Ci, respectively, commonly referred to as the mirror site and the inversion site. The substitution of Cr^{3+} in alexandrite is predominately at the mirror site Al1 [107] (Figure 9.1).

Beryllium aluminate ($BeAl_2O_4$ also known as chrysoberyl) has a structure that is isomorphous with olivine ($(Mg, Fe)_2SiO_4$). The oxygens form a distorted hexagonal close-packed array in which one-eighth of the tetrahedral interstices are occupied by beryllium, and one-half the octahedral sites are filled by aluminum. The space group of the structure, as previously determined by others, is Pnma, orthorhombic, with four molecules per unit cell. The lattice parameters are a=9.404, b=5.476, and c=4.427 A. Chromium ions replace aluminum ions in the structure.

Chromium ion-doped lasers belong to the class of tunable (or vibronic) solid-state lasers. Historically, alexandrite (Cr^{3+}-doped chrysoberyl, $Cr^{3+}:BeAl_2O_4$) was the first room temperature, tunable solid-state laser reported in the mid-1970s [108–111]. In alexandrite, the presence of broad absorption bands in the visible region makes it possible to optically excite the gain medium with flash lamps or visible diode and/or solid-state lasers. The vibronically broadened upper laser level $^4T_2$ (lifetime=6.6 μs) remains populated at room temperature, since it is close to the lower metastable level ^2E (lifetime=1.54 ms) with an energy difference of only 800 cm^{-1} (approximately 4 kBT at room temperature, where kB=Boltzmann constant and T=absolute temperature). The resulting $^4T_2 \rightarrow ^4A_2$ transition yields broadly tunable laser emission in the 701–758 nm wavelength range.

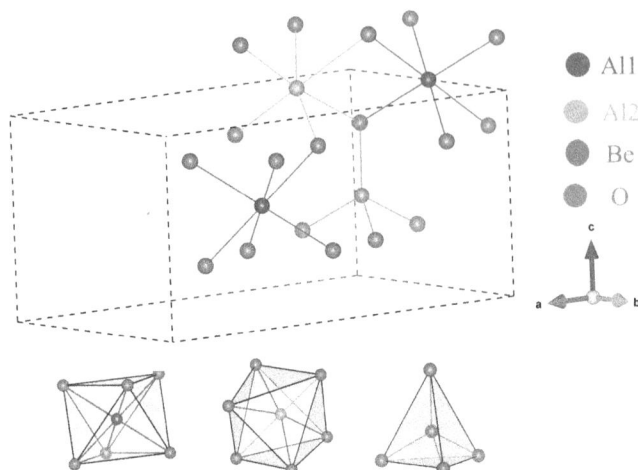

FIGURE 9.1 Figure shows unit cell coordination of BeAl2O4.

The effective upper state lifetime of alexandrite is 260 μs at room temperature, making the alexandrite gain medium particularly useful for Q-switched operation. The close proximity of the $^4T_2$ and ^2E levels further leads to temperature-dependent emission properties [1]. Lasers in blue/UV region can be obtained using frequency doubling, as this laser can be tuned from 700 to 858 nm. Moreover, phonon-assisted transitions around 750 nm permit laser operation in this region similar to R lines at 680 nm in Al_2O_3:Cr. Favorable laser properties of alexandrite include (i) high ef product, making it possible to obtain low-threshold CW operation, (ii) high intrinsic slope efficiency (η_0) of 65%, comparable to that of Ti^{3+}:sapphire laser, (iii) negligible Auger-type ETU, (iv) high thermal conductivity comparable to that of Ti^{3+}:sapphire, (v) high fracture toughness and thermal shock resistance which allows power scaling and durability against optical damage, and (vi) high nonlinear RI which can be employed to initiate Kerr-lens mode-locked operation to generate ultrashort optical pulses.

The broad emission bandwidth of the alexandrite gain medium can be utilized for the generation of ultrashort optical pulses, as well. Ghanbari et al. reported the first experimental demonstration of Kerr-lens mode-locking in alexandrite lasers, producing 170-fs pulses at 755 nm with an average power of 780 mW and pulse repetition frequency of 80 MHz [112]. Cihan et al. later used an extended, multipass-cavity design to lower the repetition rate of a KLM alexandrite laser to 5.6 MHz and generated 70-fs pulses near 750 nm [113]. The extended cavity design further eliminated the Q-switching/spectral instabilities, in agreement with previous simulations which show that chaotic behavior of alexandrite lasers depends on the resonator length [114].

Alexandrite lasers have been widely used in medical imaging (such as 3D photoacoustic mammography [115], high-contrast vascular imaging [116], dermatology [117–119], LiDAR systems [120], and ultraviolet generation, among others). In a recent experiment, Wang et al. reported efficient second harmonic generation with a diode-pumped CW alexandrite laser and obtained record of 2.55 W of second harmonic power at 378 nm [121]. Coney and Damzen reported diamond-shaped slab amplifier architecture for the development of high-energy Q-switched alexandrite lasers for LiDAR applications [122]. Fibrich et al. described a microchip laser based on alexandrite [123].

9.2.4 YAG:Nd

The YAG ($Y_3Al_5O_{12}$) possesses a number of properties that are required of a solid-state laser host. These crystals are robust, thermally stable, and have cubic structure. Various trivalent ions can be easily substituted; lanthanides at yttrium site and 3d group ions at aluminum. Laser operation of YAG:Nd was first demonstrated by Geusic et al. [124] at Bell Laboratories in 1964. They reported the continuous operation of a YAG:Nd^{3+} laser pumped at room temperature with a tungsten lamp. Intense PL emission is at 1,064 nm with lifetime 230 μs. Owing to presence of several energy levels of Nd^{3+}, light can be absorbed over wide range of wavelengths. The first oscillation of solar powered laser (SPL) was reported in 1966 [125] using YAG:Nd^{3+}, and related applications are relevant even today [126]. First, production of YAG:Nd laser crystals at industrial scale was achieved in 1965 at union carbide. Since that time, this material has been extensively used for solid-state lasers [127,128]. YAG:Nd lasers operate in both pulsed and continuous modes.

The phase diagram of Al$_2$O$_3$-Y$_2$O$_3$ is rather complicated. Several intermediate compounds such as Y$_3$Al$_5$O$_{12}$ (YAG, m.p. 1,970°C) with cubic garnet structure, YAlO$_3$ (yttrium aluminum perovskite (YAP), m.p. 1,870°C), and Y$_4$Al$_2$O$_9$ (yttrium aluminate monoclinic-YAM, 1,942°C) are well known. A metastable hexagonal phase (YAH) with the same YAP stoichiometry is also known. Usually, this phase is witnessed when soft chemical methods are employed for the synthesis. Four-fold coordination of aluminum in the YAG structure is somewhat unstable. The instability is more pronounced by Nd^{3+} doping. Therefore, the growth of YAG:Nd (Y$_3$Al$_5$O$_{12}$:Nd^{3+}) is much less straightforward than that of the pure YAG. High melting point of YAG presents difficulties in crystal growth. Small crystals have been obtained at temperatures as low as 1,250°C using flux [129].

In 1979, Caslavsky and Viechnicki [130] reported an improvement in the heat exchanger method for the growth of YAG:Nd crystals. During the entire crystal growth, a planar solid–liquid interface was maintained. This enabled elimination of the faceted growth and the resulting strained core. These single crystals were larger in weight and diameter than those produced earlier by the Czochralski technique [131] or horizontal directional solidification [132]. The cost of laser rods fabricated from larger HEM boules is lower because of more efficient materials utilization.

Ikesue et al. fabricated highly transparent YAG:Nd ceramics using powders of Al$_2$O$_3$, Y$_2$O$_3$, and Nd$_2$O$_3$ with particles smaller than 2 μm as starting materials [133]. The scattering loss (0.009 cm^{-1}) for this sample was sufficiently low to obtain laser output for the very first time. A slope efficiency of 28% was reported with a 600-mW laser-diode (LD) end-pumping scheme. For the fabrication of transparent ceramics, the hot press method, the wet chemical method, and the urea precipitation methods have been used. Recently, Konoshima Chemical Co., Ltd. fabricated YAG:Nd ceramics successfully by a new method [134,135]. Polycrystalline Nd:Y$_3$Al$_5$O$_{12}$ (YAG:Nd) ceramic laser material has several advantages. To quote, a sample with neodymium concentration exceeding 1% and large size, typically 450×10 mm, can be produced. Production on mass scale is feasible since special techniques or expensive devices are not needed during this process. Moreover, better homogeneity and composite structures can be achieved [136]. Powders to be used for ceramic fabrication can be prepared at progressively low temperatures, as low as 950°C by methods like decomposition of organic precursors [137], 900°C by cryochemical synthesis [138], 850°C by glycol route [139], or even 500°C using a green chemistry method [140] that combines advantages of a mechanically assisted metathesis reaction and the benefits of molten salts as reaction media. Various methods for ceramic fabrication of YAG and the problems involved are summed up by several authors [141–143]. Though the powders are obtained at relatively low temperatures, the transparent ceramic fabrication still required temperatures as high as 1,700°C [144]. Sintering under high pressure (8 GPa) could bring this down to 450°C [145].

Another way to overcome difficulties in growing large single crystals is to grow crystal fibers. Single-crystal fibers can be applied in devices that require high optical energy densities and high gain optical amplifiers. YAG:Nd fibers for laser applications can be drawn with relative ease using micro-pulling-down (μPD) [146] or laser-heated pedestal growth [147] method. Efficient YAG:Nd lasers were demonstrated using gradient-index crystal fibers. A laser output power of 145 mW was achieved along with a slope efficiency of 28.9%. Up to 40% higher Nd ion concentration compared to that in the source rod at the center of the crystal fiber was obtained. Measurement by low-coherence reflectometry indicated an index difference of 0.0284 between the center and the edge of the crystal fiber [148].

Apart from several direct uses of YAG:Nd lasers, several interesting indirect applications have been suggested. The most remarkable among these is the SPL for generation of electricity [149]. Solar radiations are absorbed by YAG:Nd and converted to a single wavelength around 1.064 μm. This can be used for photovoltaic conversion of solar energy. However, the NIR emission of Nd^{3+} ions has its origin in f-f transitions. Such transitions are characterized by small oscillator strength due to forbidden nature and very narrow absorption spectrum. These are inadequate to cover entire solar spectrum. A scheme to improve the laser efficiency of the YAG:Nd crystal is to add ions that can sensitize Nd^{3+} ions. Cr^{3+} has broad absorption bands in the visible region that overlap with the solar spectrum/emission spectra of flash lamps, and its use as a sensitizing ion to improve the laser output efficiency of YAG:Nd crystal was proposed as early as 1964 [150]. By and large, the optical conversion efficiency of optically pumped YAG:Nd laser is as moderate as 2%–3%. A flash lamp-pumped YAG:Nd, Cr ceramic laser operated at 1,064 nm wavelength was demonstrated with single-shot operation. A 10.4 J output energy at 1,064 nm was obtained, with a laser efficiency of 4.9% [151]. In 1988, Sun et al. [152] reported that the Nd, Ce:YAG crystal had the laser output efficiency more than that of the YAG:Nd crystal. Several such reports followed [153–155]. The Ce^{3+} absorption is about ten times more intense than that of Nd^{3+}. Moreover, Ce^{3+} absorption being in the visible region nearly eliminates the multi-phonon relaxation of Nd^{3+} in YAG arising from NIR excitations [156]. Doping by both Ce and Cr broadens the absorption spectrum further [157]. Efficiency can also be increased by partial substitution of Al by Sc [158]. Sc^{3+} being much large compared to the Al^{3+}, substitution of Al by Sc leads to lattice expansion. Such Sc^{3+}-substituted phases, most notably being the $Y_3ScAl_4O_{12}$:Nd, have higher absorption coefficient and emission intensity, as well as longer lifetimes in comparison to YAG:Nd-based amplifier media [159]. Laser space solar power system (L-SSPS) is planned to solve an energy problem by Aerospace Exploration Agency in Japan. It is a big solid laser system to collect a huge amount of solar light, to emit an ultrahigh-power laser light by the laser medium, to transmit it on the earth, and change it to an available type of energy such as electricity. Since L-SSPS is an ultrahigh-power system operated in space, the laser medium must satisfy the conditions of high-power weight ratio, high efficiency, excellent durability, and maintainability in space, thermal equilibrium operation, and excellent long-range propagation of the oscillated laser. YAG ($Y_3Al_5O_{12}$) ceramics doped doubly with Nd and Cr is considered to be a suitable laser medium for L-SSPS at present [25]. Proposals have also been considered to use solar-pumped solid-state lasers for solar-power satellites [160].

9.2.5 OTHER YAG-BASED LASERS

Apart from Nd^{3+}, Yb^{3+} ion also can give emission around 1 μm. Moreover, it is smaller in size compared to Nd^{3+}, and thus more suitable to substitute at Y^{3+} sites. $Y_3Al_5O_{12}$ and $Yb_3Al_5O_{12}$ are completely miscible over a wide range. This miscibility is due to their yttrium identical structures, there being only about a 1.5% difference in unit cell size. Moreover, their melting points differ only by 200°C [161]. YAG:Yb crystals have two strong absorption peaks located at 940 and 970 nm. It is possible to obtain higher QE by 970-nm pumping owing to the smaller energy difference between absorbed pump photons and lasing photons. The fluorescent lifetime of YAG:Yb is almost five times longer as compared to YAG:Nd. This enables YAG:Yb to store much higher energy [162]. If 940-nm light is used for pumping, the difference between emission (1,030 nm) and absorption in YAG:Yb is much smaller compared to that for YAG:Nd. As a consequence, heat generation in YAG:Yb is three times lower than in YAG:Nd. This imparts overall better laser efficiency. Because there is no additional energy structure above the $^2F_{5/2}$ level, various processes such as concentration quenching, ESA, or upconversion are missing and do not influence the YAG:Yb laser performances [163]. However, usefulness of Yb^{3+} has been limited because it has only one excited $^2F_{5/2}$ state located at approximately 10,000 cm^{-1} from the ground state. The crystal field splits the $^2F_{5/2}$ level several folds within span of about 700 cm^{-1} at room temperature. Owing to such splitting, the ytterbium lasers can be considered as quasi-three-level. The lack of adequate excited states implies that the pumping will be rather inefficient using broadband sources like flash lamps. Only a fraction of pumping light lies in the ytterbium absorption band. To keep the pumping threshold within reasonable limits, the gain medium has to be kept at low temperatures (LNT to 210 K).

The first report on stimulated emission in YAG:Yb was published by Johonson et al. in 1965 at Bell Telephone Laboratories [164]. They also studied Tm^{3+}, Ho^{3+}, and Er^{3+} ions in YAG and observed coherent oscillations. The emission being at much longer wavelengths was termed as masers. The high 325-J laser threshold and low overall efficiency, primarily due to inefficient pumping by flash lamp because of the lack of Yb^{3+} absorption bands, discouraged further immediate interest. Reinberg et al. noted that emission of GaAs:Si diode matches very well with Yb^{3+} excitation around 940 nm. Using this diode for pumping, they reported for the first time laser action at 1.029 μm [165]. Subsequently, van Der Ziel et al. [166,167] observed laser oscillations at 2.1 μm in thin films of YAG doped with Ho and sensitized by other lanthanides line Tm, Yb, and Er. From these early studies, interest and developments in YAG:Yb lasers are continuing till date [168]. In 2012, the Yb:YAG single-crystal fiber was prepared by μPD, and the continuous output power and slope efficiency reached 251 W and 53% at 1,030 nm, respectively, which is a record [169].

The concept of fully crystalline double-clad waveguide devices for bulk solid-state lasers became practicable when focus shifted from conventional round cross section clad/core structures to rectangular shapes – a planar waveguide or a channel waveguide [170]. One of the best results was witnessed for "YAG:Yb or YAG:Nd planar waveguide, directly clad-pumped by two laser diode bars. This laser demonstrated an optical-to-optical slope efficiency of 50% and was single-mode in the guided direction (~8 μm) and highly multimode in the unguided direction (~500 μm)" [171].

YAG:Er is an eye-safe laser as Er^{3+} emission around 1.6 μm is safe for eyes. The eye-safe laser radiation region is defined by the spectral response of the eye. Radiations outside the window 400–1,400 nm, both on shorter and longer wavelength side, are almost completely absorbed by tissues. These wavelengths do not penetrate the eye and inflict retinal damage. A wavelength of 1.5 mm is considered as a safe one for direct looking at the radiation beam of energy density a hundred times higher than for CO_2 laser and 2×10^5 times higher than for wavelength of YAG:Nd laser [172]. The lasers at 1.6 μm are important as the eye-safe lasers operating within an atmospheric window. Such lasers are required for atmospheric sounding and many other applications such as laser chemistry, nonlinear laser spectroscopy, coherent laser radar, laser ranging, and medicine [173]. A diode-pumped laser enjoys the advantages of being efficient, reliable, and compact [174]. White and Schleusener [175] and Zverev et al. [176] lase in YAG:Er^{3+} at room temperature. The pumping was achieved by a flash lamp, and relatively high thresholds were necessary. Stange et al. [177] obtained CW laser action of YAG:Er^{3+} at room temperature. The lasing takes place between the metastable $^4I_{3/2}$ multiplets and an upper Stark level of the $^4I_{5/2}$ ground-state manifold. To minimize the reabsorption losses, the Er^{3+} concentration has to be kept to a low value. Laser pumping at a wavelength of 647.1 nm yields lowest thresholds around 30 mW and slope efficiencies up to 9.7% for YAG:Er^{3+}. Low absorption of Er^{3+} may be remedied by using Ce^{3+} as a sensitizer [178].

The laser transitions $^4I_{11/2} \rightarrow {}^4I_{13/2}$ are self-saturating transitions. The reason is that the lifetime of the terminal level $^4I_{13/2}$ is longer than that of initial level $^4I_{11/2}$. The self-saturation effect is closely related to Er^{3+} concentrations: When Er concentrations are low, the cross-relaxation quenching process is less probable. Under such circumstances, direct laser emissions $^4I_{13/2} \rightarrow {}^4I_{15/2}$ (1.66 μm) or $^4S_{3/2} \rightarrow {}^4I_{9/2}$ (1.77 μm) are observed. For high Er concentrations, higher levels are populated by nonradiative relaxations, the population of lower Stark levels increases, and $^4I_{11/2} \rightarrow {}^4I_{13/2}$ (2.94 μm) laser transition becomes feasible.

Lasing action at 2.94 μm on the $^4I_{11/2} \rightarrow {}^4I_{13/2}$ transition in YAG:Er was reported first by Zharikov et al. [179]. The significance of this wavelength in context of applicability in medicine and biology provided incentive for rigorous investigations of YAG:Er laser properties by several research groups [180]. This emission can be induced for high concentration of the order of 30% of Er^{3+}.

9.2.6 LuAG

One of the earliest results related to using LuAG as a laser host is due to Kaminskii et al. [181]. Laser action excitation conditions were found by these authors for new transitions of Er^{3+} ions in LuAG crystals at 300 K. With the use of the deactivation method, the way of laser parameter improvement is shown for Er^{3+} ions at the generation on the self-saturating $^4I_{11/2} \rightarrow {}^4I_{13/2}$ transition (2.69–2.94 pm).

Lutetium aluminum garnet, $Lu_3Al_5O_{12}$ (LuAG), similar to the well-known $Y_3Al_5O_{12}$ (YAG) crystal, belongs to the isostructural garnet family and exhibits high thermal conductivity as well as excellent physical and chemical properties. The LuAG was reported as congruent material, so it is possible to grow undoped and RE^{3+}-doped LuAG single crystals from the melt by the μ-PD technique. From the other side, a slight difference of the lattice constant, that is 9.0075 and 11.9164 Å, respectively, for YAG and LuAG, results in a stronger crystal field for the latter and thus enhances the Stark effect of the doped active ions like Ho^{3+}. The large manifold splitting of the Ho^{3+} is favorable for branching rations of 5I7 → 5I8 transition at room temperature and beneficial for laser performance at 2.1 μm through slightly reducing the lower laser thermal population of Ho:LuAG [182].

Ho:LuAG laser operating at ambient temperature was reported. Output at 2.1 μm was obtained with a slope efficiency of 82% and a threshold energy of 4.4 mJ. A maximum absorbed energy of 17 mJ yielded 10 mJ of laser output energy [183]. The room temperature Q-switched performances of a Ho:LuAG laser operated at 2.1 μm were reported by Duan et al. At the repetition rate of 10 kHz, the maximum average output power of 9.9 W with a slope efficiency of 69.9% relative to absorbed pump power was obtained in Ho:LuAG laser. Also, the minimum pulse width of 33.0 ns was obtained, corresponding to the peak power of 30.0 kW [184]. Single-crystal fiber (SCF) Ho:LuAG laser was reported very recently. Fibers have been grown by the μ-PD technique. PL spectra of the fibers are not much different from those of the bulk crystal. A 1.9-μm laser diode was used for studying laser performances. These studies included two different pump configurations, pump guiding and free-space propagation. Laser slope efficiencies obtained with both pump configurations can be higher than 50%, and a maximum output power of 6.01 W is achieved at ~2.09 μm with the former pump [182].

Duan et al. have also given a brief account of Tm, Ho:LuAG laser. Another room temperature operation was reported for Tm:LuAg. Pulsed lasing at 2.3 μm could be accomplished in 2% Tm:LuAg. This laser operated at room temperature. The emission comes from the Tm^{3+} $^3H_4$-$^3H_5$ transition. For pumping, a pulsed alexandrite laser at 785 nm in a collinear geometry was used. The slope efficiency was 1.5% [185]. Rodriguez et al. [186] used a pulsed laser at 1.75 gm to pump the $^3F_4$ manifold in Tm:LuAG, and a laser action near 2 pm was observed at room temperature. An optical-to-optical conversion efficiency of 44% was observed with a 90% output coupler at a laser threshold of 21.5 mJ. Ho^{3+} does not have an absorption band to match the emission of commercially available laser diodes, and hence, there is a need for co-doping with thulium. Tm ions act as sensitizers to efficiently transfer the absorbed pumping energy into the Ho metastable energy state. Kushwaha et al. [187] observed CW lasing in a Ho:Tm:LuAG rod at 2.1 μm at cryogenic temperatures (77–200 K). For pumping, a 785-nm laser diode was used. Corresponding to 4 W pumping power at 77 K, 1.38 W at 2.1 μm resulted. At room temperature, laser output energy of 11.6 mJ was obtained at 1 Hz using the Ho:Tm:LuAG crystal by Sato et al. [188]. First report on SCF laser based on Tm:LuAG has also appeared recently. In a similar setup, a CW output power of 2.44 W resulted with a slope efficiency of 11.7% using a 783-nm diode for pumping. The beam quality factors M2 were 1.14 and 1.67 in the x and y directions, respectively [189].

Lasing was also achieved for Nd:LuAG. With a fiber-coupled diode laser as pump source, the CW laser action of Nd:LuAG crystal was demonstrated. The maximum output power at 1,064 nm was obtained to be 3.8 W under the incident pump power of 17.3 W, with the optical conversion efficiency 22.0% and the slope efficiency 25.7% [190].

Yb^{3+} is a quasi-three-level laser system. Such systems perform efficiently only at cryogenic temperatures. Owing to favorable temperature dependence of the thermal conductivity, LuAG is a preferred host for Yb^{3+}-based lasers. David et al. described diode-pumped cryogenic Yb:LuAG laser in the CW and pulsed modes. 140 K was found to be optimum operating temperature. A maximum output power of 10.46 W with slope efficiency values of 48.6% and 67% with respect to incident and absorbed power, respectively, was achieved for an output coupler transmission of 40% [191]. These authors have also nicely summed up work on Yb:LuAG lasers.

9.2.7 GGG

Gadolinium gallium garnet (GGG) is easy to grow and also produced at commercial scale. Initially, it was used as a substrate for magnetic devices. Subsequently, it became important as a host for solid-state laser. Gadolinium gallium garnet, $Gd_3Ga_5O_{12}$ (GGG), continues to receive attention of researchers for a number of years. Lasing in this host can be achieved for Nd^{3+} as well as Yb^{3+}. You et al. [192] fabricated a 50-mW single-longitudinal-mode microchip laser at 2.7 μm using a 600-μm-thick Er:GGG crystal. Laser action has been observed in $Gd_{2.88}Nd_{0.12}Ga_5O_{12}$ at 1.0632 μ way back in 1964 [193]. The outstanding features of garnets provided impetus for the subsequent investigations. Lasers have been constructed using Nd:GGG [124], Er:GGG [194], and Cr, Ca:GGG. Though YAG:Nd is most widely used laser, Nd:GGG also has some attractive features. "Core-free crystals of this garnet could be easily grown, it also exhibits weak concentration quenching, and it is capable of producing high average power in pulsed laser operation [195]". Nd:GGG could generate 30 kW of average output power [196]. Compare to YAG:Nd, GGG:Nd exhibits broader emission line. This makes it suitable for mode-locking laser [197]. The long fluorescent lifetime (265 μs) promotes pulse laser energy enhancement (>100 μJ) [198]. Smaller stimulated emission cross section renders it suitable for generating shorter laser pulses through passive Q-switching technique. By passive Q-switching of dual-loss modulation (electro-optic modulator/GaAs or Cr^{4+}:YAG/GaAs) or mode-locking, sub-nanosecond pulses have been achieved with Nd:GGG lasers [199]. However, the pulse width of single-loss-modulation Q-switched Nd:GGG laser still remained in a 7-ns level [200], while the pulses produced from an optical parametric oscillator, which were pumped by a passively Q-switched Nd:GGG laser, were limited to ~1.4 ns in duration [201].

An epitaxially grown ytterbium-doped gadolinium gallium garnet (Yb:GGG)-buried channel waveguide laser was prepared on an yttrium-substituted GGG substrate (Y:GGG). The effective light guiding structure is obtained as a result of the difference of refractive indices of Yb:GGG and Y:GGG. The Yb:GGG waveguide exhibits single-mode 1.025-μm lasing operation at room temperature. The incident threshold and slope efficiency of the 5-mm-long waveguide laser when pumped at 0.941 μm are 80 mW and 13.4%, respectively [202]. Chénais et al. constructed the first Yb:GGG laser [203]. Pumped by a diode, it produces 4.2 W of CW laser power. 63% slope efficiency, high gain (2.26), and broad tunability (53 nm) were the other significant features.

Tm:GGG is also considered as a promising laser crystal [204]. 2-μm laser sources are needed in applications such as eye-safe LiDAR, remote sensing of atmospheric species and medical surgery, as well as plastic processing. (Tm^{3+}) possesses states which are suitable for lasing in the spectral range 1.8–2.1 μm. The first Tm:GGG laser was introduced in 1994 [205]. It operated at the level of 40 mW. In a next step, two-mirror linear cavity configuration was constructed by Wang et al. The ensuing laser operated in CW mode. The maximum output power reaches 0.58 W with laser threshold-absorbed pump power of about 0.39 W and overall slope efficiency of about 18.4% [206].

9.2.8 MgO-based Lasers

9.2.8.1 MgO:Cr³⁺

$MgO:Cr^{3+}$ is a very-well-studied system [207]. Cr-doped MgO is interesting material owing to its high thermal conductivity, the broad emission spectrum, and the possibility of diode laser pumping. At ambient, emission comes from $^4T_2 \rightarrow {}^4A_2$ transition. This is in form of a broadband spread over 700–1,050 nm. The decay is exponential with lifetime of 30 µs. This emission is attributed to Cr^{3+} ions on orthorhombically distorted sites. Absorption spectra contain three bands ($^4A_2 \rightarrow {}^2E$), 619 nm ($^4A_2 \rightarrow {}^4T_2$), 452 nm ($^4A_2 \rightarrow {}^4T_{1a}$), and 283 nm ($^4A_2 \rightarrow {}^4T_{1b}$).Using a mirror with a transmittance of 1% at the laser wavelength of 830 nm, an output power of 48 mW and a slope efficiency of 2.1% with respect to the absorbed pump power were achieved [208]. Lasing can take place at 824, 870, and 878 nm, enabling continuous tuning. ESA at the lasing wavelength imparts only moderate efficiency. The (Ip–Iu)/Ip spectrum of Cr^{3+}:MgO is proportional to ($\sigma_{GSA+}\sigma_{SE-}\sigma_{ESA}$). Also, laser oscillation is possible only between 785 and 945 nm, where $\sigma_{SE>}\sigma_{ESA}$. For these reasons, a further improvement cannot be foreseen.

9.2.8.2 MgO:Ni²⁺

Luminescence of Ni^{2+} in MgO was studied in late sixties [209]. The incorporation of Ni^{2+} into MgO results in an infrared (1–2 µm) laser system which can be highly efficient. The transition responsible for this lasing action is between the $^3T_{2g} \rightarrow {}^3A_{2g}$ levels [210]. The emission lifetime is typically of the order of ms, and the emission QE is near unity at room temperature [211]. In principle, room-temperature laser oscillation should be possible for MgO:Ni²⁺. Despite these advantageous spectroscopic data, laser oscillation was obtained only at temperatures below 100 K. The absence of laser oscillation at room temperature can be explained by ESA, which overlaps with the spectral range of emission [212]. The ESA data can be shared in two parts: one region from 700 to 780 nm attributed to $^3T_2$ (^3F) $\rightarrow {}^1T_2$ (^1D) ESA transition and other from 560 to 640 nm attributed to $^3T_2$ (^3F) $\rightarrow {}^3T_1$ (^3P) [213]. Detailed discussions on ESA in MgO:Ni²⁺ and its consequences on laser properties are provided by Moncorge and Benyattou [214].

9.3 UPCONVERSION OXIDE LASERS

The phrase upconversion (UC) was coined to indicate one of the anti-Stokes processes. It involves combining energy of two low-energy photons to obtain one high-energy photon and originally was called as summation of photons. UC is essentially a nonlinear process, in which an electron is excited to successively higher energy states by sequentially absorbing photons. The idea of combining energy of two photons was first floated by Bloembergen in 1959 [215]. Later, Auzel [216] elaborated the UC mechanism and suggested such phosphors. Novelty of the concept attracted attention of researcher in the field of luminescence, and the list of upconverting phosphors went on expanding constantly. UC phosphors are useful for several applications such as bioimaging [217], solid-state lasers [218], solar cells [219], infrared (IR) quantum counter detectors, and temperature sensors [220].

9.3.1 UC MECHANISMS

Principle processes which make up the UC can be distinguished in four types. The most obvious one is the classical ESA). A more efficient way of achieving UC involves a sensitizer. Sequential energy transfer from a sensitizer to an activator in vicinity had been termed as ETU. There is subtle different between ETU and cooperative energy transfer (CET). In CET, two or more sensitizer ions simultaneously transfer energy to an activator. Finally, a fourth process the photon avalanche effect (PAE) is also based on sequential energy transfer, but of downconversion type (called cross-relaxation), whereas the UC step itself is due to ESA [221].

Lanthanides have been known to be most efficient upconverters. Yb^{3+}-Er^{3+} pair is most frequently used [222], for the reason that Yb^{3+}- can absorb radiation from LED emitting around 980 nm. Other pairs are Yb^{3+}-Tm^{3+} [223], Yb^{3+}-Tb^{3+} [224], Yb^{3+}-Pr^{3+} [225], and Yb^{3+}-Ho^{3+} [226]. Apart from lanthanide pairs, "organic and organo-metallic chromophores" also exhibit efficient UC [227]. In these upconverters, organic molecules with conjugated p-systems play the role of acceptor, while sensitization is achieved through organometallic complexes. In recent years, UC in these organic systems has been pursued with greater vigor [228]. Such materials have been systematically reviewed by Singh-Rachford and Castellano [229], and more recently by Healy et al. [230]. In the following, we discuss two important upconverting materials in context of lasers.

9.3.2   YAP:Er

$YAlO_3$ (YAP) possesses certain features like excellent thermal and mechanical properties, natural birefringence, and structural anisotropy. These properties make it suitable to be used as a laser host [231]. It is suitable for high-power operation and high repetition rate. This is made possible due to high mechanical strength, Mohs hardness of nearly 8.5, and thermal conductivity 11 W/m/K. Due to matching sizes of Er^{3+} (880 pm) and Y^{3+} (900 pm) ions, doping in high concentration is possible and crystals can be grown without much difficulty. Moreover, it has the high gain along the b-axis, which facilitates CW lasing operation.

The first CW upconversion lasing was achieved [232] in $Er:YAlO_3$ through ESA. The emission corresponded to $^4S_{3/2} \rightarrow {}^4I_{15/2}$ transition at 550 nm. Pumping from ground state to the excited state followed by ESA involved pumping at two different wavelengths. Entire lasing operation can be described by four steps: (i) raising the Er ion from the ground state ($^4I_{15/2}$) to the $^4I_{9/2}$ state; (ii) relaxation to the $^4I_{11/2}$ state having long lifetime; (iii) by ESA transition from $^4I_{11/2}$ state to $^4F_{5/2}$ state; and (iv) relaxation to the $^4S_{3/2}$ emitting level. Upconversion lasing using two-wavelength pumping was demonstrated in $Er:YAlO_3$ by Scheps [233]. Laser emission was produced using both sequential two-step pumping and cross-relaxation energy transfer. In addition, photon avalanche upconversion pumping was demonstrated. Selection among these pumping mechanisms is determined by the pump wavelength, and laser operation was obtained with excitation between 785 and 840 nm. The highest laser output power was achieved at 34 K, where 918 mW of pump power at 807 nm produced 121 mW of TEM_{00} emission. The optical conversion efficiency was 13%. The 550-nm emission in Er:YAP is remarkable because the lasing at these wavelengths can be achieved through three routes. The actual route that is followed is decided by the pumping wavelength. Among the three routes, cross-relaxation energy transfer leads to the highest laser power.

9.3.3   YAG:Tm

Blue emission coming from the $Tm^{3+}\,{}^1G_4 \rightarrow {}^3H_6$ transition gave rise to the laser effect at 486 nm under ESA pumping with two photons of 785 and 638 nm in the Tm:YAG bulk crystal [234]. Though low temperature is required for the operation of this laser in bulk YAG crystal, the data presented suggest that room-temperature operation in this crystal should be possible in guided-wave structures. Such structures were subsequently studied [235].

The mechanism of this laser is explained in Figure 9. 2. A photon at 785 nm excites the Tm^{3+} ion to the $^3H_4$ level. The Tm^{3+} ion then cross-relaxes by interacting with a neighboring ground-state ion to yield two ions in the metastable $^3F_4$ level. A second photon at 638 nm subsequently pumps the $^3F_4$ ion to the $^1G_4$ upper laser level. At a starting heat sink temperature of 10 K and maximum pump powers of 340 mW at 785 nm and 190 mW at 638 nm, the strongest features were identified as transitions from the lowest Stark component of the 1G4 manifold to the Z3, Z4, and Z5 components of the ground manifold. The blue emission disappeared almost completely when the dye laser beam was blocked but remained at nearly half the intensity when the Ti:sapphire laser beam was blocked. The strong excitation that was achieved with dye laser pumping alone at 638 nm can be explained by the absorption avalanche process, in which absorption builds up from a low initial level through accumulation of ions in the metastable state. When the pump beams were correctly aligned with respect to the crystal, laser threshold was reached with 340 mW at 785 nm and 100 mW at 638 nm. The laser wavelength was found to be 486.2 nm, corresponding to the transition terminating in the Z4 component of the $^3H_6$ ground manifold that lies 241 cm^{-1} above the lowest Stark component. Following this success, interaction between thulium and ytterbium ions 3,4 or between thulium and

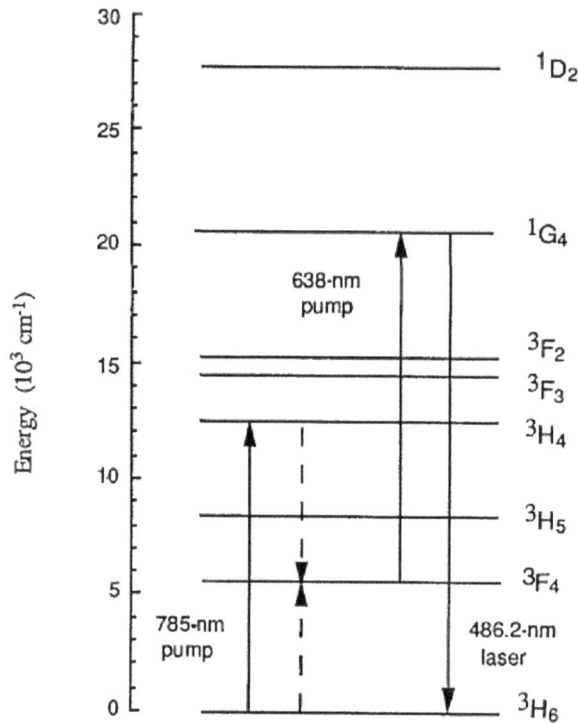

FIGURE 9.2 Figure shows dominant pumping mechanism for the blue upconversion laser in Tm:YAG.

holmium ions 5 was investigated and gave rise to upconverted blue fluorescence under single wavelength excitation around 930–970 nm Yb^{3+} $^2F_{7/2} \rightarrow {}^2F_{5/2}$ transition or 785 nm Tm^{3+} $^3H_6 \rightarrow {}^3H_4$ transition, respectively, but lasing under such configuration has never been reported.

9.4   CONCLUSIONS AND OUTLOOK

Crystals containing active metal ions were discovered as facile laser media which could substitute for the predecessor gas- and liquid-based lasers. Immediately after the discovery of lasing in ruby and CaF_2:Sm, search was instituted for similar lasing action in previously studied phosphors. Inventions of some solid-state lasers during such efforts were incidental. Later, conscious efforts were made to find laser media with better and better properties. Notwithstanding these efforts, few crystals have stood the test of practical circumstances. In fact, the gap between demonstration exercises of laser action and the engineering of practical systems is often wide and difficult to bridge.

Oxides occupy a special place among solid-state laser hosts. Materials like Al_2O_3, YVO_4, YAG, etc., stood the test of time. This is somewhat surprising in view of their high melting points and difficult crystal growth. Their stability and robustness can be one of the reasons. Large bandgap and transparency windows are another. High melting points pose crystal growth problems but these are conducive for attaining high damage threshold. Difficulties in crystal growth in several cases could be circumvented by use of ceramics and fiber lasers. In future, more and more oxides systems need to be developed using these approaches.

REFERENCES

1. Sennaroglu A., Morova Y. (2022) Divalent (Cr^{2+}), trivalent (Cr^{3+}), and tetravalent (Cr^{4+}) chromium ion-doped tunable solid-state lasers operating in the near and mid-infrared spectral regions. *Appl. Phys. B* 128, 9–34. https://doi.org/10.1007/s00340-021-07735-1.

2. Dobrovinskaya E.R., Lytvynov L.A., Pishchik V. (2009) *Sapphire Material, Manufacturing Applications.* Springer.

3. Bradley J.D.B., Pollnau M. (2011) Erbium-doped integrated waveguide amplifiers and lasers. *Laser Photon Rev.* 5(3), 368–403. https://doi.org/10.1002/lpor.201000015

4. Yang J., van Dalfsen K., Worhoff, K.A.F., Pollnau, M. (2010) High-gain Al_2O_3:Nd3+ channel waveguide amplifiers at 880 nm, 1060 nm and 1330 nm. *Appl. Phys. B: Lasers Opt.* 101(1–2), 119–127. https://doi.org/10.1007/s00340-010-4001-2

5. Bernhardi E.H., van Wolferen H.A.G.M., Wörhoff K., de Ridder R.M., Pollnau M. Highly efficient, low-threshold monolithic distributed-Bragg-reflector channel waveguide laser in Al_2O_3: Yb3+. *Opt. Lett.* 36(5), 603–605. https://doi.org/10.1364/OL.36.000603

6. Bradley J.D.B., Stoffer R., Agazzi L., Wörhoff F., Ay K., Pollnau M. (2010) Integrated Al_2O_3: Er^{3+} ring lasers on silicon with wide wavelength selectivity. *Opt. Lett.* 35(1), 73–75. https://doi.org/10.1364/OL.35.000073

7. Purnawirman N.L., Emir S.M., Singh G., Singh N., Baldycheva A., Hosseini E.S., Sun J., Moresco M., Adam T.N., Leake G., Coolbaugh D., Bradley J.D.B., Watts M.R. (2017) Ultra-narrow-line width Al_2O_3: Er^{3+} lasers with a wavelength-insensitive waveguide design on a wafer-scale silicon nitride platform. *Opt. Exp.* 25(12), 13705–13713. https://doi.org/10.1364/OE.25.013705

8. Su Z., Li N., Salih Magden E., Su Z., Li N., Salih Magden E., Byrd M., Purnawirman, Adam T.N., Leake G., Coolbaugh D., Bradley J.D.B., Watts M.R. (2016) Ultra-compact and low-threshold thulium microcavity laser monolithically integrated on silicon. *Opt. Lett.* 41(24), 5708–5711. https://doi.org/10.1364/OL.41.00570841:5708.

9. Li N., Magden E.S., Su Z., Singh N., Ruocco A., Xin M., Byrd M., Callahan P.T., Bradley J.D.B., Baiocco C., Vermeulen D., Wattset M.R. (2018) Broadband 2-μm emission on silicon chips: Monolithically integrated Holmium lasers. *Opt. Exp.* 26(3), 2220–2230. https://doi.org/10.1364/OE.26.002220

10. Denker B., Shklovsky E. (2013) *Handbook of Solid-State Lasers: Materials, Systems and Applications.* Woodhead Publishing, Cambridge.

11. Haupt F., Imamoglu A., Kroner M. (2014) Single quantum dot as an optical thermometer for Millikelvin temperatures. *Phys. Rev. Appl.* 2, 024001. https://doi.org/10.1103/PhysRevApplied.2.024001

12. Adachi S. (2021) Luminescence spectroscopy of Cr^{3+} in Al_2O_3 polymorphs. *Opt. Mater.* 114, 111000. https://doi.org/10.1016/j.optmat.2021.111000

13. Maciel G.S., Rakov N., Fokine M., Carvalho I.C.S., Carlos B. (2006) Pinheiro strong upconversion from Er_3Al_5O12 ceramic powders prepared by low temperature direct combustion synthesis. *Appl. Phys. Lett.* 89, 081109-1-3.

14. Kim Y., Lee S.M., Park C.S., Lee S.L., Lee M.Y. (1997) Substrate dependence on the optical properties of Al_2O_3 films grown by atomic layer deposition. *Appl. Phys. Lett.* 71, 3604–3606.

15. Gusev E.P., Copel M., Cartier E., Baumvol I.J.R., Krug C., Gribelyuk M.A. (2000) High resolution depth profiling in ultrathin Al_2O_3 films on Si. *Appl. Phys. Lett.* 76, 176–178.

16. Pillonnet-Minardi A., Marty O., Bovier C., Garapon C., Mugnier J. (2001) Optical and structural analysis of Eu^{3+}-doped alumina planar waveguides elaborated by the sol-gel process. *J. Opt. Mater.* 16, 9–13.

17. Rebohle L., Braun M., Wutzler R., Liu B., Sun J.M., Helm M., Skorupa W. (2014) *Appl. Phys. Lett.* 104, 251113.

18. Serna R., Afonso C.N. (1996) *Appl. Phys. Lett.* 69, 1541.

19. Maiman T. (1960) *Nature* 187, 493.

20. Boulon G. (2012) *Opt. Mater.* 34, 499–459.

21. Sewani V.K., Stöhr R.J., Kolesov R., Vallabhapurapu H.H., Simmet T., Morello A., Laucht A. (2020) Spin thermometry and spin relaxation of optically detected Cr^{3+} ions in ruby Al_2O_3. *Phys. Rev. B* 102, 104114. DOI: 10.1103/PhysRevB.102.104114.

22. Rani G., Sahare P.D. (2014) Structural and photoluminescent properties of Al_2O_3: Cr^{3+} nanoparticles via solution combustion synthesis method. *Adv. Powder Technol.* 25, 767–772

23. Wang S., Li D., Xiao Y. (2016) Insight into the effects of titanium salt and sintering temperature on the luminescence of Ti: Alpha Al_2O_3 phosphor. *Optik* 127, 8562–8569. https://doi.org/10.1016/j.ijleo.2016.06.071

24. Martinez-Martinez R., Rivera S., Yescas-Mendoza E., Alvarez E., Falcony C., Caldino U. (2010) White light generation in Al_2O_3:Ce^{3+}:Tb^{3+}:Mn^{2+} films deposited by ultrasonic spray pyrolysis. *Thin Solid Films* 518, 5724–5730.

25. Reber C., Güdel H.U. (1989) Near-infrared luminescence spectroscopy of Al_2O_3: V^{3+} and YP3O9: V^{3+}. *Chem. Phys. Lett.* 154, 425–431

26. Moulton P.F. (1986) Spectroscopic and laser characteristics of Ti: Al_2O_3. *J. Opt. Soc. Am. B* 3(1), 125–133. doi:10.1364/JOSAB.3.000125

27. Steele T.R., Gerstenberger D.C., Drobshoff A., Wallace R.W. (1991) Broadly tunable high-power operation of an all-solid-state titanium-doped sapphire laser system. *Optica Publishing Group* 16(6), 399–401. doi:10.1364/OL.16.000399

28. Withnall R. (2005) SPECTROSCOPY I Raman spectroscopy, in Guenther, Robert D. (ed.), *Encyclopedia of Modern Optics*. Elsevier, Oxford, pp. 119–134. doi:10.1016/b0-12-369395-0/00960-x

29. Spence D.E., Kean, P.N., Sibbett, W. (1991) 60-fsec pulse generation from a self-mode-locked Ti:sapphire laser. *Opt. Lett.* 16(1), 42–44. doi:10.1364/OL.16.000042

30. Strickland D., Mourou, G. (1985) Compression of amplified chirped optical pulses. *Opt. Commun.* 55, 447–449. doi:10.1016/0030-4018(85)90151-8

31. The Nobel Prize in physics 2018. www.nobelprize.org. Retrieved 2018-10-02

32. Langley A.J., Divall E.J., Hooker M.H.R., Hutchinson A.J.M, Lecot P., Marshall D., Payne M.E., Taday P.F. (2000) Central laser facility annual report sRAL-TR-2000-034, p. 196, https://www.clf.rl.ac.uk/Reports/1999-2000/pdf/86.pdf

33. Mikhailik V.B., Kraus H., Wahl D. (2005) Luminescence studies of Ti-doped Al_2O_3 using vacuum ultraviolet synchrotron radiation. *Appl. Phys. Lett.* 86, 101909. https://doi.org/10.1063/1.1880451

34. Agazzi L., Bradley J.D.B, Dijkstra M, et al. (2010) Monolithic integration of erbium-doped amplifiers with silicon-on-insulator waveguides. *Opt. Exp.* 18, 27703.

35. Mu J., Dijkstra M., García-Blanco S.M. (2019) Monolithically integrated microring lasers in silicon nitride photonics. In *Proc. Annual Symp.* IEEE Benelux Photonics Society, Amsterdam.

36. Li N., Xin M., Su Z, et al. (2020) A silicon photonic data link with a monolithic erbium-doped laser. *Sci. Rep.* 10, 1114.

37. de Goede M, Chang L, Mu J, et al. (2019) Al_2O_3: Yb^{3+} integrated microdisk laser label-free biosensor. *Opt. Lett.* 44, 5937–5940.

38. Su Z, Li N, Salih Magden E, et al. (2016) Ultra-compact and low-threshold thulium microcavity laser monolithically integrated on silicon. *Opt. Lett.* 41, 5708.

39. Hendriks W.A.P.M., Chang L., van Emmerik C.I., Mu J., de Goede M., Dijkstra M., Garcia-Blanco S.M. (2021) Rare-earth ion doped Al_2O_3 for active integrated photonics. *Adv. Phys.: X* 6, 1833753. https://doi.org/10.1080/23746149.2020.1833753

40. Pirri A., Toci G., Patrizi B., Vannini M. (2018) An overview on Yb-doped transparent polycrystalline sesquioxides laser ceramics. *IEEE J. Sel. Top. Quant. Electron* 24(5), 1–8. https://doi.org/10.1109/JSTQE.2018.2799003.

41. Baer C.R.E., Kraenkel C., Heckl O.H., Golling M., Suedmeyer T., Peters R., Petermann K., Huber G., Keller U. (2009) 227-fs pulses from a mode-locked $Yb:LuScO_3$ thin disk laser. *Opt. Exp.* 17(13), 10725–10730. https://doi.org/10.1364/OE.17.010725.

42. Zhou Z.Y., Guan X.F., Huang X.X., Xu B., Xu H.Y., Cai Z.P., Xu X.D., Liu P., Li D.Z., Zhang J., Xu J. (2017) Tm^{3+}-doped LuYO3 mixed sesquioxide ceramic laser: Effective 2.05 μm source operating in continuous-wave and passive Q-switching regimes. *Opt. Lett.* 42(19), 3781–3784. https://doi.org/10.1364/OL.42.003781.

43. Balamurugan S., Rodewald U.C., Harmening T., Wuellen L., Mohr D., Deters H., Eckert H., Poettgen R. (2010) PbO/PbF_2 flux growth of $YScO_3$ and $LaScO_3$ single crystals- structure and solid-state NMR spectroscopy. *Z. Naturf. B* 65(10), 1199–1205. https://doi.org/10.1515/znb-2010-1004.

44. Ziyu L., Akio I., Jiang L. (2021) Research progress and prospects of rare-earth doped Sesquioxide laser ceramics. *J. Eu. Cer. Soc.* 41, 3895–3910. https://doi.org/10.1016/j.jeurceramsoc.2021.02.026.

45. Hoskins R.H., Soffer B.H. (1963) Stimulated emission From $Y_2O_3:Nd^{3+}$. *Appl. Phys. Lett.* 4, 22–23. https://doi.org/10.1063/1.1723577.

46. Greskovich C., Chernoch J.P. (1973) Polycrystalline ceramic lasers. *J. Appl. Phys.* 44, 4599–4606. https://doi.org/10.1063/1.1662008.

47. Greskovich C., Chernoch J.P. (1974) Improved polycrystalline ceramic lasers. *J. Appl. Phys.* 45, 4495–4502. https://doi.org/10.1063/1.1663077.

48. Lu J., Murai T., Takaichi K., Uematsu T., Ueda K., Yagi H., Yanagitani T., Kaminskii A.A. (2002) Highly efficient CW $Nd:Y_2O_3$ ceramic laser. *Adv. SSL Pro.*, 318–320.

49. Lu J., Takaichi K., Uematsu T., Shirakawa A., Musha M., Ueda K., Yagi H., Yanagitani T, Kaminskii A.A. (2002) Yb^{3+}: Y_2O_3 Ceramics- a Novel Solid-State Laser Material. *Jpn. J. Appl. Phys.* 41, L1373–L1375. https://doi.org/10.1143/JJAP.41.L1373.

50. Kong J., Lu J., Takaichi K., Uematsu T., Ueda K., Tang D.Y., Shen D.Y., Yagi H., Yanagitani T., Kaminskii A.A. (2003) Diode-pumped Yb:Y$_2$O$_3$ ceramic laser. *Appl. Phys. Lett.* 82(16), 2556–2558. https://doi.org/10.1063/1.1569049.

51. Yin D.L., Wang J., Liu P., Zhu H.Y., Yao B.Q., Dong Z.L., Tang D.Y. (2019) Fabrication and microstructural characterizations of lasing grade Nd:Y$_2$O$_3$ ceramics. *J. Am. Ceram. Soc.* 102(12), 7462–7468. https://doi.org/10.1111/jace.16671.

52. Kong J., Tang D.Y., Lu J., Ueda K. (2004) Spectral characteristics of a Yb-doped Y$_2$O$_3$ ceramic laser. *Appl. Phys. B* 79, 449–455. https://doi.org/10.1007/s00340-004-1594-3.

53. Xie G.Q., Tang D.Y., Zhao L.M., Qian L.J., Ueda K. (2007) High-power self-mode-locked Yb:Y$_2$O$_3$ ceramic laser. *Opt. Lett.* 32(18), 2741–2743. https://doi.org/10.1364/OL.32.002741.

54. Takaichi K., Yagi H., Shirakawa A., Ueda K., Yanagitani T., Kaminskii A.A. (2005) Highly transparent Yb-doped ceramics and the laser-diode-pumped ceramic lasers. *Laser. Electro-Opt.*, 271–272.

55. Merkle L.D., Newburgh G.A., Ter-Gabrielyan N., Michael A., Dubinskii M. (2008) Temperature-dependent lasing and spectroscopy of Yb:Y$_2$O$_3$ and Yb:Sc$_2$O$_3$. *Opt. Commun.* 281(23), 5855–5861. https://doi.org/10.1016/j.optcom.2008.08.043.

56. David S.P., Jambunathan V., Yue F.X., Lucianetti A., Mocek T. (2019) Efficient diode pumped Yb:Y$_2$O$_3$ cryogenic laser. *Appl. Phys. B* 125, 137/1-5.

57. Mun J.H., Jouini A., Novoselov, A., et al. (2007) Growth and characterization of Tm-doped Y$_2$O$_3$ single crystals. *Opt. Mater.* 29(11), 1390–1393. https://doi.org/10.1016/j.optmat.2006.03.042

58. Szela J., Sloyan K.A., Parsonage T.L., Mackenzie J.I., Eason R.W. (2013) Thulium-doped yttria planar waveguide laser grown by pulsed laser deposition. In *Conference on Lasers & Electro-Optics Europe & International Quantum Electronics Conference CLEO EUROPE/IQEC*, pp. 1–1. https://doi.org/10.1109/CLEOE-IQEC.2013.6801312

59. Ryabochkina P.A., Chabushkin A.N., Kopylov Y.L., Balashov V.V., Lopukhin K.V. (2016) Two-micron lasing in diode-pumped Tm:Y$_2$O$_3$ ceramics. *Quant. Electron* 46, 597–600. https://doi.org/10.1070/QEL16084.

60. Huang H.T., Wang H., Shen D.Y. (2017) VBG-locked continuous-wave and passively QswitchedTm:Y$_2$O$_3$ ceramic laser at 2.1 μm. *Opt. Mater. Exp.* 7(9), 3147–3154. https://doi.org/10.1364/OME.7.003147.

61. Yue F.X., Jambunathan V., David S.P., Mateos X., Aguilo M., Diaz F., Sulc J., Lucianetti A., Mocek T. (2020) Spectroscopy and diode-pumped continuous-wave laser operation of Tm:Y$_2$O$_3$ transparent ceramic at cryogenic temperatures. *Appl. Phys. B* 126, 44/1–8. https://doi.org/10.1007/s00340-020-7385-7.

62. Newburgh G.A., Daniels A.W., Ikesue A., Dubinskii M. (2010) Resonantly pumped 2.1-μmHo:Y$_2$O$_3$ ceramic laser. *Laser. Electro-Opt. IEEE*. https://doi.org/10.1364/CLEO.2010.CMDD2.

63. Huang D.D., Yang Q.H., Wang Y.G., Zhang H.J., LuSh.Z., Zou Y.W., Wei Z.Y. (2013) Spectral and laser properties of Yb and Ho co-doped (YLa)2O3 transparent ceramic. *Chinese Phys. B* 22, 037801/1-4. https://doi.org/10.1088/1674-1056/22/3/037801.

64. Wang J., Zhao Y.G., Yin D., Liu P., Ma J., Wang Y., Shen D.Y., Dong Z.L., Kong L.B., Tang D.Y. (2018) Holmium doped yttria transparent ceramics for 2-μm solid state lasers. *J. Eur. Ceram. Soc.* 38(4), 1986–1989. https://doi.org/10.1016/j.jeurceramsoc.2017.9.019.

65. Ter-Gabrielyan N., Merkle L.D., Newburgh G.A., Dubinskii M. (2008) Cryogenically-cooled laser based on resonantly-pumped Er^{3+}:Y$_2$O$_3$ ceramic. In *Conference on Lasers and Electro-Optics and 2008 Conference on Quantum Electronics and Laser Science*, pp. 1–2. https://doi.org/10.1109/CLEO.2008.4551577.

66. Sanamyan T., Simmons J., Dubinskii M. (2010) Er^{3+}-doped Y$_2$O$_3$ ceramic laser at ~ 2.7 μm with direct diode pumping of the upper laser level. *Laser Phys. Lett.* 7, 206–209. https://doi.org/10.1002/lapl.200910132.

67. Ren X.J., Wang Y., Zhang J., Tang D.Y., Shen D.Y. (2017) Short-pulse-width repetitively Q-switched similar to 2.7 μmEr:Y$_2$O$_3$ ceramic laser. *Appl. Sci. B* 7(11), 1201–1207. https://doi.org/10.3390/app7111201.

68. Sanamyan T., Simmons J., Dubinskii M. (2010) Efficient cryo-cooled 2.7-μm Er^{3+}:Y$_2$O$_3$ ceramic laser with direct diode pumping of the upper laser level. *Laser Phys. Lett.* 7, 569–572. https://doi.org/10.1002/lapl.201010031.

69. Lu J., Takaichi K., Uematsu T., Shirakawa A., Musha M., Ueda K., Yagi H., Yanagitani T., Kaminskii A.A. (2002) Promising ceramic laser material: Highly transparent Nd^{3+}:Lu2O3 ceramic. *Appl. Phys. Lett.* 81(23), 4324–4326. https://doi.org/10.1063/1.1527234.

70. Hao L., Wu K., Cong H., Yu H., Zhang H., Wang Z., Wang J. (2011) Spectroscopy and laser performance of Nd:Lu2O3 crystal. *Opt. Exp.*, 1917774–17779. https://doi.org/10.1364/OE.19.017774

71. Toci G., Vannini M., Ciofini M., Lapucci A., Pirri A., Ito A., Goto T., Yoshikawa A., Ikesue A., Alombert-Goget G., Guyot Y., Boulon G. (2015) Nd³⁺-doped Lu2O3 transparent sesquioxide ceramics elaborated by the Spark Plasma Sintering (SPS) method. Part 2: First laser output results and comparison with Nd³⁺-doped Lu2O3 and Nd³⁺-Y₂O₃ ceramics elaborated by a conventional method. *Opt. Mater.* 41, 12–16. https://doi.org/10.1016/j.optmat.2014.09.033.

72. Xu C.W., Yang C.D., Zhang H., Duan Y.M., Zhu H.Y., Tang D.Y., Huang H.H., Zhang J. (2016) Efficient laser operation based on transparent Nd:Lu2O3 ceramic fabricated by spark plasma sintering. *Opt. Exp.* 24(18), 20571–20579. https://doi.org/10.1364/OE.24.020571.

73. Liu Z.Y., Toci G., Pirri A., Patrizi B., Li J. (2021) Fabrication and optical property of laser-level Nd:Lu2O3 transparent ceramics for solid-state laser applications. *J. Inorg. Mater.* 36, 210–216. https://doi.org/10.15541/jim20200143.

74. Griebner U., Petrov V., Petermann K., Peters V. (2004) Passively mode-locked Yb:Lu2O3 laser. *Opt. Exp.* 12, 3125–3130. https://doi.org/10.1364/OPEX.9.003125.

75. Takaichi K., Yagi H., Shirakawa A., Ueda K., Hosokawa S., Yanagitani T., Kaminskii A.A. (2005) Lu2O3:Yb³⁺ ceramics – A novel gain material for high-power solid-state lasers. *Phys. Status Solidi (a)* 202(1), R1–R3. https://doi.org/10.1002/pssa.200409078.

76. Tokurakawa M., Takaichi K., Shirakawa A., Ueda K., Yagi Hi, Hosokawa S., Yanagitani T., Kaminskii A.A. (2006) Diode-pumped mode-locked Yb³⁺:Lu2O3 ceramic laser. *Opt. Exp.* 14(26), 12832–12838. https://doi.org/10.1364/OE.14.012832.

77. Tokurakawa M., Shirakawa A., Ueda K., Yagi Hi, Hosokawa S., Yanagitani T., Kaminskii A.A. (2008) Diode-pumped 65 fs Kerr-lens mode-locked Yb³⁺:Lu2O3 and nondoped Y₂O₃ combined ceramic laser. *Opt. Lett.* 33(12), 1380–1382. https://doi.org/10.1364/OL.33.001380.

78. Sanghera J., Frantz J., Kim W., Villalobos G., Baker C., Sadowski B., Hunt M., Miklos F., Lutz A., Aggarwal I. (2011) 10% Yb³⁺-Lu2O3 ceramic laser with 74% efficiency. *Opt. Lett.* 36 (4), 576–578. https://doi.org/10.1364/OL.36.000576.

79. David S.P., Jambunathan V., Yue F.X., Garrec B.J.L, Lucianetti A., Mocek T. (2019) Laser performances of diode pumped Yb:Lu2O3 transparent ceramic at cryogenic temperatures. *Opt. Mater. Exp.* 9(12), 4669–4676. https://doi.org/10.1364/OME.9.004669.

80. Koopmann P., Lamrini S., Scholle K., Fuhrberg P., Petermann K., Huber G. (2011) Efficient diode-pumped laser operation of Tm:Lu2O3 around 2 µm. *Opt. Lett.* 36, 948–950. https://doi.org/10.1364/OL.36.000948

81. Antipov O.L., Golovkin S.Y., Gorshkov O.N., Zakharov N.G., Zinoviev A.P., Kasatkin A.P., Kruglova M.V., Marychev M.O., Novikov A.A., Sakharov N.V. (2011) Structural, optical, and spectroscopic properties and efficient two-micron lasing of new Tm³⁺:Lu2O3 ceramics. *Quant. Electron.* 41(10), 863–868. https://doi.org/10.1070/QE2011v041n10ABEH014653

82. Antipov O.L., Novikov A.A., Zakharov N.G., Zinoviev A.P., Yagi H., Sakharov N.V., Kruglova M.V., Marychev M.O., Gorshkov O.N., Lagatskii A.A. (2013) Efficient 2.1-µm lasers based on Tm³⁺:Lu2O3 ceramics pumped by 800-nm laser diodes. *Phys. Status Solidi C* 10(6), 969–973. https://doi.org/10.1002/pssc.201300008.

83. Lagatsky A.A., Antipov O.L., Sibbett W. (2012) Broadly tunable femtosecond Tm:Lu2O3 ceramic laser operating around 2070 nm. *Opt. Exp.* 20(17), 19349–19354.https://doi.org/10.1364/OE.20.019349.

84. Morris J., Stevenson N.K., Bookey H.T., Kar A.K., Brown C.T.A., Hopkins J.M., Dawson M.D., Lagatsky A.A. (2017) 1.9 µm waveguide laser fabricated by ultrafast laser inscription in Tm:Lu2O3 ceramic. *Opt. Exp.* 25(13), 14910–14917. https://doi.org/10.1364/OE.25.014910.

85. Baylam I., Canbaz F., Sennaroglu A. (2018) Dual-wavelength temporal dynamics of a gain-switched 2-µm Tm³⁺:Lu2O3 ceramic laser. *IEEE J. Sel. Top. Quant. Electron.* 24, 1601208. https://doi.org/10.1109/JSTQE.2018.2805825.

86. Lamrini S., Koopmann P., Schäfer M., Scholle K., Fuhrberg P., Scholle K., Petermann K., Huber G. (2011) Efficient laser operation of Ho:Lu2O3 at room temperature. In *Conference on Lasers and Electro-Optics Europe and 12th European Quantum Electronics Conference (CLEO EUROPE/EQEC)*, pp. 1–1. https://doi.org/10.1109/CLEOE.2011.5942451.

87. Koopmann P., Lamrini S., Scholle K., Schäfer M., Fuhrberg P., Huber G. (2011) Multi-watt laser operation and laser parameters of Ho-doped Lu2O3 at 2.12 µm. *Opt. Mater. Exp.* 1, 1447–1456. https://doi.org/10.1364/OME.1.001447

88. Kim W., Baker C., Bowman S., Florea C., Villalobos G., Shaw B., Sadowski B., Hunt M., Aggarwal I., Sanghera J. (2013) Laser oscillation from Ho³⁺ doped Lu2O3 ceramics. *Opt. Mater. Exp.* 3(7), 913–919. https://doi.org/10.1364/OME.3.000913.

89. Peters V., Fornasiero L., Mix E., Diening A., Petermann K., Huber G.(1998) Spectroscopic Characterization and Diode-Pumped Laser Action at 2.7 μm of Er:Lu2O3. In *CLEO/Europe Conference on Lasers and Electro-Optics*, pp. 379–379. https://doi.org/10.1109/CLEOE.1998.719598.

90. Li T., Beil K., Kränkel C., Huber G. (2012) Efficient high-power continuous wave Er:Lu2O3 laser at 2.85 μm. *Opt. Lett.* 37, 2568–2570. https://doi.org/10.1364/OL.37.002568.

91. Wang Li, Huang H., Shen D., Zhang J., Chen H., Wang Y., Liu X., Tang D. (2014) Room temperature continuous-wave laser performance of LD pumped Er:Lu2O3 and Er:Y$_2$O$_3$ ceramic at 2.7 μm. *Opt. Exp.* 22, 19495–19503. https://doi.org/10.1364/OE.22.019495.

92. Qiao X.B., Huang H.T., Yang H., Zhang L., Wang L., Shen D.Y., Zhang J., Tang D.Y. (2015) Fabrication, optical properties and LD-pumped 2.7 μm laser performance of low Er^{3+} concentration doped Lu2O3 transparent ceramics. *J. Alloys Compd.* 640, 51–55. https://doi.org/10.1016/j.jallcom.2015.03.190.

93. Wang L., Huang H.T., Shen D.Y., Zhang J., Tang D.Y. (2018) High power and short pulse width operation of passively Q-switched Er:Lu2O3 ceramic laser at 2.7 μm. *Appl. Sci. B* 8(5), 801/1-7. https://doi.org/10.3390/app8050801

94. Peters V., Mix E., Fornasiero L., Petermann K., Huber G., Basun S. (2000) Efficient laser operation of Yb^{3+} : Sc2O3 and spectroscopic characterization of Pr^{3+} in cubic sesquioxide. *Laser Phys.* 10, 417–421.

95. Lu J., Bisson J.F., Takaichi K., Uematsu T., Shirakawa A., Musha M., Ueda K., Yagi H., Yanagitani T., Kaminskii A.A. (2003) Yb^{3+}:Sc2O3 ceramic laser. *Appl. Phys. Lett.* 83(6), 1101–1103. https://doi.org/10.1063/1.1600851.

96. Takaichi K., Yagi H., Becker P., Shirakawa A., Ueda K., Bohatý L., Yanagitani T., Kaminskii A.A. (2007) New data on investigation of novel laser ceramic on the base of cubic scandium sesquioxide: Two-band tunable CW generation of Yb^{3+}:Sc2O3 with laser-diode pumping and the dispersion of refractive index in the visible and near-IR of undoped Sc2O3. *Laser Phys. Lett.* 4, 507–510. https://doi.org/10.1002/lapl.200710020.

97. Tokurakawa M., Shirakawa A., Ueda K., Yagi H., Yanagitani T., Kaminskii A.A. (2007) Diode-pumped sub 100-fs Kerr-lens mode-locked Yb^{3+}:Sc2O3 ceramic laser. *Opt. Lett.* 32(23), 3382–3384. https://doi.org/10.1364/OL.32.003382

98. Pirri A., Toci G., Vannini M. (2011) First laser oscillation and broad tunability of 1 at. % Yb-doped Sc2O3 and Lu2O3ceramics. *Opt. Lett.* 36(21), 4284–4286. https://doi.org/10.1364/OL.36.004284.

99. Toci G., Hostasa J., Patrizi B., Biasini V., Pirri A., Piancastelli A., Vannini M. (2020) Fabrication and laser performances of Yb:Sc2O3 transparent ceramics from different combination of vacuum sintering and hot isostatic pressing conditions. *J. Eur. Ceram. Soc.* 40(3), 881–886. https://doi.org/10.1016/j.jeurceramsoc.2019.10.059.

100. Ivanov V., Kaigorodov A.S., Khrustov V.R., Osipov V.V., Medvedev A.I., Murzakaev A.M., Orlov A.N. (2007) Properties of the translucent ceramics Nd:Y$_2$O$_3$ prepared by pulsed compaction and sintering of weakly aggregated nanopowders. *Glass Phys. Chem.* 33, 387–393. https://doi.org/10.1134/S108765960704013X.

101. Permin D.A., Balabanov S.S., Snetkov I.L., Palashov O.V., Novikova A.V., Klyusik O.N., Ladenkov I.V. (2020) Hot pressing of Yb:Sc2O3 laser ceramics with LiF sintering aid. *Opt. Mater.* 100, 109701/1-6. https://doi.org/10.1016/j.optmat.2020.109701.

102. Lupei V., Lupei A., Gheorghe C., Ikesue A., Stefan A., Ciupina V., Prodan A. (2007) Spectroscopic characteristics of RE^{3+}:Sc2O3 ceramics. *Proc. SPIE* 6785, 67850C. https://doi.org/10.1117/9.755967.

103. Fornasiero L., Mix E., Peters V., Petermann K., Huber G. (2000) Czochralski growth and laser parameters of RE^{3+}-doped Y$_2$O$_3$ and Sc2O3. *Ceram. Intern.* 26, 589–592. https://doi.org/10.1016/S0272-8842(99)00101-7.

104. Koopmann P., Lamrini S., Scholle K., Schäfer M., Fuhrberg P., Huber G. (2013) Holmium-doped Lu2O3, Y$_2$O$_3$, and Sc2O3 for lasers above 2.1 μm. *Opts. Exp.* 21, 3926–3931. https://doi.org/10.1364/OE.21.003926.

105. Farrell E.F., Fang J.H., Newnham R.E. (1963) Refinement of the chrysoberyl structure. *Am. Mineral* 48(7–8), 804–810.

106. Bragg W.L., Brown G.B. (1926) V. Die Kristallstruktur von Chrysoberyll (BeAl2O4). *Z. Kristallogr: Crystal. Mater.* 63(1–6), 122–143. https://doi.org/10.1524/zkri.1926.63.1.122.

107. Ching, W.Y., Xu, Y.N., Brickeen, B.K. (2001) Comparative study of the electronic structure of two laser crystals: BeAl2O4 and LiYF 4. *Phys. Rev. B* 63(11), 115101. https://doi.org/10.1103/PhysRevB.63.115101

108. Morris R.C., Cline C.F. (1976) Chromium-doped beryllium aluminate lasers. Patent US 3,997,853A.

109. Walling J.C., Peterson O., Jenssen H.P., Morris R.C., Dell E.O. (1980) Tunable alexandrite lasers. *IEEE J. Quant. Electron.* 16(6), 1302–1315. https://doi.org/doi: 10.1109/JQE.1980.1070430.

110. Walling J.C., Jenssen H.P., Morris R.C., Dell E.O., Peterson O.G. (1979) Tunable-laser performance in BeAl2O4:Cr^{3+}. *Opt. Lett.* 4(6), 182–183. https://doi.org/10.1364/OL.4.000182

111. Walling J., Peterson O. (1980) High gain laser performance in alexandrite. *IEEE J. Quant. Electron* 16s(2), 119–120. https://doi.org/10.1109/JQE.1980.1070440.

112. Ghanbari S., Akbari R., Major A. (2016) Femtosecond Kerr-lens mode-locked Alexandrite laser. *Opt. Exp.* 24(13), 14836–14840. https://doi.org/10.1364/OE.24.014836.

113. Cihan C., Muti A., Baylam I., Kocabas A., Demirbas U., Sennaroglu A. (2018) 70 femtosecond Kerr-lens mode-locked multipass-cavity Alexandrite laser. *Opt. Lett.* 43(6), 1315–1318. https://doi.org/10.1364/OL.43.001315.

114. Gadomski W., Ratajska-Gadomska B. (2000) Homoclinic orbits and chaos in the vibronic short-cavity standing-wave alexandrite laser. *J. Opt. Soc. Am. B-Opt. Phys.* 17(2), 188–197. https://doi.org/10.1364/JOSAB.17.000188.

115. Kruger R.A, Kuzmiak C.M., Lam R.B., Reinecke D.R., Del Rio S.P., Steed D. (2013) Dedicated 3D photoacoustic breast imaging. *Med. Phys.* 40(11), 113301. https://doi.org/10.1118/1.4824317.

116. Niederhauser J.J, Jaeger M., Lemor R., Weber P., Frenz M. (2005) Combined ultrasound and opto-acoustic system for real-time high-contrast vascular imaging in vivo. *IEEE Trans. Med. Imaging* 24(4), 436–440. https://doi.org/10.1109/TMI.2004.843199.

117. Alster T.S. (1995) Q-switched alexandrite laser treatment (755 Nm) of professional and amateur tattoos. *J. Am. Acad. Dermatol.* 33(1), 69–73. https://doi.org/10.1016/0190-9622(95)90013-6.

118. Gan S.D., Graber E.M. (2013) Laser hair removal: A review. *Dermatol. Surg.* 39(6), 823–838. https://doi.org/10.1111/dsu.12116.

119. Polder K.D., Landau J.M., Vergilis-Kalner I.J., Goldberg L.H., Friedman P.M., Bruce S. (2011) Laser eradication of pigmented lesions: a review. *Dermatol Surg.* 37(5), 572–595. https://doi.org/10.1111/j.1524-4725.2011.01971.x.

120. Higdon N.S., Browell E.V., Ponsardin P., Grossmann B.E., Butler C.F., Chyba T.H., Mayo M.N., Allen R.J., Heuser A.W., Grant W.B., Ismail S., Mayor S.D., Carter A.F. (1994) Airborne differential absorption LiDAR system for measurements of atmospheric water vapor and aerosols. *Appl. Opt.* 33(27), 6422–6438. https://doi.org/10.1364/AO.33.006422.

121. Song Y., Wang Z.M., Bo Y., Zhang F.F., Zhang Y.X., Zong N.S, Peng Q.J. (2021) 2.55 W continuous-wave 378nm laser by intracavity frequency doubling of a diode-pumped Alexandrite laser. *Appl. Opt.* 60(20), 5900–5905. https://doi.org/10.1364/AO.430135.

122. Coney A.T., Damzen M.J. (2021) High-energy diode-pumped alexandrite amplifier development with applications in satellite-based LiDAR. *J. Opt. Soc. Am. B* 38(1), 209–219. https://doi.org/10.1364/JOSAB.409921.

123. Fibrich M., Sulc J., Jelinkova H. (2019) Alexandrite microchip lasers. *Opt. Exp.* 27, 16975–16982. https://doi.org/10.1364/OE.27.016975.

124. Geusic F.E., Marcos H.M., van Uitert L.G. (1964) Laser oscillations In Nd-doped yttrium aluminum, yttrium gallium and gadolinium garnets. *Appl. Phys. Lett.* 4, 182–184. https://doi.org/10.1063/1.1753928.

125. Young C.G. (1966) A sun-pumped CW one-watt laser. *Appl. Opt.* 5, 993–997. https://doi.org/10.1364/AO.5.000993.

126. Hasegawa K., Ichikawa T., Mizuno S., Takeda Y., Ito H., Ikesue A., Motohiro T., Yamaga M. (2015) Energy transfer efficiency from Cr^{3+} to Nd3+ in solar-pumped laser using transparent Nd/Cr:Y3Al5O12 ceramics. *Opt. Exp.* 23, A519–A524. https://doi.org/10.1364/OE.23.00A519.

127. Takashi K. (1969) *Oyo Buturi* 38, 985.

128. (1981) High quality Nd: YAG laser rods for production laser systems, union carbide electronics division sample book. https://doi.org/10.1063/1.1754094

129. Tachibana M., Iwanade A., Miyakawa K. (2021) Distribution coefficient of rare-earth dopants in Y3Al5O12 garnet. *J. Cryst. Growth*, 568–569, 126191. https://doi.org/10.1016/j.jcrysgro.2021.126191

130. Caslavsky J.L., Viechnicki D. (1979) Melt growth of Nd:Y3Al5O12 (Nd:YAG) using the heat exchange method (HEM). *J. Cryst. Growth* 45, 601–606. https://doi.org/10.1016/0022-0248(79)90175-1

131. Khattak C.P., Schmid F. (1984) Growth of large-diameter crystals by Hem Tm for optical and laser application Proc. SPIE 0505. *Adv. Opt. Mater.* https://doi.org/10.1117/9.964616.

132. Guo H., Zhang M., Han J., Nie Y. (2013) Crystal growth, Judd-Ofelt analysis and radiative properties of YAG:Nd single crystal grown by HDS. *J. Lumin.* 140, 135–137. https://doi.org/10.1016/j.jlumin.2013.03.002.

133. Ikesue A., Kinoshita T., Kamata K., Yoshida K. (1995) Fabrication and optical properties of high-performance polycrystalline Nd:YAG ceramics for solid-state lasers. *J. Am. Ceram.* Soc. 78, 1033–1040. https://doi.org/10.1111/j.1151-2916.1995.tb08433.x.

134. Yanagitani T., Yagi H., Ichikawa M. (1998) Japan Patent 10-101333.

135. Yanagitani T., Yagi H., Hiro Y. (1998) Japan Patent 10-101411.

136. Ikesue A., Aung Y.L. (2006) Synthesis and performance of advanced ceramic lasers. *J. Am. Ceram. Soc.* 89, 1936–1944. https://doi.org/10.1111/j.1551-2916.2006.01043.x.

137. Belli Dell'Amico D., Biagini P., Bongiovanni G., Chiaberge S., Di Giacomo A., Labella L., Marchetti F., Marra G., Mura A., Quochi F., Samaritani S., Sarritzu V.A. (2018) Convenient preparation of nano-powders of Y_2O_3, Y3Al5O12 and Nd:Y3Al5O12 and study of the photoluminescent emission properties of the neodymium doped oxide. *Inorgan. Chim. Acta.* 470, 149–157. https://doi.org/10.1016/j.ica.2017.05.012

138. Zimina G.V., Novoselov A.V., Smirnova I.N., Spiridonov F.M., Pushkina G.Y., Komissarova L.N. (2010) Synthesis and study of yttrium aluminum garnets doped with neodymium and ytterbium. *Russ. J. Inorg. Chem.* 55, 1833–1836. https://doi.org/10.1134/S003602361012003X

139. Kaithwas N., Dave M., Kar S., Verma S., Bartwal K.S. (2010) Preparation of Nd:Y3Al5O12 nanocrystals by low temperature glycol route. *Crys. Rese. Tech.* 45, 1179–1182. https://doi.org/10.1002/crat.201000465

140. Mendoza-Mendoza E., Montemayor S.M., Maczka M., Marciniak M., Fuentes A.F. (2013) A facile and "green-chemistry" method to synthesize pure and Nd-doped Y3Al5O12nanopowders at low-temperatures. *Ceram. Inter.* 39, 9405–9414. https://doi.org/10.1016/j.ceramint.2013.05.057

141. Liu Q., Chen C., Dai J., Hu Z., Chen H., Li J. (2018) Effect of ammonium carbonate to metal ions molar ratio on synthesis and sintering of YAG:Nd nanopowders. *Opt. Mater.* 80, 127–137. https://doi.org/10.1016/j.optmat.2018.04.026

142. Sakar N., Gergeroglu H., AlperAkalin S., Oguzlar S., Yildirim S. (2020) Synthesis, structural and optical characterization of Nd: YAG powders via flame spray pyrolysis. *Opt. Mater.* 103, 109819. https://doi.org/10.1016/j.optmat.2020.109819.

143. Suarez M., Fernandez A., Menendez J.L., Nygren M., Torrecillas R., Zhao Z. (2010) Hot isostatic pressing of optically active YAG:Nd powders doped by a colloidal processing route. *J. Euro. Ceram. Soc.* 30, 1489–1494. https://doi.org/10.1016/j.jeurceramsoc.2009.9.006.

144. Li J., Wu Y., Pan Y., Liu W., Zhu Y., Guo J. (2008) Solid-state-reactive fabrication of Cr, YAG:Nd transparent ceramics: The influence of raw material. *J. Ceram. Soc. Jpn.* 116, 572–577. https://doi.org/10.2109/jcersj2.116.572.

145. Hreniak D., Strek W., Gluchowski P., Fedyk R., Lojkowski W. (2008) The concentration dependence of luminescence of Nd:Y3Al5O12 nanoceramics. *J. Alloys Comps.* 451, 549–552. https://doi.org/10.1016/j.jallcom.2007.04.137.

146. Chani V.I., Yoshikawa A., Kuwano Y., Hasegawa K., Fukuda T. (1999) Growth of Y3Al5O12 : Nd fiber crystals by micro-pulling-down technique. *J. Cryst. Growth* 204, 155–162. https://doi.org/10.1016/S0022-0248(99)00170-0.

147. Feigelson R.S. (1986) Pulling optical fibers. *J. Cryst. Growth* 79, 669–680. https://doi.org/10.1016/0022-0248(86)90535-X.

148. Lo C.Y., Huang P.L., Chou T.S., L.M. Lee, Chang T.Y., Huang S.L., Lin L., Yan Lin H., Ho F.C. (2002) Efficient Nd: Y3Al5O12 crystal fiber laser. *Jpn. J. Appl. Phys. Part 2: Lett.* 41, L1228–L1231. https://doi.org/10.1143/JJAP.41.L1228.

149. Shinohara Y., Kisara K., Suzuki H. (2010) Prospect of YAG laser medium for laser space solar power system and possible application of graded structure for efficient cooling system. *Mater. Sci. Forum* 9–14, 631–632. https://doi.org/10.4028/www.scientific.net/MSF.631-632.9.

150. Kiss Z., Duncan R. (1964) Cross-pumped Cr^{3+}, Nd^{3+}: YAG laser system. *Appl. Phys. Lett.* 5, 200–202. https://doi.org/10.1063/1.1723587.

151. Yagi H., Yanagitani T., Yoshida H., Nakatsuka M., Ueda K. (2007) The optical properties and laser characteristics of Cr^{3+} and Nd^{3+} co-doped Y3Al5O12 ceramics. *Opt. Laser Tech.* 39, 1295–1300. https://doi.org/10.1016/j.optlastec.2006.06.016.

152. Sun H.J., Liang Z.R., Ying Z.Q., Zhang S.X., Zhai Q.Y., Zhang S.Y., et al. (1988) Laser performance of (Nd, Ce):YAG crystals. *J. Synth. Cryst.* 317, 359. https://doi.org/10.16553/j.cnki.issn1000-985x.1988.z1.317.

153. Villars B., Hill E.S., Durfee C.G. (2015) Design and development of a high-power LED- pumped Ce: Nd: YAG laser. *Opt. Lett.* 40, 3049–3052. https://doi.org/10.1364/OL.40.003049.

154. Payziyev S., Makhmudov K., Abdel-Hadi Y.A. (2018) Simulation of a new solar Ce: Nd: YAG laser system. *Optik* 156, 891–895. https://doi.org/10.1016/j.ijleo.2017.9.071.

155. Guo Y., Huang J., Ke G., Ma Y., Quan J., Yi G. (2021) Growth and optical properties of the Nd, Ce:YAG laser crystal. *J. Lumin.* 236, 118134. https://doi.org/10.1016/j.jlumin.2021.118134.

156. Li Y., Zhou S.M., Lin H., Hou H.R., Li W.J. (2010) Intense 1064 nm emission by the efficient energy transfer from Ce^{3+} to Nd^{3+} in Ce/Nd co-doped YAG transparent ceramics. *Opt. Mater.* 32, 1223–1226. https://doi.org/10.1016/j.optmat.2010.04.003.

157. Endo M. (2010) Optical characteristics of Cr^{3+} and Nd^{3+} codoped Y3Al5O12 ceramics. *Opt. Laser Tech.* 42, 610–616. https://doi.org/10.1016/j.optlastec.2009.10.09.

158. Carreaud J., Boulesteix R., Maitre A., Rabinovitch Y., Brenier A., Labruyère A., Couderc V. (2013) From elaboration to laser properties of transparent polycrystalline Nd-doped Y3Al5O12 and Y3ScAl4O12 ceramics: A comparative study. *Opt. Mater.* 35, 704–711. https://doi.org/10.1016/j.optmat.209.07.021.

159. Sato Y., Saikawa J., Taira T., Ikesue A. (2007) Characteristics of Nd^{3+} doped Y3ScAl4O12 ceramic laser. *Opt. Mater.* 29, 1277–1282. https://doi.org/10.1016/j.optmat.2006.01.032.

160. Endo M. (2007) Feasibility study of a conical-toroidal mirror resonator for solar-pumped thin disk lasers. *Opt. Exp.* 15, 5482–5493. https://doi.org/10.1364/OE.15.005482.

161. Xu X., Zhao Z., Song P., Zhou J., Xu J., Deng P. (2004) Structural, thermal, and luminescent properties of Yb-doped Y3Al5O12 crystals. *J. Opt. Soc. Am. B* 21, 543–547. https://doi.org/10.1364/JOSAB.21.000543.

162. Lacovara P., Choi H.K., Wang C.A., Aggarwal R.L., Fan T.Y. (1991) Room-temperature diode-pumped Yb:YAG laser. *Opt. Lett.* 16, 1089–1091. https://doi.org/10.1364/OL.16.001089

163. Dascalu T., Pavel N., Taira T. (2003) 10 W continuous-wave diode edge-pumped microchip composite Yb:Y3Al5O12 laser. *Appl. Phys. Lett.* 83, 4086–4088. https://doi.org/10.1063/1.1627960.

164. Johnson L.F., Geusic J.E., Van Uitert L.G. (1965) Coherent oscillations from Tm^{3+}, Ho^{3+}, Yb^{3+} and Er^{3+} ions in yttrium aluminum garnet. *Appl. Phys. Lett.* 7, 127–129. https://doi.org/10.1063/1.1754339.

165. Reinberg A.R., Riseberg L.A., Brown R.M.,. Wacker R.W, Holton W.C. (1971) GaAs:Si LED pumped Yb-doped YAG laser. *Appl. Phys. Lett.* 19, 11–13. https://doi.org/10.1063/1.1653723.

166. van der Ziel J.P., Bonner W.A., Kopf L., Van Uitert L.C. (1972) Coherent emission from Ho^{3+} ions in epitaxially grown thin aluminum garnet films. *Phys. Lett.* 42A, 105–106. https://doi.org/10.1016/0375-9601(72)90049-7.

167. van der Ziel J.P., Bonner W.A., Kopf L., Singh S., Van Uitert L.C. (1973) Laser oscillation from Ho^{3+} and Nd^{3+} ions in epitaxially grown thin aluminum garnet films. *Appl. Phys. Lett.* 22, 656–657. https://doi.org/ 10.1063/1.1654543.

168. Gao P., Zhang Le, Yao Q., Ma Y., Shao C., Zhou T., Liu M., Zhu L., Chen H., Cheng X., Yang H. (2021) A novel route to fabricate Yb:YAG ceramic fiber and its optical performance. *J. Eur. Cer. Soc.* 41, 4598–4608. https://doi.org/10.1016/j.jeurceramsoc.2021.03.019.

169. Delen X., Piehler S., Didierjean J., Aubry N., Voss A., Ahmed M.A., Graf T., Balembois F., George. (2012) P.250 W single-crystal fiber Yb:YAG laser. *Opt. Lett.* 37, 2898–2900. https://doi.org/10.1364/OL.37.002898.

170. Ter-Gabrielyan N., Fromzel V., Mu X., Meissner H., Dubinskii M. (2012) High efficiency, resonantly diode pumped, double-clad, Er:YAG-core, waveguide laser. *Opt. Exp.* 20, 25554–25561. https://doi.org/10.1117/9.1518452.

171. Beach R.J., Mitchell S.C., Meissner H.E., Meissner O.R., Krupke W.F., McMahon J.M., Bennett J.W., Shepherd D.P. (2001) Continuous-wave and passive Q-switched cladding-pumped planar waveguide lasers. *Opt. Lett.* 26, 881–883. https://doi.org/10.1364/OL.26.000881

172. Geeraets W.J., Berry E.R. (1968) Ocular spectral characteristics as related to hazards from lasers and other light sources. *Am. J. Ophthalmol.* 61, 15–20. https://doi.org/10.1016/0002-9394(68)91780-7.

173. Yu Y., Wu Z., Zhang S. (2000) Concentration effects of Er ion in YAG:Er^{3+} laser crystals. *J. Alloys Compd.* 302, 204–208. https://doi.org/10.1016/S0925-8388(99)00630-1.

174. Spariosu K., Birnbaum M., Viana B. (1994) Er^{3+}:Y3Al5O12 laser dynamics: Effects of upconversion. *J. Opt. Soc. Am. B: Opt. Phys.* 11, 894–900. https://doi.org/10.1364/JOSAB.11.000894.

175. White K.O., Schleusener S.A. (1972) Coincidence of Er: YAG laser emission with methane absorption at 1645.1 nm. *Appl. Phys. Lett.*, 21419–21420. https://doi.org/10.1063/1.1654438.

176. Zverev G.M., Garmash V.M., Onishchenko A.M., Pashkov V.A., Semenov A.A., KolbetskovYu M., Smirnov A.I. (1974) *Zh. Prikl. Spektrosk.* 21, 821–823. [English transl.: *J. Appl. Spectrosc. (USSR)* 21, 1467–1469 (1974)].

177. Stange H., Petermann K., Huber G., Duczynski E.W. (1989) Continuous wave 1.6 μm laser action in Er doped garnets at room temperature. *Appl. Phys. B* 49, 269–273. https://doi.org/10.1007/BF00714646.

178. Yu Y., Zhang S., Tie S., Song M. (1995) Spectral characteristics of Er^{3+}activated and Ce^{3+} sensitized yttrium aluminium garnet laser crystals. *J. Alloys Compd.* 217, 148–150. https://doi.org/10.1016/0925-8388(94)01308-5.

179. Zharikov E.V., Zhekov V.I., Kulevskii L.A., Murina T.M., Osiko V.V., Prokhorov A.M., Savelev A.D., Smirnov V.V., Starikov B.P., Tomoshechkin M.I. (1975) Stimulated emission from Er^{3+}ions in yttrium aluminum garnet crystals at $\lambda = 2.94$ μ. *Sov. J. Quant. Electron.* 4, 1039–1041. https://dx.doi.org/10.1070/QE1975v004n08ABEH011147.

180. Pollack S.A., Chang D.B., Birnbaum M., Kokta M. (1991) Upconversion-pumped 2.8-2.9-μm lasing of Er^{3+} ion in garnets. *J. Appl. Phys.* 70, 7227–7239. https://doi.org/10.1063/1.349767.

181. Kaminskii A.A., Butaeva T.I., Fedorov V.A., BagdasarovKh S., Petrosyan A.G. (1977) Absorption, luminescence, and stimulated emission investigations in Lu3Al5O12-Er^{3+} Crystals. *Physica Status Solidi (a)* 39(2), 541–548. https://doi.org/10.1002/pssa.2210390222.

182. Liu J., Song Q., Wang Z., Wang Y., Dong J., Xu J., Liu P., Li D., Wang J., Wang C., Lebbou K. (2022) Ho: LuAG single crystal fiber: Growth, spectroscopy and laser characteristics. *Opt. Exp.* 30(4), 5826–5834. https://doi.org/10.1364/OE.449790

183. Hart D.W., Jani M., Barnes N.P. (1996) Room-temperature lasing of end-pumped Ho:Lu3Al5O9. *Opt. Lett.* 21(10), 728–730. https://doi.org/10.1364/OL.21.000728.

184. Duan X., Yao B., Li G., Youlun Ju, Wang Y., Zhao G. (2009) High efficient actively Q-switched Ho:LuAG laser. *Opt. Exp.* 17(24), 21691–2169. https://doi.org/10.1364/OE.17.021691.

185. Sudesh V., Piper J.A. (2000) Spectroscopy, modeling, and laser operation of thulium-doped crystals at 2.3μm. *IEEE J. Quant. Electron.* 36(7), 879–884. https://doi.org/10.1109/3.848362

186. Rodriguez W.J., Naranjo F.L., Barnes N.P., Kokta M.K. (1994) *Tech. Dig. Conf. on Laser and Electro-Optics*, Anaheim. CA 174.

187. Kushawaha V., Chen Y., Yan Y., Major L. (1996) High-efficiency continuous-wave diode-pumped Tm:Ho:LuAG laser at 2.1μm. *App. Phys. B: Lasers Opt.* 62(1), 109–111. https://doi.org/10.1007/BF01081257.

188. Sato H., Shimamura K., Sudesh V., Masahiko Ito, Machida H., Fukuda T. (2002) Growth and characterization of Tm, Ho-codoped Lu3Al5O12 single crystals by the Czochralski technique. *J. Cryst. Growth* 234(2–3), 463–468. https://doi.org/10.1016/S0022-0248(01)01724-9.

189. Liu J., Dong J., Wang Y., Yuan H., Song Q., Xue Y., Xu J., Liu P., Li D., Lebbou K., Wang Z., Zhao Y., Xu X., Xu J. (2021) Laser operation of Tm: LuAG single-crystal fiber grown by the micro-pulling down method. *Crystals* 11(8), 898. https://doi.org/10.3390/cryst11080898.

190. Xu X.D., Wang X.D., Meng J.Q., Y Cheng D Z Li., Cheng S.S., F Wu Zhao Z.W., Xu J. (2009) Crystal growth, spectral and laser properties of Nd:LuAG single crystal. *Laser Phys. Lett.* 6(9), 678–681. https://doi.org/10.1002/lapl.200910053.

191. Paul David S., Jambunathan V., Yue F., Lucianetti A., Mocek T. (2021) Diode pumped cryogenic Yb:Lu3Al5O12 laser in continuous-wave and pulsed regime. *Opt. Laser Tech.* 135, 106720. https://doi.org/10.1016/j.optlastec.2020.106720

192. You Z.Y., Wang Y., Xu J.L., Zhu Z.J., Li J.F., Wang H.Y., Tu C.Y. (2015) Single-longitudinal-mode Er:GGG microchip laser operating at 2.7μm. *Opt. Lett.* 40(16), 3846–3849. https://doi.org/10.1364/OL.40.003846.

193. Linares R.C. (1964) Growth of garnet laser crystals. *Solid State Commun.* 2(8), 229–231. https://doi.org/10.1016/0038-1098(64)90369-2.

194. Osiko V.V., Sigachev V.B., Strelov V.I., Timoscechkin M.I. (1991) Erbium gadolinium gallium garnet crystal laser. *Sov. J. Quant. Electron.* 21(2), 159–161. https://doi.org/10.1070/QE1991v021n02ABEH003740.

195. Yoshida K., Yoshida H., Kato Y. (1988) Characterization of high average power Nd:GGG slab lasers. *IEEE J. Quant. Electron.* 24(6), 1188–1192. https://doi.org/10.1109/3.243.

196. Rotter M.D., Dane C.B., Fochs S., Fortune K.L., Merrill R., Yamamoto B. (2004) Solid-state heat-capacity lasers: Good candidates for the marketplace. *Photon. Spectra* 278(38), 44–52. https://doi.org/10.1364/ASSP.2004.278.

197. Qin L.J., Tang D.Y., Xie G.Q., Luo H., Dong C.M., Jia Z.T., Yu H.H., Tao X.T. (2008) Diode-end-pumped passively mode-locked Nd:GGG laser with a semiconductor saturable mirror. *Opt. Commun.* 281(18), 4762–4764. https://doi.org/10.1016/j.optcom.2008.06.011.

198. Qin L.J., Tang D.Y., Xie G.Q., Dong C.M., Jia Z.T., Tao X.T. Tao. (2008) High-power continuous wave and passively Q-switched laser operations of a Nd:GGG crystal. *Laser Phys. Lett.* 5. 100–103. https://doi.org/10.1002/lapl.200710097.

199. Qiao J., Zhao J., Yang K., Zhao S., Li Y., Li G., Li D., Qiao W., Li T., Chu H. (2015) Diode-pumped Nd:Gd3Ga5O12---> KTiOPO4 green laser doubly passively Q-switched mode-locked by GaAs and Cr⁴⁺:YAG saturable absorbers. *Jpn. J. Appl. Phys.* 54(3), 032701. https://doi.org/10.7567/JJAP.54.032701.

200. Shen J.P., Ding C.F. (2012) Narrow pulse width and high power passively Q-switched Nd³⁺:Gd3Ga5O12 laser with Cr⁴⁺:YAG saturable absorber. *Laser Phys.* 22, 1659–1663. https://doi.org/10.1134/S1054660X12110126.

201. Li Z., Huang H., He J., Zhang B., Xu J. Li Z.Y., Huang H.T., He J.L., Zhang B.T., Xu J.L. (2010) High peak power eye-safe intracavity optical parametric oscillator pumped by a diode-pumped passively Q-switched Nd:GGG laser. *Laser Phys.* 20, 1302–1306. https://doi.org/10.1134/S1054660X10110125.

202. Shimokozono M., Sugimoto N., Tate A., Katoh Y. (1996) Room-temperature operation of an Yb-doped Gd3Ga5O12 buried channel waveguide laser at 1.025 μm wavelength. *Appl. Phys. Lett.* 68(16), 2177. https://doi.org/10.1063/1.116004.

203. Chénais S., Druon F., Balembois F., Georges P., Brenier A., Boulon G. (2003) Diode-pumped Yb:GGG laser: Comparison with Yb:YAG. *Opt. Mater.* 22(2), 99–106. https://doi.org/10.1016/S0925-3467(02)00353-1.

204. Zhou H., Ma X., Chen G., Wancong L., Yan W., Zhenyu Y., Jianfu Li., Zhu Z., Tu C. (2009) Tm³⁺ doped Gd3Ga5O12 crystal: A potential tunable laser crystal at 2.0 μm. *J. Alloys Compd.* 475(1–2), 555–559. https://doi.org/10.1016/j.jallcom.2008.07.084.

205. Moncorge R., Manaa H., Koselja M., Boulon G., Madej C., Souriau J.C., Borel C., Wyon C. (1994) Comparative optical study and 2 μm laser performance of the Tm³⁺ doped oxide crystals: Y3Al5O12, YAlO3, Gd3Ga5O12, Y2SiO5, SrY4(SiO4)3O. *J. Phys. IV Fr.* 04(4), 377–379. https://doi.org/10.1051/jp4:1994490.

206. Wang Y., Lan J., Zhou Z., Xiaofeng G., Bin Xu, Huiying Xu, Zhiping Cai, Wang Y., Tu C. (2017) Continuous-wave laser operation of diode-pumped Tm-doped Gd3Ga5O12 crystal, *Opt. Mater.* 66, 185–188. https://doi.org/10.1016/j.optmat.2017.02.014.

207. Henry M.O., Larkin J.P., Imbusch G.F. (1976) Nature of the broadband luminescence center in MgO:Cr³⁺. *M Phys. Rev. B.* 13(5), 1893. https://doi.org/10.1103/PhysRevB.13.1893.

208. Kuck S., Heumann E., Karner T., Maaroos A. (1999) Continuous-wave room-temperature laser oscillation of Cr³⁺:MgO. *Opt. Lett.* 24(14), 966–968. https://doi.org/10.1364/OL.24.000966.

209. Ralph J.E., Townsend M.G. (1968) Near-infrared fluorescence and absorption spectra of Co²⁺ and Ni²⁺ in MgO. *J. Chem. Phys.* 48(1), 149–154. https://doi.org/10.1063/1.1664462.

210. Mooradian A. (1979) Laser Focus 15, 28. *Rep. Prog. Phys.* 42, 1533.

211. Iverson M.V., Windscheif J.C., Sibley W.A. (1980) Optical parameters for the MgO:Ni²⁺ laser system. *Appl. Phys. Lett.* 36(3), 183–184. https://doi.org/10.1063/1.91439

212. Moulton P.F. (1982) *Handbook of Lasers Science and Technology I, Lasers and Masers*, Weber M. J. (ed.). Chemical Rubber, Boca Raton, p. 60

213. Moncorge R., Benyattou T. (1988) Excited-state absorption measurements in the Ni²⁺ doped MgO and MgF2 vibronic laser systems. *J. Lumin.* 40–41(2), 105–106. https://doi.org/10.1016/0022-2313(88)90109-3

214. Moncorge R., Benyattou T. Excited-state absorption of Ni²⁺ in MgF2 and MgO. *Phys. Rev. B* 37(16), 9186–9196. https://doi.org/10.1103/PhysRevB.37

215. Bloembergen N. (1959) Solid state infrared quantum counters. *Phys. Rev. Lett.* 2(3), 84–85. https://doi.org/10.1103/PhysRevLett.2.84.

216. Auzel F. (1966) Computeurquantique par transfertd'energie entre de Yb³⁺ a Tm³⁺ dans un tungstate mixte et dans verre germinate. *C. R. Acad. Sci.* 263, 819–821.

217. Haase M., Schafer H. (2011) Upconverting nanoparticles. *Angew. Chem. Int. Ed.* 50(26), 5808–5829. https://doi.org/10.1002/anie.201001159.

218. Bowman S.R., Shaw L.B., Feldman B.J., Ganem J. (1996) A 7-/spl mu/m praseodymium-based solid-state laser. *IEEE J. Quant. Electron.* 32, 646–649. https://doi.org/10.1109/3.488838

219. Wang H.Q., Batentschuk M., Osvet A., Pinna L., Brabec C.J. (2011) Rare earth ion doped upconversion materials for photovoltaic applications. *Adv. Mater.* 23(22–23), 2675–2680. https://doi.org/10.1002/adma.201100511.

220. Joubert M.F. (1999) Photon avalanche upconversion in rare earth laser materials. *Opt. Mater.* 11(2–3), 181–203. https://doi.org/10.1016/S0925-3467(98)00043-3.

221. Khare A. (2020) A critical review on the efficiency improvement of upconversion assisted solar cells. *J. Alloys Compd.* 821, 153214. https://doi.org/10.1016/j.jallcom.2019.153214.

222. Wang S., Zhang B., Li Y., Zhang J., Chen B. (2021) High quality upconversion luminescence and excellent thermal stability of Yb³⁺, Er³⁺codoped Sr3NaY(PO4)3F phosphors. *Mater. Res. Bull.* 140, 111306. https://doi.org/10.1016/j.materresbull.2021.111306.

223. Dogan A., Erdem M. (2021) Investigation of the optical temperature sensing properties of up-converting TeO2-ZnO-BaO activated with Yb^{3+}/Tm^{3+} glasses. *Sensors Actuators A: Phys.* 322, 112645. https://doi.org/10.1016/j.sna.2021.112645.

224. Deshmukh P., Kumar Deo R., Ahlawat A., Khan A.A., Singh R., Karnal A.K., Satapathy S. (2021) Spectroscopic investigation of upconversion and downshifting properties LaF3:Tb^{3+}, Yb^{3+}: A dual mode green emitter nanophosphor. *J Alloys Compd.* 859, 157857. https://doi.org/10.1016/j.jallcom.2020.157857.

225. Wang W., J.T., Dong J., Xue Y., Hu D., Hou W., Tang H., Wang Q., Xu X., Xu J. (2021) Growth spectroscopic properties and up-conversion of Yb, Pr co-doped CaF2 crystals. *J. Lumin.* 233, 117931. https://doi.org/10.1016/j.jlumin.2021.117931.

226. Lim C.S., Aleksandrovsky A., Atuchin V., Molokeev M., Oreshonkov A. (2020) Microwave-employed sol-gel synthesis of Scheelite-type microcrystalline AgGd(MoO4)2:Yb^{3+}/Ho^{3+} upconversion yellow phosphors and their spectroscopic properties. *Crystals* 10, 1000–1014. https://doi.org/10.3390/cryst10111000.

227. Ansari A.A., Nazeer.uddin M.K., Tavakoli M.M. (2021) Organic-inorganic upconversion nanoparticles hybrid in dye-sensitized solar cells. *Coord. Chem. Rev.* 436, 213805. https://doi.org/10.1016/j.ccr.2021.213805

228. Wild J. de, Meijerink A., Rath J.K., van Sark W.G.J.H.M., Schropp R.E.I. (2010) Enhanced near-infrared response of a-Si:H solar cells with β-NaYF4:Yb^{3+} (18%), Er^{3+} (2%) upconversion phosphors. *Sol. Energy Mater. Sol. Cells* 94(12), 2395–2398. https://doi.org/10.1016/j.solmat.2010.08.024.

229. Singh-Rachford T.N., Castellano F.N. (2010) Photon upconversion based on sensitized triplet-triplet annihilation. *Coord. Chem. Rev.* 254(21–22), 2560–2573. https://doi.org/10.1016/j.ccr.2010.01.003.

230. Healy C., Hermanspahn L., Kruger P.E. (2021) Photon upconversion in self-assembled materials. *Coord. Chem. Rev.* 432, 213756. https://doi.org/10.1016/j.ccr.2020.213756.

231. Quan C., Sun D., Luo J., Zhang H., Fang Z., Zhao X., Hua L., Cheng M., Zhang Q., Yin S. (2018) Growth, structure and spectroscopic properties of Er, Pr:YAP laser crystal. *Opt. Mater.* 84, 59–65. https://doi.org/10.1016/j.optmat.2018.06.049

232. Silversmith A.J., Lenth W., Macfarlane R.M. (1987) Green infrared pumped erbium upconversion laser. *Appl. Phys. Lett.* 51, 1977–1979. https://doi.org/10.1063/1.98316.

233. Scheps R. (1994) Er^{3+}:YAlO3 upconversion laser. *IEEE J. Quant. Electron.* 30, 2914–2924. https://doi.org/10.1109/3.362717.

234. Scott B.P., Zhao F., Chang R.S.F., Djeu N. (1993) Upconversion-pumped blue laser in Tm:YAG. *Opt. Lett.* 18, 113–115. https://doi.org/10.1364/OL.18.000113.

235. Szachowicz M., Joubert M.F., Moretti P., Couchaud M., Ferrand B., Borca C., Boudrioua A. (2008) H+ implanted channel waveguides in buried epitaxial crystalline YAG:Nd, Tm layers and infrared-to-blue upconversion characterization. *J. Appl. Phys.* 104, 113104. https://doi.org/10.1063/1.2976303.

10 Metal Oxides
Antimicrobial Techniques

V. R. Raikwar

Ramniranjan Jhunjhunwala College of Arts, Science and Commerce

CONTENTS

10.1 INTRODUCTION

Many bacteria, fungi, and viruses are becoming resistant to existing antibacterial agents. In addition, new microbial species appear for unknown reasons. Diseases caused by microorganisms have become a significant threat to the human race. This could be the result of inefficiencies in various forms affecting sustainable development. The use of antimicrobial drugs today or as a result of unsustainable development harms nature and ecosystems. The development of antimicrobial agents is necessary for the event of new diseases. Although some infectious diseases have been eradicated, many more are still creating havoc in society. Infectious diseases are the leading cause of death worldwide.

From the biblical plagues and plagues of ancient Athens to the Black Death of the Middle Ages, to the 1918 "Spanish flu" pandemic, and more recently the HIV/AIDS pandemic and COVID-19, infectious diseases have continued to appear and re-emerge in ways that defy accurate predictions. The emergence of the resurgence of some diseases like tuberculosis and cholera and the emergence of new variants creating endemic in all parts of the world have been the cause of global health threats [1,2]. The pathogenic bacteria develop resistance to the antibiotic drugs, which has been a cause of worry ever since the first antibiotic, penicillin, was developed. It is a need of the hour to develop materials that have an antimicrobial activity that can be used effectively against the strain.

As the threat of emerging pathogens and drug resistance emerges, materials that are biocompatible, economic, and effective against the bacterial strains are needed to be developed. One of the promising alternatives that exhibit remarkable antibacterial activities against several bacteria, including pathogens, is engineered nanomaterials specifically, metal and metal oxide nanoparticles that show remarkable antimicrobial activity [3–5]. They have an excellent antimicrobial activity, which highlights the role of these nanoparticles in several biomedical products [6,7].

10.2 METHODS OF SYNTHESIS OF METAL OXIDES FOR ANTIMICROBIAL ACTIVITY

The method plays a very important role in the synthesis of any material for a particular application. The selection of synthesis method determines many structural and physiochemical properties, such as crystal structure, morphology, size, shape, types of defects, and dispersity of the metal oxides. The metal oxides are synthesized in the following ways:

10.2.1 BIOSYNTHESIS METHOD

Biosynthesis is important since biocompatibility is one of the most important requirements for the nanomaterial to be used in any medicine. Plant-mediated synthesis of nanoparticles (NPs) is a revolutionary technique that has a wide range of applications in agriculture, the food industry, and medicine. Nanoparticles synthesized using conventional methods have limited uses in the medical domain due to their toxicity. Due to the physico-chemical properties of plant-based nanoparticles, the limitations of conventional chemical and physical methods of nanoparticle synthesis have been considerably overcome. ZnO nanoparticles were synthesized from leaf extracts [8–11]. Raliya et al. [12] have synthesized ZnO, MgO, and TiO_2 nanoparticles by using fungus. The major drawbacks of nanoparticle synthesis using biological routes are the less control on the size and shape of nanoparticles, the process can be slow, and after a certain period of time, degradation of synthesized metal oxides can take place. By varying parameters like microorganism type and strain, its growth phase, culture growth medium, pH, substrate concentrations, temperature, reaction time, the addition of non-target ions, and a source compound of the wanted nanoparticle, it is possible to control the size of the particle and their monodispersity [13].

10.2.2 CHEMICAL METHOD

In the chemical method, the solvent with a high boiling point, other precursors, and surfactants are heated slowly to a high temperature. The major steps in this process are monomer formation and accumulation, nucleation, and growth. The growth process and the boiling point of the solvent control the size and diameter of the particles [14]. CeO_2 nanoparticles were synthesized by chemical method resulting in 25–50 nm size particles. The as-synthesized metal oxide particles were tested against gram-negative and gram-positive bacteria like E. coli, P. aeruginosa, S. pneumoniae, Proteus mirabilis, Klebsiella, and Haemophilus influenzae, S. epidermidis, Enterobacter aerogenes, Citrobacter freundii, and Enterobacter cloacae [15]. ZnO and Vanadium pentoxide (V_2O_5) nanowires with H_2O_2 and Br^- nanoparticles synthesized by this method were tested against gram-negative bacteria like E. coli [16–18].

10.2.3 MICROWAVE METHOD

The precursors are dissolved in deionized water and stirred to mix well at room temperature till it is jellified. This gel is then irradiated with microwave energy and is cooled at room temperature.

The resulting precipitate is vacuum filtered, washed with deionized water and alcohol, and dried in a vacuum at 80°C for 1 hour. CaO nanoparticles are synthesized by this method resulting in 16 nm particle size and found that the antibacterial effect was more prominent against gram-positive than gram-negative strains [19]. Magnesium oxide nanowires were synthesized using the microwave method and tested for E. coli and Bacillus sp. Bacterial activities showed decreased bacteriostatic activity against E. coli and Bacillus sp. with increasing MgO nanowire concentration [20].

10.2.4 PYROLYTIC METHOD

This method is used to prepare metal oxide nanoparticles by mixing the atomized solution with soluble polymers and preparing a porous matrix before heating the precursor molecules. Due to these steps, the aggregation or agglomeration of nanoparticles is prevented. Cobalt oxide (Co_3O_4) prepared by the pyrolytic method showed antibacterial activity on two bacteria, S. aureus and E. coli [21].

10.2.5 HYDROTHERMAL METHOD

This method uses an autoclave or steel reactor, which can be kept in an oven/furnace at 120°C–200°C under 2,000 psi pressure. The aqueous solvents can be dissolved and recrystallized in these operating conditions. Different forms of nanoparticles can be obtained by the hydrothermal method by controlling the operating parameters. Quantum dot CdTe conjugated with ZnO was prepared by hydrothermal method and impeded microbial growth of B. subtilis and E. coli and deterred biofilm formation of P. aeruginosa through photocatalytic action [22]. Few layered graphene sheets decorated by ZnO nanoparticles (FLG/ZnO composite) having 20–40 nm size were prepared by Hummers method (graphene oxide) hydrothermal followed by reduction (FLG/ZnO composite). This material exhibited significant antibacterial activity against S. Typhi than E. coli [23]. ZnO nanorods were prepared by hydrothermal to check their antimicrobial activity against S. aureus, E. coli, and A. niger microorganisms and inhibited their growth [24].

10.2.6 SONOCHEMICAL METHOD

In the sonochemical method, the solution of the starting material (e.g. metal salts) is subjected to an enhanced ultrasonic vibration to break the chemical bonds of compounds. The ultrasonic waves traveling through a solution cause a change in compression and relaxation. This leads to sound cavitation, which results in the continuous formation, growth, and collapse of foam in the liquid. In addition, the change in pressure creates microscopic bubbles that explode violently leading to the appearance of shock waves in the gas phase deflating the bubble. The effect of the accumulation of millions of deflated bubbles produces a large amount of energy released in the solution. Transient temperature ~5,000 K, pressure ~1,800 atm, and cooling flow greater than 1,010 K/s were recorded at the localized cavitational implosion hotspots [25]. The advantages of the sonochemical method are uniform size distribution, a higher surface area, faster reaction time and improved phase purity of the metal oxide nanoparticles. Using this method, many metal oxides ZnO [26], TiO_2 [27], Fe_3O_4 [28], Mn-doped Fe_2O_3 [29], $PbWO_4$ [30] were synthesized.

10.2.7 MICROSYNTHESIS METHOD

In this method, the precursors are added to a fungal filtrate and by stirring continuously a white precipitate will be formed from a yellowish-brown precipitate. Mycogenesis of cerium oxide nanoparticles using Aspergillus niger culture filtrate resulted in a fine powder that can inhibit the growth of Streptococcus pneumonia, B. subtilis, Proteus Vulgaris, and E. coli. The precipitate of metal oxide nanopowder was obtained using this method and tested against the bacteria. CeO_2 nanoparticles are made by this method [31].

TABLE 10.1
Methods of Synthesis of Metal Oxide Nanoparticles and Some Materials Prepared Using Them

Sr No.	Method of Synthesis	Metal Oxides	References
1	Biosynthesis	ZnO, MgO, TiO_2	[8–13]
2	Microwave	CaO, MgO	[19,20]
3	Pyrolytic	CO_3O_4	[21]
4	Hydrothermal	ZnO, FLG/ZnO	[22–24]
5	Sonochemical	TiO_2, ZnO, $PbWO_4$, Fe_2O_3, Fe_3O_4	[25–30]
6	Mycosynthesis	CeO_2	[31]
7	Combustion/sol-gel combustion	$GdAlO_3$:Eu^{3+}, ZnO, CuO, Fe_2O_3, SAOED/ RB-CF, Ag+/CaTiO3: Pr3+	[35,36,74,77]
8	Chemical	CeO_2, ZnO, V_2O_5, Fe_3O_4, Fe_2-xAgxO_3, CaO,	[15,37–39]

10.2.8 COMBUSTION METHOD

The combustion synthesis method is used to prepare inorganic phosphors at relatively low temperatures using organic fuels and metal nitrates as oxidizers. The method has many important features like the use of relatively simple equipment, formation of high-purity products, stabilization of metastable phases, and formation of virtually any size and shape of products. These inorganic metal oxides have applications in optoelectronic devices and displays [32–34]. These materials can be used as antimicrobial agents. $GdAlO_3$:Eu^{3+} was synthesized by the combustion method, and it was shown that incorporating Li, K, and Na cations in the host matrix increased the photoluminescence. These materials showed antimicrobial activities against pathogenic microorganisms [35]. A comparative study of CuO, ZnO, and Fe_2O_3 synthesized by the sol-gel combustion method was done. It was found that ZnO nanoparticles were the smallest in size as compared to the other two materials, thereby having the highest surface-to-volume ratio. The antibacterial activity was compared, and it was concluded that the antibacterial activity increased with an increase in surface-to-volume ratio due to a decrease in the size of nanoparticles. The ZnO nanoparticles are more effective against gram-positive bacterial strains compared with gram-negative bacterial strains [36]. Table 10.1 summarizes some of the metal oxides prepared using various methods that are tested for antimicrobial activity.

10.3 MECHANISMS OF METAL OXIDE NANOPARTICLES FOR ANTIMICROBIAL ACTIVITY

Due to the unique physical, chemical, electrical, magnetic, optical, and biological properties, metal oxide nanoparticles are very important to scientists as an antibacterial agent. The mechanism of the antibacterial activity of metal oxide nanoparticles is still in its infancy. Elucidation of the molecular mechanism of metal oxide nanoparticles is now a subject of rigorous research. To investigate possible mechanisms, many studies have been undertaken. The antibacterial activity of metal oxide nanoparticles revealed by recent studies shows that the mechanism is not exhaustive, but research has revealed it as bacterial growth delay/killing. By combining one or more mechanisms, the nature and chemistry of metal oxide nanoparticles can be decided. The main activity is the production of reactive oxygen species (ROS) (Figure 10.1). The other activities are electrostatic interaction, disturbance in metal/metal ion homeostasis, protein and enzyme dysfunction, genotoxicity, and photokilling.

The mechanism of metal oxide nanomaterials interacting with living cells and microorganisms has been widely explored and explained in various ways: The common one is ROS formation. The other mechanisms are interaction with the cell membrane, particle internalization, and binding with specific targets such as proteins or DNA. In oxides, which are made from poly metals, the

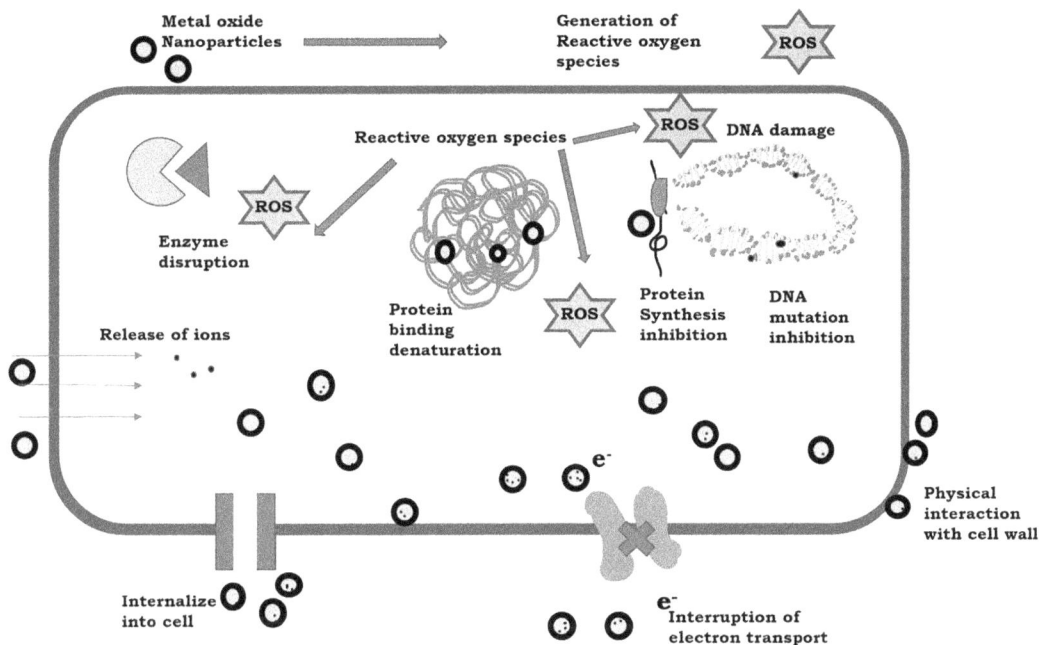

FIGURE 10.1 The plausible mechanisms of metal oxide nanoparticle mediated antimicrobial activities.

physicochemical parameters of the respective components are altered, which can lead to a new reaction in the living organism. Some polymetallic oxide nanoparticles showed little tendency to agglomerate in solutions and biological fluids, resulting in increased antibacterial activity and high biocompatibility for their components.

10.3.1 BACTERIAL CELL WALL INTERACTION AND CELL PENETRATION

When bacterial cells are exposed to nanoparticles, it can lead to membrane damage caused by the adsorption of nanoparticles. This is followed by penetration into the cell [40–42]. Many studies suggested that adsorption on the cell wall following its disintegration is the primary mechanism of toxicity [43,44]. Cell wall depolarization takes place and adsorption of NPs makes the cell wall become more permeable. The study by et al. revealed that the bacterial cell wall becomes blurry, indicating cell wall degradation as confirmed by a laser scanning confocal microscope [45].

10.3.2 OXIDATIVE STRESS AND ROS

Oxidative stress has been confirmed as a key contributor to changing the permeability of the cell membrane, which can result in bacterial cell membrane damage. ROS are highly reactive chemicals formed from O_2. Some of the examples are hydrogen peroxide, hydroxyl radicals, singlet oxygen, and superoxide. This group of molecules has many functions, including antimicrobial activity. ROS are created from metal oxide nanoparticles and help in antibacterial activity. The membrane of bacteria contains phospholipids that can be oxidized by ROS resulting in the destruction of the cell membrane. ROS can damage DNA/RNA and proteins in the cells of bacteria causing the death of bacteria [46–48]. It was discussed in detail that increasing ROS levels in the biosphere represented growing stress levels and thus shaped the evolution of species. If ROS is more than the cell's capacity, oxidative stress comes into the picture. The injuries through oxidative stress in cells are diverse. The damage caused by oxidative stress on the different steps of the central dogma of molecular biology in bacteria is shown in Figure 10.2.

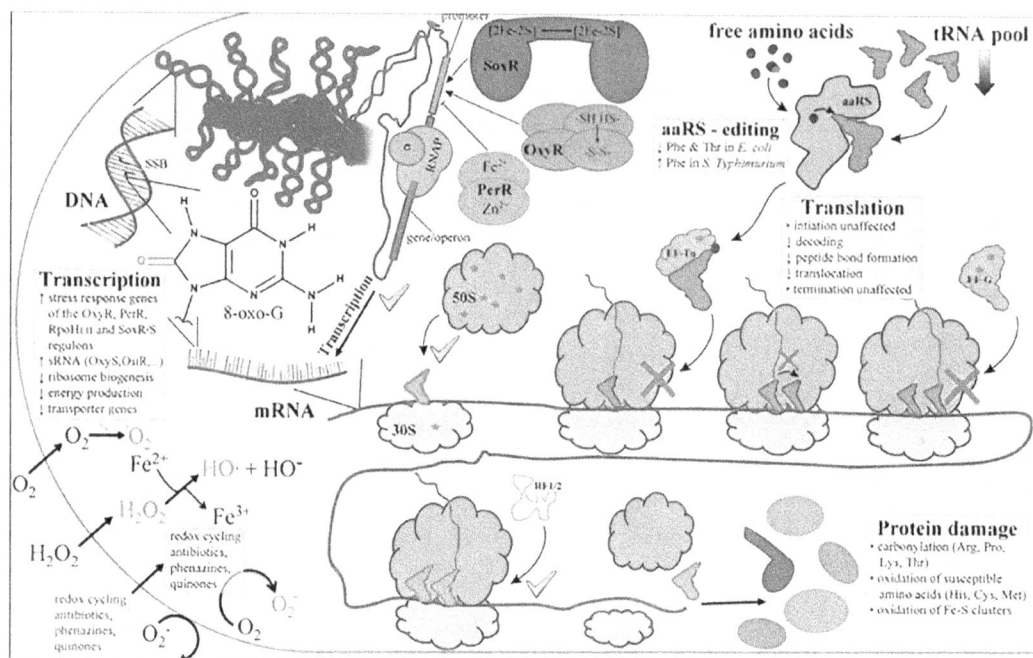

FIGURE 10.2 Damage caused by oxidative stress in bacteria. (Copyright © 2021 Fasnacht and Polacek under Creative Commons Attribution License.)

It was concluded that the central dogma of molecular biology is nonetheless susceptible to oxidative damage, but the pathogens are having their myriad defense mechanisms against ROS [49]. Rai et al. showed that ZnO nanoparticles produced ROS species on the membrane of bacteria cells and showed antibacterial activity against both gram-positive and gram-negative bacteria [50]. The silver nanoparticles and TiO$_2$ nanoparticles were effectively used as antibacterial agents by Besinis et al. against S. Mutans. They showed that silver nanoparticles have strong antimicrobial activity than chlorhexidine [51]. Xie et al. studied the antibacterial activity of ZnO nanoparticles against Campylobacter jejuni (C. jejuni). They proposed the mechanism of disruption of the cell membrane and oxidative stress in C. jejuni. Their results are significant as the ZnO nanoparticles caused morphological changes, membrane leakage that can be measurable and up to a 52-fold increment in oxidative stress gene expression in C. jejuni [52]. Foster et al. in their work showed that titanium dioxide nanoparticles can adhere to the surface of bacterial cells to produce ROS. These ROS ultimately damage the composition and structure of the cell membrane, in this way interfering with the function of the cell membrane and causing leakage of cellular contents, resulting in bacterial death [53]. MgO nanoparticles are tested against gram-positive and gram-negative bacteria by Sawai et al. They showed that the presence of active oxygen such as superoxide on the surfaces of MgO nanoparticles was the key factor that affects their antibacterial activity [54]. The mixed nanostructure of ZnO-MgO was evaluated by Vidic et al. against both gram-positive and gram-negative bacteria and confirmed that ZnO nanoparticles showed high antimicrobial activity against both, MgO showed moderate and ZnO-MgO mixed phase nanoparticles showed high antimicrobial activity against gram-positive bacteria. They suggested the use of a mixed phase as a new therapeutic for bacterial infection [55].

10.3.3 ELECTROSTATIC INTERACTION

The metal cations separated from metal oxides may get attracted by the negatively charged cell membrane of bacteria and bound to it by electrostatic interactions after which they insert into the hydrophobic core and perturb the structure. This results in an increase in bacterial membrane permeability that induces leakage of cytoplasmic components and consequently the death of bacteria. In this regard, adsorption of metal ions to cell walls is the potential process in destructing bacteria. The metal ions released in the surrounding media get bound with the negatively charged functional groups of the bacterial cell membrane such as phosphate and carboxyl groups. This process is known as biosorption. Different metal ions have shown affinity to different sites; for example, zinc ions showed high affinity towards the –SH groups of proteins. The closely spaced highly ordered cell membranes become confused and dispersed, which destroys their inherent function and leads to bacterial death [56]. Gram-negative bacteria have more negative charge than gram-positive bacteria [57]. Therefore, electrostatic interaction will be stronger in gram-negative bacteria. Lipopolysaccharide (LPS) in the outermost leaflet of the lipid bilayer is the cause of having more charge per unit, thereby making them highly negative in charge [58]. The surface area that is available for interaction also affects the electrostatic force of attraction. The metal oxide nanoparticles are characterized by small size thereby having a large surface area. This feature of metal oxide nanoparticles makes them potential candidates for imparting cytotoxicity. The attachment of a large number of metal oxide nanoparticles alters the structure and permeability of the cell membrane and the strong bond with the membranes results in the disorganization of cell walls. The pores on the membrane are of the order of nanometers and the size of bacteria is of the order of micrometers. The metal oxide nanoparticles enter the cell structure and cause leakage of intracellular contents resulting in the destruction of bacteria [59].

10.3.4 PHOTOKILLING

When metal oxide nanoparticles in contact with bacteria are exposed to light, photokilling process takes place. Generation of O_2^- is the main reason, whereas the generation and release of the electron occur through light. The released electrons are grabbed by metal oxide nanoparticles leading to the generation of more ROS. Thus, the effect of light will result in photochemical alteration of the cell membrane, damage to proteins, and abnormal cell division. The photokilling was studied using Fe_3O_4 nanoparticles, and the growth of bacteria was inhibited [60]. UV light was used by another group for photokilling microorganisms using TiO_2 nanoparticles [61]. Pt-dispersed TiO_2 thin film showed excellent photokilling after exposure of UV light for 5 minutes (Figure 10.3) [62].

10.3.5 PROTEIN AND ENZYME DYSFUNCTION

The Dysfunctioning of enzymes of a bacterial cell is another mechanism through which the metal oxide nanoparticles carry out antimicrobial activity. Protein-bound carbonyls are the result of catalyzing the oxidation of amino acid side chains by metal oxide nanoparticles. This protein carbonylation will reduce the catalytic activity of enzymes and result in protein degradation. These ions also

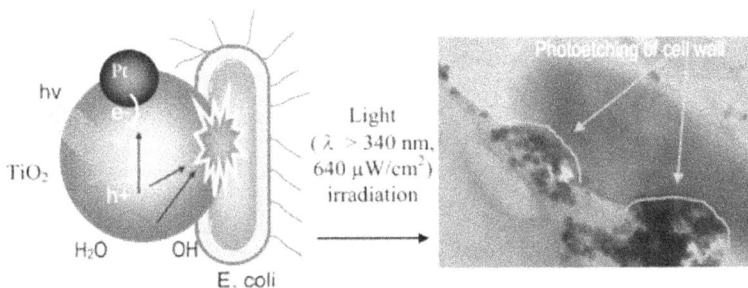

FIGURE 10.3 Photokilling of bacteria using TiO_2 nanoparticles.

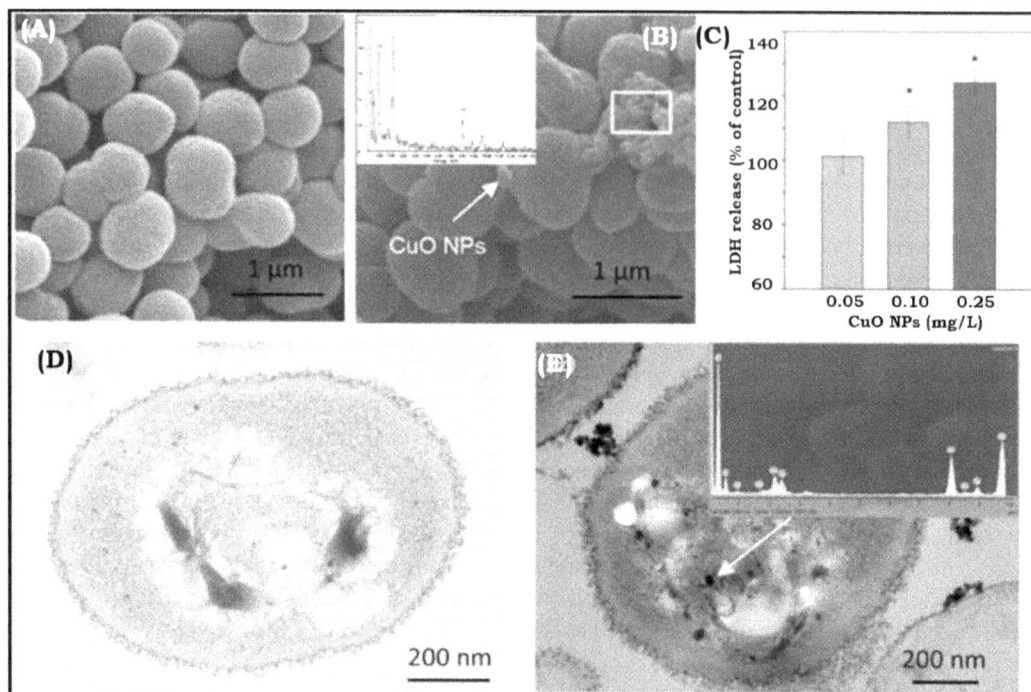

FIGURE 10.4 Effect of CuO NPs on cellular structure of *P. denitrificans*.

react with -SH groups of many proteins and enzymes to make them inactive [63–65]. The influence of CuO nanoparticles on bacterial denitrification was studied by Su et al. as shown in Figure 10.4.

They observed that CuO nanoparticles can alter significantly the expression of key proteins. The proteomic bioinformatics analysis showed that CuO nanoparticles caused the regulation of proteins involved in nitrogen metabolism, electron transfer, and substance transport [66]. This further deteriorates the cell's inner functionalization. The crossing of nanoparticles from the bacteria cell membrane and then damaging the vital enzymes of gram-positive and gram-negative bacteria by highly stabilized CuO nanoparticles were studied [67,68].

10.3.6 DISORDER IN METAL ION HOMEOSTASIS

For microbial survival, metal ion homeostasis is necessary wherein the metabolic functions are regulated by coenzymes, cofactors, and catalysts. If metal ions are excess in number, it will produce disorder in metabolic functions. As shown in Figure 10.1, when metal ions bind with DNA, its helical structure is disturbed. The copper oxide nanoparticles release copper ions that carry a positive charge, which neutralizes the charges on the LPS and increase the permeability of the outer membrane. The disordered membrane allowed more metal cations to accumulate in the bacterial cell which is reduced to metal atoms by thiol groups (-SH) in enzymes and proteins and hence inactivating the essential proteins blocking, respiration and this leads to cell death [69]. Long-chain polycations were coated on the cell surfaces that kill both gram-positive and gram-negative bacteria [70].

10.4 IMPORTANT AND NEW ERA ANTIMICROBIAL APPLICATIONS

Nanosized metal oxides are widely utilized in skincare merchandise because they have a large spectrum of antimicrobial properties and decreased pores and skin irritation. These nanoparticle-based sunscreens or creams for dermatological packages have already been commercialized. The first sunscreen containing nano-TiO_2 turned commercialized in 1989, and later on, in 1991, ZnO turned into additionally introduced. Furthermore, 70% of TiO_2 sunscreens and 30% of zinc sunscreens contained those ingredients withinside the nanoform [71]. Lancôme Soleil from L'Oréal is one such instance of a nano-based sunscreen. Nanoparticles of ZnO and TiO_2 are frequently utilized in cosmetics; further to the robust antimicrobial hobby, additionally they exhibit physicochemical properties that are beneficial for those cosmetics. The antimicrobial property enables to reflow of odors and protects the pores and skin from colonization via way of means of pathogens. However, permeation through the pores and skin is one of the main worries related to such sunscreens. Various studies have been carried out to determine the penetration capacity of nanoparticles through pores and skin. Permeation of gold and iron oxide nanoparticles particularly with suggest smaller sizes turned into located in a few research. However, the nanoparticles of TiO_2 and ZnO are typically categorized as secure substances for dermal packages. Modern sunscreens contain titanium dioxide (TiO_2) or zinc oxide (ZnO) nanoparticles that are insoluble, colorless, and reflect/disperse ultraviolet (UV) rays more effectively than larger particles. Most of the available theoretical and experimental evidence suggests that insoluble NPs do not penetrate into or through normal or damaged human skin. Oral and topical toxicity data indicate that TiO_2 and ZnO NPs have low systemic toxicity and are well tolerated in the skin [72]. The use of inorganic luminescent particles is being explored for new application areas like the textile industry. For the development of self-disinfecting photodynamic textiles with enhanced antibacterial activity, a facile method has been reported wherein the long-persistent phosphor $SrAl_2O_4$:Eu^{2+}, Dy^{3+} (SAOED) and the photosensitizer rose bengal (RB) onto cotton fabric (CF) are applied by knife coating and photocrosslinking methods (Figure 10.5). The prepared composite material, termed SAOED/RB-CF, showed superior antibacterial efficacy capable of 99.999% (5 log units, detection limit) and 99.986% (3.9 log units) photoinactivation against Staphylococcus aureus and methicillin-resistant S. aureus (MRSA), respectively, under visible light illumination (Xenon lamp). This study suggested that the light emitted from the phosphorescence of the photoexcited SAOED could be absorbed by the rose bengal photosensitizer under

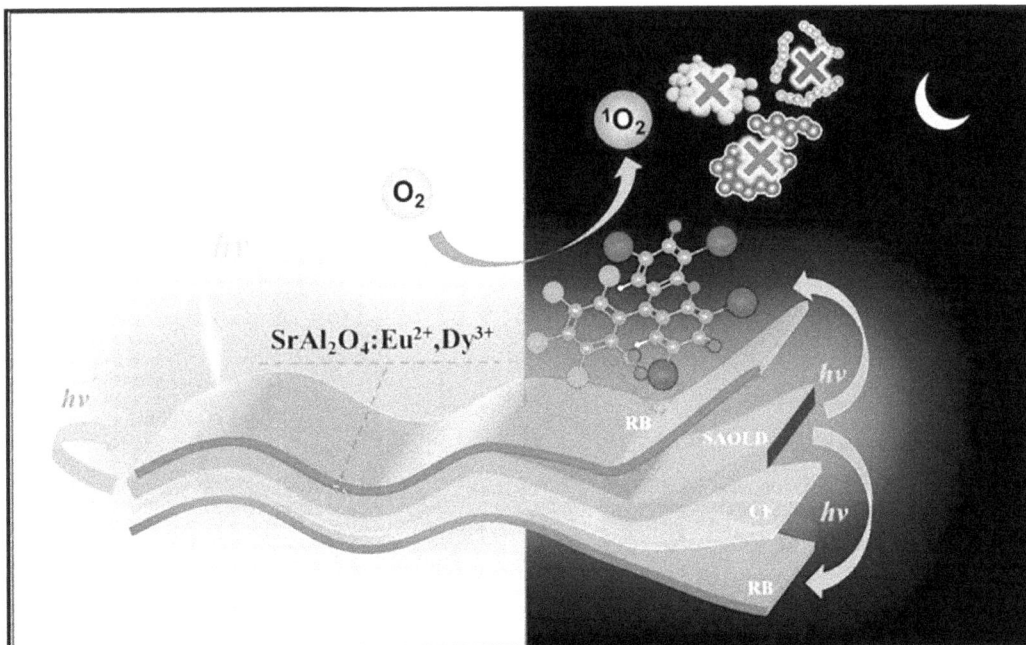

FIGURE 10.5 SAOED/RB-CF shows synergistic antibacterial efficacy by generating singlet oxygen even under dark condition.

dark conditions, resulting in effective synergistic pathogen inactivation [73]. The use of PEGylated palladium-doped CeO_2 for cancerous cell destruction shows a promising application in the field of antimicrobial study. The fluorescent staining results confirmed the cell death in A549 cells through cellular damage by oxidative stress.

The reported material exhibited higher antibacterial activity against bacterial pathogens. Figure 10.6 shows the synthesis, biocompatibility, anticancer activity, and antibacterial activity of PEGylated palladium-doped CeO_2 [74]. A new composite based on aluminum oxide nanoparticles and borosiloxane polymers has been prepared, and its mechanical and physicochemical properties have been studied. The composite inhibits bacteria growth and does not affect the growth of host animal cells. The increase in the concentration of Al_2O_3 nanoparticles showed an increase in cytotoxicity. The reported composite is capable of generating ROS such as hydroxyl radicals and hydrogen peroxide (Figure 10.7). The rate of ROS generation increases threefold with an increase in the concentration of Al_2O_3. It has been shown that the composite can act as a key biomarker of oxidative stress as 8-oxoguanine in DNA [75]. A multifunctional material $CaTiO_3$:Pr^{3+} coated with silver has been prepared and studied for its luminescence and antibacterial properties by Liu et al. They used the sol-gel combustion method for preparing this material. They showed that the emission intensity decreases with the increase in the concentration of Ag^+ in the sample. The reason given for this effect was the special absorptive capacity of Ag^+. When the silver compounds (Band gap 1.46 eV) are irradiated with light having energy more than their band gaps, a separation of valence band electron and conduction band holes is generated. The separated electron hole pairs absorb energy and the energy of Pr^{3+} ions get decreased. Hence, as the activator's energy decreases in luminescent solids, their photoluminescence decreases. They also studied the effect of temperature on the antimicrobial property. The optimized temperature was 800°C. They showed that the antibacterial property increased with increase in the addition of Ag^+ coating. The Ag^+ coated $CaTiO_3$:Pr^{3+} red emitting phosphor excited by ultraviolet radiation thus can be used as a red phosphor in the

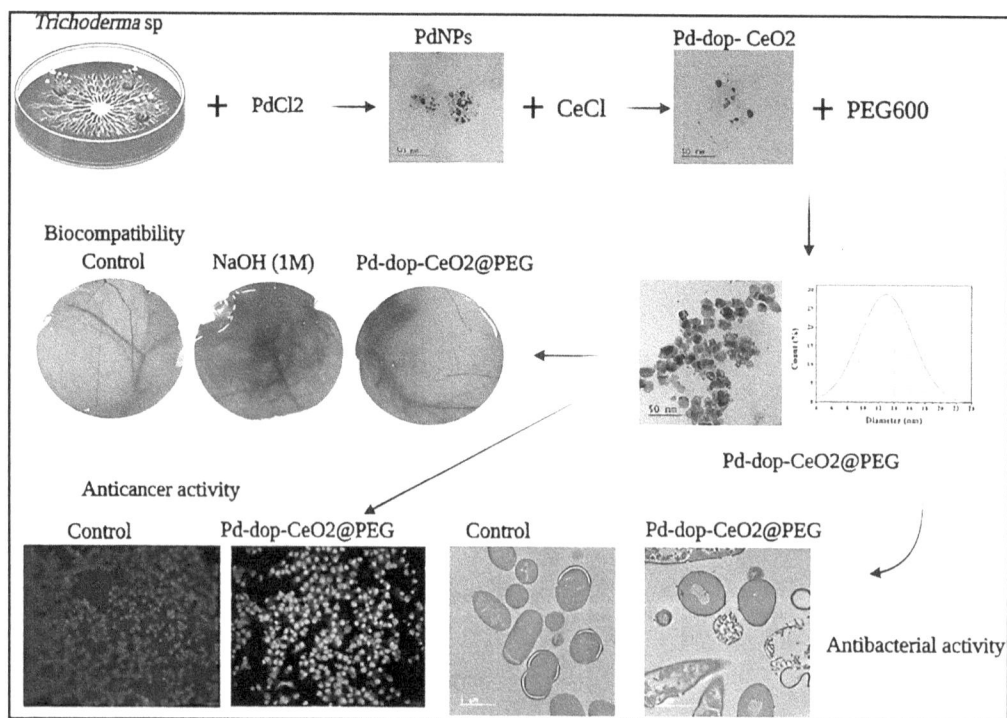

FIGURE 10.6 The synthesis, biocompatibility, anticancer activity, and antibacterial activity of PEGylated palladium-doped CeO_2.

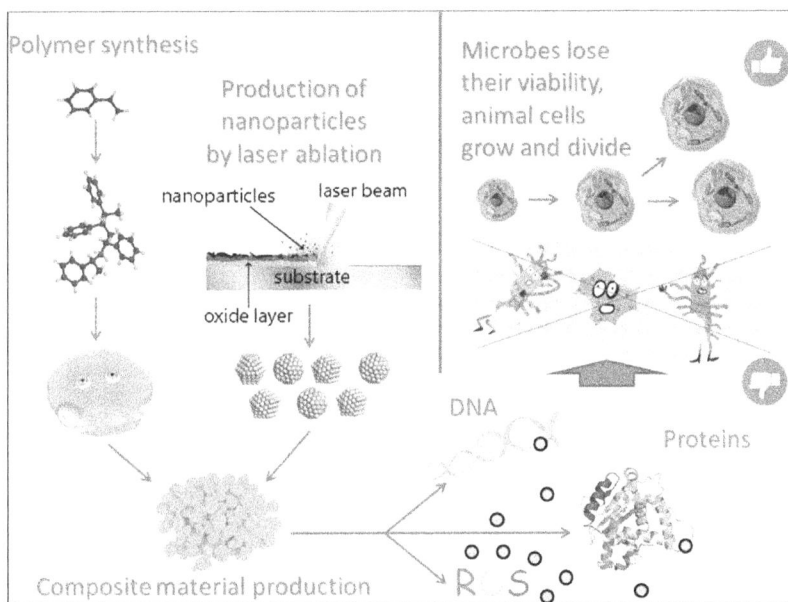

FIGURE 10.7 The composite of polymer and metal oxide nanoparticles inhibiting bacteria growth and not affecting growth of host animal cells by generating ROS.

manufacturing of white light-emitting diodes and antibacterial material as well. This multifunctional material having luminescence and antimicrobial property can be used in medical devices [76]. The facile preparation of traffic warning clothing from long-persistent photoluminescent cotton fabrics has been reported by Khattab et al. They spray-coat a binder with strontium aluminum oxide phosphor (SrAl$_2$O$_4$:Eu, Dy) in a water environment on cotton fabric. They studied the breathability, flexibility, and comfortability of the coated cotton fabrics and showed that the long-lasting luminescent cotton fabric has the added advantage of being readily visible to the naked eye and its luminescence can be located in the dark. Morphology, surface, and elemental composition, as well as color fastness and luminescence properties of the spray-coated cotton substrates, were explored. They studied the antibacterial and antifungal properties of spray-coated cotton substrates against two gram-negative and gram-positive bacteria, including E. coli and S. aureus, respectively, and Candida albicans fungus by applying the plate agar count methodology, and the antibacterial/antifungal reduction percentage caused by the spray-coated cotton fabrics were calculated. It was observed that the blank cotton fabric showed no inhibition influence on the reduction percentage, but spray-coated fabric showed significant resistance to the pathogens. They rated this microbial resistance ranging from poor to excellent with increasing the concentration of strontium aluminate phosphor. Thus, the reported cotton fabric having antimicrobial properties with the added advantage of glowing in the dark proves additional safety [77].

10.4.1 ANTIMICROBIAL BLUE LIGHT APPLICATIONS

Preservation technologies important in the agricultural and medical industries. In this context, light-emitting diode (LED) illumination is widely applied for microbial inactivation. In recent years, LEDs with blue wavelengths are used as an alternative technology to ultraviolet irradiation. Minimally preserved foods such as freshly cut fruits and vegetables that are eaten raw are more prone to contamination by microorganism. This leads to various diseases and food wastage. Moreover, it loses its taste and texture. Consumers prefer the fresh look and smell of minimally processed foods that are preservative-free. To address all these issues in the food industry, thermal

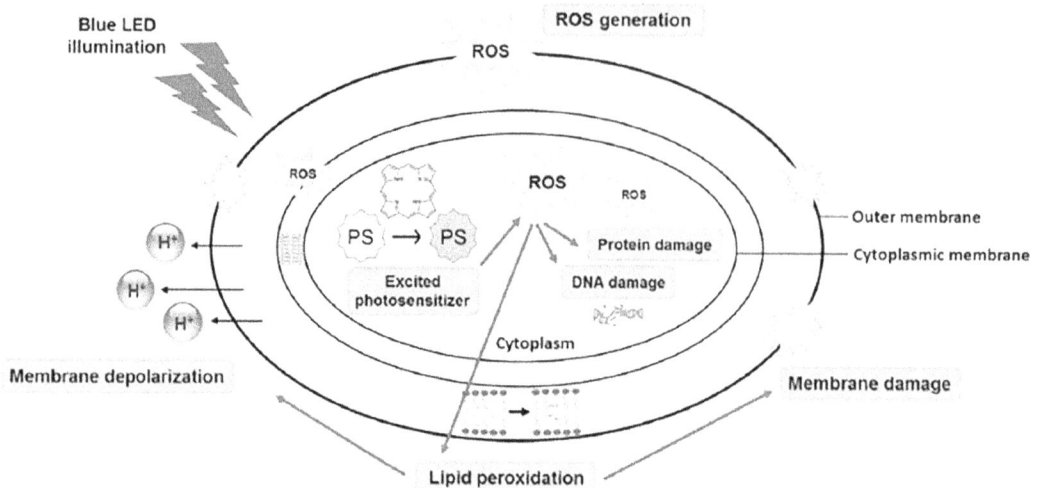

FIGURE 10.8 Antibacterial activity caused by LED illumination.

technology is primarily used for antimicrobial treatment. But the thermal treatment causes loss of nutrition, taste, and texture of fresh produce [78]. To avoid such unwanted effects, nonthermal treatments are explored and are being used for antibacterial treatment for the last 35 years. Among these, light-based technologies have promising potential for foodborne pathogens. The ultraviolet light is generally used for sterilization and in the food packaging industry for its antibacterial effect. But it has a harmful effect on humans for long exposure time and therefore the need arises for the development of safe way for food as well as for people working in the food and agricultural industry [79,80].

LEDs are semiconductor-based light source with a narrowband of emission wavelengths. It can be engineered to emit various colors of light of which the visible range spreads from 400 to 700 nm. The blue, green, and red wavelength regions are used in food industry. LED lights are advantages like more life span, low energy consumption, compact size to name a few as compared to conventional light sources. The LED has been widely used in many applications in the field of postharvest storage, delaying of fruit ripening, cultivation of crops, food preservation, medical therapy and lighting (indoor and outdoor) [81–86]. Blue LEDs exhibit outstanding electrical and optical properties such as high light output power, high brightness, low forward driving voltage, and high internal quantum efficiency (IQE). Figure 10.8 shows the mechanism of LED-driven antibacterial activity.

When the LED light is allowed to fall on the food, it generates ROS, which creates oxidative stress in the membrane of the bacterial cell wall. The possible antibacterial mechanism is that light rays from LEDs excite endogenous photosensitizers such as cytochromes, flavins, NADH, and porphyrins that are naturally present in the cells. This generally results in the formation of ROS, like hydrogen peroxide, hydroxyl radicals, superoxide anions, and singlet oxygen. This process occurs after the absorption of light energy in the presence of oxygen. The cellular components, such as DNA, lipids, and protein, are damaged in this process and eventually lead to the destruction of bacterial cells [87,88]. Also, lipid peroxidation can affect ion pumps in cell membranes and thus increase membrane permeability or depolarization [89]. Table 10.2 shows the effect of blue LED on the killing of microbes studied by various groups in recent years. Along with the illumination time, temperature and intensity of light decide the rate of inhibition of microbes.

10.5 CONCLUSION

Ultraviolet (UV) radiation is a well-known tool for disinfecting food and food surfaces. But, under some conditions, visible light has been proven to have bactericidal characteristics that

TABLE 10.2
Effect of LED Illumination on Various Wavelengths on Pathogenic Bacteria in Foods

Pathogenic Bacteria	Wavelength (nm)	Food	Treatment Time	Temp. (°C)	Result	Reference
S. Montevideo	460	Orange juice	4.9–13.6 hours	4	3.3 log reduction at 4°	[90]
S. Newport				12	3.6 log reduction at 12°	
S. Saintpaul				20	4.8 log reduction at 20°	
S. Gaminara	405	Chicken skin	7.5 hours	10	1.7–2.1 log reduction	[91]
S. Typhimurium						
C. Jejuni						
C. coli	405	Skim milk	90 minutes	5	3.0–5.9 log reduction at 405 nm	[92]
E. coli						
	433			10	2.4–4.1 log reduction at 433 nm	
	460			15	2.8–3.7 log reduction at 460 nm	
S. Montevideo	460	Fresh-cut pineapples	8.7–24 hours	7	0.6–1.0 log reduction at 7°C	[93]
S. Newport				16	0.7–1.7 log reduction at 16°C	
S. Saintpaul				25		
S. Agona	405	Fresh-cut papaya	48 hours	4	1.0–1.2 log reduction at 4°C	[94]
S. Newport				10	0.3–1.3 log reduction at 10°C	
S. Saintpaul				20		
S. Typhimurium						
S. Enteritidis	405	Cooked chicken	48 hours	4	1.0–1.6 log reduction at 4°C	[95]
E. coli O157:H7	405	Fresh-cut mango	48 hours	4	1.0–1.2 log reduction at 4°C	[96]
S. Typhimurium				10		
L. Monocytogenes				20	1.2 log reduction In S. Typhimurium	
E. coli O157:H7	405	Chicken purge	10–20 minutes	10	0.7–1.0 log reduction on chicken purge	[97]
L. Monocytogenes		Chicken skin				
S. Aureus					0.2–0.4 log reduction on chicken skin	
S. saprophyticus	460–470	Sliced cheese	168 hours	4	3.6–5.1 log reduction at 4°C	[98]
Salmonella spp.						
L. Monocytogenes						
P. fluorescens			48 hours	25	1.9–2.0 log reduction at 25°C	
L. Monocytogenes	460	Smoked salmon	–	10	1.0 log reduction at 10°C	[99]
E. coli O157:H7	460–470	Fresh-cut apples and cherry tomatoes	5 days	4	<1.48 log reduction at 4°C	[100]

allow it to play an important role in food preservation. In food production, as well as in agriculture and horticulture, the visible light has proven its prime role as it stimulates photosynthesis, which is essential for plant growth and development. The use of visible light is the need of the hour, as the application of low light levels helps the crops retain postharvest quality by reducing senescence and enhancing phytochemical and nutritional content in a variety of species. In the agriculture and food industry, artificial light treatments are being used to disinfect water and food, as well as to enhance plant health and development by employing light energy of various wavelengths. The use of UV and visible range of light spectrum could be the future of non-toxic, chemical-less media for protecting the postharvest produce. The idea could be further implemented in developing the novel applications like germ-free fabrics, food containers, furniture, and everyday gadgets such as home appliances, displays, and mobile phones. For this, there is an urgent need of developing light-emitting metal oxides that can produce ROS that further leads to the inactivation of pathogens. Moreover, the rapid growth and progress in computer-based technology can be integrated with material development, and modern germ-free gadgets can be devised in the near future.

10.6   CHALLENGES AND FUTURE PERSPECTIVES

Due to the innumerable advantages of metal oxide nanoparticles, researchers around the world are making a concerted effort to elucidate the interactions of metal oxides with organisms. But, the mechanism of human toxicity of metal oxide nanoparticles remains to be elucidated. The following characteristics pose a threat to the use of metal oxide nanoparticles as antibacterial agents in humans and other organisms: structure; solubility; the ability to overcome various barriers; synthesis in the system; produces oxidants; the ability to penetrate the nucleus; exposure time; dosage, etc. Before being marketed as an antibacterial agent, it is necessary to assess the risks and side effects. The toxicity of some metal oxide nanoparticles varies with dose and duration of exposure, in addition to cell and tissue types. The primary concern is the prooxidant nature, followed by its potential to damage biomolecules and cell membranes. The microbiological properties of metal oxide nanoparticles depend not only on their physicochemical properties but also on the type of bacterial strain. For this reason, they hold promise as excellent antimicrobial agents for the future and find their place for wide application in nanomedicine. Metal oxides exhibiting antibacterial activities indicate a positive intention to use them as antibacterial agents in medicine as soon as possible. Although technology is developing at a rapid pace, the discovery of new antibiotics is lagging and the worry is that bacteria may become resistant to the drugs at a faster rate than the discovery of new antibiotics. The worldwide spread of MDR bacteria has occurred at a rapid pace as globalization has accelerated in recent years. Various metal oxide nanoparticles exhibiting antibacterial activities against different strains of bacteria and fungi have been extensively reported in recent years.

The rate of development of antibiotic resistance is much faster than the discovery of new antibiotics; To combat the emergence of antibiotic resistance, researchers are constantly searching for new materials such as antibiotic agents. Metal oxide nanoparticles hold great promise in overcoming this microbial resistance. The next obstacle to using metal oxide nanoparticles as antibacterial agents is their potential toxicity to humans. Although they can be an alternative to conventional antibiotics, disruption of beneficial microflora is a concern in addition to systemic toxicity, and effective use needs to be explored. Now it is a matter of knowing the interactions of metal oxide nanoparticles with human cells and organs, such as their ability to cross the blood-brain barrier and the blood-testis barrier. The knowledge gained through research in the coming years will explore metal oxide nanoparticles in agriculture, environmental science, food science, medicine, and veterinary medicine. Bacteria eliminate common and resistant antibiotics via the ATP-binding cassette (ABC) transporter. The entry of divalent metal ions into the cell via the ABC transporter has not been explored by scientists yet. ABC transporters are essential for cell viability and pathogenicity because they participate in transmembrane translocation of essential substrates as well as RNA translation and DNA repair; Any change in these protein systems cannot counteract unwanted changes inside the cell. This requires further investigation of the interaction of metal oxide nanoparticles with the ABC transporter. Nanotechnologists, microbiologists and toxicologists need to find ways to use metal oxide nanoparticles and their assays against specific bacteria. A better understanding of the mechanistic pathways and their safe use will help to modify metal oxide nanoparticles, not only to produce excellent antimicrobials, less toxic to humans, but also to prevent or reduce the undesirable effects on the ecosystem.

The field of nanotechnology should be used wisely for the well-being of mankind without damage to our environment. The adverse effects of use of technology on the surroundings can lead us towards unprecedented situations, which can be avoided through proper precautions and following the safety norms. The use of metal oxide nanoparticles in medicines should be tested in vivo rigorously and should be chosen carefully. The reproducibility with safety concern should be assessed and the toxicity assays should be reported. The clinical studies for various metal oxides should be drafted and accurately evaluated for long term so that critical assessment of the medicine and their effects can be studied. The mechanisms and role of various metal oxides and their applications in every field of life should be monitored and after rigorous testing, they should be used in any product.

REFERENCES

1. Morens, D. M., Folkers, G. K., Fauci, A. S. (2004). The challenge of emerging and re-emerging infectious diseases. *Nature*, 430(6996), 242–249. https://doi.org/10.1038/nature02759

2. Fauci, A. S., Touchette, N. A., Folkers, G. K. (2005). Emerging infectious diseases: A 10-year perspective from the national Institute of Allergy and Infectious Diseases. *International Journal of Risk and Safety in Medicine*, 17(3, 4), 157–167. doi: 10.3201/eid1104.041167

3. Dizaj, S. M., Lotfipour, F., Barzegar-Jalali, M., Zarrintan, M. H., Adibkia, K. (2014). Antimicrobial activity of the metals and metal oxide nanoparticles. *Materials Science and Engineering C*, 44, 278–284. https://doi.org/10.1016/j.msec.2014.08.031

4. Huh, A. J., Kwon, Y. J. (2011). "Nanoantibiotics": A new paradigm for treating infectious diseases using nanomaterials in the antibiotics resistant era. *Journal of Controlled Release*, 156(2), 128–145. https://doi.org/10.1016/j.jconrel.2011.07.002

5. Khan, M., Khan, S. T., Khan, M., Adil, S. F., Musarrat, J., Al-Khedhairy, A. A., Al-Warthan, A., Siddiqui, M. R., Alkhathlan, H. Z. (2014). Antibacterial properties of silver nanoparticles synthesized using Pulicaria glutinosa plant extract as a green bioreductant. *International Journal of Nanomedicine*, 9, 3551–3565. https://doi.org/10.2147/IJN.S61983

6. Dreaden, E. C., Alkilany, A. M., Huang, X., Murphy, C. J., El- Sayed, M. A. (2012). The golden age: Gold nanoparticles for biomedicine. *Chemical Society Reviews*, 41(7), 2740–2779. https://doi.org/10.1039/C1CS15237H

7. Heath, J. R. (2015). Nanotechnologies for biomedical science and translational medicine. *Proceedings of the National Academy of Sciences of the USA*, 112(47), 14436–14443. https://doi.org/10.1073/pnas.1515202112

8. Rajiv, P., Rajeshwari, S., Venckatesh, R. (2013). Bio-fabrication of zinc oxide nanoparticles using leaf extract of Parthenium hysterophorus L. and its size-dependent antifungal activity against plant fungal pathogens. *Spectrochim Acta Part A: Molecular and Biomolecular Spectroscopy.*, 112, 384–387. https://doi.org/10.1016/j.saa.2013.04.072

9. Dejen, K. D., Zereffa, E. A., Murthy, H. A., Merga, A. (2020). Synthesis of ZnO and ZnO/PVA nanocomposite using aqueous Moringa oleifeira leaf extract template: Antibacterial and electrochemical activities. *Reviews on Advanced Materials Science*, 59(1), 464–476. https://doi.org/10.1515/rams-2020-0021

10. Ramesh, M. A., Viruthagiri, G. (2015). Green synthesis of ZnO nanoparticles using Solanum nigrum leaf extract and their antibacterial activity. *Spectrochimica Acta Part A: Molecular and Biomolecular Spectroscopy*, 136, 864–870. https://doi.org/10.1016/j.saa.2014.09.105

11. Abdelsattar, A. S., Farouk, W. M., Gouda, S. M., Safwat, A., Hakim, T. A., El-Shibiny, A. (2022). Utilization of Ocimum basilicum extracts for zinc oxide nanoparticles synthesis and their antibacterial activity after a novel combination with phage. *Materials Letters*, 309, 131344. https://doi.org/10.1016/j.matlet.2021.131344

12. Raliya, R., Tarafdar, J. (2014). Biosynthesis and characterization of zinc, magnesium and titanium nanoparticles: An eco-friendly approach. *International Nano Letters*, 4, 1–10. https://doi.org/10.1007/s40089-014-0093-8

13. Li, X., Xu, H., Chen, Z. S., Chen, G. (2011). Biosynthesis of nanoparticles by microorganisms and their applications. *Journal of Nanomaterials*, 2011, 1–16. https://doi.org/10.1155/2011/270974

14. Kwon, S. G., Piao, Y., Park, J., et al. (2007). Kinetics of monodisperse iron oxide nanocrystal formation by 'heating-up' process. *Journal of the American Chemical Society*, 129, 12571–12584. https://doi.org/10.1021/ja074633q

15. Masadeh, M. M., Karasneh, G. A., Al-Akhras, M. A., Albiss, B. A., Aljarah, K. M., Al-Azzam, S. I., Alzoubi, K. H. (2015). Cerium oxide and iron oxide nanoparticles abolish the antibacterial activity of ciprofloxacin against gram positive and gram negative biofilm bacteria. *Cytotechnology*, 67(3), 427–435. https://doi.org/10.1007/s10616-014-9701-8

16. Gordon, T., Perlstein, B., Houbara, O., Felner, I., Banin, E., Margel, S. (2011). Synthesis and characterization of zinc/iron oxide composite nanoparticles and their antibacterial properties. *Colloids and Surfaces A: Physicochemical and Engineering Aspects*, 374(1–3), 1–8. https://doi.org/10.1016/j.colsurfa.2010.10.015

17. Narayanan, P. M., Wilson, W. S., Abraham, A. T., Sevanan, M. (2012). Synthesis, characterization, and antimicrobial activity of zinc oxide nanoparticles against human pathogens. *BioNanoScience*, 2(4), 329–335. https://doi.org/10.1007/s12668-012-0061-6

18. Natalio, F., André, R., Hartog, A. F., Stoll, B., Jochum, K. P., Wever, R., Tremel, W. (2012). Vanadium pentoxide nanoparticles mimic vanadium haloperoxidases and thwart biofilm formation. *Nature Nanotechnology*, 7(8), 530–535. https://doi.org/10.1038/nnano.2012.91

19. Roy, A., Gauri, S. S., Bhattacharya, M., Bhattacharya, J. (2013). Antimicrobial activity of CaO nanoparticles. *Journal of Biomedical Nanotechnology*, 9(9), 1570–1578. https://doi.org/10.1166/jbn.2013.1681

20. Al-Hazmi, F., Alnowaiser, F., Al-Ghamdi, A. A., Al-Ghamdi, A. A., Aly, M. M., Al-Tuwirqi, R. M., El-Tantawy, F. (2012). A new large-scale synthesis of magnesium oxide nanowires: Structural and antibacterial properties. *Superlattices and Microstructures*, 52(2), 200–209. https://doi.org/10.1016/j.spmi.2012.04.013

21. Ghosh, T., Dash, S. K., Chakraborty, P., Guha, A., Kawaguchi, K., Roy, S., Chattopadhyay, T., Das, D. (2014). Preparation of antiferromagnetic Co_3O_4 nanoparticles from two different precursors by pyrolytic method: In vitro antimicrobial activity. *RSC Advances*, 4(29), 15022–15029. DOI: 10.1039/c3ra47769j

22. Patil, A. B., Bhanage, B. M. Green methodologies in the synthesis of metal and metal oxide nanoparticles. In: Kharisov, B. I., Kharissova, O. V., Dias, H. V. R., editors. *Nanomaterials for Environmental Protection* (p. 293–311). Hoboken, NJ: John Wiley & Sons, Inc.; 2014. https://doi.org/10.1002/9781118845530.ch18

23. Bykkam, S., Narsingam, S., Ahmadipour, M., Dayakar, T., Rao, K. V., Chakra, C. S., Kalakotla, S. (2015). Few layered graphene sheet decorated by ZnO nanoparticles for anti-bacterial application. *Superlattices and Microstructures*, 83, 776–784. https://doi.org/10.1016/j.spmi.2015.03.063

24. Jaisai, M., Baruah, S., Dutta, J. (2012). Paper modified with ZnO nanorods-antimicrobial studies. *Beilstein Journal of Nanotechnology*, 3(1), 684–691. https://doi.org/10.3762/bjnano.3.78

25. Suslick, K. S. (1990). Sonochemistry. *Science*, 247(4949), 1439–1445. https://doi.org/10.1126/science.247.4949.1439

26. Jung, S. H., Oh, E., Lee, K. H., Yang, Y., Park, C. G., Park, W., Jeong, S. H. (2008). Sonochemical preparation of shape-selective ZnO nanostructures. *Crystal Growth and Design*, 8(1), 265–269. https://doi.org/10.1021/cg070296l

27. Jimmy, C. Y., Yu, J., Ho, W., Zhang, L. (2001). Preparation of highly photocatalytic active nano-sized TiO2 particles via ultrasonic irradiation. *Chemical Communications*, (19), 1942–1943. https://doi.org/10.1039/B105471F

28. Kumar, R. V., Koltypin, Y., Xu, X. N., Yeshurun, Y., Gedanken, A., Felner, I. (2001). Fabrication of magnetite nanorods by ultrasound irradiation. *Journal of Applied Physics*, 89(11), 6324–6328. https://doi.org/10.1063/1.1369408

29. Lai, J., Shafi, K. V., Loos, K., Ulman, A., Lee, Y., Vogt, T., Estournes, C. (2003). Doping γ-Fe2O3 nanoparticles with Mn (III) suppresses the transition to the α-Fe2O3 structure. *Journal of the American Chemical Society*, 125(38), 11470–11471. https://doi.org/10.1021/ja035409d

30. Geng, J., Zhu, J. J., Lu, D. J., Chen, H. Y. (2006). Hollow PbWO4 nanospindles via a facile sonochemical route. *Inorganic Chemistry*, 45(20), 8403–8407. https://doi.org/10.1021/ic0608804

31. Gopinath, K., Karthika, V., Sundaravadivelan, C., Gowri, S., Arumugam, A. (2015). Mycogenesis of cerium oxide nanoparticles using Aspergillus niger culture filtrate and their applications for antibacterial and larvicidal activities. *Journal of Nanostructure in Chemistry*, 5(3), 295–303. https://doi.org/10.1007/s40097-015-0161-2

32. Patil, K. C., Aruna, S. T., Ekambaram, S. (1997). Combustion synthesis. *Current Opinion in Solid State and Materials Science*, 2(2), 158–165. https://doi.org/10.1016/S1359-0286(97)80060-5

33. Aruna, S. T., Mukasyan, A. S. (2008). Combustion synthesis and nanomaterials. *Current Opinion in Solid State and Materials Science*, 12(3–4), 44–50. https://doi.org/10.1016/j.cossms.2008.12.002

34. Hedaoo, V. P., Bhatkar, V. B., Omanwar, S. K. (2015). PbCaB2O5 doped with Eu3+: A novel red emitting phosphor. *Optical Materials*, 45, 91–96. https://doi.org/10.1016/j.optmat.2015.02.037.

35. Venkatesh, R., Dhananjaya, N., Sateesh, M. K., Shabaaz Begum, J. P., Yashodha, S. R., Nagabhushana, H., Shivakumara, C. (2018). Effect of Li, Na, K cations on photoluminescence of GdAlO3:Eu3+ nanophosphor and study of Li cation on its antimicrobial activity. *Journal of Alloys and Compounds*, 732, 725–739. https://doi.org/10.1016/j.jallcom.2017.10.117.

36. Azam, A., Ahmed, A. S., Oves, M., Khan, M. S., Habib, S. S., Memic, A. (2012). Antimicrobial activity of metal oxide nanoparticles against Gram-positive and Gram-negative bacteria: A comparative study. *International Journal of Nanomedicine*, 7, 6003–6009. https://doi.org/10.2147%2FIJN.S35347

37. Mukherjee, M., De, S. (2015). Reduction of microbial contamination from drinking water using an iron oxide nanoparticle-impregnated ultrafiltration mixed matrix membrane: Preparation, characterization and antimicrobial properties. *Environmental Science: Water Research & Technology*, 1(2), 204–217. https://doi.org/10.1039/C4EW00094C

38. Bhushan, M., Muthukamalam, S., Sudharani, S., Viswanath, A. K. (2015). Synthesis of α-Fe 2− x Ag x O 3 nanocrystals and study of their optical, magnetic and antibacterial properties. *RSC Advances*, 5(40), 32006–32014. https://doi.org/10.1039/C4RA17259K

39. Tang, Z. X., Yu, Z., Zhang, Z. L., Zhang, X. Y., Pan, Q. Q., Shi, L. E. (2013). Sonication-assisted preparation of CaO nanoparticles for antibacterial agents. *Química Nova*, 36(7), 933–936. Doi: 10.1590/S0100-40422013000700002

40. Pelletier, D. A., Suresh, A. K., Holton, G. A., McKeown, C. K., Wang, W., Gu, B., Mortensen N P, Allison D. P., Joy D. C., Allison M. R., Brown S. D., Phelps T. J., Doktycz, M. J. (2010). Effects of engineered cerium oxide nanoparticles on bacterial growth and viability. *Applied and Environmental Microbiology*, 76(24), 7981–7989. https://doi.org/10.1128/AEM.00650-10

41. Thill, A., Zeyons, O., Spalla, O., Chauvat, F., Rose, J., Auffan, M., Flank, A. M. (2006). Cytotoxicity of CeO2 nanoparticles for Escherichia coli. Physico-chemical insight of the cytotoxicity mechanism. *Environmental Science & Technology*, 40(19), 6151–6156. https://doi.org/10.1021/es060999b

42. Stoimenov, P. K., Klinger, R. L., Marchin, G. L., Klabunde, K. J. (2002). Metal oxide nanoparticles as bactericidal agents. *Langmuir*, 18(17), 6679–6686. https://doi.org/10.1021/la0202374

43. McQuillan, J. S., Groenaga Infante, H., Stokes, E., Shaw, A. M. (2012). Silver nanoparticle enhanced silver ion stress response in Escherichia coli K12. *Nanotoxicology*, 6(8), 857–866. https://doi.org/10.3109/17435390.2011.626532

44. Leung, Y. H., Ng, A. M., Xu, X., et al. (2014). Mechanisms of antibacterial activity of MgO: Non-ROS mediated toxicity of MgO nanoparticles towards Escherichia coli. *Small*, 10(6), 1171–1183. https://doi.org/10.1002/smll.201302434

45. Mukha, I. P., Eremenko, A. M., Smirnova, N. P., Mikhienkova, A. I., Korchak, G. I., Gorchev, V. F., Chunikhin, A. Y. (2013). Antimicrobial activity of stable silver nanoparticles of a certain size. *Applied Biochemistry and Microbiology*, 49(2), 199–206. https://doi.org/10.1134/S0003683813020117

46. Fones, H., Preston, G. M. (2012). Reactive oxygen and oxidative stress tolerance in plant pathogenic Pseudomonas. *FEMS Microbiology Letters*, 327(1), 1–8. https://doi.org/10.1111/j.1574-6968.2011.02449.x

47. Vatansever, F., de Melo, W. C., Avci, P., Vecchio, D., Sadasivam, M., Gupta, A., Chandran R., Karimi M., Parizotto N. A., Yin R., Tegos G. P., Hamblin, M. R. (2013). Antimicrobial strategies centered around reactive oxygen species-bactericidal antibiotics, photodynamic therapy, and beyond. *FEMS Microbiology Reviews*, 37(6), 955–989. https://doi.org/10.1111/1574-6976.12026

48. Ong, K. S., Cheow, Y. L., Lee, S. M. (2017). The role of reactive oxygen species in the antimicrobial activity of pyochelin. *Journal of Advanced Research*, 8(4), 393–398. https://doi.org/10.1016/j.jare.2017.05.007

49. Fasnacht, M., Polacek, N. (2021). Oxidative stress in bacteria and the central dogma of molecular biology. *Frontiers in Molecular Biosciences*, 8, 392. https://doi.org/10.3389/fmolb.2021.671037

50. Ravishankar Rai, V., Jamuna Bai, A. Nanoparticles and their potential applications as antimicrobials, Science against microbial pathogens: Communicating current research and technological advances. In: Méndez-Vilas, A., editor. *Formatex* (Vol. 1, p. 197–209), Microbiology Series, No. 3; 2011. Spain.

51. Besinis, A., De Peralta, T., Handy, R. D. (2014). The antibacterial effects of silver, titanium dioxide and silica dioxide nanoparticles compared to the dental disinfectant chlorhexidine on Streptococcus mutans using a suite of bioassays. *Nanotoxicology*, 8(1), 1–16. https://doi.org/10.3109/17435390.2012.742935

52. Xie, Y., He, Y., Irwin, P. L., Jin, T., Shi, X. (2011). Antibacterial activity and mechanism of action of zinc oxide nanoparticles against Campylobacter jejuni. *Applied and Environmental Microbiology*, 77(7), 2325–2331. https://doi.org/10.1128/AEM.02149-10

53. Foster, H. A., Ditta, I. B., Varghese, S., Steele, A. (2011). Photocatalytic disinfection using titanium dioxide: Spectrum and mechanism of antimicrobial activity. *Applied Microbiology and Biotechnology*, 90(6), 1847–1868.https://doi.org/10.1007/s00253-011-3213-7

54. Sawai, J., Kojima, H., Igarashi, H., Hashimoto, A., Shoji, S., Sawaki, T., Hakoda, A., Kawada, E., Kokugan, T., Shimizu, M. (2000). Antibacterial characteristics of magnesium oxide powder. *World Journal of Microbiology and Biotechnology*, 16(2), 187–194. https://doi.org/10.1023/A:1008916209784

55. Vidic, J., Stankic, S., Haque, F., Ciric, D., Le Goffic, R., Vidy, A., Jupille, J., Delmas, B. (2013). Selective antibacterial effects of mixed ZnMgO nanoparticles. *Journal of Nanoparticle Research*, 15(5), 1–10. https://doi.org/10.1007/s11051-013-1595-4

56. Padmavathy, N., Vijayaraghavan, R. (2011). Interaction of ZnO nanoparticles with microbes - a physio and biochemical assay. *Journal of Biomedical Nanotechnology*, 7(6), 813–822. https://doi.org/10.1166/jbn.2011.1343

57. Chung, Y. C., Su, Y. P., Chen, C. C., Jia, G., Wang, H. L., Wu, J. G., Lin, J. G. (2004). Relationship between antibacterial activity of chitosan and surface characteristics of cell wall. *Acta Pharmacologica Sinica*, 25(7), 932–936. PMID: 15210068.

58. Beveridge T. J. (1999). Structures of Gram-negative cell walls and their derived membrane vesicles. *Journal of Bacteriology*, 181, 4725–4733. https://doi.org/10.1128/JB.181.16.4725-4733.1999

59. Liu, Y. J., He, L. L., Mustapha, A., Li, H., Hu, Z. Q., Lin, M. S. (2009). Antibacterial activities of zinc oxide nanoparticles against Escherichia coli O157: H7. *Journal of Applied Microbiology*, 107(4), 1193–1201. https://doi.org/10.1111/j.1365-2672.2009.04303.x

60. Chen, W. J., Chen, Y. C. (2010). Fe3O4/TiO2 core/shell magnetic nanoparticle-based photokilling of pathogenic bacteria. *Nanomedicine*, 5(10), 1585–1593. https://doi.org/10.2217/nnm.10.92

61. Tsuang, Y. H., Sun, J. S., Huang, Y. C., Lu, C. H., Chang, W. H. S., Wang, C. C. (2008). Studies of photokilling of bacteria using titanium dioxide nanoparticles. *Artificial Organs*, 32(2), 167–174. https://doi.org/10.1111/j.1525-1594.2007.00530.x

62. Pal, B., Singh, I., Angrish, K., Aminedi, R., Das, N. (2012). Rapid photokilling of gram-negative Escherichia coli bacteria by platinum dispersed titania nanocomposite films. *Materials Chemistry and Physics*, 136(1), 21–27. https://doi.org/10.1016/j.matchemphys.2012.06.001

63. Lynch, I., Dawson, K. A. (2008). Protein-nanoparticle interactions. *Nano Today*, 3(1–2), 40–47. https://doi.org/10.1016/S1748-0132(08)70014-8

64. Aggarwal, P., Hall, J. B., McLeland, C. B., Dobrovolskaia, M. A., McNeil, S. E. (2009). Nanoparticle interaction with plasma proteins as it relates to particle biodistribution, biocompatibility and therapeutic efficacy. *Advanced Drug Delivery Reviews*, 61(6), 428–437. https://doi.org/10.1016/j.addr.2009.03.009

65. Hatchett, D. W., White, H. S. (1996). Electrochemistry of sulfur adlayers on the low-index faces of silver. *The Journal of Physical Chemistry*, 100(23), 9854–9859. https://doi.org/10.1021/jp953757z

66. Su, Y., Zheng, X., Chen, Y., Li, M., Liu, K. (2015). Alteration of intracellular protein expressions as a key mechanism of the deterioration of bacterial denitrification caused by copper oxide nanoparticles. *Scientific Reports*, 5(1), 1–11. https://doi.org/10.1038/srep15824

67. Mahapatra, O., Bhagat, M., Gopalakrishnan, C., Arunachalam, K. D. (2008). Ultrafine dispersed CuO nanoparticles and their antibacterial activity. *Journal of Experimental Nanoscience*, 3(3), 185–193. https://doi.org/10.1080/17458080802395460

68. Ahamed, M., Alhadlaq, H. A., Khan, M. A., Karuppiah, P., Al-Dhabi, N. A. (2014). Synthesis, characterization, and antimicrobial activity of copper oxide nanoparticles. *Journal of Nanomaterials*, 2014(17), 1–4. https://doi.org/10.1155/2014/637858

69. Smetana, A. B., Klabunde, K. J., Marchin, G. R., Sorensen, C. M. (2008). Biocidal activity of nanocrystalline silver powders and particles. *Langmuir*, 24(14), 7457–7464. https://doi.org/10.1021/la800091y

70. Tiller, J. C., Liao, C. J., Lewis, K., Klibanov, A. M. (2001). Designing surfaces that kill bacteria on contact. *Proceedings of the National Academy of Sciences*, 98(11), 5981–5985. https://doi.org/10.1073/pnas.111143098

71. Faunce, T., Murray, K., Nasu, H., Bowman, D. (2008). Sunscreen safety: The precautionary principle, the Australian therapeutic goods administration, and nanoparticles in sunscreens. *NanoEthics*, 2(3), 231–240. https://doi.org/10.1007/s11569-008-0041-z

72. Nohynek, G. J., Lademann, J., Ribaud, C., Roberts, M. S. (2007). Grey goo on the skin? Nanotechnology, cosmetic and sunscreen safety. *Critical Reviews in Toxicology*, 37(3), 251–277. https://doi.org/10.1080/10408440601177780

73. Jin, F., Liao, S., Wang, Q., Shen, H., Jiang, C., Zhang, J., Wei, Q., Ghiladi, R. A. (2022). Dual-functionalized luminescent/photodynamic composite fabrics: Synergistic antibacterial activity for self-disinfecting textiles. *Applied Surface Science*, 587152737. https://doi.org/10.1016/j.apsusc.2022.152737

74. Saravanakumar, K., Sathiyaseelan, A., Priya, V. V., Wang, M. H. (2022). PEGylated palladium doped ceria oxide nanoparticles (Pd-dop-CeO2-PEG NPs) for inhibition of bacterial pathogens and human lung cancer cell proliferation. *Journal of Drug Delivery Science and Technology*, 103367. https://doi.org/10.1016/j.jddst.2022.103367

75. Astashev, M. E., Sarimov, R. M., Serov, D. A., et al. (2022). Antibacterial behavior of organosilicon composite with nano aluminum oxide without influencing animal cells. *Reactive and Functional Polymers*, 170, 105143. https://doi.org/10.1016/j.reactfunctpolym.2021.105143

76. Liu, Z., Qiu, K., Tang, Q., Wu, Y., Wang, J. (2019). Synthesis of Ag+/CaTiO3: Pr3+ with luminescence and antibacterial properties. *Advanced Powder Technology*, 30(1), 23–29. https://doi.org/10.1016/j.apt.2018.10.003

77. Khattab, T. A., Fouda, M. M. G., Abdelrahman, M. S., et al. (2019) Development of illuminant glow-in-the-dark cotton fabric coated by luminescent composite with antimicrobial activity and ultraviolet protection. *Journal of Fluorescence*, 29, 703–710. https://doi.org/10.1007/s10895-019-02384-2

78. Ma̅nas, P., Pag̅an, R. (2005). Microbial inactivation by new technologies of food preservation-a review. *Journal of Applied Microbiology*, 98(6), 1387–1399. https://doi.org/10.1111/j.1365-2672.2005.02561.x

79. Elmnasser, N., Guillou, S., Leroi, F., Orange, N., Bakhrouf, A., Federighi, M. (2007). Pulsed-light system as a novel food decontamination technology: A review. *Canadian Journal of Microbiology*, 53(7), 813–821. https://doi.org/10.1139/W07-042

80. Hijnen, W. A. M., Beerendonk, E. F., Medema, G. J. (2006). Inactivation credit of UV radiation for viruses, bacteria and protozoan (oo)cysts in water: A review. *Water Research*, 40, 3–22. https://doi.org/10.1016/j.watres.2005.10.030

81. Craford, M. G. LEDs for solid state lighting and other emerging applications: Status, trends, and challenges. In: *Fifth International Conference on Solid State Lighting* (Vol. 5941, p. 594101). International Society for Optics and Photonics; 2005, September. https://doi.org/10.1117/12.625918

82. D'Souza, C., Yuk, H. G., Khoo, G. H., Zhou, W. (2015). Application of light-emitting diodes in food production, postharvest preservation, and microbiological food safety. *Comprehensive Reviews in Food Science and Food Safety*, 14(6), 719–740. https://doi.org/10.1111/1541-4337.12155

83. Amin, R. M., Bhayana, B., Hamblin, M. R., Dai, T. (2016). Antimicrobial blue light inactivation of Pseudomonas aeruginosa by photo-excitation of endogenous porphyrins: In vitro and in vivo studies. *Lasers in Surgery and Medicine*, 48, 562–568. https://doi.org/10.1002/lsm.22474

84. Kunz, D., Wirth, J., Sculean, A., Eick, S. (2019). In- vitro-activity of additive application of hydrogen peroxide in antimicrobial photodynamic therapy using LED in the blue spectrum against bacteria and biofilm associated with periodontal disease. *Photodiagnosis and Photodynamic Therapy*, 26, 306–312. https://doi.org/10.1016/j.pdpdt.2019.04.015

85. Setlur, A. A. (2009). Phosphors for LED based lighting. *Electrochemical Society Interface*, 18(4), 32. https://doi.org/10.1149/2.F04094IF

86. Shirai, A., Kawasaka, K., Tsuchiya, K. (2022). Antimicrobial action of phenolic acids combined with violet 405-nm light for disinfecting pathogenic and spoilage fungi. *Journal of Photochemistry and Photobiology B: Biology*, 229, 112411. https://doi.org/10.1016/j.jphotobiol.2022.112411

87. Luk̕siene, Z., Zukauskas, A. (2009). Prospects of photosensitization in control of pathogenic and harmful micro-organisms. *Journal of Applied Microbiology*, 107(5), 1415–1424. https://doi.org/10.1111/j.1365-2672.2009.04341.x

88. Bumah, V. V., Masson-Meyers, D. S., Cashin, S., Enwemeka, C. S. (2013). Wavelength and bacterial density influence the bactericidal effect of blue light on methicillin-resistant Staphylococcus aureus (MRSA). *Photomedicine and Laser Surgery*, 31(11), 547–553. https://doi.org/10.1089/pho.2012.3461

89. Hyun, J. E., Lee, S. Y. (2020). Blue light-emitting diodes as eco-friendly non-thermal technology in food preservation. *Trends in Food Science & Technology*, 105, 284–295. https://doi.org/10.1016/j.tifs.2020.09.008

90. Ghate, V., Kumar, A., Zhou, W., Yuk, H. G. (2016). Irradiance and temperature influence the bactericidal effect of 460-nanometer light-emitting diodes on Salmonella in orange juice. *Journal of Food Protection*, 79, 553–560. https://doi.org/10.4315/0362-028X.JFP-15-394

91. Gunther IV, N. W., Phillips, J. G., Sommers, C. (2016). The effects of 405-nm visible light on the survival of Campylobacter on chicken skin and stainless steel. *Foodborne Pathogens and Disease*, 13(5), 245–250. https://doi.org/10.1089/fpd.2015.2084

92. Srimagal, A., Ramesh, T., Sahu, J. K. (2016). Effect of light emitting diode treatment on inactivation of Escherichia coli in milk. *LWT-Food Science and Technology*, 71, 378–385. https://doi.org/10.1016/j.lwt.2016.04.028

93. Ghate, V., Kumar, A., Kim, M. J., Bang, W. S., Zhou, W., Yuk, H. G. (2017). Effect of 460 nm light emitting diode illumination on survival of Salmonella spp. on fresh-cut pineapples at different irradiances and temperatures. *Journal of Food Engineering*, 196, 130–138. https://doi.org/10.1016/j.jfoodeng.2016.10.013

94. Kim, M. J., Bang, W. S., Yuk, H. G. (2017). 405±5 nm light emitting diode illumination causes photodynamic inactivation of Salmonella spp. on fresh-cut papaya without deterioration. *Food Microbiology*, 62, 124–132. https://doi.org/10.1016/j.fm.2016.10.002

95. Kim, M. J., Ng, B. X. A., Zwe, Y. H., Yuk, H. G. (2017). Photodynamic inactivation of Salmonella enterica Enteritidis by 405±5-nm light-emitting diode and its application to control salmonellosis on cooked chicken. *Food Control*, 82, 305–315. https://doi.org/10.1016/j.foodcont.2017.06.040

96. Kim, M. J., Tang, C. H., Bang, W. S., Yuk, H. G. (2017). Antibacterial effect of 405±5 nm light emitting diode illumination against Escherichia coli O157:H7, Listeria monocytogenes, and Salmonella on the surface of fresh-cut mango and its influence on fruit quality. *International Journal of Food Microbiology*, 244, 82–89. https://doi.org/10.1016/j.ijfoodmicro.2016.12.023

97. Sommers, C., Gunther IV, N. W., Sheen, S. (2017). Inactivation of Salmonella spp., pathogenic Escherichia coli, Staphylococcus spp., or Listeria monocytogenes in chicken purge or skin using a 405-nm LED array. *Food Microbiology*, 64, 135–138. https://doi.org/10.1016/j.fm.2016.12.011

98. Hyun, J. E., Lee, S. Y. (2020). Antibacterial effect and mechanisms of action of 460-470 nm light-emitting diode against Listeria monocytogenes and Pseudomonas fluorescens on the surface of packaged sliced cheese. *Food Microbiology*, 86, 103314. https://doi.org/10.1016/j.fm.2019.103314

99. Kim, M. J., Lianto, D. K., Koo, G. H., Yuk, H. G. (2021). Antibacterial mechanism of riboflavin-mediated 460 nm light emitting diode illumination against Listeria monocytogenes in phosphate-buffered saline and on smoked salmon. *Food Control*, 124, 107930. https://doi.org/10.1016/j.foodcont.2021.107930

100. Hyun, J. E., Moon, S. K., Lee, S. Y. (2022). Application of blue light-emitting diode in combination with antimicrobials or photosensitizers to inactivate Escherichia coli O157: H7 on fresh-cut apples and cherry tomatoes. *Food Control*, 131, 108453. https://doi.org/10.1016/j.foodcont.2021.108453

11 Metal Oxides
Bioimaging and Biomedical Applications

V. G. Thakare

Shri Shivaji Art's, Commerce and Science College

CONTENTS

11.1 INTRODUCTION

Due to their physicochemical characteristics, metal oxide nanoparticles (MONPs) are particularly important in many chemical and physical fields [1]. Metal oxide nanoparticles' diverse physicochemical properties enable them to perform a variety of biological tasks based on their antibacterial [2], anti-diabetic [3], anti-inflammatory [4], drug delivery [5,6], antioxidant [7,8], anticancer [9], antifungal [10] and bioimaging capabilities [11]. Metal oxide nanoparticles are adaptable and can be used in a wide range of biomedical applications, including targeted drug administration, bioimaging, biosensors, tissue engineering, antibacterial use, photosensitisers for photodynamic therapy (PDT), photothermal therapy (PTT), and many more [12,13]. The morphological and physicochemical qualities of metal oxide particles determine the majority of their usable capabilities; however, their chemical properties can vary greatly depending on the manufacturing process's preparation and processing steps. Metal oxide nanoparticles can be created using a variety of manufacturing processes [14]. Methods for precipitating different substances [15–17] include co-precipitation method [18], combustion synthesis [19], mechanochemical process [20], surfactant precipitation [21], sol-gel method [22,23], simple ammonia precipitation [24], electrochemical method [25], solvo-thermal, hydrothermal synthesis [26], colloid solutions [27], emulsion methods [28,29], and green synthesis [30].

Nanotechnology has recently garnered more attention as a result of its distinctive structural, behav-ioural, and varied functional characteristics, such as size, high chemical stability with enhanced surface area, functionalisable surface with a variety of molecules, and biocompatibility with a wide variety of cell types applications [31]. These characteristics are largely responsible for the phenom-enon. It has been demonstrated that nano-platforms are good tools for biomedical applications [32]. Metal oxide nanoparticles [MONPs] are biocompatible and have a distinctive structure, intriguing surface characteristics, a large surface area, and good mechanical stability. These factors have led to a great deal of interest in metal oxide nanoparticles in the areas of biomedical treatments, bio-imaging, and biosensing. So, for the past few decades, scientists have been actively designing and creating diverse metal oxide nanoparticles. Magnetic iron oxide (Fe_3O_4), titanium oxide (TiO_2), alu-minium oxide (Al_2O_3), zinc oxide (ZnO), cobalt oxide, silver, gold, and more recently ceria (CeO_2) nanomaterials are of great interest to researchers for various biomedical applications [33]. These nanoparticles are the most commonly used nanoparticles. These materials are suitable for a range of diverse biomedical applications due to their fascinating catalytic, antioxidant, and bactericidal activ-ity, mechanical stability, and biocompatibility. For instance, ZnO nanoparticles are biodegradable and have low toxicity. Chemically, ZnO has an abundance of –OH groups on its surface, which can easily be functionalised by different surface-decorating molecules. If the surface is in close contact with the solution, ZnO can slowly dissolve in both acidic (such as in the tumour cells and tumour microenvironment) and strong basic conditions. ZnO nanoparticles have drawn a lot of interest in biomedical applications due to their favourable characteristics [34]. Due to their unusual magnetic properties, low cost, chemical stability, and biocompatibility, iron oxide nanoparticles have garnered a lot of interest. Iron oxide nanoparticles are frequently used for excellent biomedical applications due to their special characteristics. Because of its exceptional chemical, biological, and magnetic features, such as chemical stability, nontoxicity, biocompatibility, high saturation magnetisation, and high magnetic susceptibility, iron oxide nanoparticles have received a lot of attention in the litera-ture [35]. The preferred material for medical implants is titanium because it offers a biocompatible surface for cell adhesion and growth. Targeted medication delivery, magnetic resonance imaging (MRI), and cell labelling and separation have all made extensive use of magnetic iron oxides. Due to its catalytic and antioxidant abilities, ceria has recently attracted a lot of attention [36].

The most frequently used metal oxides, characterisation methods for NP metal oxides, and cur-rent developments in a few specific biomedical applications are the main topics of this chapter. The biomedical uses of NPs metal oxides are covered in the final section of this chapter.

11.2   CHARACTERISATION TECHNIQUES

Research done in the past has shown that the morphological properties of nanoparticles and NPs have an effect on the variety of applications for which they can be used. X-ray diffraction, ultra-violet visible spectroscopy, scanning tunnelling microscope, atomic force microscope, scanning electron microscope, dynamic light scattering, transmission electron microscope, differential scan-ning calorimetry, Fourier transmission infrared spectroscopy, energy dispersive X-ray, and particle size analysis were the methods that were utilised to provide evidence of this [37]. In this article, several different approaches of characterisation are investigated in depth in order to investigate the morphological and biological features of metal oxide nanoparticles.

11.2.1   Scanning Electron Microscopy

The surface topography, crystalline structure, chemical makeup, and electrical behaviour of materi-als are all studied using scanning electron microscopy (SEM) [38]. There are typically two different types of microscopy accessible. Scanning electron microscopy (SEM) and optical microscopy (OM) (SEM). The OM is the most traditional and has been in use for the past two centuries. It is a straight-forward tool with few functions. Additionally, it is also known as light microscopy. Following are some characteristics and traits where SEM and OM diverge.

a. OM is based on light, but the basic working principle of SEM is electron emission.
b. Compound OM has two lenses, whereas OM just has one lens. The use of lenses that enlarge images by bending light.
c. In comparison with SEM, which can magnify objects up to 300,000 times, current OM only magnifies objects 400–1,000 times their original size.
d. While solid material and living cells can both be studied by OM, relatively few small organic particles and small solid fragments can be seen. SEM offers finer-grained field and grayscale pictures.

SEM is therefore more beneficial for studying materials and living things, but it is more expensive and difficult to maintain. SEM is one method for visualising surface magnetic microstructure on the 10 nm length scale [39–41].The SEM technique is mostly employed to observe the surfaces of cells, tissues, and organs, with internal features receiving less consideration [42]. Through the use of electron high-energy beam scanning, the sample's surface is imaged. SEM uses electrons to magnify the image as opposed to the light that traditional microscopes employ.

When nanoparticles are prone to agglomerating, there may not always be a clear way to distinguish between them in the sample. The way the specimen interacts with the electron beam determines the resolution of the SEM image.

Atoms interact with one another as a result of the collision of an input electron beam with the surface of a sample. This results in the creation of secondary electrons, backscattered electrons, diffracted electrons, and specific X-rays that are highly dependent on the topography, morphology, and chemical composition of the surface. Secondary electron imaging (SEM) produces a picture that has a very high resolution and can detect details that are as small as 5 nm. In addition to using specific X-rays and backscattered electrons from the sample, scanning electron microscopy (SEM) is utilised in order to produce an image and determine the component parts that make up the sample. Due to the high-resolution capabilities of SEM, it may be used to identify the nanoscale features of nanostructured materials. These features are significant to the properties and applications of the nanostructured materials.

11.2.2 Transmission Electron Microscopy

Ernst Ruska and Max Knoll created the transmission electron microscope in 1931. The resolution, contrast, and signal-to-noise ratio (S/N) of an image are indicators of its quality. When compared to extremely early models, which had resolutions of about 100 nm, modern electron microscopes have resolutions of 0.1 nm (1) and even better [43]. It is a useful technique for the full physicochemical characterisation of freshly generated nanoparticulates; in fact, over 15,000 articles on nanoparticles have employed transmission electron microscopy (TEM) as one of the methodologies in the previous 10 years. On a total of around 6,500 research articles on this topic published over the past 10 years, 2,400 (or 37%) were published from 2014 to 2016 [44]. TEM is also recently emerging as a technique of choice to investigate the impact of nanocomposites on biological systems. The method of microscopy known as TEM involves passing an electron beam through an extremely thin material while allowing it to interact with the material along the way. A picture is formed as a result of the interaction of electrons as they move through the specimen. This image is then expanded and focused onto an imaging device such as a fluorescent screen, a layer of photographic film, or a sensor such as a CCD camera. The extremely short de Broglie wavelength of the electron enables transmission electron microscopes (TEMs) to provide images with a resolution that is significantly superior to that of light microscopes. This makes it possible to view excellent detail, even as small as a single column of atoms, which is approximately 10,000 times smaller than the smallest thing that can be seen using a light microscope. The TEM is an important analytical tool that is used in a wide number of scientific fields, including the biological and physical sciences. Utilising TEMs is beneficial to the study of a wide variety of topics, including nanotechnology, cancer, viruses, materials science, and semiconductors. The ability of the material to absorb electrons as a function of both its thickness and composition is what causes the contrast in TEM pictures when the magnification level is set to a lower value. At higher magnifications, the intensity of the picture is regulated by more sophisticated wave interactions; as a result, analysis of the acquired images requires the assistance of an expert. Through the use of multiple modes of operation, the TEM can investigate changes in chemical identity, crystal orientation, electronic structure, and sample-induced electron phase shift in addition to the standard absorption-based imaging.

A specialist must prepare the specimens for the TEM technique. For powder materials, tiny particles are mixed with ethanol or distilled water to form a suspension, which is then dropped onto a layer of carbon that has been placed on a copper grid. The solvent is then removed [45].

11.2.3 FLUORESCENCE TECHNOLOGY

The technique of fluorescence microscopy is utilised so that structures, substances, or proteins that are located inside of cells can be identified. After taking in light of a lower wavelength, fluorescent molecules then give off light of a longer wavelength. When fluorescent molecules absorb a specific wavelength of light that corresponds to the electron's absorption, the electron in the given orbital will migrate to a higher energy level, known as the excited state. These electrons are currently in a condition that is unstable; as a result, they will eventually return to their ground state, at which point they will release energy in the form of light and heat. The scientific term for this type of energy release is fluorescence. Because of the reduction in energy caused by the loss of heat, the light that is emitted has a longer wavelength than the light that is absorbed, also known as excitation light [46].

In fluorescence microscopy, a cell is first stained with a dye, and then the cell is illuminated by light that has been filtered to match the wavelength at which the dye absorbs light. After that, an observation is made of the light given off by the dye by employing a filter that lets through only the wavelength that the dye emits. The dye shines brightly against a dark background due to the fact that the eyepieces and camera port of the microscope are designed to let just the wavelength of light that is emitting light through. The epi-illumination system is used in the construction of the vast majority of microscopes. When epi-illumination excitation is being performed, light must pass through the objective lens in order to illuminate the target. The light that is emitted by the specimen is focused by the same objective lens that was used to take the image. In this particular research project, a fluorescence microscope was utilised to investigate the morphology of both the cells and the scaffolds [47]. The optical components of a fluorescence microscope are shown in Figures 11.1 and 11.2, respectively (FLoid Cell Imaging Station).

11.2.4 TOXICITY STUDY

Metallic nanoparticles (NPs) must be nontoxic, biocompatible, and stable in biological mediums in order to be employed in therapy and diagnosis. Numerous mechanisms of toxicity, including oxidative stress, genotoxicity, cytotoxicity, and inflammation, have been discovered in recent investigations on the toxicology of metal oxides [48]. NPs are widely used in various applications in contemporary civilisation and have permeated daily life. However, it is important to design efficient toxicity testing models for these NPs because of their toxicity to species, notably humans. The method and length of exposure also affect a nanomaterial's toxicity. The four most common exposure pathways that are frequently encountered by researchers, workers in industrial facilities, and ultimately end users are inhalation, ingestion, injection, and skin contact [49]. Therefore, a

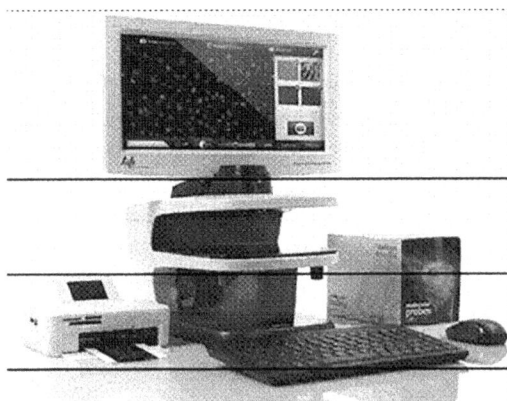

FIGURE 11.1 Fluorescence microscopy (FLoid Cell Imaging Station).

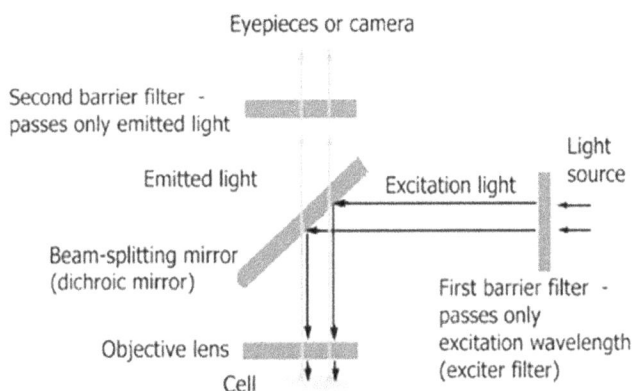

FIGURE 11.2 The optical system of fluorescence microscope.

detailed investigation of the important biological reactions at each route of exposure is required [50]. Various metal oxides were the subject of a cytotoxic investigation, according to Jeng and Swanson. When examined at high concentrations, they discovered that ZnO was extremely poisonous, whereas Al_2O_3 was only moderately toxic when compared to Fe_3O_4 and TiO_2. It's interesting to note that CrO_3 wasn't hazardous at the tested [51]. According to the research, green synthesis is safer and more environmentally friendly than physical and chemical synthesis [52]. Zhang reported that distinct cytotoxicity was linked to exposure to ZnO, TiO_2, SiO_2, and Al_2O_3 nanoparticles, and this suggests that these nanomaterials should be used with extreme caution [53]. On the other hand, metal oxides are finding more and more uses in biomedicine, including gene delivery, medication delivery, biosensors, and bioimaging.

11.2.5 BIOCOMPATIBILITY

Biocompatibility is a constant challenge for metal oxide nanomaterial applications in biomedicine. A significant experiment for screening a chemical of potential pharmacological use is the cytotoxicity assay. A test called a cytotoxicity assay is used to characterise and assess any materials' possible hazardous or detrimental effects on living things or cell cultures. To test the cytotoxic effect, several kinds of molecules, plant extracts, and NPs have been employed [54]. The ZnONPs demonstrated antiproliferative capabilities against the MDAMB-231 breast cancer cell lines, according to cytotoxicity studies. The most frequent uses of ZnONPs result in numerous severe adverse effects during chemotherapy and radiotherapy for cancer. Although the FDA has given ZnO approval for cosmetic purposes, its exact toxicological profile and cytotoxicity mechanism are still poorly understood. The biocompatibility of ZnO nanoparticles without surface coatings or modifications has been the subject of numerous investigations. Other studies, however, have demonstrated that ZnO nanoparticles are both nontoxic and preferentially harmful to bacteria or cancer cells, which may be helpful for applications involving cancer therapy.

It was also claimed that, in contrast to malignant cells, nanoceria can shield normal cells from radiation harm. Although radiation therapy is a frequently used method of treating cancer, it can seriously harm healthy cells that are near to the treatment site. The production of free radicals during treatment is thought to be the root of this harm. Nanoceria can offer radioprotection to healthy human breast cells but not to cancerous human breast cells (human breast tumour cell line MCF-7) according to an invitro investigation [55,56].

11.2.6 Biodegradability

For the sake of biomedical applications, the development of nanomaterials that are biocompatible, biodegradable, and functionalised remains a tremendously active and fruitful area of study. Nanoparticles with a paramagnetic core, quantum dots (QDs), nanoshells, and carbon nanotubes (CNTs) are among the many additional nanomaterials with a variety of biomedical applications that have been thoroughly researched. Zinc oxide (ZnO), which has unique features such as semiconducting, piezoelectric, and optical, might be exhibited by a wide range of different nanostructures. As a result, nanomaterials based on ZnO are currently under consideration for a vast array of applications, including energy storage, nanosensors, cosmetic items, nano-optical devices, and nanoelectronic devices. ZnO nanoparticles have a number of important properties, including biodegradability and low toxicity, which are among the most important (Table 11.1).

11.3 METAL OXIDES: BIOMEDICAL APPLICATIONS

Metal oxides have several applications in the field of biomedicine; some of these applications are shown in Figure 11.3: targeted drug delivery, bioimaging, biosensors, tissue engineering, antibacterial application, photosensitiser agents for photodynamic therapy (PDT), and photothermal therapy (PTT).

11.3.1 Targeted Drug Delivery

Because it has the potential to improve medication selectivity towards the cells that are intended to be affected, the utilisation of nanostructured materials is an exciting strategy for the administration of drugs [70]. This is because it can be used to administer drugs. Recent studies have revealed that nano-clusters that mimic fullerenes are capable of piercing cell barriers, transporting drugs to their destinations, and releasing them at those locations. This ability allows the nano-clusters to act as drug delivery vehicles. Nanoparticles that are composed of metal oxides are becoming an increasingly important component of one of the most recent groups of materials that are used in the pharmaceutical sector as well as in other applications that are related to health. The fast growing fields of medicine, medical engineering, and other fields allied to medicine have all helped to contribute to the discovery of innovative ways of administering medications. Some examples of these fields include the following: The process of developing methods and tools that make it possible to distribute medical chemicals in a manner that is both more effective and less harmful is referred to as the creation of a "drug delivery system," or DDS for short. This process is referred to by the acronym DDS [71]. It is hypothesised that the pharmacological carriers that are used satisfy a number of conditions for the physicochemical qualities of the pharmaceuticals that they carry. This is based on the fact that these conditions are necessary for the carriers to be used. They should

TABLE 11.1

Nanoparticles Characterisations, Related Biocompatibility and Toxicity Mechanisms

Nanoparticles	Synthesis Method	Physiochemical Characterisations	Biocompatibility Assays	Cells	Results	Ref
CeO_2	Colloid solution method	Shape: isotropic Size: 1.5–2.5 nm	MTT assays	NCCTC clone L9L29	No cytotoxic effect on cell and do not change their morphology	[57]
ZnO	Leaf extracts of Calotropis gigantea were biosynthesised using Rivina humilis as a starting material (L.)	Shape: circular morphology, crystalline in naturecrystalline nature	MTT assays	Neuro 2a neuroblastoma cells	Biogenic ZnONPs exhibit powerful antioxidant, antibacterial, and antimicrobial properties, in addition to phytochemical and cytotoxic ones.	[58]
Al_2O_3	Biosynthesis leaf extracts of lemongrass	Size: 34.5 nm	MTTassays	Cancer cell lines ofMDAMB-231	No cytotoxic effect on cell	[59]
Ag	Biosynthesis of silver nanoparticles from the leaves of Artemisia vulgaris, an environmentally friendly process	Shape: spherical Size: 25 nm	MTTassays	HeLa and MCF-7 cell lines	Compared to AVLE alone, the green produced nanoparticles showed effective antibacterial efficacy against harmful microorganisms. Silver nanoparticles (AV-AgNPs) showed promising antioxidant capabilities in invitro antioxidant experiments.	[60]
Zn-doped CeO_2-NPs	Sol-gel	Shape: spherical morphology Size: between about 15 and 30 nm	MTTassays	Neuro2A cells	Exhibited a negligible harmful effect at high nanoparticle concentrations, making them a suitable candidate for a variety of biological and medicinal applications.	[61]
Cerium oxide/ polyallylamine Nanoparticles	Sol–gel synthesis	Shape: spherical aggregations' Size: 46.24, 28.58, and 45.52 nm	MTT assay and RBC hemolysis assay	(MCF7, HeLa, and erythrocyte	CeO_2-NPs exhibited anticancer effects on the viability of MCF7 cells, with half-maximal inhibitory concentration (IC50) values decreasing by a significant amount each time. There was a correlation between a decrease in the viability of cancer cells and the 50% hemolytic concentration (HC50)	[62]
Cd doped $CoFe_2O_4$	Microwave-modified Pechini sol–gel method	Shape: spherical and rod Size: 18–39 nm.	MTT assay	3T3 fibroblast cells	All materials were shown to be nontoxic by assay and may be used in biological applications.	[63]
ZnO nano-crystals	Microwave-assisted combustion.	Shape: different shapes Size:~27 to ~85 nm	MTT assay and cell viability	3T3 cells	The low-toxicity, as-synthesised nanoparticles shown favourable antioxidant and antibacterial properties.	[15]

(Continued)

TABLE 11.1 (Continued)

Nanoparticles Characterisations, Related Biocompatibility and Toxicity Mechanisms

Nanoparticles	Synthesis Method	Physiochemical Characterisations	Biocompatibility Assays	Cells	Results	Ref
ZnO nanoparticles	Green synthesis	Shape: rod Size:20–130 nm	MCF-7 breast cancer cells	MTT assay	Action against breast cancer cells that is cytotoxic. The results showed that increasing the concentration of ZnO NPs from 10 to 50 g/mL significantly reduced the viability of MCF-7 breast cancer cells.	[64]
Chitosancoated iron oxide nanoparticles	Cross linking method	Shape: spherical morphology Size:18 nm	Cancer cell lines (SiHa, HeLa)	MTT assay	The superparamagnetic CS MNPs. On cancer cell lines, these nanoparticles were discovered to be non-cytotoxic (SiHa, HeLa). Magnetic resonance imaging (MRI), magnetic hyperthermia, and targeted medication administration are just a few of the biomedical uses for the produced MNPs.	[65]
Cerium-doped bioactive glass/chitosan/polyethylene oxide composite	Electrospinning technique	Shape: nanofibres Size: 200–850 nm	MTT assay	Fibroblast cells	Tissue engineering and cell culture were in agreement with the findings of the MTT assay, which demonstrated that the presence of 8Ce-BG had a positive influence on the adherence of cells to one another, the multiplication of cells, and the vitality of the cells.	[66]
CeO$_2$-NPs	Biosyntheses by sol-gel method.	Size: ranging from 18 to 60 nm in diameter, with a typical size of approximately 34 nm.	MTT assay	WEHI 164 cancer cell line	CeO$_2$-NPs were found to be low toxic and nontoxic in a range of concentrations ranging from 0.97 to 250 g/mL, as determined by the results of an in vitro cytotoxicity experiment performed on the WEHI 164 cell line. The investigation was carried out on the cell line.	[67]
(Doxorubicin) DOX-loaded ZnO nanoparticles	Co-precipitation	Size:476.4–2.51 nm in	MTT assay	Cancer cells (MCF-7)	Beneficial for the administration of anticancer medication	[68]
Iron-oxide nanoparticles	Chemical precipitation method	Shape: crystalline nature Size:15 nm	MTT assay	A-549 cells	Studies on cell viability indicated the presence of cytotoxic effects caused by PDT in cells that had been treated with PS-CA-SPIONs and then exposed to light. According to the findings, PS-CA-SPIONs have the potential to serve as delivery agents in photodynamic therapy.	[69]

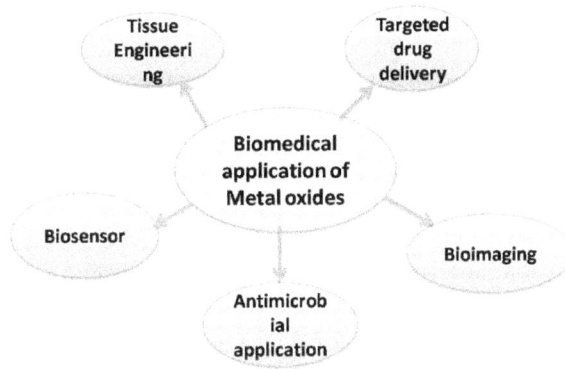

FIGURE 11.3 Biomedical application of metal oxides.

be characterised by having an activity that lasts for an extended period of time, in addition to being stable over time and having a regulated release of the active ingredient. In addition to this, the active ingredient should have a regulated release from the capsules. In addition to this, the carriers for active compounds shouldn't result in any negative side effects, they should be biocompatible with living creatures, and they should be stable over time. In addition to this, the ease of production of drug carriers as well as the cost-effectiveness of these drug carriers has to be taken into consideration as important criteria. The implementation of these systems should not result in the occurrence of any unexpected side effects. Both the capability to deliver the active substance precisely to the destination and the capability to keep a steady concentration of the drug in the blood throughout the course of pharmacotherapy are two extremely essential factors that need to be taken into consideration. It is imperative that these factors be taken into account. In addition to those qualities, the implementation of this kind of product should be uncomplicated and uncomplicated at all times. One of the strategies that is regarded to be one of the most preferred ways is the use of metallic and oxide nanoparticles as carriers for the delivery of active compounds. [This] strategy is one of the approaches that is considered to be one of the most promising approaches. Light is one of the more interesting external stimuli for releasing a drug from a delivery system among the new strategies for treating cancer because of its capability to provide both spatial and temporal control over the release of anticancer agents. This is because light can provide control over the release of anticancer agents. This is because light is an external stimulus, and external stimuli have an effect on internal processes. Additionally, there is a lot of interest in the use of TiO_2 nanostructures as potential carriers for photoactive medicines. In addition to its photoactivity, nanostructures based on TiO_2 offer a wide range of other features that are quite important. Because of these qualities, which include a high surface area, stability, availability, and the possibility for surface modification, they are appropriate carriers for the attachment of a wide variety of different medications [72].

The fundamental objective of the design process that goes into the production of a DDS is to ensure that the medicine is discharged in a controlled manner and delivered to the intended location. Investigations into DDS began with an emphasis on the kinetics of drug release, as well as the sensitivity of polymers to pH and temperature, nasal delivery, and oral drug delivery. These were the primary areas of research in the early days. In recent years, one of the areas that have attracted the attention of a substantial number of researchers is the possibility of utilising nanomaterials in conjunction with various types of DDSs. It is believed that these materials offer a number of benefits, including greater therapeutic qualities and safety, lower toxicity, biocompatibility, and suitability for accurate transport to the cells that are the focus of the treatment. Some of these benefits are listed below. When it comes to the process of formulating a nanofluid for the administration of a medication, the system that is used must be capable of providing drug loading and release qualities, in addition to having a longer shelf life and being biocompatible. One of the most effective cancer therapies involves the utilisation of magnetic NPs, the placement of a magnet in close proximity to the tumour, and the subsequent injection of the medication into the vessels that are located geographically closest to the malignant growth. It is possible that the dynamics of these NPs are affected by the peristaltic motion of the waves that are created in the cone-shaped asymmetric channel walls. The channel walls are responsible for the generation of these waves. When it comes to the process of treating malignant tissues, doing research into the flow of nanofluids, which is influenced by this type of movement, can be of significant benefit. When it comes to therapeutic applications, the most frequent type of magnetic nanoparticle used for magnetic medication administration is known as superparamagnetic iron oxide nanoparticles (SPIOs). Recent studies have provided light on the various applications of SPIOs that can be used in the diagnosis, screening, and treatment of cancer [73]. Extensive research has been conducted in the field of targeted drug delivery on a wide variety of nanoparticles, including CeO_2NPs, IO, AgNPs, AuNPs, TiO_2NPs, and ZnONPs [74,75].

11.3.2 BIOIMAGING

A variety of imaging modalities, such as magnetic resonance imaging (MRI), computed tomography (CT), positron emission tomography (PET), and ultrasound, are utilised in the process of identifying and diagnosing medical conditions. These methods are non-invasive, and some of them can create high-resolution images of the organs inside the body. Contrast chemicals are typically utilised in these bioimaging techniques in order to distinguish between healthy and sick tissue, as well as to identify the organ or tissue of interest.

Imaging with magnetic resonance (often known as MRI) is a strong technology used in medical diagnostics and imaging. It offers excellent resolutions both spatially and temporally and, in contrast to other technologies such as CT scanning, emits no ionising radiation and does not require any sort of intrusive procedure. In actuality, the resolution can be high enough to make it possible to follow the movement of individual cells while they are still in vivo. Additionally, MRI enables three-dimensional imaging on several planes; as a result, a large number of tissue layers can be examined in a single session. There are a number of reviews available on the topic of SPION synthesis for MRI [76]. Contrast agents known as superparamagnetic iron oxide nanoparticles (SPIONs) are widely employed in clinical settings and are considered to be largely nontoxic. SPIONs are beneficial for magnetic resonance imaging (MRI) applications due to their superparamagnetism as well as their high magnetic response in the presence of an applied magnetic field. It is crucial to surface-modify SPIONs with a biocompatible polymer such as dextran, PEG, alginate, starch, chitosan, or d-mannose in order to stable them in solution and to increase their in vivo biocompatibility. Dextran, which has been used in clinical contrast agent formulations, is one of the polymers that is utilised most frequently for this objective and serves as an example. Contrast agents can fall into one of two categories: (i) paramagnetic materials (such as lanthanides and gadolinium), which increase the signal intensity by lowering the T1 relaxation and producing a brighter image; or (ii) superparamagnetic iron oxide nanomaterials, which have an effect on the T2 relaxation [77]. Coating the particles with dextran, PEG, or chitosan can be accomplished either by hydrostatic interactions or through the formation of covalent bonds to the particle surface via an amino-silane intermediate. The SPIONs are often entrapped or encapsulated within a polymer matrix when the matrix is coated with alginate, starch, or another polysaccharide coating. The various encapsulant materials that were described endow the composite particles with a wide variety of colloidal, biological, and chemical properties. These materials also have an effect on the magnetic and MRI properties of the SPION [78]. In the field of medical imaging, magnetic iron oxide nanoparticles/QDs can be used as a dual-mode probe, specifically as an integrated contrast agent for fluorescence imaging and magnetic resonance imaging [78,79]. For instance, based on the magnetic nanoparticles as MRI contrast agents, one of the most exciting applications is the multimodal imaging, which integrates optical imaging (for instance, dyes, QDs, and Au nanoparticles) or positron emission tomography (PET) imaging (for instance, isotopes) with MRI [80, 81].

11.3.3 Biosensors

It is a piece of analytical equipment that is put to use in the process of examining biological samples. A chemical, biological, or biochemical response is converted into an electrical signal via it. There are three fundamental parts that make up a biosensor. (i) the bioreceptor, which typically consists of nucleic acids, enzymes, cells, antibodies, and tissues; (ii) the transducer, which can be electrochemical, optical, electronic, piezoelectric, pyroelectric, or gravimetric; and (iii) the electronic unit, which typically contains the amplifier, the processor, and the display. $CoFe_2O_4$ nanoparticles, in addition to having a magnetic susceptibility that is inherent to the material, have a high surface area, which allows for the possibility of surface functionalisation with antibodies and receptors, and possess excellent absorption capability due to the porous structure of the nanoparticles. The surface of complementary biomolecules makes it easier to perform a wide variety of studies, and their capacity for surface adsorption makes them a strong option for use in bioseparation [82].

11.3.4 Tissue Engineering

Tissue engineering is a multidisciplinary science that combines the principles of engineering, cell biology, and medicine in order to achieve its ultimate goal, which is the regeneration of specific cells and/or functioning tissues. The primary objective of tissue engineering is to create microstructured scaffolds that are biocompatible and can either be two or three dimensional. These scaffolds serve as a foundation upon which cells can adhere, grow, differentiate, and proliferate. Ex vivo or in vivo cell colonisation of these scaffolds is possible. Ex vivo cell colonisation allows for subsequent grafting into the body, while in vivo cell colonisation allows for the regeneration of damaged organs or tissues. It is noteworthy to note that tissue grafts have been advocated for use in the field of drug discovery as a means of reducing the number of animals required for testing the effects of various treatments. This is one way that the number of animals needed for testing can be cut down significantly. The majority of the time, the cells that are going to be cultured are taken from a tissue biopsy, cultured in vitro, and then planted into a porous three-dimensional scaffold. This process is repeated several times. The challenge at hand is to achieve successful cell seeding while also making certain that the cell-cell interactions, in particular those that take place between heterotypic cells and between cell layers, are of a functional nature. In addition, in order to generate a tissue construct that works in a manner that is comparable to how it would under in vivo settings, a three-dimensional structure is required. In order to achieve this objective, a technique that has been dubbed "magnetic force-based tissue engineering" (sometimes abbreviated as "Mag-TE") has been described. In order to endow the cells with magneto-responsive properties, this technique takes use of magnetic nanodevices (such as MCLs, magnetic gelatin NPs, MNP-loaded hydroxyapatite and collagen, etc.) [83]. In green synthesis of metal or metal oxide nanoparticles, the use of gums and naturally occurring carbohydrate polymers leads to the development of polymer-coated NPs that are less toxic and more effective in their ability to inhibit the growth of microorganisms. Green synthesis is also known as sustainable synthesis. Green chemistry may be found here, and it can be used to increase the biocompatibility of nanometals that are intended for use in biomedical applications, such as the treatment of cells and the regeneration of tissue. These applications include cellular treatment and tissue regeneration [84].

11.3.5 Antimicrobial Application

Nanoparticles made of metal oxides have recently gained attention as potential options because to their low toxicity, increased durability, excellent biocompatibility, and increased stability when compared to organic nanoparticles. Metal oxide nanoparticles (NPs) are known to possess powerful antibacterial capabilities [85,86]. Some examples of these nanoparticles are silver oxide (Ag_2O), titanium dioxide (TiO_2), silicon (Si), copper oxide (CuO), zinc oxide (ZnO), gold (Au), calcium oxide (CaO), and magnesium oxide (MgO) [87]. Studies conducted in vitro demonstrated that metal nanoparticles reduced the growth of a variety of microbiological species. Because of its antibacterial activity, zinc oxide is well suited for use in topical treatments such as wound dressings and coatings for a variety of medical equipment to prevent the formation of biofilms [88]. It was reported by Jiang that ZnO NPs that were prepared by biosynthetic and environmentally friendly technology exhibited strong potential for biomedical applications. This is because it possesses excellent anticancer and antibacterial activity, and it is more cost-effective (low priced), nontoxic, and biocompatible than chemical and physical methods [89]. Both gram-negative and gram-positive bacterial strains, such as S. aureus and E. coli, were susceptible to ZnO nanoparticles' potent antibacterial action when they were synthesised using the straightforward and cost-effective sol-gel approach. The ZnO-NPs' ability to inhibit the growth of bacteria was demonstrated by means of agar diffusion experiments. The findings of this study may make it easier to comprehend the mechanism of ZnO NPs [90]. TiO_2 nanoparticles that are green generated and those that are derived chemically are equally effective in killing microorganisms; however, the antibacterial activity of biologically derived NPs is superior. The plant extracts, which act as capping agents, are responsible for the outstanding antibacterial properties of these compounds. Nadeem and his fellow workers [91]. TiO_2 nanoparticles have been created using (i) the use of extract from the Hibiscus flower as a capping agent and (ii) the chemical approach. It has been discovered that the TiO_2 nanoparticles that were capped and stabilised by phenolic and amine moieties of flower extract are smaller in size and more dispersed than the TiO_2 nanoparticles that were prepared by a chemical method. This was discovered by comparing the two types of TiO_2 nanoparticles. In addition, the floral extract-stabilised TiO_2 nanoparticles had significant antibacterial efficacy against pathogenic bacteria, which was on par with that of a conventional antibiotic. When we compare the flower extract-stabilised TiO_2 nanoparticles to the chemically synthesised TiO_2 nanoparticles, we find that the flower extract-stabilised TiO_2 nanoparticles have improved dispersibility, stability, and surface coatings [92,93]. This leads us to the conclusion that the flower extract-stabilised TiO_2 nanoparticles may have potential biomedical applications. Ansari is the first person to report using the leaf extract of lemon grass in the production of Al_2O_3-NPs [94]. This discovery was made by Ansari. The produced NPs appear to be cost-effective, nontoxic, and environmentally benign. Additionally, they demonstrated excellent antibacterial activity against MDR strains of P. aeruginosa, which demonstrates their compatibility for pharmaceutical and other biomedical applications. The current study came to a successful conclusion after optimising the procedure of production of iron oxide nanoparticles from an extract of L. inermis (Henna) that was functionalised by L-tyrosine. The effect of various parameters on the creation of nanoparticles was investigated by altering the concentrations of plant extracts, as well as the pH, temperature, and salt concentrations. Monitoring the absorbance allowed for the recording of the changes. The iron oxide nanoparticles are extremely stable at room temperature, and there is no oxidation that takes place even after they have been stored for a whole month. These nanoparticles have antibacterial action against pathogenic bacteria, and as a result, they have the potential to be used as a powerful source of nanomedicine. In addition, they have a wide variety of additional possible applications in medical and environmental biotechnology [95].

Because secondary metabolites can be discovered in a wide variety of plant extracts, there is the possibility that NPs can be synthesised using a technique that is based on biological processes. Because of the reduction in, and subsequent stabilisation of, the efficiency of secondary metabolites, this potential was made attainable. It is possible to achieve the goals of increasing the quality of ZnO NPs and optimising the reaction conditions in order to achieve improvements in the production of ZnO NPs on a large scale as a result of the extensive information on the formation mechanism of ZnO NPs that has been gathered. This can be done with the help of the information that has been gathered. ZnO NPs can improve agricultural outcomes such as fertility efficiency, an increase in the rate of germination, the development of root systems, the size of fruits, and the sugar and protein content of crops. The amount of biomolecules that are present in the plant extract that is used in the synthesis of ZnO NPs determines, to a large extent, the extent to which ZnO NPs that are synthesised from plant extracts can improve agricultural outcomes. It is able to quickly enter the cell walls of bacteria and generate reactive oxygen species (ROS), which makes it a promising therapeutic agent for the treatment of cancer as well as microbial infections [96,97]. The presence of phytochemicals in the leaf extract itself helps to the development of metal oxide nanoparticles in a number of different ways. One of these ways is by inducing a reaction that involves both oxidation and reduction. Amines and alkanes are two examples of the functional groups of phytochemicals that play a role in the production of nanoparticles. Amines and alkanes are frequently discovered in secondary metabolites such as terpenoids, flavonoids, and alkaloids, to name a few examples of these types of metabolites [98].

Pterocarpus marsupium heartwood extract was used in the phytoassisted synthesis of magnesium-oxide nanoparticles by Ammulu and colleagues. The results showed that the magnesium-oxide nanoparticles have exceptional antioxidant, antibacterial, anti-diabetic, and anti-inflammatory capabilities. The inhibition of beta-amylase was used as a test for the anti-diabetic action of phytoassisted MgO-NPs, and the IC50 value was determined to be 56.32. The albumin denaturation method was used to investigate the anti-inflammatory activity of MgO-NPs; the IC50 value was reported to be 81.69 [99].

11.3.6 PHOTOSENSITISER AGENTS FOR PHOTODYNAMIC THERAPY (PDT)

The photodynamic therapy, often known as PDT, offers significant potential for use in the treatment of cancer [100]. This is an alternative to the conventional method of surgery. It is a novel treatment for a variety of ailments that is activated by light from the outside and involves three primary components: (i) the photosensitiser, often known as PS; (ii) light, typically in the form of a laser; and (iii) oxygen in the tissue. After systemic, local, or topical injection of PS followed by an adequate incubation period, selective illumination of the region of interest is carried out using light of the appropriate wavelength and power. After being exposed to light, PS is able to transfer the absorbed photon energy to the oxygen molecules in its immediate environment, thereby producing ROS such as singlet oxygen or free radicals, which can cause cell death and tissue destruction without causing any harm to the healthy tissue that is adjacent to the affected area. On the other hand, the majority of PSs have drawbacks such as prolonged cutaneous photosensitivity, low water solubility, and insufficient selectivity. Upconversion nanoparticles are the most cutting-edge method, since they are able to transform low-energy radiation into higher-energy emissions, making it possible to perform PDT on deeper tissue layers. Imaging tumours and treating cancer with PDT are two possible clinical applications for such nanoparticles. We may finally be able to bring PDT to the forefront of oncological diagnosis and intervention if we are successful in developing appropriate nanoparticles. Particle beam therapy (PDT) makes excellent use of a wide variety of nanomaterials, including gold, silver, copper, and other nanoparticles based on metals, so that these problems can be solved. In the treatment of a variety of malignancies, including those that can be found in the head and neck, the lungs, and the bladder, as well as certain skin cancers, photodynamic therapy is a technique that is frequently utilised. It has also been used successfully in the treatment of non-cancerous illnesses such as age-related macular degeneration, psoriasis, and atherosclerosis, and it has demonstrated some usefulness in anti-viral therapy including the treatment of herpes. The need for delicate surgery and extended periods of recuperation are reduced, along with the creation of scar tissue and deformity, when a patient undergoes photodynamic therapy. These benefits are enjoyed by both the patient and the treating physician. PDT does have some downsides, the most significant of which is the general photosensitisation of the patient's skin tissue, which occurs as a side effect of the treatment [101]. PDT is unquestionably a path that holds a great deal of potential for the treatment of cancer. However, the use of PDT is complicated by the fact that traditional PSs have a low water solubility and a shallow light penetration, both of which limit the effectiveness of the treatment. Therefore, photodynamic therapy is not yet acknowledged as a primary treatment option [102]. The use of nanoparticles in PDT, and more specifically nanoparticles based on metals, is a very promising strategy for the breakthrough of traditional approaches in the not too distant future. EPR effect can be used to deliver hydrophobic PSs to the locations of certain malignancies using metal-based nanoparticles as carriers. [These nanoparticles] can be employed as carriers of hydrophobic PSs [103].

11.3.7 PHOTOTHERMAL THERAPY (PTT)

Photothermal therapy, often known as PTT, is a technology that is now being investigated for its potential to treat cancer. PTT has received a significant amount of attention in recent years as a result of the unique advantages it offers in comparison with more conventional cancer treatments such as surgery, chemotherapy, and radiotherapy. These advantages include a high level of specificity, minimal invasiveness, and precise spatial–temporal selectivity [104]. A PTT agent is applied during the process of PTT to absorb near-infrared (NIR) light and generate heat in order to destroy cancer cells [105]. In recent years, a great number of novel PTT agents have been produced, including carbon-based compounds, sulphides, and selenides, as well as nanostructures made of gold. The applications of these inorganic nanomaterials in clinical trials are restricted, despite the fact that they have a high photothermal conversion efficiency. This is mostly due to the biotoxicity, which has the ability to cause damage to cells, tissues, and organs in the human body [106]. Titanium oxides have not been documented for use in PTT applications, primarily because the common titanium oxide materials have a low NIR absorption. TiO_2, the most well-known of these types of materials, has been the subject of extensive research for use in electronics, as a catalyst, and in solar-cell applications. Despite this, TiO_2 materials typically have a bandgap that is greater than 3.0 eV and can only absorb ultraviolet (UV) light. Doping and defect engineering were recently used to successfully tailor the bandgap of TiO_2 to visible light; nevertheless, such materials are unable to absorb NIR light and have not been explored for photothermal applications [107].

Experiments in vitro and in vivo using nanoparticles of different morphologies of iron oxides have been very successful. These nanoparticles have shown the potential to kill cancer cells in vitro and to considerably reduce the size of a tumour in vivo. Recent research has investigated the applicability of iron oxide nanoparticles in PTT, where it has been found to have a number of distinct advantages over AuNPs [108].

11.4 LIMITATIONS, CHALLENGES, AND FUTURE SCOPE

In order to decide how much accuracy is required in sample characterisations and product consistency during the process of formulation development, it will be important to know the mechanism behind the toxicity of metal oxide nanoparticles, which is caused by changes in the shape, size, and surface properties of the particles [109]. Other obstacles that need to be addressed include issues relating to biocompatibility, toxicological, and immunological characteristics [110].

Therefore, future research should concentrate on the simple fabrication of low-toxicity anisotropic nanoparticles with different chemical compositions. This will allow the physicochemical properties to be optimised, which will in turn reduce the required dosage not only of the nanoparticles but also of the therapeutic drugs, as well as the side effects associated with them [111]. The study of metal oxide nanoparticles in the future would therefore be fascinating due to the attractive properties that it possesses, particularly in light of the fact that other areas of biomedical science are currently undergoing development. It is possible that the most important biomedical applications for these particles will be in the field of healthcare for older people, specifically in the treatment of conditions that affect the musculoskeletal system. Many of these conditions are characterised by severe inflammation, incapacity, and discomfort. Improving the ability to control inflammation is an important aim, and magnetic nanoparticles may contribute to this goal to a significant level [112]. Nevertheless, additional research is required into a number of significant concerns about metal oxide nanoparticles. These concerns include the following: (i) there is no comparative analysis of its biological advantages in comparison to those of other metal nanoparticles; (ii) the limitations of metal oxide nanoparticles' toxicity towards biological systems continue to be a contentious subject in recent research; (iii) There has not been enough evidence-based, randomised study done to particularly investigate therapeutic functions in enhancing anticancer, antibacterial, anti-inflammatory, and anti-diabetic effects [113].

11.5 CONCLUSION

Metal oxides have been considered as suitable nanomaterials because they have been found to be applicable in various aspects of biomedical science and biotechnology. This is in consideration of the fact that the field of biomedical science is expanding at a rapid rate, which necessitates the use of a new class of materials. When it comes to the production of metal oxide NPs, the most difficult challenge that still needs to be overcome is adapting these NPs, in terms of shape, size, and surface modification, so that they are suitable for use in biomedical applications while also ensuring that they are biocompatible and nontoxic. Despite this, there is a significant possibility that metal oxide nanoparticles may find widespread usage in the field of biomedicine.

REFERENCES

1. Nikolova M. P., and Chavali M. S. (2020). Metal oxide nanoparticles as biomedical materials biomimetics, 5, 27–74, doi:10.3390/biomimetics5020027.
2. Oun A., Shankar S., and Rhim J. W. (2019). Multifunctional nanocellulose/metal and metal oxide nanoparticle hybrid nanomaterial. *Critical Reviews in Food Science and Nutrition*, 60, 1–26, doi:10.10 80/10408398.2018.1536966.
3. Barui A. K., Kotcherlakota R., and Patra C. R. (2018). *Biomedical Applications of Zinc Oxide Nanoparticles*. Elsevier Inc.
4. Vijayakumar N., Bhuvaneshwari V. K., Ayyadurai G. K., Jayaprakash R., Gopinath K., Nicoletti M., Alarifi S., and Govindarajan M. (2022). Green synthesis of zinc oxide nanoparticles using Anoectochilus elatus, and their biomedical applications. *Journal of Biological Sciences*, 29(4), 2270–2279, doi:10.1016/j.sjbs.2021.11.065.
5. Rasmussen J. W., Martinez E., Louka P., and Wingett D. G. (2010). Zinc oxide nanoparticles for selective destruction of tumor cells and potential for drug delivery applications. *Expert Opinion in Drug Delivery*, 7(9), 1063–1077, doi:10.1517/17425247.2010.502560.
6. Hassanpour P. P., Yunes E. K., Abbas A., Abolfazl D. S., Nasibova A. N., Khalilov R. and Taras K. (2018). Biomedical applications of aluminium oxide nanoparticles. *Micro & Nano Letters*, 13(9), 1227–1231, doi:10.1049/mnl.2018.5070.
7. Liua Y., and Shib J. (2019). Review antioxidative nanomaterials and biomedical applications. *Nano Today*, 27, 1–32, doi:10.1016/j.nantod.2019.05.008.
8. Wang Z., and Tang M., (2020). Research progress on toxicity, function, and mechanism of metal oxide nanoparticles on vascular endothelial cells. *Journal of Applied Toxicology*, 4121, 1–18, doi:10.1002/jat.4121.
9. Waris A., Din M., Ali A., Ali M., Afridi S., Baset A., & Khan A. U. (2020). A comprehensive review of green synthesis of copper oxide nanoparticles and their diverse biomedical applications. *Inorganic Chemistry Communications*, 123, 108369–108409, doi:10.1016/j.inoche.2020.108369.
10. Foggi C. C., Fabbro M. T., Santos L. P. S., De S., Yuri V. B., Vergani C. E., Machado A. L., Cordoncillo E., Andres J., and Longo E. (2017). Synthesis and evaluation of α-Ag2 WO4 as novel antifungal agent. *Chemical Physics Letters*, 674, 125–129, doi:10.1016/j.cplett.2017.02.067
11. Namara K. M., and Syed A. M. (2016). To fail, nanoparticles in biomedical applications. *Advances in Physics: X*, 2(1), 54–88, doi:10.1080/23746149.2016.1254570.
12. Teow S. Y., Wong M. M., Yap H., Peh S., and Shameli K. (2018). Bactericidal properties of plants-derived metal and metal oxide nanoparticles (NPs). *Molecules*, 23, 1366–1391, doi:10.3390/molecules23051366.
13. Laurent S., Boutry S., and Muller R.N. (2018). *Iron Oxide Nanoparticles for Biomedical Applications Metal Oxide Particles and Their, Prospects for Applications*. Elsevier Ltd.

14. Kapoor P. N., Bhag A. K., Mulukutla R. S., and Klabunde K. J. (2008). Mixed metal oxide nanoparticles. *Encyclopedia of Nanoscience and Nanotechnology*, 6(2), 2280–2290, doi:10.1201/NOE0849396397. ch197.

15. Seabra A. B., and Duran N. (2015). Nanotoxicology of metal oxide nanoparticles metals, 5, 934–975, doi:10.3390/met5020934.

16. Murthy S., Effiong P., and Fei C. C. (2020). *Metal Oxide Powder Technologies, Metal Oxide Nanoparticles in Biomedical Applications*. Elsevier Inc.

17. Mirzaei H., and Darroudi M. (2016). Zinc oxide nanoparticles: Biological synthesis and biomedical application. *Ceramic International*, 8842(16), 31814–31838, doi:10.1016/j.ceramint.2016.10.051.

18. Hamed N., Marziyeh S., Kheiri M. H., Hossein D., and Soodabeh D., (2017). Preparation of magnetic albumin nanoparticles via a simple and one-pot desolvation and co-precipitation method for medical and pharmaceutical applications. *International Journal of Biological Macromolecules*, 108, 909–915, doi:10.1016/j.ijbiomac.2017.10.180.

19. Azizi S., Mohamad R., and Shahri M. M., (2017). Green microwave-assisted combustion synthesis of zinc oxide nanoparticles with Citrullus colocynthis (L.) Schrad: Characterization and Biomedical Applications. *Molecules*, 22, 301–313, doi:10.3390/molecules22020301.

20. Francesca P., Mariangela B., Aurelio L. B., Franco P., Mariateresa M., Alessandra P., Giuseppa G. M., Giorgio L., Elisa N., Carlo C., and Paola G. (2014). Biodistribution and acute toxicity of a nanofluid containing manganese iron oxide nanoparticles produced by a mechanochemical process. *International Journal of Nanomedicine*, 9, 1919–1929, doi:10.2147/IJN.S56394.

21. Kim D.K., Zhang Y., Voit W., Rao K.V., Kehr J., Bjelke B., and Muhammed M. (2001). Superparamagnetic iron oxide nanoparticles for bio-medical applications, 44, 1713–1717, doi:10.1016/s1359-6462(01)00870-3.

22. Shafiee P., Nafchi R., Eskandarinezhad M., Mahmoudi S., and Ahmadi E. (2021). Sol-gel zinc oxide nanoparticles: Advances in synthesis and applications. *Synthesis and Sintering*, 1, 242–254, doi:10.53063/synsint.2021.1477.

23. Spanhel L., and Anderson M. A. (1991). Semiconductor clusters in the sol-gel process-quantized aggregation, gelation, and crystal-growth in concentrated ZnO colloids. *Journal of the American Chemical Society*, 113, 2826–2833, doi:10.1021/ja00008a004.

24. Renuka N. K. (2012). Structural characteristics of quantum-size ceria nano particles synthesized via simple ammonia precipitation. *Journal of Alloys and Compounds*, 513, 230–235, doi:10.1016/j.jallcom.2011.10.027.

25. Rheima A. M., Anber A. A., Abdullah H. I., and Ismail A. H. (2021). Synthesis of alpha-gamma aluminum oxide nanocomposite via electrochemical method for antibacterial activity. *Nano Biomedicine and Engineering*, 13, 1–5, doi:10.5101/nbe.v13i1.p1-5.

26. Hamrayev H., Shameli K., and Yusefi M. (2021). Preparation of zinc oxide nanoparticles and its cancer treatment effects: A review paper. *Journal of Advance Research in Micro and Nano Engineering*, 2, 1–11, https://akademiabaru.com/submit/index.php/armne/article/view/2962.

27. Ivanov V. K., Polezhaeva O. S., Shaporev A. S., Baranchikov A. E., Shcherbakov A. B., and Usatenko A. V. (2010). Synthesis and thermal stability of nanocrystalline ceria sols stabilized by citricand polyacrylic acids. *Russian Journal of Inorganic Chemistry*, 55(3), 328–332, doi:10.1134/s0036023610030046.

28. Akintelu S. A., and Folorunso A. F. (2020). A review on green synthesis of zinc oxide nanoparticles using plant extracts and its biomedical applications. *Bio Nano Science*, 1, 1–13, doi:10.1007/s12668-020-00774-6.

29. Noqta O. A., Aziz A. A., Usman I. A., and Bououdina M. (2019). Recent advances in iron oxide nanoparticles (IONPs): Synthesis and surface modification for biomedical applications. *Journal of Superconductivity and Novel Magnetism*, 32, 779–795, doi:10.1007/s10948-018-4939-6.

30. Hamed M., and Majid D., (2017). Zinc oxide nanoparticles: Biological synthesis and biomedical applications. *Ceramics International*, 43, 907–914, doi:10.1016/j.ceramint.2016.10.051.

31. Anik M. I., Hossain M. K., Hossain I., Mahfuz A. M. U. B. Rahman M. T., and Ahmed I. (2021). Recent progress of magnetic nanoparticles in biomedical applications: A review. *Nano Select*, 2, 1146–1186, doi:10.1002/nano.202000162.

32. Saifi M. A., Seal S., and Godugu C. (2021). Nanoceria the versatile nanoparticles: Promising biomedical applications. *Journal of Controlled Release*, 338, 164–189, doi:10.1016/j.jconrel.2021.08.033.

33. Andreescu S., Ornatska M., Erlichman J. S., Estevez A., and Leiter J. C. (2012). *Biomedical Applications of Metal Oxide Nanoparticles, Fine Particles in Medicine and Pharmacy*. Springer, Matijević (ed.).

34. Zhang Y., Nayak T. R., Hong H., and Cai W. (2013). Biomedical applications of zinc oxide nanomaterials. *Current Molecular Medicine*, 13(10), 1633–1645, doi:10.2174/1566524013666131111130058.

35. Bloemen M., Brullot W., Luong T. T., Geukens N., and Gils A. (2012). Improved functionalization of oleic acid-coated iron oxide nanoparticles for biomedical applications. *Journal of Nanoparticle Research*, 14, 1100–1110, doi:10.1007/s11051-012-1100-5.

36. García M. C., Torres J., Antonella V., Córdoba D., Longhi M., and Uberman P. M. (2022). *Metal Oxides for Biomedical and Biosensor Application, Drug Delivery Using Metal Oxide Nanoparticles*. Elsevier.

37. Akintelu S. A., and Folorunso A. S. (2020). A review on green synthesis of zinc oxide nanoparticles using plant extracts and its biomedical applications. *Bio Nano Science*, doi:10.1007/s12668-020-00774-6.

38. Mutukwa D., Taziwa R., and Khoteseng L. E. (2022). A review of the green synthesis of ZnO nanoparticles utilising Southern African indigenous medicinal plants. *Nanomaterials*, 12(19), 3456, doi: 10.3390/nano12193456.

39. Scheinfein, M. R., Unguris, J. Kelley, M. H., Pierce, D. T., and Celotta, R. J. (1990). Scanning electron microscopy with polarization analysis (SEMPA), 61, 2501–2529, doi:10.1063/1.1141908

40. Oatley C. W. (1982). The early history of the scanning electron microscope. *Journal of Applied Physics*, 53, 1–14. doi:10.1063/1.331666.

41. Seiler H. (1983). Secondary electron emission in the scanning electron microscope. *Journal of Applied Physics*, 54, 1–19, doi:10.1063/1.332840.

42. Tánaka K. (1989). High resolution scanning electron microscopy of the cell. 65(2), 1–10, doi:10.1111/j.1768-322X.1989.tb00777.x.

43. Franken L. E., Kay G., Egbert J B., and Stuart C. A. (2020). A technical introduction to transmission electron microscopy for soft matter: Imaging, possibilities, choices, and technical developments. *Small*, 16, 1906198–1906213, doi:10.1002/smll.201906198-13.

44. Malatesta M. (2016). Transmission electron microscopy for nanomedicine: Novel applications for long-established techniques. *European Journal of Histochemistry*, 60(4), 2751–2756. doi:10.4081/ejh.2016.2751.

45. De Aza P. N., Luklinska Z. B., Anseau M., Guitian F., and De Aza S. (1996). Morphological studies of pseudowollastonite for biomedical application. *Journal of Microscopy*, 182(1), 24–31, doi:10.1111/j.1365-2818.1996.tb04794.x.

46. Harris D. C., and Bertolucci M.D. (1989). *Symmetry and Spectroscopy, an Introduction to Vibrational and Electronic Spectroscopy*. Dover Publications, Inc., New York.

47. Lee W., Reece P., Marchington R., Metzger N., and Dholakia K. (2007). Construction and calibration of an optical trap on a fluorescence optical microscope. *Nature Protocols*, 2, 3226–323, doi:10.1038/nprot.2007.446.

48. Kogan M. J,. Ivonne O., Leticia H., Ariel G. R., Cruz, Luis J., and Fernando A. (2007). Peptides and metallic nanoparticles for biomedical applications. *Nanomedicine*, 2(3), 287–306, doi:10.2217/17435889.2.3.287.

49. Murthy S. (2020). *Metal Oxide Powder Technologie Metal Oxide Nanoparticles in Biomedical Applications*. Elsevier.

50. Remya N. S., Syama S., Sabareeswaran A., and Mohanan P. V. (2016). Toxicity, toxicokinetics and biodistribution of dextran stabilized iron oxide nanoparticles for biomedical applications. *International Journal of Pharmaceutics*, 511, 586–98, doi: 10.1016/j.ijpharm.2016.06.119.

51. Jeng H. A., and Swanson J. (2006). Toxicity of metal oxide nanoparticles in mammalian cells. *Journal of Environmental Science and Health*, 41, 2699–2711, doi:10.1080/10934520600966177.

52. Kalpana V. N., and Rajeswari D. V. (2018). A review on green synthesis, biomedical applications, and toxicity studies of ZnO NPs. *Bioinorganic Chemistry and Applications*, 1–12, doi:10.1155/2018/3569758.

53. Zhang X. Q., Yin L. H., Tang M., and Pu Y. P. (2011). ZnO, TiO_2, SiO_2, and Al_2O_3 nanoparticles-induced toxic effects on human fetal lung fibroblasts. *Biomedical and Environmental Sciences*, 24, 661-669, doi:10.3967/0895-3988.2011.06.011.

54. Waris A., Din M., Ali A., Afridi S., Baset A., Khan A. U., and Ali M. (2021). Green fabrication of Co and Co3O4 nanoparticles and their biomedical applications: A review. *Open Life Sciences*, 16, 14–30, doi:10.1515/biol-2021-0003.

55. Karakoti A. S., Monteiro-Riviere N. A., Aggarwal R., Davis J. P., Narayan R. J., Self W. T., Ginnis J. M., and Seal S. (2008). Nanoceriaas antioxidant: Synthesis and biomedical applications, 60(3), 33–37, doi:10.1007/s11837-008-0029-8.

56. Caputo F., Mameli M., and Sienkiewicz A. (2017). A novel synthetic approach of cerium oxide nanoparticles with improved biomedical activity. *Scientific Reports*, 7, 4636–4659, doi:10.1038/s41598-017-04098-6

57. Ivanova O. S., Shekunova T. O., Ivanov V. K., Shcherbakov A. B., Popov A. L., Davydova G. A., Selezneva I. I., Kopitsa G. P., and Tretyakov Y. D. (2011). One stage synthesis of ceria colloid solutions for biomedical use. *Chemistry*, 437(2), 103–106, doi:10.1134/S0012500811040070.

58. Annapoorani A., Koodaligam A., Beulaja M., Gowrikumar S., Chitra P. Stephen A., Prabhu N. M., Janarthanan S. G., and Manikandan R. (2021). Eco-friendly synthesis of zinc oxide nanoparticles using Rivina humilis leaf extract and their biomedical applications. *Process Biochemistry*, 112, 192–202, doi:10.1016/j.procbio.2021.11.022

59. Rajashekara S., Shrivastava A., Sumhitha S., and Kumari S. (2020). Biomedical applications of biogenic zinc oxide nanoparticles manufactured from leaf extracts of Calotropis gigantea (L.) Dryand. *BioNanoScience*, 10, 654–671, doi:10.1007/s12668-020-00746-w.

60. Rasheed T., Bilal M., Iqbal H. M. N., and Li C. (2017). Green biosynthesis of silver nanoparticles using leaves extract of Artemisia vulgaris and their potential biomedical applications. *Bionterfaces*, 158, 408–415, doi:10.1016/j.colsurfb.2017.07.020.

61. Alireza A., Mansoureh K., Danial T., Arezoo R., Khorsand Z. A., and Majid D. (2017). Zinc-doped cerium oxide nanoparticles: Sol-gel synthesis, characterization, and investigation of their in vitro cytotoxicity effects. *Journal of Molecular Structure*, 1149, 771–776, doi:10.1016/j.molstruc.2017.08.069.

62. Motaharesadat H., Issa A., Mohammad M., and Masoud M. (2020). Gel synthesis, physico-chemical and biological characterization of cerium oxide/polyallylamine nanoparticles. *Polymers*, 12(7), 1444–1458, doi:10.3390/polym12071444

63. Gharibshahian M., Nourbakhsh M. S., and Mirzaee O. (2018). Evaluation of the superparamagnetic and biological properties of microwave assisted synthesized Zn & Cd doped CoFe2O4 nanoparticles via Pechini sol-gel method. *Journal of Sol-Gel Science and Technology*, 85, 1–9, doi:10.1007/s10971-017-4570-1.

64. Devaraj B., and Bhuvaneshwari V. (2019). Synthesis of zinc oxide nanoparticles (ZnO NPs) using pure bioflavonoid rutin and their biomedical applications: Antibacterial, antioxidant and cytotoxic activities. *Research on Chemical Intermediates*, 45, 2065–2078, doi:10.1007/s11164-018-03717-9.

65. Unsoy G., Yalcin S., Khodadust R., Gunduz G., and Gunduz U. (2012). Synthesis optimization and characterization of chitosan-coated iron oxide nanoparticles produced for biomedical applications. *Journal of Nanoparticle Research*, 14, 964–977, doi:10.1007/s11051-012-0964-8.

66. Alieza S., Razaghian A. A., Amirhossein M., and Masoud M. (2020). Synthesis and characterization of electrospun cerium-doped bioactive glass/chitosan/polyethylene oxide composite scaffolds for tissue engineering applications. *Ceramics International*, 47, 1–12, doi:10.1016/j.ceramint.2020.08.130.

67. Nourmohammadi E., Oskuee K. R., Hasanzadeh L., Mohajeri M., Hashemzadeh A., Rezayi M., and Darroudi M. (2018). Cytotoxic activity of greener synthesis of cerium oxide nanoparticles using carrageenan towards a WEHI 164 cancer cell line, 44(16), 19570–19575,

68. Sharma H., Kumar, Choudhary K., Mishra C., Pawan K., and Vaidya B. (2014). Development and characterization of metal oxide nanoparticles for the delivery of anticancer drug. *Artificial Cells, Nanomedicine, and Biotechnology*, 44, 672–679, doi:10.3109/21691401.2014.978980.

69. Kumar P., Agnihotri S., and Roy I. (2016). Synthesis of photoactive SPIONs doped with visible light activated photosensitizer.

70. Xiaoying C., Zhangping S., Huanran Z., and Saeid O. (2019). Effect of metal atoms on the electronic properties of metal oxide nanoclusters for use in drug delivery applications: A density functional theory study. *Molecular Physics*, 118, 1–9, doi:10.1080/00268976.2019.1692150.

71. Xiaoying C., Zhangping S., Huanran Z., and Saeid O. (2019). Effect of metal atoms on the electronic properties of metal oxide nanoclusters for use in drug delivery applications: A density functional theory study. *Molecular Physics*, 118, 1–9, doi:10.1080/00268976.2019.1692150.

72. Jafari S., Mahyad B., Hashemzadeh H., Janfaza S., Gholikhani T., and Tayebi L. (2020). Biomedical Applications of TiO$_2$ Nanostructures: Recent Advances. *International Journal of Nanomedicine*, 15, 3447–3470, doi:10.2147/IJN.S249441.

73. Mojgan S., Mohadeseh A., Alibakhsh K., Ali R. R., and Zahra T. (2020). Role of nanofluids in drug delivery and biomedical technology: Methods and applications. *Nanotechnology, Science and Applications*, 13, 47–59, doi:10.2147/NSA.S260374.

74. Liying H. E., Yumin S. U., Lanhong J., and Shikao S. H. I. (2015). Recent advances of cerium oxide nanoparticles in synthesis, luminescence and biomedical studies: A review, *Journal of Rare Earths*, 33, 791–799, doi:10.1016/S1002-0721(14)60486-5.

75. Canaparo R., Foglietta F., Limongi T., and Serpe L. (2020). Biomedical applications of reactive oxygen species generation by metal nanoparticles. *Materials*, 14, 53–64. doi:10.3390/ma14010053

76. Lodhia J., Mandarano G., Ferris N.J., Eu P., and Cowell S.F. (2010). Development and use of iron oxide nanoparticles (Part 1): Synthesis of iron oxide nanoparticles for MRI. *Biomedical Imaging and Intervention Journal*, 6, 1–11, doi:10.2349/biij.6.2.e12.

77. Reddy K. R., Reddy P. A., Reddy C. V., Shetti N. P., Babu B., Ravindranadh K., Shankare M.V., Reddy M. C., Sonig S. K., and Naveen S. (2019). Functionalized magnetic nanoparticles/biopolymer hybrids: Synthesis methods, properties and biomedical applications, 46, 227–254, doi:10.1016/bs.mim.2019.04.005 1-28.

78. Serge Y., Tim L., Perry E., and Frank G. (2013). Superparamagnetic iron oxide nanoparticles (SPIONs): Synthesis and surface modification techniques for use with MRI and other biomedical applications. *Current Pharmaceutical Design*, 19, 493–509, doi:10.2174/138161213804143707.

79. Liu F., Laurent S., Fattahi H., Elst, L. V., and Muller R. N. (2011). Superparamagnetic nanosystems based on iron oxide nanoparticles for biomedical imaging, 6, 519–528, doi:10.2217/nnm.11.16.

80. Oh J. K., and Park J. M. (2011). Iron oxide-based superparamagnetic polymeric nanomaterials: Design, preparation, and biomedical application, 36, 168–189, doi:10.1016/j.progpolymsci.2010.08.005.

81. Gao J., Gu H., and Xu B. (2009). Multifunctional magnetic nanoparticles: Design, synthesis, and biomedical applications. *Accounts of Chemical Research*, 42, 1097–1107, doi:10.1021/ar9000026.

82. Srinivasan S. Y., Paknikar K. M, Gajbhiye D., and Bodas V. (2018). Applications of cobalt ferrite nanoparticles in biomedical nanotechnology. *Nanomedicine*, 13, 1221–1238, doi:10.2217/nnm-2017-0379 10.

83. Reddy H. L., Arias J. L., Nicolas J., and Couvreur P. (2012). Magnetic nanoparticles: Design and characterization, toxicity and biocompatibility, pharmaceutical and biomedical applications. *Chemical Reviews*, 112(11), 5818–5878, doi:10.1021/cr300068p.

84. Makvandi P., Yu W. C., Zare E. N., Borzacchiello A., Niu, L.N., and Tay F. R. (2020). Metal-based nanomaterials in biomedical applications: Antimicrobial activity and cytotoxicity aspects. *Advanced Functional Materials*, 30, 1–40, doi:10.1002/adfm.201910021.

85. Sadiq I. M., Chowdhury B., Chandrasekaran N., and Mukherjee A. (2009). Antimicrobial sensitivity of Escherichia coli to alumina nanoparticles. *Nanomedicine: Nanotechnology, Biology and Medicine*, 5, 282–286, doi:10.1016/j.nano.2009.01.002.

86. Mishra P. K., Mishra H., Ekielski A., Talegaonkar S., and Vaidya B. (2017). Zinc oxide nanoparticles: A promising nanomaterial for biomedical applications. *Drug Discovery Today*, 22, 1825–1834, doi:10.1016/j.drudis.2017.08.006.

87. Dizaj S. M., Lotfipour F., Jalali M. B., Zarrintan M. H., and Adibkia K. (2014). Antimicrobial activity of the metals and metal oxides nanoparticles. *Materials Science & Engineering C*, 44, 278–284, doi: 10.1016/j.msec.2014.08.031.

88. Denisa F., Ficai D., Oprea O., Ficai A., and Maria H. (2014). Metal oxide nanoparticles: Potential uses in biomedical applications. *Current Proteomics*, 11, 139–149. doi:10.2174/1570164611102140917122838.

89. Jiang J., Pi J., and Cai J., (2018). The advancing of zinc oxide nanoparticles for biomedical applications. *Bioinorganic Chemistry and Applications*, 1–18, doi:10.1155/2018/1062562.

90. Albukhaty S., Al-Karagoly H., and Dragh M. A. (2020). Synthesis of zinc oxide nanoparticles and evaluated its activity against bacterial isolates. *Journal of Biotech Research*, 11, 47–53.

91. Muhammad N., Duangjai T., Christophe H., Haider A. B., Hashmi S. S., Waqar A., and Adnan Z. (2018). The current trends in the green syntheses of titanium oxide nanoparticles and their applications. *Green Chemistry Letters and Reviews*, 11, 492–502, doi:10.1080/17518253.2018.1538430.

92. Kumar P. S. M., Francis A. P., and Devasena T. Biosynthesized and chemically synthesized titania nanoparticles: Comparative analysis of antibacterial activity. *Journal of Environmental Nanotechnology*, 3, 73–81, doi:10.13074/jent.2014.09.143098.

93. Rajeshkumar S., Santhoshkumar J., Jule L. T., and Ramaswamy K. (2021). Phytosynthesis of titanium dioxide nanoparticles using king of bitter Andrographis paniculata and its embryonic toxicology evaluation and biomedical potential. *Bioinorganic Chemistry and Applications*, Article ID 6267634, 11 pages, doi:10.1155/2021/6267634.

94. Ansari M. A., Khan H. M., Alzohairy M. A., Jalal M., Ali S. G., Pal R., and Musarrat J. (2015). Green synthesis of Al_2O_3 nanoparticles and their bactericidal potential against clinical isolates of multi-drug resistant Pseudomonas aeruginosa. *World Journal of Microbiology and Biotechnology*, 31, 153–164, doi:10.1007/s11274-014-1757-2.

95. Chauhan S., and Upadhyay L. S. B. (2019). Biosynthesis of iron oxide nanoparticles using plant derivatives of Lawsonia inermis (Henna) and its surface modification for biomedical application. *Nanotechnology for Environmental Engineering*, 4, 8–12, doi:10.1007/s41204-019-0055-5.

96. Adewale A. S., and Folorunso A. S. (2020). A review on green synthesis of zinc oxide nanoparticles using plant extracts and its biomedical applications. *BioNanoScience*, 10, 848–863, doi:10.1007/s12668-020-00774-6.

97. Ansari M. A., Baykal A. A., and Rehman S. S. (2018). Synthesis and characterization of antibacterial activity of spinel chromium-substituted copper ferrite nanoparticles for biomedical application. *Journal of Inorganic and Organometallic Polymers and Materials*, 28, 2316–2327, doi:10.1007/s10904-018-0889-5.

98. Santhoshkumar J., Kumar S V., and Rajeshkumar S. (2017). Synthesis of zinc oxide nanoparticles using plant leaf extract against urinary tract infection pathogen. *Resource-Efficient Technologies*, 3, 459–465. doi:10.1016/j.reffit.2017.05.001.

99. Manne A. A., Viswanath K. V., Giduturi A. K., Vemuri P. K., Mangamuri U., and Poda S. (2021). Phytoassisted synthesis of magnesium oxide nanoparticles from Pterocarpus marsupium rox.b heartwood extract and its biomedical applications. *Journal of Genetic Engineering and Biotechnology*, 19, 21–49, doi:10.1186/s43141-021-00119-0.

100. Chen Z. A., Kuthati Y., Kankala R. K., Chang Y. C., Liu C. L., Weng C. F., Mou C. Y., and Lee C. H. (2015). Encapsulation of palladium porphyrin photosensitizer in layered metal oxide nanoparticles for photodynamic therapy against skin melanoma. *Science and Technology of Advanced Materials*, 16, 54205–54213, doi:10.1088/1468-6996/16/5/054205.

101. Bechet D., Couleaud P., Frochot C., Viriot M. L., Guillemin F., and Muriel B. H. (2008). Nanoparticles as vehicles for delivery of photodynamic therapy agents. *Trends in Biotechnology*, 26, 612–621, doi:10.1016/j.tibtech.2008.07.007.

102. Josefsen L. B., and Boyle R. W. (2008). Photodynamic therapy and the development of metal-based photosensitisers. *Metal-Based Drugs*, 1–23, doi:10.1155/2008/276109.

103. Jingyao S., Semen K., Ying L., Yao H., Daming W., and Zhaogang Y. (2018). Recent progress in metal-based nanoparticles mediated photodynamic therapy. *Molecules*, 23, 1704–1727, doi:10.3390/molecules23071704.

104. Jiang W., Fu Q., and Wei H. (2019). TiN nanoparticles: Synthesis and application as near-infrared photothermal agents for cancer therapy. *Journal of Materials Science*, 54, 5743–5756, doi:10.1007/s10853-018-03272-z.

105. Ji X., Shao R., Elliott A.M., Stafford R.J., Esparza-Coss E., Bankson J.A., Liang G., Luo Z.P., Park K., and Markert J.T. (2007). Bifunctional gold nanoshells with a superparamagnetic iron oxide-silica core suitable for both MR imaging and photothermal therapy. *Journal of Physical Chemistry C: Nanomaterials and Interfaces*, 111, 6245–6251, doi:10.1021/jp0702245.

106. Hu W., Shi J., Lv W., Jia X., and Ariga K. (2022). Nanomaterials and their composite scaffolds for photothermal therapy and tissue engineering applications. *Science and Technology of Advanced Materials*, 23, 393–412, doi:10.1080/14686996.2021.1924044.

107. Ou G., Li Z., Li D., Cheng L., Liu Z., and Wu H. (2016). Photothermal therapy by using titanium oxide nanoparticles. *Nano Research*, 9, 1236–1243, doi:10.1007/s12274-016-1019-8.

108. Mona L. P., Songca S. P., and Ajibade P. A. (2022). Synthesis and encapsulation of iron oxide nanorods for application in magnetic hyperthermia and photothermal therapy. *Nanotechnology Reviews*, 11, 176–190, doi:10.1515/ntrev-2022-0011.

109. Gupta A. K., Naregalkar R. R., Vaidy V. D., and M. Gupta, (2007). *Nanomedicine*, 2, 23–39, doi:10.2217/17435889.2.1.23.

110. Arias L. S., Pessan J. P., Vieira A. P. M., Lima T. M. T., Botazzo A. C., and Monteito D. R. (2018). Iron oxide nanoparticles for biomedical applications: A perspective on synthesis. *Drugs, Antimicrobial Activity, and Toxicity, Antibiotics*, 7, 46–78, doi:10.3390/antibiotics7020046.

111. Andrade R. G. D., Veloso S. R. S., and Castanheira E. M. S. (2020). Shape anisotropic iron oxide-based magnetic nanoparticles: Synthesis and biomedical applications. *International Journal of Molecular Sciences*, 21, 16–25, doi:10.3390/ijms21072455.

112. Guptaa A. K., and Gupta M. (2005) Synthesis and surface engineering of iron oxide nanoparticles for biomedical applications. *Biomaterials*, 26, 3995–4021, doi:10.1016/j.biomaterials.2004.10.012.

113. Jiang J., Pi J., and Cai J. (2018) The advancing of zinc oxide nanoparticles for biomedical applications, 1–18, doi:10.1155/2018/1062562.

12 Metal Oxides
Anti-Counterfeiting

V. S. Singh and S. R. Dhakate
CSIR – National Physical Laboratory

CONTENTS

12.1 INTRODUCTION

Duplicating something genuine with an intention to pirate, counterfeit or replace and reduce the value of the legal one comes under counterfeiting issues. Usually, the word counterfeit is used for fake banknotes and documents. To counter such forgeries, various technological ways of ANTC are implemented these days.

12.1.1 TYPES OF COUNTERFEITING

1. **Fake Brand Names and Trademarks:** Imitation of branded products such as clothes, shoes, bags, medicines and food items including household plasticwares may bear the duplicate names or marks, of questionable quality. Duplications of such branded and costly products are done due to lack of availability of enough money and appropriate quality of products to economically backward countries in the world. Figure 12.1a shows the photo

DOI: 10.1201/9781003366232-12

of one such popular brand counterfeited by producing almost same type of logo in order to make an elusive version of the original one.

Such frauds in the field of pharmaceuticals are also gaining potential every day, posing serious threats to millions of life.

2. **Fake Currency, Stamps and Documents:** Forgery of government money generally of paper notes and of confidential documents such as intellectual property rights, patents and copyrights agreements are done worldwide. Figure 12.1b shows the copied or a fake sample of the original Rs. 500 note. Further, illegal postage stamps applied to these government documents encourage the impulse of such type of fraudulent activities even more.

3. **Art works:** similarly as above, famous artworks (such as the one shown in Figure 12.1c) have been copied several times for fulfilling profitable interests.

12.1.2 Solutions to Counter-Counterfeiting through Various Technological Methods

Recently, EUIPO and European Observatory on Infringements of Intellectual Property Rights announced guidelines to technologically counter these types of forgeries. They involve electronic, chemical, physical, mechanical and digital technologies. Basically, an ANTC technology addresses the problem of lack of understanding solutions for advanced forgeries challenging the authentication of the original products. ANTC could be executed in two ways: either directly by human inspection or by using a separate device to detect it. But all the methods share common characteristics for

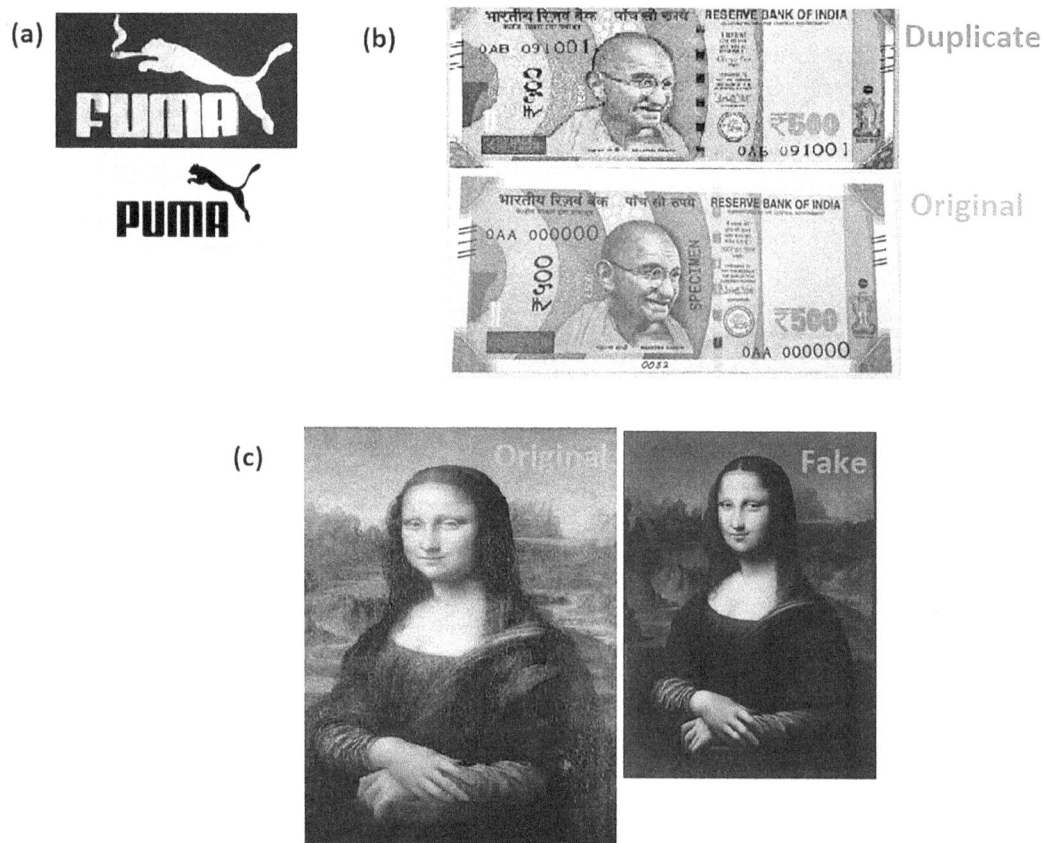

FIGURE 12.1 Counterfeiting of a popular brand logo (a), Indian currency paper note of Rs. 500 (b) and famous portrait of Mona Lisa painting (c).

Technological Methods

Electronic
—Electronic Seals
—Magnetic Strips
—Contact Chips
— RFID

Chemical
— DNA Coding
— Chemical Tracing
— Glow Coding

Physical
Security Inks
—UV-Sensitive Ink
—IR-Sensitive Ink
—Magnetic Ink
—Optical Variable Ink
— Thermally-sensitive Inks
— Reactive Inks
— Penetrating Inks
— Optical Memories
— Barcodes
— Security Holograms
— Watermarks

Mechanical
— Labels
— Tags
— Laser Engraving
— Seals
— Security Threads
— Security Films

Digital
—Digital Watermarks
—Fingerprinting
— Hashing

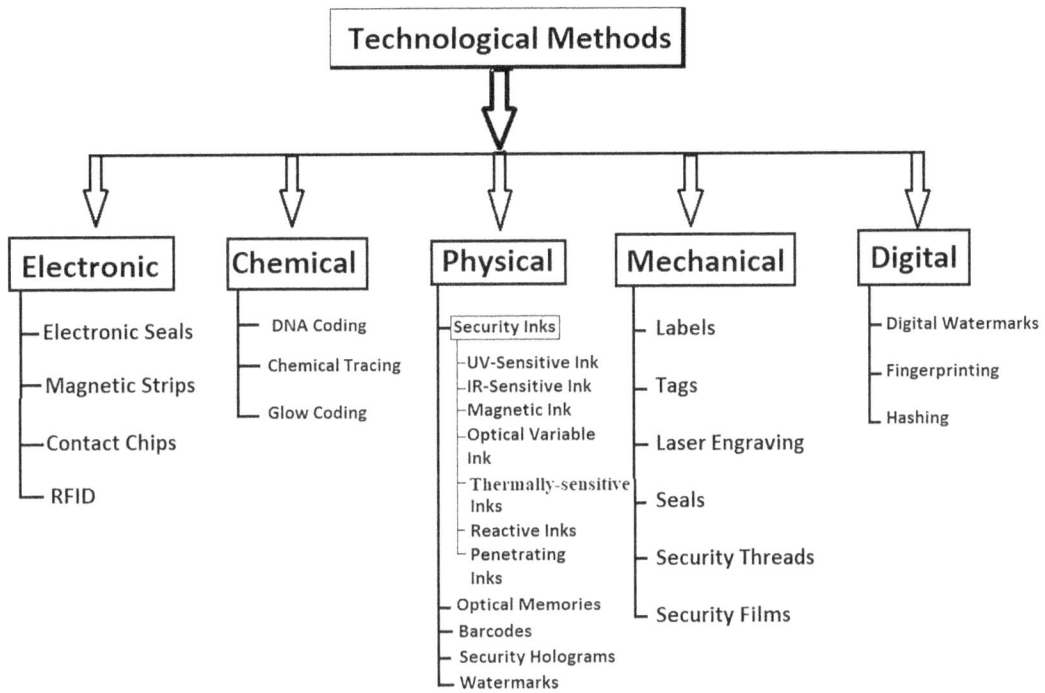

FIGURE 12.2   Flowchart diagram explaining the various technological measures introduced to ANTC system.

identifying the original, which are the specific type of markers. These marks drawn on the secret codes are attached to the authentic products that allow the modern technologies to counter forgery effectively (Figure 12.2).

Among these, the most common and popular way comprises of secret security inks. Due to being optically intuitive, these can be grouped under physical methods, although EUIPO classified them as marking methods [1].

12.1.3 Measures to Cease Counterfeiting Using Luminescent Materials (Phosphors)

Mostly ink-based ANTC technologies relay on development of phosphors. Luminescent paints consist of mainly 'phosphors'. These paints are easy to use, read and apply to various environments. ANTC inks could be either visible or invisible. Both help not only in evaluating the authentication of the product, but also in tracking/tracing the theft. Identification directly through vision is the easiest technique, which also results in producing a cost-efficient technology. Combining both visible and invisible inks together gives a higher level of protection against counterfeiting.

There are different types of ink technologies depending on the type of ink used, discussed as follows:

1. UV-sensitive inks: Inks which show different colors (comes under visible range) on application of light under the UV range comes under this category. Security codes/labels painted with such inks appear different during daylight and emits another visible color under UV light.
2. IR -sensitive inks: IR inks work similar to UV inks, but differ in range of excitation and emission wavelengths. This implies that IR inks are usually invisible and transparent under daylight and become visible under IR light. This technique applied to secret codes/writing is very useful in preventing unauthorized reproduction of goods and documents.

3. Magnetic inks [2,3]: These types of inks are mainly assigned to the paints which are luminescent as well as have magnetic properties. On application of magnets, encoded information becomes visible on the authentic document. This technique is usually applied to banknotes and cheques.

4. Optical variable inks (OVI): This technique uses inks which show different colors (drawings or writings) when viewed from different angles. Usually, this can be seen in various labels and hologram stickers on products and goods used in daily life. This is the simplest technique since it can be perceived directly by human eyes even though it is not so easy to replicate the original one.

5. Thermally sensitive inks: These inks change colors on increasing or decreasing temperature known as temperature-dependent inks. Some thermo-luminescent types of paints can also be applicable to this technology, although thermo-luminescence follows a different basic principle, but colors become visible only on stimulation by heating processes. Some light sensitive materials shows luminescence only at low-temperatures, can also find applications under this category. These color changes can be temporary or permanent in nature, i.e., the paint can regain its actual color on returning to its normal temperature.

6. Reactive inks: As the name suggests, these inks become visible or show color changes due to chemical reactions. These inks when in contact with certain types of solvents or solutions result in exchanging of ions. These inks could be either erasable due to water-soluble properties or could result in the color alterations.

7. Mechanical measures: Inks changing colors or becoming visible on application of stress or strain, i.e., implying mechanical pressure, are categorized under mechanical methods. This is another simpler route of identification of genuine products.

Therefore, various kinds of luminescent inks/paints are used in ANTC processes in identifying the theft. Luminescent phosphors make an indigenous part of today's advanced technological ANTC measures. These inks are embedded into the secret codes written or drawn in the form of images on labels and seals/stamps, etc. assigned to the product's tags/logos (Figure 12.3).

Although holograms depend on a different theory, images containing luminescent paints also work somewhat in a similar way, and sometimes, these elusive inks are used in holograms for better security.

Security provided against the reproduction, duplication or copying the particular goods or information depends upon the characteristics of the chemical and physical composition of the ink used. These security inks are further painted/printed and applied to barcodes/QR codes and labels, etc. affixed to the products or authorized documents.

Generally, barcodes consist of 2D/3D black and white lines or patches (of different sizes and shapes) such as in QR codes. But sometimes colored luminous paints are also applied in order to meet higher order of security demands. For example, two-dimensional barcodes are read with the help of surface-enhanced Raman scattering (SERS) probes, which provides higher level of technological security against counterfeiting due to the involvement of much difficult routes of fabrication and detection [4]. Figure 12.4 displays the working of two & three-dimensional QR codes in real time.

There are many more other ways of identifying the authenticity that EUIPO has summarized into its guidelines, but we have discussed only few which employ luminescence techniques as measures against false deeds.

12.2 METAL OXIDES

Metals such as Ti, Fe, Ni, Cu, Mo, Si, Sn and Sb combine with oxygen to form their monoxides/dioxides TiO_2 (titanium(IV) oxide), MnO_2 (manganese dioxide), CeO_2 (cerium(IV) oxide), CuO (copper(II) oxide), SnO_2 (tin(IV) oxide), ZnO (zinc oxide), NiO (nickel oxide), Fe_2O_3 (ferric oxide), Y_2O_3 (yttrium oxide), In_2O_3 (indium oxide), WO_3 (tungsten oxide) and even ternary-oxides, for example $CaTiO_3$ and $NiCo_2O_4$.

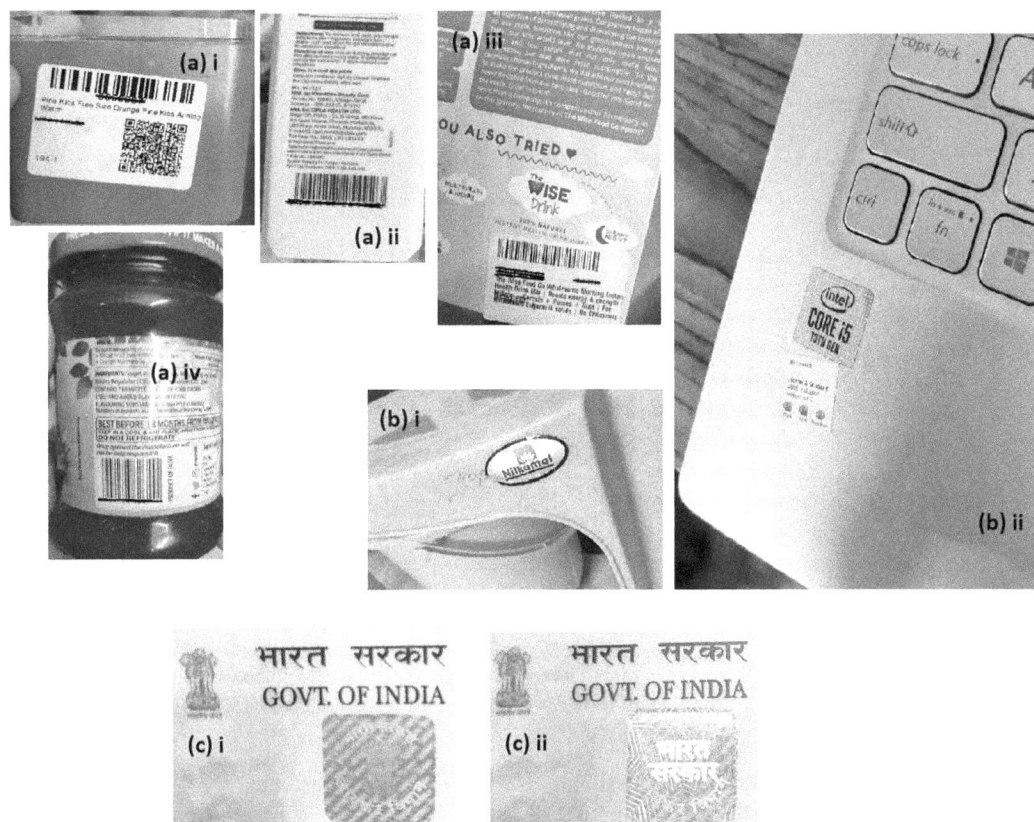

FIGURE 12.3 Some examples of technologically advanced ANTC measures. (a) i–iv show 2D and 3D barcodes (QR code) on different household items, (b) i & ii show company's logos on a plasticware and on a laptop, revealing the originality of the products, (c) i & ii government holograms on an identity card displaying different colors and patterns on seeing from different angles.

There are many metal oxide-based technological advances in recent years as already have been described in the present book. Apart from these wonderful properties of high chemical resistivity, temperature stability and low refractive index, metal oxides have a great quality of being luminescent hosts when doped with rare-earth ions. Hence, this ensures plenty of applications of metal oxides in the modern world and plays an indistinguishable role in our daily life. Thermal sensing [5,6], pH sensing [7], bio-medicine [8,9], gas sensing [10–12], energy storage [13,14], semiconductors in electronics [15], catalysts in chemical industry [16,17] and detection of heavy-metal ions for the removal from wastewater [18,19] are some of the many advantages and applications of metal oxides [20].

12.3 LUMINESCENCE AND LANTHANIDES

Luminescence attained due to the addition of rare earths as impurities to the host lattice matrix is known to be lanthanide (Ln^{3+}) luminescence. Lanthanides are popular due to their simple synthesis procedures, high performance, narrow bands and strong emission intensities, mainly characteristics to a particular rare-earth element [21,22]. There are various luminescence-established ANTC procedures depending on numerous mechanisms: photoluminescence (PL) [23,24], temperature-dependent (thermal-sensitive) luminescence [25], mechanoluminescence (ML) [26], chemiluminescence (CL) [27] and electroluminescence (EL) [28]. Some of these are discussed here:

The position of ML emission wavelength due to most of the lanthanide ions (Tb^{3+}, Pr^{3+}, Ho^{3+}, Er^{3+}, Dy^{3+}, Sm^{3+}, Eu^{3+}, Tm^{3+}, Nd^{3+} and Yb^{3+}) have been collectively reported by Du et al. [22]. The corresponding PL emissions for respective ions also produce similar spectra. These are characteristic

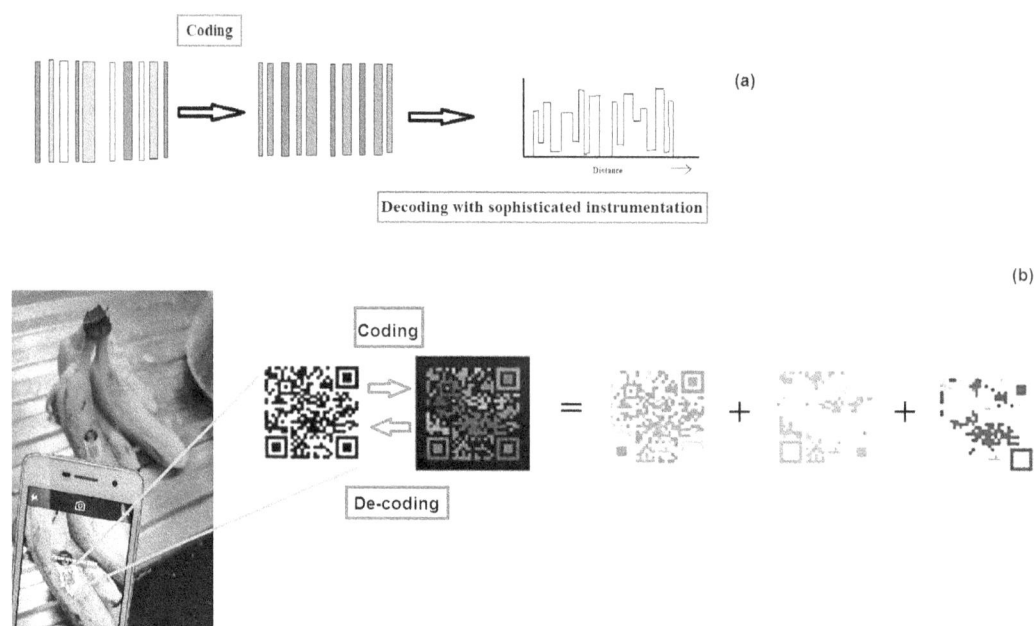

FIGURE 12.4 (a) Coding of 2D barcode (due to its length and breadth on 2D surface) through highly sophisticated techniques such as SERS, (b) 3D QR code consisting of red, green & blue coding (patches) under visible/UV light and decoding after scanning through a mobile phone.

emissions and give similar forms of emission in every host. Kaczmarek et al. observed wavelength-dependent luminescence when white color Eu^{3+}-codoped powder sample turns bright reddish under UV light, whereas temperature-dependent luminescence was reported for white-color Sm^{3+}-codoped same sample turning orange on slight variation in temperature [25]. Various kinds of luminescence phenomena can be understood with the help of the following figure (Figure 12.5).

Luminescent materials consist of many kinds of optical materials ranging from organic/inorganic complexes, metallic nanoparticles, quantum dots (QDs), polymer dots (PDs), carbon dots (Cds) and lanthanides-activated compounds [29,30]. Each one of these is having certain advantages over the others (Table 12.1). Among these, lanthanides-activated compounds are the most valuable materials due to their unique ability to exhibit multiple optical properties such as upconversion, downconversion and downshifting. Upconversion phosphors in nano-form (UCNPs) provide high-level of protection against counterfeiting. The combination of mechanism of upconversion and downshifting lays the basis of most of the ANTC methods.

12.4 TYPES OF ANTC

12.4.1 Wavelength-Dependent ANTC

Upconversion (UCL), downconversion (DCL or DL) and downshifting luminescence (DSL) are the three major phenomena responsible for luminescent emissions. These are influenced by and originate on absorption of energy in the radiation form. These can be explained as follows:

Upconversion: is a process in which two or more photons of lower energies emit a photon of higher energy.

Downconversion: is a process in which a photon (of comparatively higher energy) gets converted into two photons of lower energies such that the energy during the whole process remains conserved.

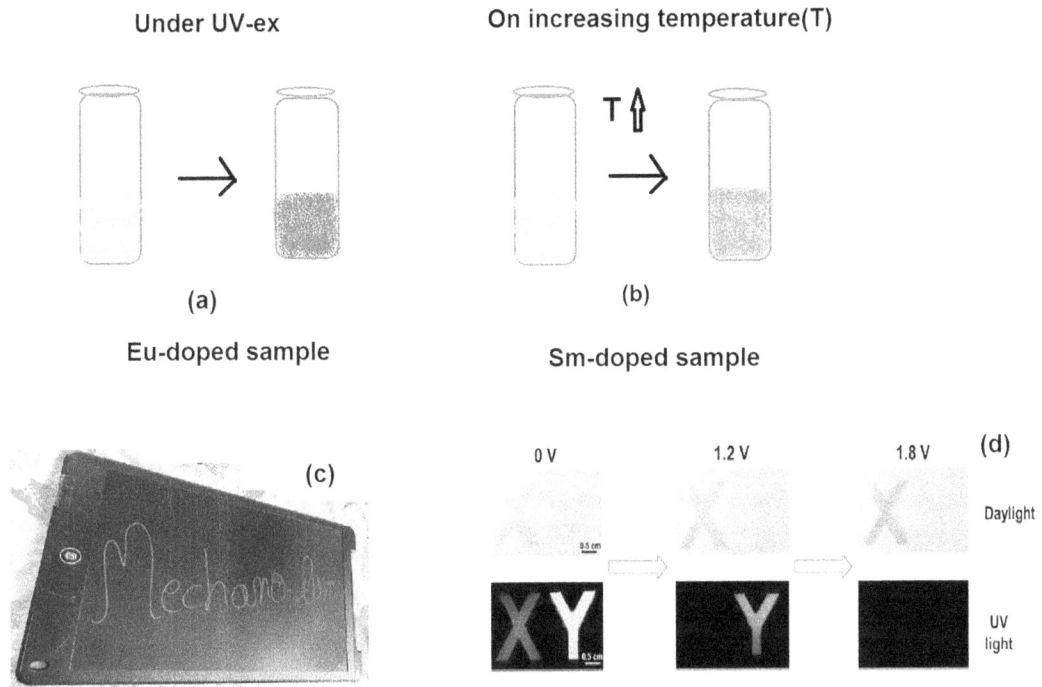

FIGURE 12.5   Wavelength-dependent luminescence – white color Eu^{3+}-codoped sample turns red under UV light (a), temperature-dependent luminescence – on slightly increasing the temperature makes white color Sm^{3+}-codoped same sample orange (b), an electronic slate could be a newest example of displaying an effect of mechanoluminescence (c) and an example of electroluminescence: on increasing the voltage (V) under UV light, letters 'X' and 'Y' appear, disappear and even change colors (d).

TABLE 12.1

Type of Materials used in Illumination with their Advantages, Disadvantages and Mechanism Responsible for the Emission of Light, Respectively [29]

Type of Luminescent Materials	Examples	Advantages	Disadvantages	Mechanism
Metallic nanoparticles	AuNPs/AgNPs	High stability, easy syntheses processes	Not visible to naked eye, needs sophisticated instruments for readouts	Absorption spectrum, scattering
Quantum dots (QDs)	CdSe/CdS/CdZnS & ZnCdS/CdZnS	Tunable excitation & emission wavelengths, narrow emission bandwidths	Cytotoxicity	Semiconductor
Carbon dots (Cds)	CD/SA, CD-calcein/SA, CD-CDTe/SA, CD-RhB/PMMA (CDs mixed in sodium alginate (SA) and PMMA matrixes	Good quantum efficiencies, cost-efficient, multi-mode & tunable emissions, long persistence	Depends on specific substrates	Semiconductor

(Continued)

TABLE 12.1 (*Continued*)
Type of Materials Used in Illumination with Their Advantages, Disadvantages and Mechanism Responsible for the Emission of Light, Respectively [29]

Type of Luminescent Materials	Examples	Advantages	Disadvantages	Mechanism
Silicon-nanoparticles	SiNPs [31,32]	Extracted from bio-waste, environment friendly, abundance of silicon tends to low cost	Emission peaks mainly lie in visible, not very suitable for harsh conditions	Stokes emission
Perovskites (ABX$_3$ where X = Cl, Br, I)	CsPbBr$_3$, CsPbI$_3$	Multi-mode tunable emissions, narrow emission bands	Unstable in humid environment	Stokes & anti-Stokes shift
Ln-doped metal-oxides	CaAl$_2$Si$_2$O$_8$:Ce^{3+}, Tb^{3+}, Yb^{3+}[33], Lu$_2$GeO$_5$:Bi^{3+}, Yb^{3+}[34], Ca$_2$MgWO$_6$:Cr^{3+}, Yb^{3+}[35], ZnO:Eu^{3+}, Yb^{3+}[36], Gd$_2$O$_3$:Bi^{3+}, Nd^{3+} [37]	Less reactivity & high stability, color tunable, infinite codes	Cumbersome syntheses processes involving high temperatures	Stokes & anti-Stokes emission
Polymer dots	P(TPE-TPA), P(DTDPE-TPA) [28]	Eco-friendly	Poor morphology	Stokes & anti-Stokes emission

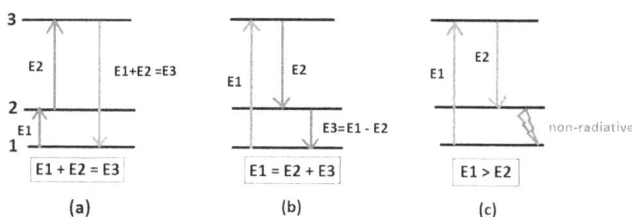

FIGURE 12.6   Upconversion (a), downconversion (b) and downshifting luminescence (c) mechanisms.

Downshifting: is usually termed for the normal luminescence in which a photon of high energy emits a photon of lower energy inefficiently. A part of the incident photon energy gets wasted into the non-radiative emissions (phonon vibrations in the crystal lattices) (Figure 12.6).

In the upconversion process, energy E1 of the first photon adds up with the energy E2 of another photon to emit a photon of higher energy E3, which is equal to the sum of E1 and E2. Similarly, but not exactly, downconversion is the reverse process of UCL. In the downconversion process, two or more photons of lower energies E2 and E3 are emitted on absorption of one photon E1 of higher energy.

Downshifting is termed for normal luminescence which obeys simple Stokes shift, where the energy emitted E2 is always less than the energy absorbed E1 because a part of the initial energy gets lost as non-radiative decays responsible for phonon energies.

The above mechanisms are the domains of photoluminescence and lie under the UV, visible and near infrared (NIR) ranges. The advantage of emission in the visible range is that it can be recognized directly through human eyes, but sometimes, phosphors emitting in UV and NIR can also prove fruitful where higher security is concerned. These emitters will need additional devices to probe into. All the three mechanisms form the backbone of modern-day ANTC methods.

12.4.1.1 Metal-Oxides as Luminescent Hosts for Effective ANTC

There are numerous luminescent metal oxides, but only few were applied to ANTC operations so far. Some are summarized and others with possible applications in the field of ANTC are discussed here.

12.4.1.1.1 Single/First-Stage ANTC

Processes involving only one or either any one of the three (upconversion/downconversion/down-shifting) luminescence mechanisms (i.e., where only wavelength-dependent ATNC is involved) can be considered as single or first-level ANTC. Generally, singly-doped phosphors are used in first-stage ANTC measures.

12.4.1.1.1.1 Single-Dopant Effect

Shivananjaiah et al. [38] doped La^{3+} ions in ZnO nanoparticles and applied for latent fingerprint (LFP) identification with two prominent emission peaks around 390 and 500 nm for 327-nm excitation. 390-nm emission occurs due to $^5D_1 \rightarrow {}^7F_0$ whereas 500-nm emission is comparatively weaker one assigned to the $^5D_3 \rightarrow {}^7F_1$ transition. Similarly, few more such types of singly doped oxide phosphors used in ANTC are listed in Table 12.2.

Typically, inorganic phosphors consist of a host compound lattice which is doped with impurity ions known as 'activators' in very small proportions. Activator ions hold energy levels that can be populated by directly absorbing energy from the excitation radiation or in an indirect way of transfer of energy from one ion to another ion in the crystal lattice. This results in the luminescence emissions. Generally, there are two types of activator ions that can be distinguished. The first type

---

TABLE 12.2
Various Metal Oxides (Singly, Doubly & Triply Doped) with Excitation and Emission Wavelengths along with their Respective Transitions and the Mode in Which they Can be applied to ANTC Measures are given in the following Table

Oxides	Excitation (λ)	Emission (λ)	Transitions	Color	Possible Mechanism & Possible Mode of ANTC	Application in ANTC
ZnO:La^{3+} [38]	327	390	$^5D_1 \rightarrow {}^7F_0$	UV-blue	Downshifting (dual-mode)	LFP identification
		500	$^5D_3 \rightarrow {}^7F_1$	Green		
NaYGeO$_4$:Tb^{3+} (0.01%)	254	621	$^5D_4 \rightarrow {}^7F_3$	Red	Cross-relaxation (multi-mode)	Phosphor mixed in PDMS
NaYGeO$_4$:Tb^{3+} (0.5%)		370–480	$^5D_3 \rightarrow {}^7F_{3,4,5,6}$	Blue		
NaYGeO$_4$:Tb^{3+} (5%) [66]		554	$^5D_4 \rightarrow {}^7F_5$	Green		
TiO$_2$:Eu^{3+} [68]	395	612	$^5D_0 \rightarrow {}^7F_2$	Red	Downshifting (single-mode)	LFPs identification
Sr$_3$Ga$_4$O$_9$: Tb^{3+} Sr$_3$Ga$_4$O$_9$:Bi^{3+} [69]	254	549	$^5D_4 \rightarrow {}^7F_5$	Green Yellow	Downshifting (dual-mode)	Long persistent luminescent (LPL)
Bi$_{0.50}$Gd$_{0.45}$Eu$_{0.05}$PO$_4$ [70]	254 365	615 580	$^5D_0 \rightarrow {}^7F_J$ (J = 1, 2, 3, 4)	Orange-red Yellow-green	Downshifting (dual-mode)	Identification of authentic documents
La$_2$Mo$_2$O$_9$:Er^{3+}, Tm^{3+} [71]	980 1,550	534, 553 660	$^2H_{11/2}/^4S_{3/2}$ $\rightarrow {}^4I_{15/2}$ $^4F_{9/2} \rightarrow {}^4I_{15/2}$	Green Red	Upconversion (dual-mode)	Optical temperature sensing & possible application in ANTC

of ions are those that show weak interaction with their host lattices. That is, their energy levels are properly shielded from their outer environment. All the lanthanides 'Ln^{3+}' come under this group and possess 'line emission' as their characteristics.

Mn^{2+}, Eu^{2+} and Ce^{3+} are the second type of activator ions that strongly interact with the crystal fields and get influenced by the host lattice's energy shells. This type of transitions takes place when the 'd' level electrons come into play. Sometimes, even s^2 ions such as Pb^{2+} or Sb^{3+} and more complex anions such as MoO$_4$$^{2+}$ or NbO$_4$$^{3+}$ also take part in transitions and result in complicated coupling interactions. These give broad-band emissions.

Lanthanides exist in the most stable 'Ln^{3+}' trivalent oxidation state and display many interesting optical properties. Lanthanide luminescence occurs due to electron transition within 4f^n electronic levels and contains complex spin–orbit coupling interactions. Usually, these f–f transitions are parity-forbidden, and inter-mixing of opposite configurations result into breakdown of selection rules. Due to asymmetry produced in the crystal field, the probability of electric dipole transition greatly increases. As a result, symmetry around the lanthanide ion gets reduced and becomes responsible for luminescent intensity. Crystal structures of lanthanide-doped lattices get disturbed and are modified in a way to enhance upconversion emissions on introduction of different sizes of ions influenced by dramatic crystal field splitting (Figure 12.7) [39,40].

For an efficient upconversion process, host lattice must have less loss to phonon energies in order to reduce non-radiative transitions and increase radiative emission. Lanthanide-doped materials with high phonon vibrations provide a limited number of applications in the field of lighting and illumination. It is a well-known fact that the d–d transition metal ions are good at tuning the excited state properties of lanthanides and hence has a potential to control the quenching effects due to phonon energies. This leads to improvement in the emission efficiency. Incorporation of transition elements into hosts helps in intensifying the upconversion process through controlled energy transfers from one ion to another. For example, Dong et al. established enhancement of several orders in intensities in oxide host materials including Yb$_2$Ti$_2$O$_7$, Al$_2$O$_3$, TiO$_2$, Gd$_2$O$_3$ and Yb$_3$Al$_5$O$_{12}$ when doped with molybdenum (Mo^{3+}) ions [41,42]. This efficient energy transfer was achieved through energy transfer from the excited state of the Yb^{3+}- MoO$_4$$^{2-}$ sensitizer to the Er^{3+} activator. Similar outcomes were noticed when manganese (Mn^{2+}) was doped in oxide hosts, through Yb^{3+}–Mn^{2+} sensitization in Yb$_3$Al$_5$O$_{12}$ [43].

12.4.1.1.1.2 Codopant Effect   Luminescence occurs in various steps such as energy absorption, energy transfer and energy emission within the crystal lattices. Energy transfer can happen by re-absorption of energy emitted by another activator or sensitizer ion. Therefore, sensitizers are used to increase the quantum yield of a particular phosphor. Hence, the necessity of codoping via another dopant comes into picture.

The Li$^+$ ion gets easily incorporated into the host lattices due to its small size. Earlier, Chen et al. had reported around 25- and 60-fold increase in the intensities of respective green and violet emission of Y$_2$O$_3$:Yb/Er nano-phosphors through Li$^+$ ions doping [44,45]. Later on, many research

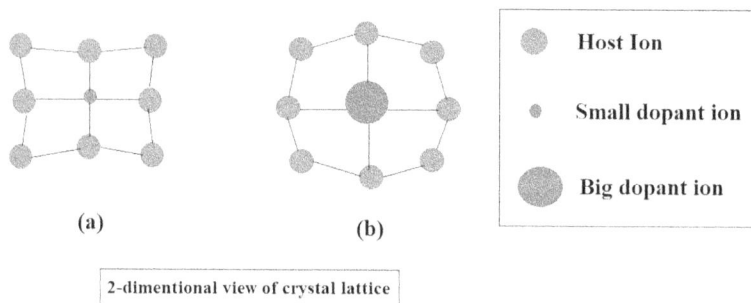

FIGURE 12.7   Crystal lattice shrinks on introduction of smaller-size dopant ion (a) and expands on bigger-size dopant ion (b) at the center.

groups have used this property of lithium ion and attained enhancement of luminescence in various oxide hosts such as TiO_2 [46], $BaTiO_3$ [47], ZrO_2 [48], ZnO [49], $ZnWO_4$ [50], $Y_3Al_5O_{12}$ [51], $CaMoO_4$ [52] and $GdVO_4$ [53].

12.4.1.1.1.3 Tri-dopant Effect Pushpendra et al. [54] studied the use of tri-doping in ANTC. They reported orange-reddish emission for $GdPO_4$:1% Er^{3+}, 10% Yb^{3+} on 980-nm NIR excitation, but also observed it on 395-nm UV light after adding 5% Eu^{3+} as a tri-dopant into its lattice. Singly doped $GdPO_4$ by Eu^{3+} ion emits in the pure red region on 395-nm excitation and hence can be used in LFP identification using powder dusting technique (Figure 12.8) [55].

It is becoming difficult day by day to meet the ever-increasing demand of complex ANTC only through first-level luminescence. Hence, other mechanisms such as mechanoluminescence, electroluminescence can also be exploited for appropriate use in ANTC.

12.4.2 TEMPERATURE-DEPENDENT/STIMULATED LUMINESCENCE

Y_2WO_6:Yb^{3+}/Er^{3+} is known to emit three colors, blue, green and yellow-green, on 254-, 365- and 980-nm excitations, respectively. Its downconversion lifetime is also helpful in temperature measurements, and therefore, it can act as an excellent multi-mode luminescent phosphor [56].

12.4.3 PRESSURE-DEPENDENT/STIMULATED LUMINESCENCE (CAN BE TERMED UNDER MECHANOLUMINESCENCE)

Ma et al. [57] cut $Y_3Al_5O_{12}$:Ce^{3+}/polydimethylsiloxane (YAG:Ce^{3+}/PDMS) elastomers into strips with different widths and then placed them into a Petri dish full of PDMS precursor cured at 353

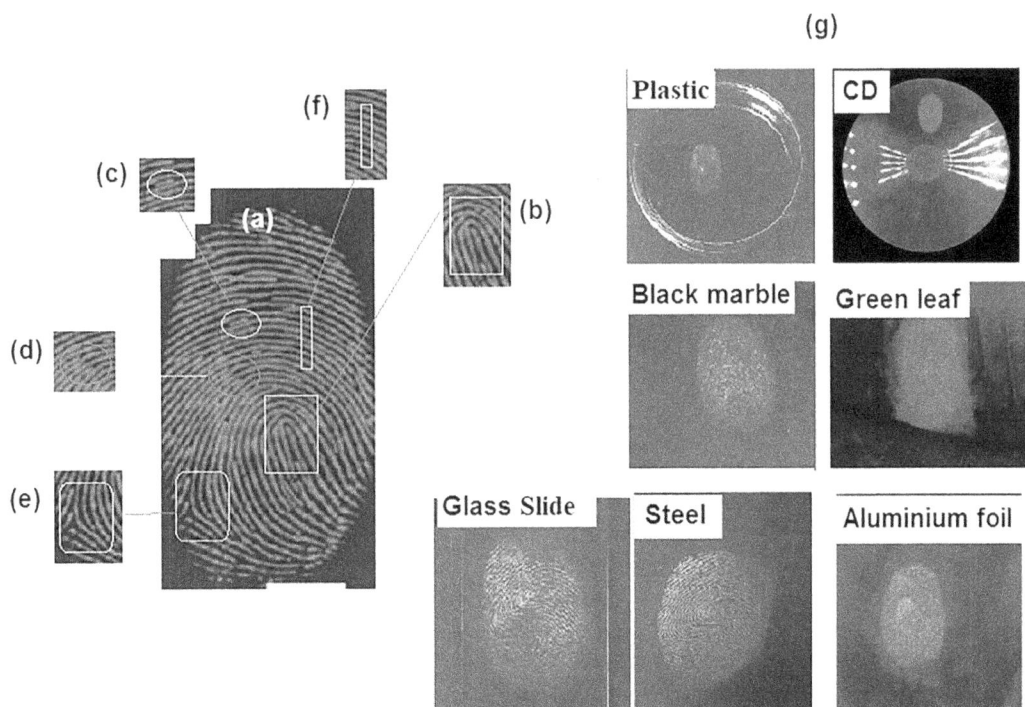

FIGURE 12.8 A fingerprint on a substance (a), fingerprint recognition by identifying unique features: loop (b), island (c), ending (d), delta (e), linear (f), etc. and fingerprints on various substances (g) under 395-nm UV light.

K for 2 hours. Then, an ANTC device was prepared with bar codes. After stimulated by folding, the strips composed of YAG:Ce^{3+} could emit yellow ML due to phonon-assisted mechanical energy transfer processes.

12.4.4 Voltage-Dependent/Stimulated Luminescence (Can Be Termed under Electroluminescence)

In 2011, Hao et al. [58] developed an electric-field-induced enhancement of upconversion luminescence. In their study, a ferroelectric host material (BaTiO$_3$) was chosen and doped with Yb^{3+} and Er^{3+} to produce a BTO:Yb/Er film. They then varied the structural symmetry of the BTO host by applying a direct current bias voltage. The green upconversion emission was increased by the factor of 2.7 times, but did not show any appreciable change in its consecutive red emission intensity, indicating the ideal voltage-dependent as well as wavelength-dependent nature of the emission simultaneously.

12.4.4.1 Second-Stage ANTC

ANTC procedures involving any two among the upconversion and/or downconversion and/or downshifting (or even with longer lifetimes) are categorized under the second stage or dual-mode measures. Sometimes, even other luminescence processes such as temperature-dependent/mechano-/electro-luminescence can also find applications in second- stage ANTC methods.

The presence of simple luminescence is not enough in second-level ANTC. PL lifetime, i.e., how persistent (glow after excitation ceases) the luminescence is, can also be useful against counterfeiting. SrAl$_2$O$_4$:Eu^{2+}, Dy^{3+} (SAED) is a well-known long persistent green phosphor. More recently, Moudam et al. [59] and Katumo et al. [60] both demonstrated SAED applications for various ANTC techniques separately as shown in the figure below (Figure 12.9).

Katumo reported long lifetime (from 0.5 to 11.7 seconds) for SAED on annealing at various temperatures, which is very useful in ANTC effects. The green emission is mainly due to the Eu^{2+} transitions 4f^7 ($^8S_{7/2}$)$\rightarrow$4f^6 5d^1, demonstrating strong crystal field interactions. Earlier studies on a similar compound Sr$_4$Al$_{14}$O$_{25}$:Eu^{2+}, Dy^{3+} [61] reports emission at 495 nm for 365-nm excitation with efficient afterglow characteristics useful in LFPs recognition.

12.4.4.2 Multi-stage/Multi-mode ANTC

ANTC procedures may use all of three (upconversion, downconversion and downshifting) processes together to introduce higher level of security, which makes it almost impossible to make copies of such complicated authentic documents.

La$_4$GeO$_8$(LGO):Eu^{2+}, Er^{3+} material is reported, which can emit red, purple, baby blue and green light under the increased excitation wavelength from 250 to 380 nm. The phosphor also shows green upconversion luminescence under the NIR (980 and 808 nm) laser irradiation and hence emits different colors on exposure to different excitation wavelengths. Also, emission intensity of this phosphor increases with excitation power as well [62]. Another good example for multi-level ANTC could be Na$_2$CaGe$_2$O$_6$ (NCG):Ln^{3+}, Yb^{3+}, Tb^{3+} (Ln^{3+}=Er^{3+}, Ho^{3+}, Tm^{3+}), an outstanding UC, DC and persistent phosphor recently reported by Jin et al. [63]. NCG:Er^{3+}, Yb^{3+} gives upconversion green, Ho^{3+}, Yb^{3+} in yellow and Tm^{3+}, Yb^{3+} in blue on NIR 980-nm excitation. The UC emission spectra are found to be laser power dependent and increase with an increase in intensity of the excitation source. I$\propto$P^n, where n represents the number of absorbed photons, P is the power of excitation, and I corresponds to emission intensity. Also, the spectra show efficient red emission when it is singly doped by Eu^{3+} ions, whereas tri-doping of Tb^{3+} exhibits long persistent luminescence (LPL) in the same lattice host. Therefore, some of its unique aspects are summed up in Figure 12.10.

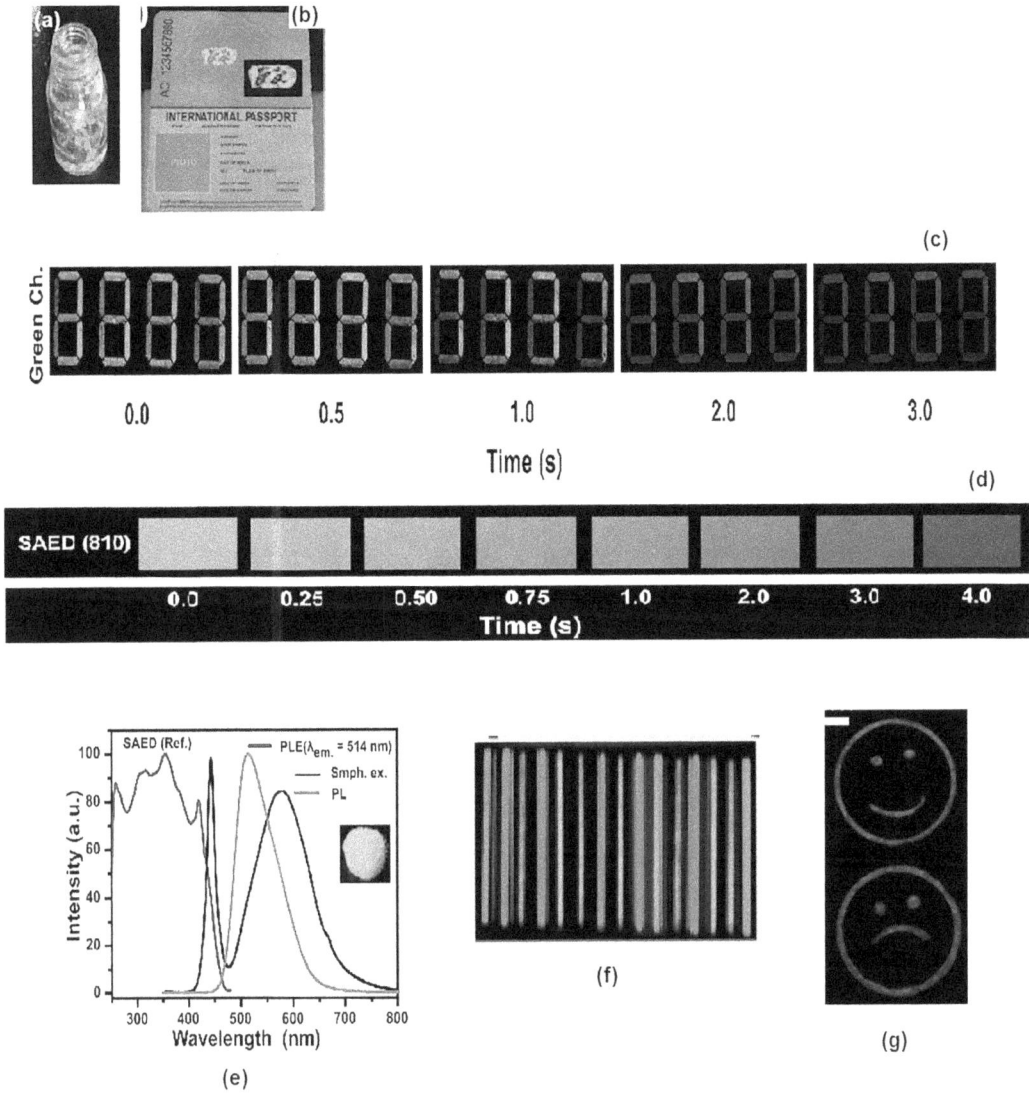

FIGURE 12.9 SAED stored in glass bottle (a), SAED (mixed in polyvinyl butyral (PVB)) screen-printed on a demo passport-type document (b) observed under visible light after UV irradiation (325 nm), showing long persistent green-afterglow green Ch. Displaying long persistence of the SAED green phosphor from time 0–3 seconds on a seven-segment digital panel (c), long persistence of SAED annealed at 810°C (d), absorption spectra (250–450 nm) of SAED overlapping with the excitation by flashlight of smartphone at 514-nm emission wavelength (e), applied to color barcodes (including other red phosphor) (f), happy and sad emoji's glowing green in dark (g).

More recently, Haider et al. [64] confirmed photochromism in $TiO_2:Yb^{3+}$, Er^{3+}. White $TiO_2:Yb^{3+}$, Er^{3+} turns pink-grayish in color on 405-nm excitation, but retains the white color on bleaching with 808-nm radiation, hence showing photon-induced reversible change in color.

Moreover, a new phosphor $Ba_2Zr_2Si_3O_{12}:Eu^{2+}$, Er^{3+} [65] was found to have both spatial and temporal luminescence effects. That is, it exhibits photo- and thermoluminescence (by UV irradiation), long persistent and upconversion on 980-nm excitation and becomes most magnificent candidate for multi-level ANTC till date.

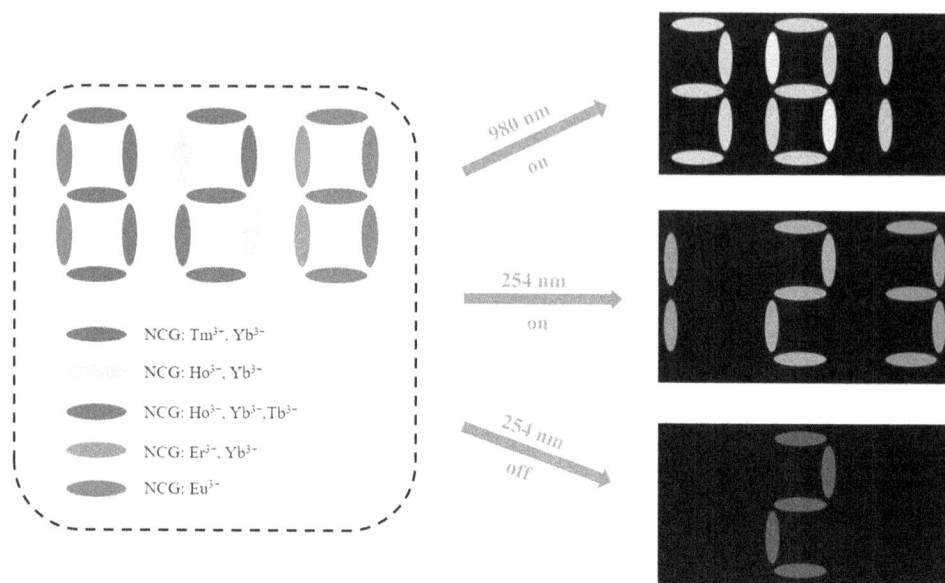

FIGURE 12.10 Displays dual-mode ANTC prototype seven-segment colored digits by using NCG:Tm^{3+}, Yb^{3+}; NCG: Ho^{3+}, Yb^{3+}; NCG:Tm^{3+}, Yb^{3+}; NCG:Ho^{3+}, Yb^{3+}, Tb^{3+} and NCG:Eu^{3+} phosphors.

Sometimes, even singly doped phosphor displaying multi-color features can also be used in multi-level ANTC (e.g., $NaYGeO_4$:Tb^{3+}) [66]. However, a unique phosphor $NaBaScSi_2O_7$:Eu^{2+} exhibiting several PL mechanisms (such as Long-persistent luminescence (LPL) and Photo-stimulated luminescence (PSL)) was reported to meet multiple-ANTC measures with single dopant 'Eu^{2+}' (Figures 12.11 and 12.12) [67].

Apart from those given in the above table, there are many more (e.g., Ca_2MgWO_6:Er^{3+}/Yb^{3+}/Eu^{3+} [72], Zn_2GeO_4:Mn^{2+}/Cr^{3+} [73], $SrGa_2Si_2O_8$:Eu^{3+} [74], $SrTiO_3$:Pr^{3+}:Li^+ [75]) oxides-based phosphors which can be useful in ANTC.

Multilevel luminescence ANTC changes more than two times under the mode of excitation light, luminescence lifetime, heat stimulation, chemical reactions, mechanical pressure and so on. Various stages of ANTC (**single mode → double/dual mode → multi-mode luminescence**) are effectively explained in Figure 12.13 in a tabular form:

Figure 12.14 shows the region on the EM radiation wavelength spectrum where human eye is most sensitive. Maximum emission should lie within the region under the curve; otherwise, eye sensitivity falls sharply before and beyond this region. Emissions in red appear more benefiting due to less scattering property of longer wavelength radiation.

12.4.4.3 Rare-Earth Metal Oxides

There are many rare-earth ion-based oxides, possessing enormous potential for ANTC solutions. Many oxides of rare-earth ions directly or on doping with other rare-earth ions show luminescence characteristics. Some of them have been already applied and are tabulated distinctly as follows (Table 12.3 and Figure 12.15):

In recent years, new application with the emergence of nanotechnology graced with the development of materials in the nano-powder form, finds a way to meet inventive demands for the better society. ANTC is one such necessity.

12.4.4.4 Nano-(Core/Shell)Phosphors

Compared to bulk materials, powders in the nano-form face more quenching effects due to their high surface/volume ratio. Energy loss mechanism dominates in small-size particles. This results

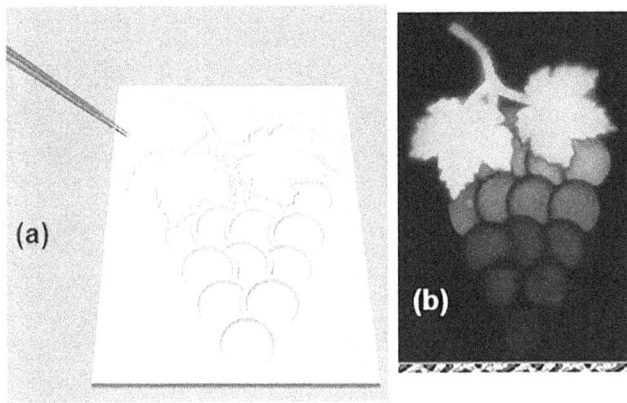

FIGURE 12.11 Image painted on paper with the use of NaYGeO$_4$:Tb^{3+} (different concentrations) phosphor appears white in daylight (a) and colored under 254-nm UV light (b).

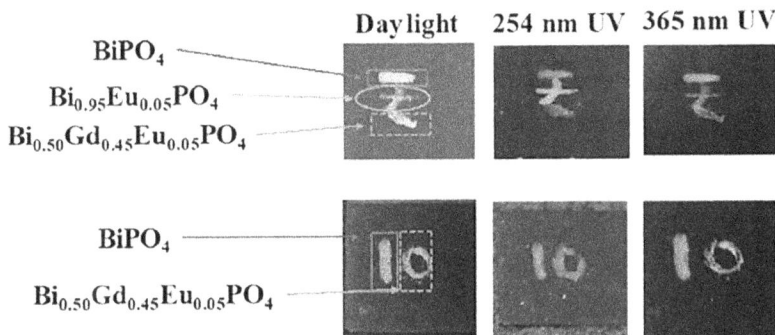

FIGURE 12.12 Indian currency symbol and number '10' appears transparent in daylight (a), red in 254-nm (b) and yellowish-green in 365-nm UV light (c) using Bi$_{0.50}$Gd$_{0.45}$Eu$_{0.05}$PO$_4$ as security ink.

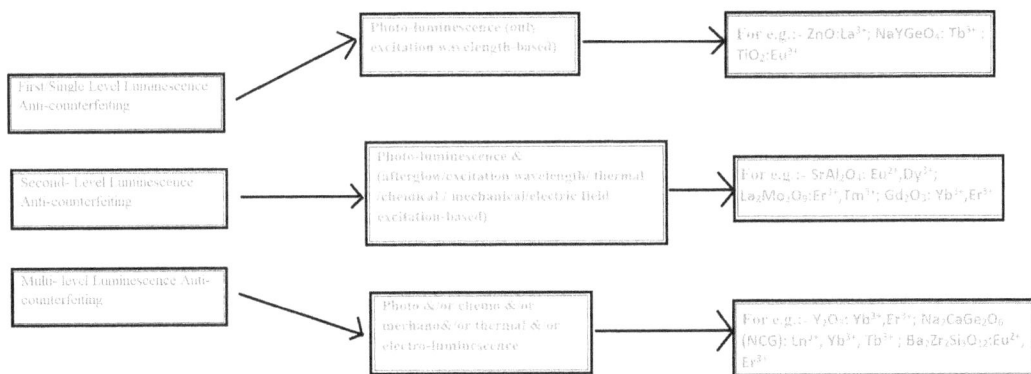

FIGURE 12.13 Flowchart-type diagram representing different stages of ANTC with their examples.

in the non-radiative processes related to the surface-induced quenching effects responsible for the low quantum yield. Lu et al. coated Y$_2$O$_3$:Yb/Tm nanoparticles with an amorphous silica layer and demonstrated that the intensity of the nanoparticles can be controlled by changing the thickness of the silica shell. In a core–shell structure, dopant ions are embedded into different layers with definite concentrations [86]. Also, SiO$_2$-based nanocomposites greatly reduce the toxicity of these inorganics in the environment.

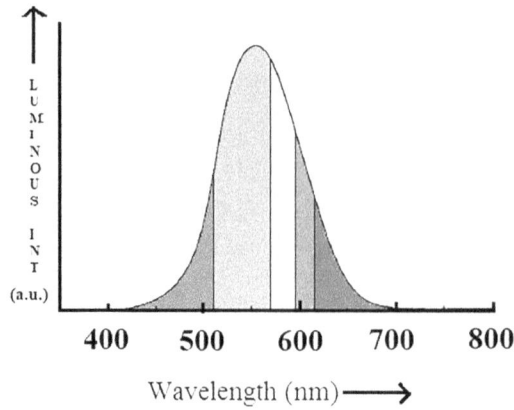

FIGURE 12.14 Luminous efficiency sensitive to the human eye with respect to wavelength 'λ' (nm). The band of colors displays the range of respective visible color on the EM radiation, visible to human eyes.

TABLE 12.3

Rare-Earth Metal Oxides Applied to ANTC with their Respective Excitation and Emission Wavelengths, Possible Mechanism Associated with them and the Colors they emit in the Visible Region

Rare-Earth Metal Oxide	Dopant	Mechanism	Excitation Wavelength (nm)	Emission Wavelength (nm)	Color
Lu_2O_3 [76]	Yb^{3+},	Upconversion	980	477 & 490	Blue
	Tm^{3+}	Upconversion	980	540 & 565	Green
	Er^{3+}	Upconversion	980	662	Red
	Yb^{3+}/Er^{3+}				
Lu_2O_3 [77]	Er^{3+}	Upconversion	804	450–500	Blue
		Upconversion	804	525 & 550	Green
		Upconversion	804	660	Red
		Upconversion	650	460, 495 & 475	Blue
		Upconversion	650	525 & 550	Green
		Normal	488	500–580	Green
		photoluminescence	488	640–690	Red
		(downshifting)	488	785–825	NIR
		Downshifting	488	840–870	NIR
		Downshifting			
Y_2O_3 [78–80]	Er^{3+}	Downshifting	488	520–570	Green
	Ce^{3+}	Downshifting	488	640–700	Red
	Tb^{3+}	Upconversion	815	525 & 550	Green
	Eu^{3+}	Upconversion	815	660	Red
	Yb^{3+}, Er^{3+}	Downshifting	381	438	Blue
		Downshifting	305	541	Green
		Downshifting	254	611	Red
		Downshifting	379	562	Green
		Downshifting	980	1127	IR
		Upconversion	980	562	Green
		Upconversion	1550	460–730	Green
		Upconversion	1550	750–1050	NIR

(Continued)

TABLE 12.3 (*Continued*)

Rare-Earth Metal Oxides Applied to ANTC with Their Respective Excitation and Emission Wavelengths, Possible Mechanism Associated with Them and the Colors They Emit in the Visible Region

Rare-Earth Metal Oxide	Dopant	Mechanism	Excitation Wavelength (nm)	Emission Wavelength (nm)	Color
YAG [81,82]	Ce^{3+}	Downshifting	470	532	Yellow
Gd_2O_3 [83]	Yb^{3+}, Er^{3+}	Upconversion	980	565	Green
	Yb^{3+},	Upconversion	980	660	Red
	Tm^{3+}	Upconversion	980	477	Blue
YBO_3 [84]	Eu^{3+}	Downshifting	254	591	Red
$LuVO_4$ [85]	Eu^{3+}	Downshifting	312	618	Red

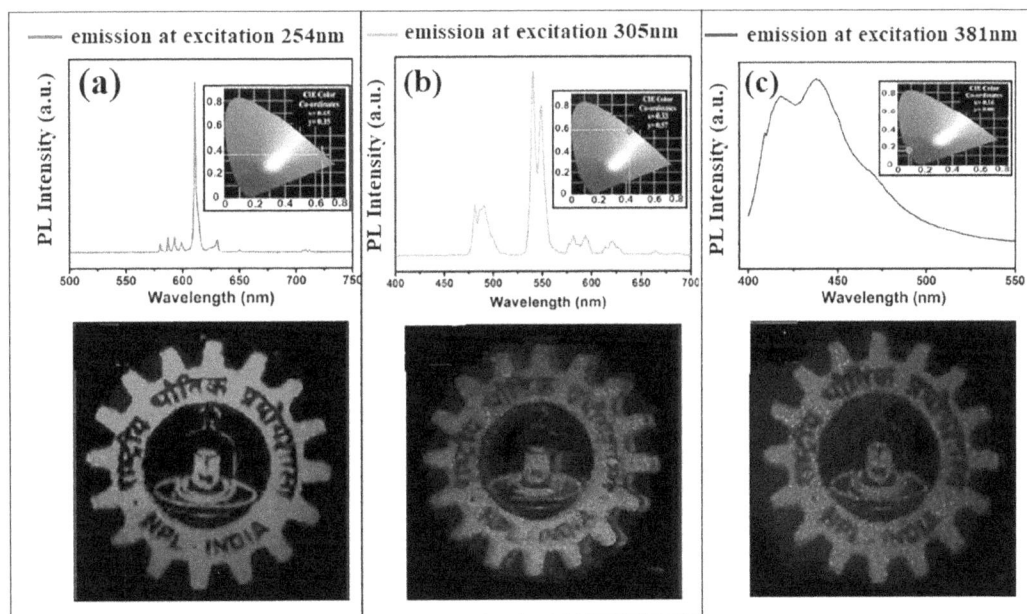

FIGURE 12.15 PL emission spectrum of Y_2O_3:Eu^{3+} with corresponding NPL logo screen printed on white paper appearing red in color (a), PL emission spectrum of Y_2O_3:Tb^{3+} with corresponding NPL logo appearing green (b), PL emission spectrum of Y_2O_3:Ce^{3+} with corresponding NPL logo appearing blue (c) under the UV excitation.

In 2010, Zhang et al. [87] and, in 2013, Ge et al. [88] gave an idea of using metal-core configuration for efficiency enhancement. They observed green emission at 549 nm from Au@SiO_2@Y_2O_3:Yb/Er nanoparticles. Afterward, other core/shell combinations with metals as shell or as core are designed and applied [40]. Au/Ag both can be used as metal in these combinations.

Very recently, the Y_2O_3:Eu@SiO_2/Y_2O_3:Er(Tm/Yb) nano-composite [89] was made, and it exhibited dual-mode co-excitation (980/405 nm) contributing to both upconversion and downshifting, which has been described in detail.

In 2013, Saif [90] introduced Ln^{3+}:$Y_2Zr_2O_7$/SiO_2 (Eu^{3+} & Tb^{3+} preferably) core–shell composite nano-phosphor used in LFP identification technique. Similarly, SiO_2@Y_2O_3:Eu^{3+}, Li^+ was used in

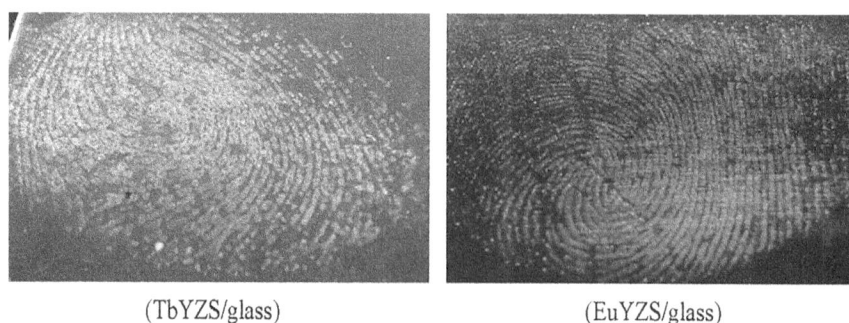

(TbYZS/glass)   (EuYZS/glass)

FIGURE 12.16 Fingerprints in green due to Tb-YZS (a) and in red due to Eu-YZS (b) on glass substance under UV light.

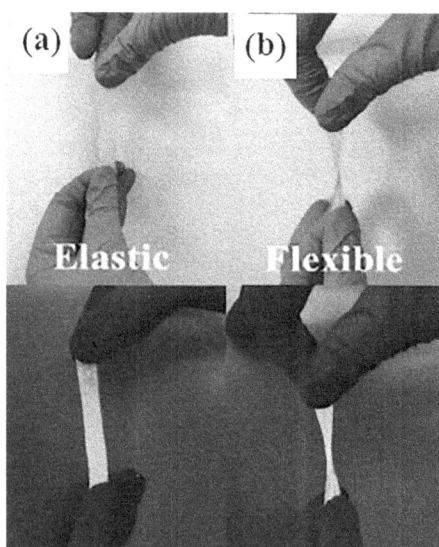

FIGURE 12.17 SAOE@TPU film gives green glow in the dark on stretching (a) and twisting (b) under UV excitation. The film can be stretched or twisted and applied to various ANTC techniques.

2017 by Venkatachalaiah et al. [91]. Both reported efficient downconversion phenomena through 254-nm UV excitation (Figure 12.16).

Eu^{3+} emission appears red whereas Tb^{3+}-doped nano-composites display green fluorescence, making the fingerprints clearly visible on UV irradiation.

Also, core–shell $SiO_2@SrTiO_3:Eu^{3+}$, Li^+ nanopowders can be used in LFP findings [92]. Likewise, $YVO_4:Er^{3+}$, $Yb^{3+}@YPO_4:Eu^{3+}$ [93] is also one of many examples of oxide-based core/shell nanoparticles.

12.4.4.5 Oxide Phosphors as Thin Films and Fibers

Efficient $SrAl_2O_4:Eu^{2+}$ (SAOE) can be drawn into nano-fibers and flexible thin films by electro-spinning it together with organic TPU. Photoluminescence was recorded at 525-nm emission for 365-nm excitation (Figure 12.17) [94].

Likewise, nano-fibers of $Bi_2Ti_2O_7:Yb^{3+}/Eu^{3+}$ through electro-spinning have been prepared, and their upconversion luminescence characteristics were studied under 980-nm excitation [95].

Both the organic and inorganic luminescent materials can prove efficient ANTC paints, but inorganics are enriched with better chemical and thermal stabilities.

12.5 CONCLUSIONS

With the dawn of technological developments, new ways of threat, theft and smuggling, etc. are also found. Every day, there is millions of dollars of loss all around the world due to counterfeiting products. Therefore, there is an ever-challenging demand to counter false duplicating. For this, many measures at the level of technology itself have been taken. Such measures come under an umbrella of anti-counterfeiting solutions to catch, destroy and judge these counterfeited units on authentication bases. Luminescence markers are creating an unprecedented pathway to counter these unjustifiable acts. Different luminescence phenomena are used in many ANTC applications. Metal oxides prove to be among the most efficient luminescent hosts doped with various rare earth metals as impurity ions used in ANTC and have been discussed in detail.

Various ANTC techniques useful in counterfeiting detection are analyzed and understood using appropriate examples.

ACKNOWLEDGMENT

Author Vartika Singh is greatly thankful to **CSIR**, Government of India, for providing research Grant of No. **31/0001(11936)/2021-EMR-I** to support the present work.

REFERENCES

1. Linsner, Bristows LLP-Marc. (2021). EUIPO observatory publishes anti-counterfeiting technology guide. Lexology. www.lexology.com. Retrieved 2021-03-18. Doi:10.2814/665780
2. Arppe, R., Sørensen, T. J. (2017). Physical unclonable functions generated through chemical methods for anti-counterfeiting. *Nat. Rev. Chem.*, 1, 31–44. https://doi.org/10.1038/s41570-017-0031.
3. Puddu, M., Paunescu, D., Stark, W. J., Grass, R. N. (2014). Magnetically recoverable, thermostable, hydrophobic DNA/silica encapsulates and their application as invisible oil tags. *ACS Nano*, 8(3), 2677–2685. https://doi.org/10.1021/nn4063853.
4. Li, D., Tang, L., Wang, J., Liu, X., Ying, Y. (2016). Multidimensional SERS barcodes on flexible patterned plasmonic metafilm for anticounterfeiting applications. *Adv. Opt. Mater.*, 4(10), 1475–1480. https://doi.org/10.1002/adom.201600247.
5. Jahanbazi, F., Mao, Y. (2021). Recent advances on metal oxide-based luminescence thermometry. *J. Mater. Chem. C*, 9, 16410–16439. https://doi.org/10.1039/D1TC03455C.
6. Wang, X., Liu, Q., Bu, Y., Liu, C-S., Liu, T., Yan, X. (2015). Optical temperature sensing of rare-earth ion doped phosphors. *RSC Adv.*, 5, 86219–86236. https://doi.org/10.1039/C5RA16986K.
7. Manjakkala, L., Szwagierczakb, D., Dahiya, R. (2020). Metal oxides based electrochemical pH sensors: Current progress and future perspectives. *Prog. Mater. Sci.*, 109(1–31), 100635. https://doi.org/10.1016/j.pmatsci.2019.100635.
8. Ahmad, F., Al-Douri, Y., Kumar, D., Ahmad, S. (2020). *Metal Oxide Powder Technologies: Fundamentals, Processing Methods and Applications: Metal-Oxide Powder Technology in Biomedicine.* Elsevier. https://doi.org/10.1016/C2018-0-02252-5
9. Murthy, S., Effiong, P., Fei, C. C. (2020). *Metal Oxide Powder Technologies: Fundamentals, Processing Methods and Applications: Metal Oxide Nanoparticles in Biomedical Applications.* Elsevier. https://doi.org/10.1016/C2018-0-02252-5.
10. Zhyrovetsky, V. M., Popovych, D. I., Savka, S. S., Serednytski, A. S. (2017). Nanopowder metal oxide for photoluminescent gas sensing. *Nanoscale Res. Lett.,* 12(1–5), 132. https://doi.org/10.1186/s11671-017-1891-5
11. Mirzaei, A., Majhi, S. M., Kim, H. W., Kim, S. S. (2021). *Metal Oxide-Based Nanofibers and Their Applications: Metal Oxide-Based Nanofibers and Their Gas-Sensing Applications.* Elsevier. https://doi.org/10.1016/C2019-0-01876-6
12. Bovhyra, R. V., Mudry, S. I., Popovych, D. I., Savka, S. S., Serednytski, A. S., Venhryn, Yu, I. (2019). Photoluminescent properties of complex metal oxide nanopowders for gas sensing. *Appl. Nanosci.*, 9, 775–780. https://doi.org/10.1007/s13204-018-0697-9
13. Mahmood, A. S., Al-Samarai, R. A., Al-Douri, Y. (2020). *Metal Oxide Powder Technologies: Fundamentals, Processing Methods and Applications: Metal Oxides Powder Technology in Energy Technologies.* Elsevier. https://doi.org/10.1016/C2018-0-02252-5

14. Monaca, A. L., Campanella, D., Paolella, A. (2021). *Metal Oxide-Based Nanofibers and Their Applications: Synthesis of One-Dimensional Metal Oxide-Based Crystals as Energy Storage Materials.* Elsevier. https://doi.org/10.1016/C2019-0-01876-6.

15. Odeh, A. A., Al-Douri, Y., (2020). *Metal Oxide Powder Technologies: Fundamentals, Processing Methods and Applications: Metal Oxides in Electronics.* Elsevier. https://doi.org/10.1016/C2018-0-02252-5.

16. Herrero, Y. R., Ullah, A. (2020). *Metal Oxide Powder Technologies: Fundamentals, Processing Methods and Applications: Metal Oxide Powder Technologies in Catalysis.* Elsevier. https://doi.org/10.1016/C2018-0-02252-5.

17. Kang, C.-S., Evans, E. A., Chase, G. G. (2021). *Metal Oxide-Based Nanofibers and Their Applications: Metal Oxide Nanofiber for Air Remediation Via Filtration, Catalysis, and Photocatalysis.* Elsevier. https://doi.org/10.1016/C2019-0-01876-6.

18. Alias, N., Rosli, S. A., Sazalli, N. A. H., Hamid, H. A., Arivalakan, S., Umar, S. N. H., Khim, B. K., Taib, B. N., Keat, Y. K., Razak, K. A., Yee, Y. F., Hussain, Z., Bakar, E. A., Kamaruddin, N. F., Manaf, A. Abd., Uchiyama, N., Kian, T. W., Matsuda, A., Kawamura, G., Sawada, K., Matsumoto, A., Lockman, Z. (2020). *Metal Oxide Powder Technologies: Fundamentals, Processing Methods and Applications: Metal Oxide for Heavy Metal Detection and Removal.* Elsevier. https://doi.org/10.1016/C2018-0-02252-5

19. El-Samak, A. A., Rahman, H., Ponnamma, D., Hassan, M. K., Zaidi, S. J., Al-Maadeed, M. A. A. (2021). *Metal Oxide-Based Nanofibers and Their Applications: Role of Metal Oxide Nanofibers in Water Purification.* Elsevier. https://doi.org/10.1016/C2019-0-01876-6.

20. Yoon, Y., Truong, P. L., Lee, D., Ko, S. H. (2022). Metal-oxide nanomaterials synthesis and applications in flexible and wearable sensors. *ACS Nanosci. Au*, 2, 64–92. https://doi.org/10.1021/acsnanoscienceau.1c00029.

21. Eliseeva, S. V., Bünzli, J.-C. G. (2010). Lanthanide luminescence for functional materials and biosciences. *Chem. Soc. Rev.*, 39, 189–227. https://doi.org/10.1039/B905604C.

22. Du, Y., Jiang, Y., Sun, T., Zhao, J., Huang, B., Peng, D., Wang, F. (2018). Mechanically excited multicolor luminescence in lanthanide ions. *Adv. Mater.*, 31(7), 1807062(1–8). https://doi.org/10.1002/adma.201807062.

23. Fan, Y., Jin, X., Wang, M., Gu, Y., Zhou, J., Zhang, J., Wang, Z. (2020). Multimode dynamic photoluminescent anticounterfeiting and encryption based on a dynamic photoluminescent material. *Chem. Eng. J.*, 393(1–8), 124799. https://doi.org/10.1016/j.cej.2020.124799.

24. Liu, J., Rijckaert, H., Zeng, M., Haustraete, K., Laforce, B., Vincze, L., Van Driessche, I., Kaczmarek, A. M., Van Deun, R. (2018). Simultaneously excited downshifting/upconversion luminescence from lanthanide-doped core/shell fluoride nanoparticles for multimode anticounterfeiting. *Adv. Funct. Mater.*, 28(1–10), 1707365. https://doi.org/10.1002/adfm.201707365.

25. Kaczmarek, A. M., Liu, Y.-Y., Wang, C., Laforce, B., Vincze, L., Van Der Voort, P., Van Hecke, K., Van Deun, R. (2017). Lanthanide "chameleon" multistage anti-counterfeit materials. *Adv. Funct. Mater.*, 27(20), 1700258(1–5). https://doi.org/10.1002/adfm.201700258.

26. Xiong, P., Peng, M., Yang, Z. (2021). Near-infrared mechanoluminescence crystals: A review. *iScience*, 24(1), 101944(1–14). https://doi.org/10.1016/j.isci.2020.101944.

27. Francis, P. S., Hogan, C. F. (2008). Chapter 13: Luminescence. *Comprehensive Analytical Chemistry*, 54, 343–373. https://doi.org/10.1016/S0166-526X(08)00613-2.

28. Lu, L., Wang, K., Wu, H., Qin, A., Tang, B. Z. (2021). Simultaneously achieving high capacity storage and multilevel anti-counterfeiting using electrochromic and electro-fluorochromic dual-functional AIE polymers. *Chem. Sci.*, 12, 7058–7065. https://doi.org/10.1039/D1SC00722J.

29. Ren, W., Lin, G., Clarke, C., Zhou, J., Jin, D. (2019). Optical nanomaterials and enabling technologies for high-security-level anticounterfeiting. *Adv. Mater.*, 32(18), 1901430(1–15). https://doi.org/10.1002/adma.201901430.

30. Abdollahi, A., Roghani-Mamaqani, H., Razavi, B., Salami-Kalajahi, M. (2020). Photoluminescent and chromic nanomaterials for anticounterfeiting technologies: Recent advances and future challenges. *ACS Nano*, 14, 11, 14417–14492. https://doi.org/10.1021/acsnano.0c07289.

31. Song, B., Wang, H., Zhong, Y., Chu, B., Su, Y., He, Y. (2018). Fluorescent and magnetic anti-counterfeiting realized by biocompatible multifunctional silicon nanoshuttles-based security ink. *Nanoscale*, 10, 1617–1621. https://doi.org/10.1039/C7NR06337G

32. Wu, Y., Zhong, Y., Chu, B., Sun, B., Song, B., Wu, S., Su, Y., He, Y. (2016). Plants-derived fluorescent silicon nanoparticles featuring excitation wavelength-dependent fluorescence spectra for anticounterfeiting application. *Chem. Commun.*, 52, 7047–7050. https://doi.org/10.1039/C6CC02872A.

33. Dai, W. B., Zhou, J., Huang, K., Hu, J., Xu, S., Xu, M. (2019). Investigation on structure and optical properties of down-conversion aluminosilicate phosphors CaAl2Si2O8: Ce/Tb/Yb. *J. Alloys Compd.*, 786, 662–667. https://doi.org/10.1016/j.jallcom.2019.02.021.

34. Luo, H., Zhang, S., Mu, Z., Wu, F., Nie, Z., Zhu, D., Feng, X., Zhang, Q. (2019). Near-infrared quantum cutting via energy transfer in Bi3+, Yb3+ codoped Lu2GeO5 down-converting phosphor. *J. Alloys Compd.*, 784, 611–619. https://doi.org/10.1016/j.jallcom.2019.01.060.

35. Xu, D., Zhang, Q., Wu, X., Li, W., J. Meng, J. (2019). Synthesis, luminescence properties and energy transfer of Ca2MgWO6:Cr3+, Yb3+ phosphors. *Mater. Res. Bull.*, 110, 135–140. https://doi.org/10.1016/j.materresbull.2018.10.023.

36. David, P. S., Panigrahi, P., Nagarajan, G. S. (2019). Enhanced near IR downconversion luminescence in Eu3+ -Yb3+ co-doped V activated ZnO host: An effort towards efficiency enhancement in Si-solar cells. *Mater. Lett.*, 249, 9–12. https://doi.org/10.1016/j.matlet.2019.03.046.

37. Liu, G.-X., Zhang, R., Xiao, Q.-L., Zou, S.-Y., Peng, W.-F., Cao, L.-W., Meng, J.-X. (2011). Efficient Bi3+ → Nd3+ energy transfer in Gd2O3:Bi3+, Nd3+. *Opt. Mater.*, 34(1), 313–316. https://doi.org/10.1016/j.optmat.2011.09.003.

38. Shivananjaiah, H. N., Sailaja Kumari, K., Geetha, M. S. (2020). Green mediated synthesis of lanthanum doped zinc oxide: Study of its structural, optical and latent fingerprint application. *J. Rare Earths*, 38(12), 1281–1287. https://doi.org/10.1016/j.jre.2020.07.012.

39. Dou, Q., Zhang, Y. (2011). Tuning of the structure and emission spectra of upconversion nanocrystals by alkali ion doping. *Langmuir*, 27, 21, 13236–13241. https://doi.org/10.1021/la201910t.

40. Han, S., Deng, R., Xie, X., Liu, X. (2014). Enhancing luminescence in lanthanide-doped upconversion nanoparticles. *Angew. Chem.*, 53(44), 11702–11715. https://doi.org/10.1002/anie.201403408.

41. Dong, B., Cao, B., He, Y., Liu, Z., Li, Z., Feng, Z. (2012). Temperature sensing and in vivo imaging by molybdenum sensitized visible upconversion luminescence of rare-earth oxides. *Adv. Mater.*, 24(15), 1987–1993. https://doi.org/10.1002/adma.201200431.

42. Cao, B. S., He, Y. Y., Feng, Z. Q., Li, Y. S., Dong, B. (2011). Optical temperature sensing behavior of enhanced green upconversion emissions from Er-Mo:Yb2Ti2O7 nanophosphor. *Sens. Actuators B: Chemical*, 159(1), 8–11. https://doi.org/10.1016/j.snb.2011.05.018.

43. Li, Z. P., Dong, B., He, Y. Y., Cao, B. S., Feng, Z. Q. (2012). Selective enhancement of green upconversion emissions of Er3+:Yb3Al5O12 nanocrystals by high excited state energy transfer with Yb3+-Mn2+ dimer sensitizing. *J. Lumin.*, 132(7), 1646–1648. https://doi.org/10.1016/j.jlumin.2012.02.034.

44. Chen, G., Liu, H., Liang, H., Somesfalean, G., Zhang, Z. (2008). Upconversion emission enhancement in Yb3+/Er3+-codoped Y2O3 nanocrystals by tridoping with Li+ ions. *J. Phys. Chem. C*, 112, 31, 12030–12036. https://doi.org/10.1021/jp804064g.

45. Chen, G. Y., Liu, H. C., Liang, H. J., Somesfalean, G., Zhang, Z. G. (2008). Enhanced multiphoton ultraviolet and blue upconversion emissions in Y2O3: Er3+ nanocrystals by codoping with Li+ ions. *Solid State Commun.*, 148(3–4), 96–100. https://doi.org/10.1016/j.ssc.2008.08.001.

46. Cao, B. S., Feng, Z. Q., He, Y. Y., Li, H., Dong, B. (2010). Opposite effect of Li+ codoping on the upconversion emissions of Er3+-doped TiO2 powders. *J. Sol-Gel Sci. Technol.*, 54(1), 101–104. https://doi.org/10.1007/s10971-010-2163-3.

47. Sun, Q., Chen, X., Liu, Z., Wang, F., Jiang, Z., Wang, C., (2011). Enhancement of the upconversion luminescence intensity in Er3+ doped BaTiO3 nanocrystals by codoping with Li+ ions. *J. Alloys Compd.*, 509(17), 5336–5340. https://doi.org/10.1016/j.jallcom.2010.12.212.

48. Liu, L., Wang, Y., Zhang, X., Yang, K., Bai, Y., Huang, C., Han, W., Li, C., Song, Y. (2011). Efficient two-color luminescence of Er3+/Yb3+/Li+:ZrO2 nanocrystals. *Opt. Mater.*, 33(8), 1234–1238. https://doi.org/10.1016/j.optmat.2011.02.019.

49. ai, Y., Wang, Y., Yang, K., Zhang, X., Peng, G., Song, Y., Pan, Z., Wang, C. H. (2008). The effect of Li on the spectrum of Er3+ in Li- and Er-codoped ZnO nanocrystals. *J. Phys. Chem. C*, 112, 12259–12263. https://doi.org/10.1021/jp803373e.

50. Luo, X., Cao, W. (2008). Upconversion luminescence properties of Li+-doped ZnWO4:Yb. *Er. J. Mater. Res.*, 23(8), 2078–2083. https://doi.org/10.1557/JMR.2008.0275.

51. Lopez, O., McKittrick, J., Shea, L. (1997). Fluorescence properties of polycrystalline Tm3+-activated Y3Al5O12 and Tm3+-Li+ co-activated Y3Al5O12 in the visible and near IR ranges. *J. Lumin.*, 71(1), 1–11. https://doi.org/10.1016/S0022-2313(96)00123-8.

52. Chung, J. H., Lee, S. Y., Shim, K. B., Kweon, S.-Y., Ur, S.-C., Ryu, J. H. (2012). Blue upconversion luminescence of CaMoO4:Li+/Yb3+/Tm3+ phosphors prepared by complex citrate method. *Appl. Phys. A*, 108, 369–373. https://doi.org/10.1007/s00339-012-6893-7.

53. Mahalingam, V., Naccache, R., Vetrone, F., Capobianco, J. A. (2012). Enhancing upconverted white light in Tm3+/Yb3+/Ho3+-doped GdVO4 nanocrystals via incorporation of Li+ ions. *Opt. Express,* 20(1), 111–119. https://doi.org/10.1364/OE.20.000111.

54. Pushpendra, Suryawanshi, I., Srinidhi, S., Singh, S., Kalia, R., Kunchala, R. K., Mudavath, S. L., Naidu, B. S. (2021). Downshifting and upconversion dual mode emission from lanthanide doped GdPO4 nanorods for unclonable anti-counterfeiting. *Mater. Today Commun.,* 26, 102144(1–10). https://doi.org/10.1016/j.mtcomm.2021.102144.

55. Pushpendra, Suryawanshi, I., Kalia, R., Kunchala, R. K., Mudavath, S. L., Naidu, B. S. (2022). Detection of latent fingerprints using luminescent Gd0.95Eu0.05PO4 nanorods. *J. Rare Earths,* 40(4), 572–578. https://doi.org/10.1016/j.jre.2021.01.015.

56. Luo, Y., Zhang, L., Liu, Y., Heydari, E., Chen, L., Bai, G. (2022). Designing dual-mode luminescence in Er3+ doped Y2WO6 microparticles for anticounterfeiting and temperature measurement. *J. Am. Ceram. Soc.,* 105(2), 1375–1385. https://doi.org/10.1111/jace.18153.

57. Ma, Z., Zhou, J., Zhang, J., Zeng, S., Zhou, H., Smith, A. T., Wang, W., Sun, L., Wang, Z. (2019). Mechanics-induced triple-mode anticounterfeiting and moving tactile sensing by simultaneously utilizing instantaneous and persistent mechanoluminescence. *Mater. Horiz.,* 6, 2003–2008. https://doi.org/10.1039/C9MH01028A.

58. Hao, J., Zhang, Y., Wei, X. (2011). Electric-induced enhancement and modulation of upconversion photoluminescence in epitaxial BaTiO3:Yb/Er thin films. *Angew. Chem.,* 123, 7008–7012; *Angew. Chem. Int. Ed.,* 50(30), 6876–6880. https://doi.org/10.1002/anie.201101374.

59. Moudam, O., Lakbita, O. (2021). Potential end-use of a europium binary photoluminescent ink for anti-counterfeiting security documents. *ACS Omega,* 6(44), 29659–29663. https://doi.org/10.1021/acsomega.1c03949.

60. Katumo, N., Li, K., Richards, B. S., Howard, I. A. (2022). Dual-color dynamic anti-counterfeiting labels with persistent emission after visible excitation allowing smartphone authentication. *Sci. Rep.,* 12, 2100(1–14). https://doi.org/10.1038/s41598-022-05885-6.

61. Sharma, V., Das, A., Kumar, V., Ntwaeaborwa, O. M., Swart, H. C. (2014). Potential of Sr4Al14O25: Eu2+, Dy3+ inorganic oxide-based nanophosphor in Latent fingermark detection. *J. Mater. Sci.,* 49, 2225–2234. https://doi.org/10.1007/s10853-013-7916-2.

62. Pei, P., Wei, R., Wang, B., Su, J., Zhang, Z., Liu, W. (2021). An advanced tunable multimodal luminescent La4GeO8: Eu2+, Er3+ phosphor for multicolor anticounterfeiting. *Adv. Funct. Mater.,* 31(31) 2102479 (1–9). https://doi.org/10.1002/adfm.202102479.

63. Jin, X., Wang, Z., Wei, Y., Fu, Z. (2022). Dual-mode multicolor luminescence based on lanthanide-doped Na2CaGe2O6 phosphor for anticounterfeiting application. *J. Lumin.,* 249, 118937(1–8). https://doi.org/10.1016/j.jlumin.2022.118937.

64. Haider, A. A., Cun, Y., Bai, X., Xu, Z., Zi, Y., Qiu, J., Song, Z., Huang, A., Yang, Z. (2022). Anti-counterfeiting applications by photochromism induced modulation of reversible upconversion luminescence in TiO2:Yb3+, Er3+ ceramic. *J. Mater. Chem. C,* 10, 6243–6251. https://doi.org/10.1039/D2TC00859A.

65. Zhu, X., Wang, T., Liu, Z., Cai, Y., Wang, C., Lv, H., Liu, Y., Wang, C., Qiu, J., Xu, X., Ma, H., Yu, X. (2022). A temporal and space anti-counterfeiting based on the four-modal luminescent Ba2Zr2Si3O12 phosphors. *Inorg. Chem.,* 61, 3223–3229. https://doi.org/10.1021/acs.inorgchem.1c03712.

66. Ma, J., Li, Y., Hu, W., Wang, W., Zhang, J., Yang, J., Wang, Y. (2020). A terbium activated multicolour photoluminescent phosphor for luminescent anticounterfeiting. *J. Rare Earths,* 38, 1039–1043. https://doi.org/10.1016/j.jre.2019.09.008.

67. Liu, Z., Zhao, L., Chen, W., Fan, X., Yang, X., Tian, S., Yu, X., Qiu, J., Xu, X. (2018). Multiple anti-counterfeiting realized in NaBaScSi2O7 with a single activator of Eu2+. *J. Mater. Chem. C,* 6, 11137–11143. https://doi.org/10.1039/C8TC04018D.

68. Venkatesha Babu, K. R., Renuka, C. G., Basavaraj, R. B., Darshan, G. P., Nagabhushana, H. (2019). One pot synthesis of TiO2:Eu3+ hierarchical structures as a highly specific luminescent sensing probe for the visualization of latent fingerprints. *J. Rare Earths,* 37(2), 134–144. https://doi.org/10.1016/j.jre.2018.05.019.

69. Wang, J., Chen, W., Peng, L., Han, T., Liu, C., Zhou, Z., Qiang, Q., Shen, F., Wang, J., Liu, B. (2022). Long persistent luminescence property of green emitting Sr3Ga4O9: Tb3+ phosphor for anti-counterfeiting application. *J. Lumin.,* 250, 119066(1–8). https://doi.org/10.1016/j.jlumin.2022.119066.

70. Pushpendra, Singh, S., Srinidhi, S., Kunchala, R. K., Kalia, R., Achary, S. N., Naidu, B. S. (2021). Structural and excitation-dependent photoluminescence properties of Bi0.95-xGdxEu0.05PO4 (0≤x≤0.95) solid solutions and their anticounterfeiting applications. *Cryst. Growth Des.*, 21, 4619–4631. https://doi.org/10.1021/acs.cgd.1c00467.

71. Xiao, Q., Yin, X., Wu, X., Fan, Y., Lv, L., Dong, X., Zhou, N., Liu, K., Luo, X. (2022). High-sensitive thermometry and tunable dual-color luminescence based on effective energy transfer pathways of Er, Tm system. *J. Lumin.*, 250, 119057(1–9). https://doi.org/10.1016/j.jlumin.2022.119057.

72. Cheng, H., Jiang, Z., Lai, F., Wang, H., Xiao, Z., Sun, J., You, W. (2022). Simultaneous achievement of sensitivity enhancement and dual-mode luminescence through co-doping Eu3+ in Ca2MgWO6:Er3+/Yb3+ phosphor. *J. Lumin*, 246, 118804(1–9). https://doi.org/10.1016/j.jlumin.2022.118804.

73. Barbosa, R., Gupta, S. K., Srivastava, B. B., Villarreal, A., De Leon, H., Peredo, M., Bose, S., Lozano, K. (2021). Bright and persistent green and red light-emitting fine fibers: A potential candidate for smart textiles. *J. Lumin.*, 231 (2021) 117760(1–8). https://doi.org/10.1016/j.jlumin.2020.117760.

74. Xu, Z., Zhu, Y., Luo, Q., Liu, X., Li, L. (2020). Luminescence, lattice occupancy and application of a new anti-counterfeiting for SrGa2Si2O8: Eu3+. *J. Lumin.*, 219, 116894(1–7). https://doi.org/10.1016/j.jlumin.2019.116894.

75. Yeshodamma, S., Sunitha, D. V., Basavaraj, R. B., Darshan, G. P., Prasad, B. D., Nagabhushana, H. (2019). Monovalent ions co-doped SrTiO3:Pr3+ nanostructures for the visualization of latent fingerprints and can be red component for solid state devices. *J. Lumin.*, 208, 371–387. https://doi.org/10.1016/j.jlumin.2018.12.044.

76. Yang, J., Zhang, C., Peng, C., Li, C., Wang, L., Chai, R., Lin, J. (2009). Controllable red, green, blue (RGB) and bright white upconversion luminescence of Lu2O3:Yb3+/Er3+/Tm3+ nanocrystals through single laser excitation at 980 nm. *Chem. Eur. J.*, 15, 4649–4655. https://doi.org/10.1002/chem.200802106.

77. Capobianco, J. A., Vetrone, F., Boyer, J. C., Speghini, A., Bettinelli, M. (2002). Visible upconversion of Er3+ doped nanocrystalline and bulk Lu2O3. *Opt. Mater.*, 19, 259–268. https://doi.org/10.1016/S0925-3467(01)00188-4.

78. Capobianco, J. A., Vetrone, F., D'Alesio, T., Tessari, G., Speghini, A., Bettinelli, M. (2000). Optical spectroscopy of nanocrystalline cubic Y2O3:Er3+ obtained by combustion synthesis. *Phys. Chem. Chem. Phys.*, 2, 3203–3207. https://doi.org/10.1039/B003031G.

79. Kumar, P., Nagpal, K., Gupta, B. K. (2017). Unclonable security codes designed from multicolour luminescent lanthanide doped Y2O3 nanorods for anti-counterfeiting. *ACS Appl. Mater. Interfaces*, 9, 16, 14301–14308. https://doi.org/10.1021/acsami.7b03353.

80. Kumar, P., Dwivedi, J., Gupta, B. K. (2014). Highly-luminescent dual mode rare-earth nanorods assisted multi-stage excitable security ink for anti-counterfeiting applications. *J. Mater. Chem. C*, 2, 10468–10475. https://doi.org/10.1039/C4TC02065K.

81. Pan, Y., Wu, M., Su, Q. (2004). Tailored photoluminescence of YAG:Ce phosphor through various methods. *J. Phys. Chem. Solids*, 65(5), 845–850. https://doi.org/10.1016/j.jpcs.2003.08.018.

82. Justel, T., Nikol, H., Ronda, C. (1998). New developments in the field of luminescent materials for lighting and displays. *Angew. Chem.*, 110, 3250–3271; Angew. *Chem. Int. Ed.*, 37(22), 3084–3103. https://doi.org/10.1002/(SICI)1521-3773(19981204)37:22<3084::AID-ANIE3084>3.0.CO;2–W.

83. Wu, M., Guan, G., Yao, B., Teng, C- P., Liu, S., Tee, S. Y., Ong, B. C., Dong, Z. L., Han, M.-Y. (2019). Upconversion luminescence of Gd2O3 :Ln3+ nanorods for white emission and cellular imaging via surface charging and crystallinity control. *ACS Appl. Nano Mater.*, 2, 3, 1421–1430. https://doi.org/10.1021/acsanm.8b02315.

84. Gangwar, A. K., Nagpal, K., Kumar, P., Singh, N., Gupta, B. K. (2019). New insight into printable europium doped yttrium borate luminescent pigment for security ink application. *J. Appl. Phys.*, 125, 074903(1–8). https://doi.org/10.1063/1.5027651.

85. Liang, L., Chen, C., Lv, Z., Xie, M., Yu, Y., Liang, C., Lou, Y., Li, C., Shi, Z. (2019). Microwave-assisted synthesis of highly water-soluble LuVO4:Eu nanoparticles as anti-counterfeit fluorescent ink. *J. Lumin.*, 206, 560–564. https://doi.org/10.1016/j.jlumin.2018.10.088.

86. Lu, Q., Guo, F. Y., Sun, L., Li, A. H., Zhao, L. C. (2008). Silica-/titania-coated Y2O3:Tm3+, Yb3+ nanoparticles with improvement in upconversion luminescence induced by different thickness shells. *J. Appl. Phys.*, 103(12), 123533(1–10). https://doi.org/10.1063/1.2946730.

87. Zhang, F., Braun, G. B., Shi, Y., Zhang, Y., Sun, X., Reich, N. O., Zhao, D., Stucky, G. (2010). Fabrication of Ag@SiO2@Y2O3:Er nanostructures for bioimaging: Tuning of the upconversion fluorescence with silver nanoparticles. *J. Am. Chem. Soc.*, 132, 2850–2851. https://doi.org/10.1021/ja909108x.

88. Ge, W., Zhang, X. R., Liu, M., Lei, Z. W., Knize, R. J., Lu, Y. (2013). Distance dependence of gold-enhanced upconversion luminescence in Au/SiO2/Y2O3:Yb3+, Er3+ nanoparticles. *Theranostics*, 3(4), 282–288. doi:10.7150/thno.5523.

89. Han, Q., Liang, Y., Li, Z., Song, Y., Wang, Y., Zhang, X. (2022). Tunable multicolor emission based on dual-mode luminescence Y2O3: Eu@SiO2/Y2O3: Er(Tm/Yb) composite nanomaterials. *J. Lumin.*, 241, 118541(1–11). https://doi.org/10.1016/j.jlumin.2021.118541.

90. Saif, M. (2013). Synthesis of down conversion, high luminescent nano-phosphor materials based on new developed Ln3+:Y2Zr2O7/SiO2 for latent fingerprint application. *J. Lumin.*, 135, 187–195. https://doi.org/10.1016/j.jlumin.2012.10.022.

91. Venkatachalaiaha, K. N., Nagabhushanac, H., Darshand, G. P., Basavaraj, R. B., Prasad, B. D. (2017). Novel and highly efficient red luminescent sensor based SiO2@Y2O3:Eu3+, M+ (M+=Li, Na, K) composite core-shell fluorescent markers for latent fingerprint recognition, security ink and solid state lightning applications. *Sensors Actuators B*, 251, 310–325. https://doi.org/10.1016/j.snb.2017.05.022.

92. Sandhyarani, A., Kokila, M. K., Darshan, G. P., Basavaraj, R. B., Prasad, B. D., Sharma, S. C., Lakshmi, T. K. S., Nagabhushana, H. (2017). Versatile core-shell SiO2@SrTiO3:Eu3+, Li+ nanopowders as fluorescent label for the visualization of latent fingerprints and anti-counterfeiting applications. *Chem. Eng. J.*, 327, 1135–1150. https://doi.org/10.1016/j.cej.2017.06.093

93. Lin, F., Sun, Z., Jia, M., Zhang, A., Fu, Z., Sheng, T. (2020). Core-shell mutual enhanced luminescence based on space isolation strategy for anti-counterfeiting applications. *J. Lumin.*, 218, 116862(1–6). https://doi.org/10.1016/j.jlumin.2019.116862.

94. Gao, W., Ge, W., Shi, J., Tian, Y., Zhu, J., Li, Y. (2021). Stretchable, flexible, and transparent SrAl2O4:Eu2+@TPU ultraviolet stimulated anti-counterfeiting film. *Chem. Eng. J.*, 405, 126949(1–10). https://doi.org/10.1016/j.cej.2020.126949.

95. Xu, M., Ge, W., Shi, J., Li, J., Tian, Y. (2021). Color-tunable and multiple upconversion luminescence of Bi2Ti2O7:Yb3+/ Eu3+ nanofibers via electrospinning process. *J. Lumin.*, 237, 118135(1–4). https://doi.org/10.1016/j.jlumin.2021.118135.

Index

For Product Safety Concerns and Information please contact our EU
representative GPSR@taylorandfrancis.com
Taylor & Francis Verlag GmbH, Kaufingerstraße 24, 80331 München, Germany